Finite Element Implementation

Finite Element Implementation

Y. K. Cheung
S. H. Lo
A. Y. T. Leung
The University of Hong Kong

Blackwell
Science

Editorial Offices:
Osney Mead, Oxford OX2 OEL
25 John Street, London WC1N 2BL
23 Ainslie Place, Edinburgh EH3 6AJ
238 Main Street, Cambridge
Massachusetts 02142, USA
54 University Street, Carlton
Victoria 3053, Australia

Other Editorial Offices:
Arnette Blackwell SA
1, rue de Lille, 75007 Paris
France

Blackwell Wissenschafts-Verlag GmbH
Kurfürstendamm 57
10707 Berlin, Germany

Feldgasse 13, A-1238 Wien
Austria

First published 1996

Set by Pure Tech India Ltd, Pondicherry
Printed and bound in Great Britain by Hartnolls Ltd, Bodmin, Cornwall

DISTRIBUTORS
Marston Book Services Ltd
PO Box 87
Oxford OX2 0DT
(*Orders*: Tel: 01865 791155
Fax: 01865 791927
Telex: 837515)

USA
Blackwell Science, Inc.
238 Main Street
Cambridge, MA 02142
(*Orders*: Tel: 800 215–1000
617 876–7000
Fax: 617 492–5263)

Canada
Oxford University Press
70 Wynford Drive
Don Mills
Ontario M3C 1J9
(*Orders*: Tel: 416 441 2941)

Australia
Blackwell Science Pty Ltd
54 University Street
Carlton, Victoria 3053
(*Orders*: Tel: 03 347–0300
Fax: 03 349–3016)

A catalogue record for this title is available from the British Library

ISBN 0–632–03937–X

Library of Congress
Cataloging-in-Publication Data

Cheung, Y. K.
Finite element implementation / Y. K. Cheung, S. H. Lo, A. Y. T. Leung.
p. cm.
Includes bibliographical references and index.
ISBN 0–632–03937–X
1. Finite element method. I. Lo, S. H. II. Leung, A. Y. T. III. Title.
TA347.F5C48 1996
620'.001'51535–dc20 95–1810
CIP

Contents

PART III

Preface

More than 25 years ago the classic text of *The Finite Element Method in Structural and Continuum Mechanics* was published as a medium-size reference book with 272 pages. A few years ago the fourth edition came out with 1455 pages, and is therefore a vastly expanded version of the first edition. Most of the contents are concerned with the theoretical development and application examples of the finite element method, and the authors have not been able to emphasize the implementation aspects. There is no doubt, however, that the time is ripe for a more extensive text on the practical aspects of the finite element method and it is with the implementation purpose in mind that the present book is written.

The book can be used by final year project students, postgraduate students, and practising engineers with some basic knowledge in computer programming and computational mechanics. In contrast to most of the existing texts, discussions on the theoretical development are restricted to only what is absolutely necessary for the understanding of the present text, while extensive reference is made to other reference texts such as the one by Zienkiewicz and Taylor. Great emphasis is paid to presenting the detailed formulation of a number of commonly used displacement type elements, to explaining to readers the advanced techniques of computation which can reduce the amount of computational efforts and increase the efficiency, to showing readers the practical problems associated with automatic mesh generation and node renumbering, and to producing a well-documented, user-friendly software package for the use of the reader for a variety of problems.

The book comprises eight chapters (a brief review of the chapters can be found in Section 1.9 of Chapter 1). It is divided into three parts: Part I (Chapters 1–4) contains elementary materials and should be read by undergraduate and postgraduate students, Part II (Chapters 5–7) deals with advanced topics and would be useful to senior postgraduate students and practising engineers working in computer analysis of engineering problems and, finally, Part III (Chapter 8) deals with computer implementation of materials presented in the first four chapters and also gives the listing of the finite element package FACILE. This last part would be useful to all readers.

We are grateful to Professor F. K. Kong of Nanyang Technological University in Singapore for his suggestion that we should write a book of this nature. We are obliged to Mr C. K. Lee, our graduate student, for drawing part of the diagrams, and to Mrs Vivian Ho, for typing part of the manuscript.

Y. K. Cheung
S. H. Lo
A. Y. T. Leung

FACILE program disks

The authors of this book are willing to provide the listing of the finite element analysis program FACILE on standard floppy diskettes on payment of the material cost and a small handling services charge. Please write a simple application note to Dr SH Lo at:

Department of Civil and Structural Engineering
The University of Hong Kong
6/F Haking Wong Building, Pokfulam Road,
HONG KONG

Tel: (852) 2859 1977

or Fax to: (852) 2559 5337
or e-mail to: hreclsh@hkucc.hku.hk

Notation

The principal symbols used in this book are presented below for easy reference. It should be noted that this is not an exhaustive list; other symbols are defined in the text where they first occur. Occasionally, the same symbol may be used to represent different physical objects, and a non-uniqueness arises. Nevertheless, it is hoped that appropriate text explanation will clarify the intended meaning of the symbols. Vectors and matrices will be denoted by bold symbols, e.g. $\mathbf{v}$, $\mathbf{K}$; however, vectors may also be indicated by a small arrow, e.g. $\vec{\psi}$, $\vec{\mathrm{p}}$.

$\mathbb{R}$	set of real numbers
$\cup$	set union
$\cap$	set intersection
$\emptyset$	empty set
$\in$	is a member of
$\notin$	is not a member of
$\subset$	is a subset of
$\forall$	for all, for each
$\|\|\mathbf{v}\|\|$	modulus or magnitude of vector $\mathbf{v}$
∇	spatial differential operator
δ_{ij}	Kronecker delta
Ω	problem domain
$\partial\Omega$ or Γ	boundary of domain Ω
Ω_e	domain of a finite element e
Δ	area of a triangle or reference triangular domain
$\square$	standard 2×2 square domain
$\mathcal{R}$	element reference domain
$\mathbf{J}$	Jacobian matrix
$\det(\mathbf{J})$	determinant of matrix $\mathbf{J}$
$\hat{\mathrm{n}}$	unit normal vector
$\hat{\mathrm{t}}$	unit tangent vector
i, j, k, l	Latin indices ranging from 1 to 3
I, J, K, L	block letter indices ranging from 1 to 2
α, β	Greek indices ranging from 1 to N, number of element nodes
ψ	arbitrary scalar function
$\vec{\psi} = (\psi_1, \psi_2, \psi_3)$	arbitrary vector function
W_k	weight at integration point k
(x_1, x_2, x_3) or (x, y, z)	Cartesian coordinates
(r, z, θ)	polar coordinates
$\xi, \eta, \varphi, \zeta$	element natural coordinates

a_1, a_2, a_3	constants related to the coordinates
b_1, b_2, b_3	of the corner points of a triangle
c_1, c_2, c_3	
$2a, 2b$	horizontal and vertical dimensions of a rectangular element
R_{ij}	components of rotation matrix **R**
$\mathbf{u} = (u_1, u_2, u_3)$	displacement vector
$= (u, v, w)$	
G_α	shape function for node α
H_α	interpolation function for nodc α
$H_{\alpha,i}$	differentiation of H_α with respect to x_i
λ, μ	Lamé constants
E	Young's modulus
ν	Poisson's ratio
$\lambda_1, \lambda_2, \lambda_3, \lambda_4, \lambda_5$	elastic constants
C_{ijkl}	components of elasticity tensor **C**
k_{ij}	components of conductivity matrix **K**
ε_{ij}	components of strain tensor $\boldsymbol{\varepsilon}$
σ_{ij}	components of stress tensor $\boldsymbol{\sigma}$
ε_0, σ_0	initial strain and stress
t, h	element thickness
$\vec{\mathrm{r}}$	initial stress(strain) force vector
$\vec{\mathrm{b}}$	body force vector
$\vec{\mathrm{s}}$	surface force vector
$\vec{\mathrm{p}}$	concentrated nodal force vector
$\vec{\mathrm{f}}$	global force vector
$\vec{\mathrm{r}}_e, \vec{\mathrm{b}}_e, \vec{\mathrm{s}}_e$	element force vectors
$\vec{\mathrm{r}}_\alpha, \vec{\mathrm{b}}_\alpha, \vec{\mathrm{s}}_\alpha$	element forces at node α
$\mathrm{r}_i^\alpha, \mathrm{b}_i^\alpha, \mathrm{s}_i^\alpha$	components of element forces at node α
$\mathbf{K}$	global stiffness matrix
$\mathbf{K}_e$	element stiffness matrix
$\mathbf{K}_e^{\alpha\beta}$	element sub-matrix relating nodes α and β.
$\mathbf{K}_{ij}^{\alpha\beta}$	components of element sub-matrix.
$\vec{\gamma}, \vec{\varphi}$	body force per unit volume
$\vec{W}$	total element body force
$\vec{q}, \vec{t}$	surface traction force per unit area
Q	strength of source per unit volume
$\vec{q}$	rate of flow
γ_1, γ_2	shear strains
θ_1, θ_2	fibre rotations
$\mathbf{x}, \mathbf{u}$	position and displacement vectors at a general point
$\bar{\mathbf{x}}, \bar{\mathbf{u}}$	position and displacement vectors of a point on reference surface
$\tilde{\mathbf{x}}, \tilde{\mathbf{u}}$	vector and displacement along the fibre line
$(\hat{e}_1, \hat{e}_2, \hat{e}_3)$	orthogonal basis on laminar surface
$(\tilde{e}_1, \tilde{e}_2, \tilde{e}_3)$	orthogonal basis along fibre line
m_{IJ}	moment per unit length
q_I	transverse shear force for unit length

$V_{\alpha\beta}$	$= \int_{\Omega_e} H_\alpha H_\beta \, d\Omega$
$A_{\alpha\beta}$	$= \int_{\Gamma_e} H_\alpha H_\beta \, da$
$h_{\alpha j}$	$= \int_{A_e} t H_{\alpha,j} \, dA$
$I_{ij}^{\alpha\beta}$	$= \int_{\Omega_e} H_{\alpha,i} H_{\beta,j} \, d\Omega$

1 Introduction

1.1 THE FINITE ELEMENT METHOD

In essence, the finite element method is a numerical technique which solves the governing equations of a complicated system through a discretization process. The system of interest can be either physical or mathematical. The domain of the system can be well-defined or subject to continual changes (moving boundary problems such as transient free surface water flow, large deformation problems, etc.). The boundary conditions can be well-defined in terms of prescribed loads and displacements, or sometimes less well-defined as in fluid-structure interaction or contact problems. The governing equations can be given in differential form or be expressed in terms of variational integrals.

Before an analysis is carried out, the whole system has to be divided into a number of individual subsystems or components, whose behaviour is readily understood. The basic units of the discretized subsystems are called *finite elements*, which should neither overlap nor have gaps between each other. The finite elements used for a domain need not be of the same type and the properties can also vary. Figure 1.1 shows how a smoothly varying surface, as defined by function ϕ, is modelled by elements of various types. When 3-node triangular elements are used, the ϕ surface is approximated by flat triangular facets. Whereas the 4-node and 8-node quadratic elements are able to represent warped and curved surfaces, and can thus better

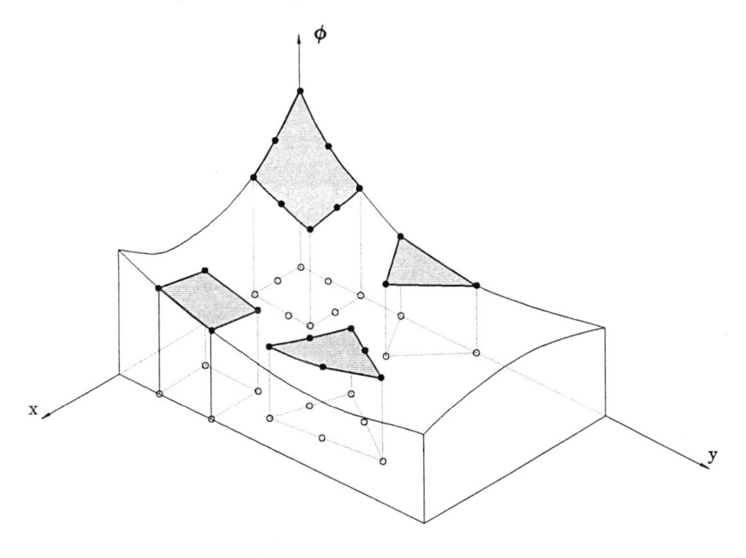

Fig. 1.1 Smooth surface approximated by finite elements

approximate the actual function. Obviously, the approximation can also be improved by using more elements instead of increasing the order of the interpolation polynomial. This sketch illustrates the basic ideas of the finite element method: piecewise approximation of a smooth function by means of simple polynomials, each of which is defined over a small region (element) and expressed in terms of the values of the functions at the element nodes.

1.2 Brief historical background

For the past one hundred years, structural mechanics has been applied to analyze building frames and trusses [1]. Stiffness matrices of bar and beam elements are derived from the theory of elementary strength of materials. With the introduction of the direct stiffness method, the element matrices can be assembled to form the global matrices relating the displacements to the forces at the nodes and the reactions at the supports of the whole structure. Beginning in 1906 and thereafter, researchers proposed the *lattice analogy* to solve continuum problems [2–4], in which the continuum is approximated by a regular mesh of elastic bars. In a paper published in 1943, Courant [5] suggested piecewise polynomial interpolation over triangular subregions as a way to get approximate numerical solutions. He recognized this approach as a Rayleigh–Ritz solution of a variational problem. This is the basis of the finite element method as we know it today.

None of the previous work was found practical at the time as there were no computers to do the tedious calculations. By 1953, engineers were writing stiffness equations in matrix notation and solving the equations with digital computers [6]. The finite element concept was introduced in the classical paper by Turner, Clough, Martin and Topp [7] in 1956. With this paper together with many others, an explosive development of the finite element method in engineering applications began. The name *finite element* was first introduced by Clough [8] in 1960. In the early sixties, the method was viewed as sound and versatile, and it became a respectable area of study in the academic circle. Research was pursued simultaneously in various parts of the world in several directions. A wide variety of elements was developed including bending elements [9], curved shell elements and the isoparametric concept was also introduced [10]. The method was soon recognized as a general method of solution for partial differential equations; its applicability to non-linear and dynamic structural problems was amply demonstrated; extension in other domains, such as soil mechanics, fluid mechanics, thermodynamics, electromagnetism, etc., produced solution to engineering problems hitherto intractable [11–21]. A mathematical basis for the method was also established using the concept of functional analysis [22–23]. By 1976, after development for two decades, the cumulative total publications on the finite element method exceeded 7000 [24].

1.3 Scope of applications and important characteristics

The finite element method has been widely used in many branches of science and engineering. The most common applications are found in mechanics – solid mechanics, fluid mechanics, heat transfer and thermal stress analysis, coupled problems, etc. We shall restrict our attention to solid mechanics and field problems in our discussions. Obviously, in a text of this size, it would not be possible to deal with all the linear and non-linear, conservative and non-conservative, static, dynamic and stability problems in solid mechanics, nor would it be practical to deal with all the steady-state, transient, moving boundary and free surface analysis of the field problems. We shall therefore concentrate on linear elastostatics in solid mechanics and steady-state field problems. It is our experience that a solid understanding of the linear analysis of the finite element method is very useful in its own right, and it makes the subject of non-linear finite element analysis much more accessible.

Generality and versatility are perhaps the most outstanding features of the finite element method. As mentioned in the last paragraph, the present-day application of the method includes almost all physical problems that are governed by differential equations. Several advantageous properties of the method have contributed to its extensive use. There follows some of the more important characteristics:

- The material properties of adjacent elements need not be the same. This allows the method to be applied to bodies composed of several types of materials.
- Irregular shaped boundaries can be approximated using elements of straight edges or matched exactly using elements with curved boundaries. The method is therefore not limited to regular nice domains with easily defined boundaries.
- Various shapes, sizes and types of elements can be employed within the same region. This flexibility allows the method to make optimal use of the finite elements available.
- Boundary conditions such as discontinuous surface loading present no difficulties for the method. Mixed boundary conditions can be handled easily in a natural way.
- The size of the elements can be made smaller in regions where the unknown parameter is expected to vary rapidly. In fact, very accurate finite element solutions can be obtained by an automatic adaptive refinement procedure using graded meshes [25], in which small elements are used in areas of high stress concentration.
- All the above features can be incorporated into one single computer program for a particular class of problems. In general, very little manipulation is necessary for different boundary and loading conditions. Hence it would also be economical to solve the same problem with different boundary and loading conditions by just forming the system stiffness matrix once only.

A major inconvenience of the finite element method has been the heavy workload in the preparation of large amounts of input data and the interpretation of voluminous outputs. The absence of control during executions of the various phases of a finite element analysis and the lack of visualization aids to monitor what is being processed in the computer are the major problems yet to be resolved. Today, the trend is to design software with facilities for model generation, man–machine interaction and graphical display capabilities.

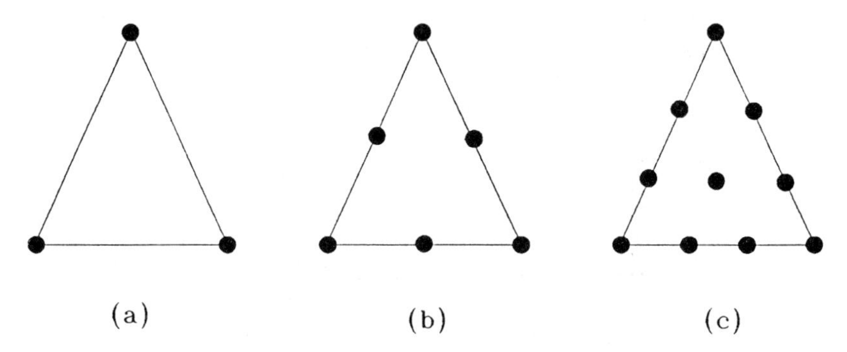

Fig. 1.2 Triangular elements with corner, edge and interior nodes

1.4 FINITE ELEMENT PROCEDURE

The most widely used finite element formulation in solid mechanics is the displacement approach. The displacement field within the element is defined in terms of assumed functions (interpolation functions) and unknown parameters at the nodes which are either displacements or displacement related quantities such as slopes and curvatures. As shown in Fig. 1.2, an element will usually have a number of corner nodes, sometimes also edge nodes, and in rare cases internal nodes.

For each finite element, a displacement function in terms of the element coordinates (x, y, z) and the nodal displacement parameters (e.g. w, θ_x, θ_y at each node of a plate bending element) is chosen to represent the displacement field, and thereby the strain and stress, within the element. A stiffness matrix relating the nodal forces to the nodal displacements can be derived through the application of the principle of virtual work or the principle of minimum total potential energy. The stiffness matrices of all the elements in the domain can be assembled to form the overall stiffness matrix for the system. After modifying the global stiffness matrix in accordance with the boundary conditions and establishing the force vector, the system of equations can be solved to yield firstly the nodal displacements, and then subsequently the stresses at any point in each individual element.

The steps of a finite element analysis can be summarized as follows:

1. Discretization of the problem domain into finite elements.
2. Selection of nodal displacement parameters and element interpolation functions.
3. Evaluation of individual element properties.
4. Assembly of system stiffness matrix.
5. Introduction of boundary conditions.
6. Formation of global force vector.
7. Solution of system matrix equations for nodal displacements.
8. Additional calculation for stresses and other parameters.

1.5 FINITE ELEMENT DISCRETIZATION

The first step of a finite element analysis is to divide the continuum or problem domain into valid finite elements. A variety of elements may be used and different sizes and shapes of elements can be employed in the same solution region. Indeed, when analyzing an elastic structure that has different types of members, such as plates, bars and beams, it is not only desirable but also necessary to use different types of elements in the analysis.

The discretization of the problem domain involves a decision on the number, size and shape of subregions used to model the real structure. The element should be small enough to give useful results and large enough to reduce the computational effort. The actual discretization process can be divided into two parts – the division of the system into elements and the labelling of the elements and nodes. The latter sounds quite simple, but is in reality quite complicated for structures of irregular boundary owing to the desire to improve computational efficiency.

The discretization of the problem domain into finite elements is a painstaking process and is also prone to errors if carried out manually. In this text, a comprehensive discussion on automatic mesh generation will be presented and the aforementioned difficulties can be minimized. The discretization of various types of problem domains by an automatic mesh generation process is illustrated with examples in the following paragraph.

Figure 1.3 is a perspective view of a four-storey structural frame, for which the position of the nodal points can be easily identified, and the whole structure can be discretized into column and beam elements in a more or less natural manner. The discretization of a two-dimensional continuum is less trivial. Figure 1.4 demonstrates how a deep beam with an opening is partitioned into triangular and quadrilateral elements. This mesh is taken from the result of an adaptive refinement analysis

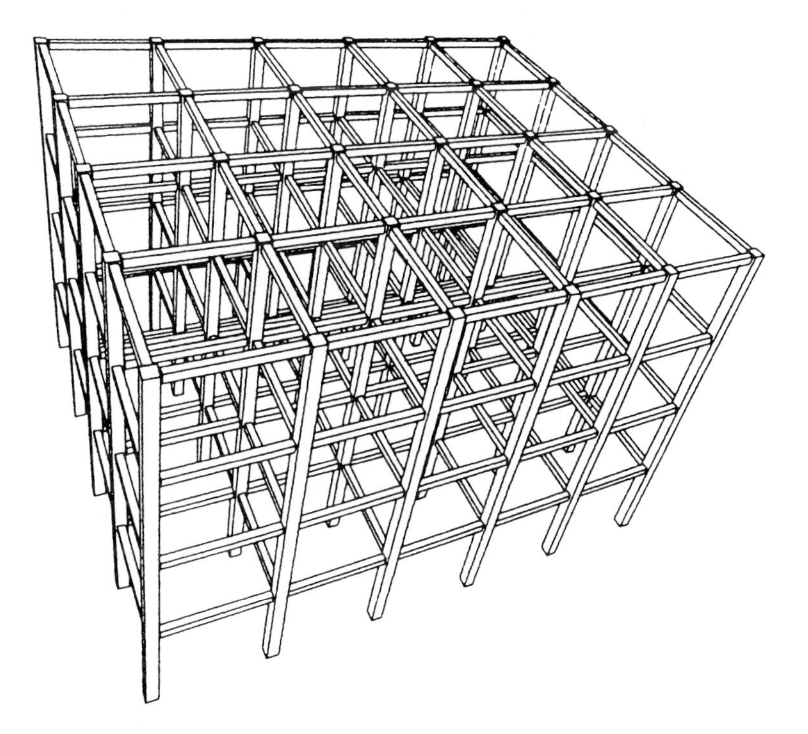

Fig. 1.3 A four-storey structural frame

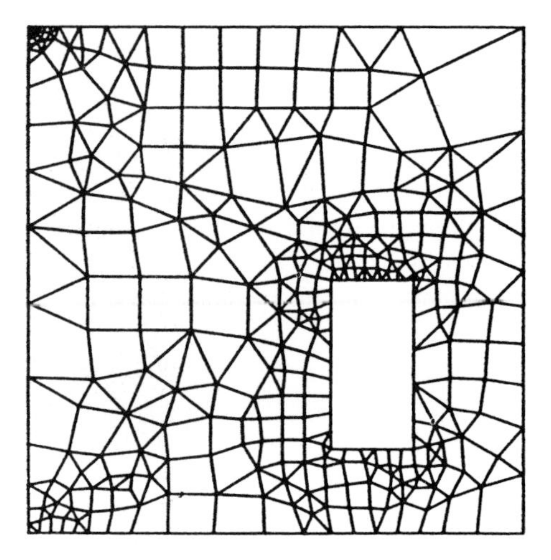

Fig. 1.4 The finite element discretization of a deep beam

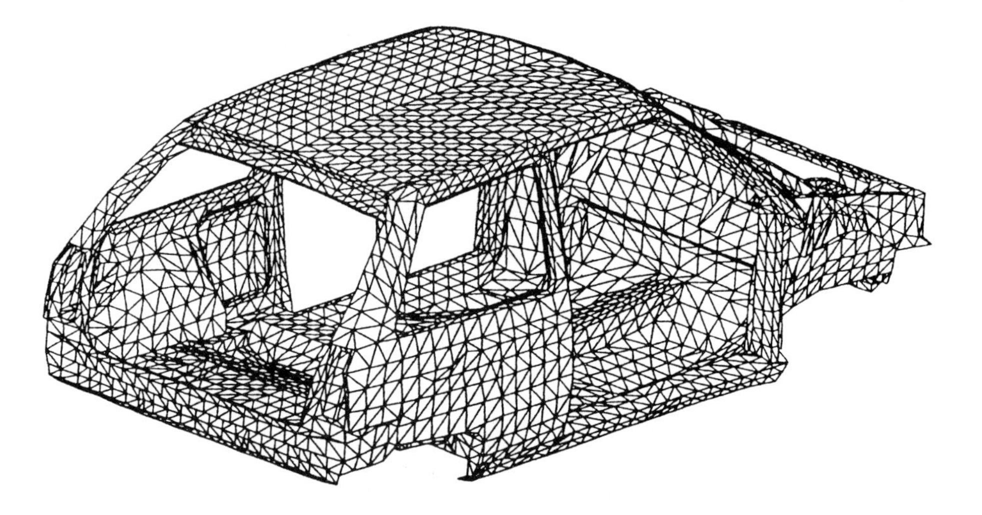

Fig. 1.5 The finite element model of a car

[26]. It can be seen that small elements are used at corner points of high stress concentration. Figure 1.5 shows the isometric view of a finite element model of a car body, which has been divided into 11 732 triangular shell elements. The last example is the discretization of a solid object, in which a champagne glass is discretized into 1117 tetrahedral elements, as shown in Fig. 1.6.

1.6 Displacement functions

There are three common types of displacement functions, and they will be discussed briefly below.

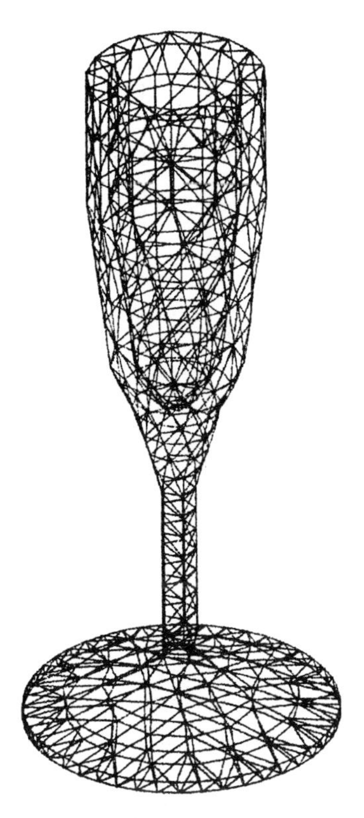

Fig. 1.6 A glass divided into tetrahedral elements

(a) Displacement function given as a simple polynomial
As an example, the displacement function of a two-dimensional element can be expressed as

$$\phi = a_1 + a_2x + a_3y + a_4x^2 + a_5xy + a_6y^2 + \cdots \tag{1.1}$$

in which the a_is are undetermined coefficients which can however be expressed in terms of the nodal displacement parameters later.
(b) Displacements given in terms of interpolation functions
Interpolation functions are usually osculatory polynomials (for one-dimensional elements) or products of osculatory polynomials (for two- and three-dimensional elements), yielding unit value of the displacement or its partial derivatives at the relevant node but zero values at all other nodes in the element. In effect, the function gives the shape of the displacement field within the element due to a unit displacement (slope) at the relevant node. Hence, the displacement function is given as

$$\phi = H_1(x, y)\phi_1 + H_2(x, y)\phi_2 + H_3(x, y)\phi_3 + \cdots \tag{1.2}$$

in which the ϕ_is are the nodal displacement parameters, and the H_is are interpolation functions.

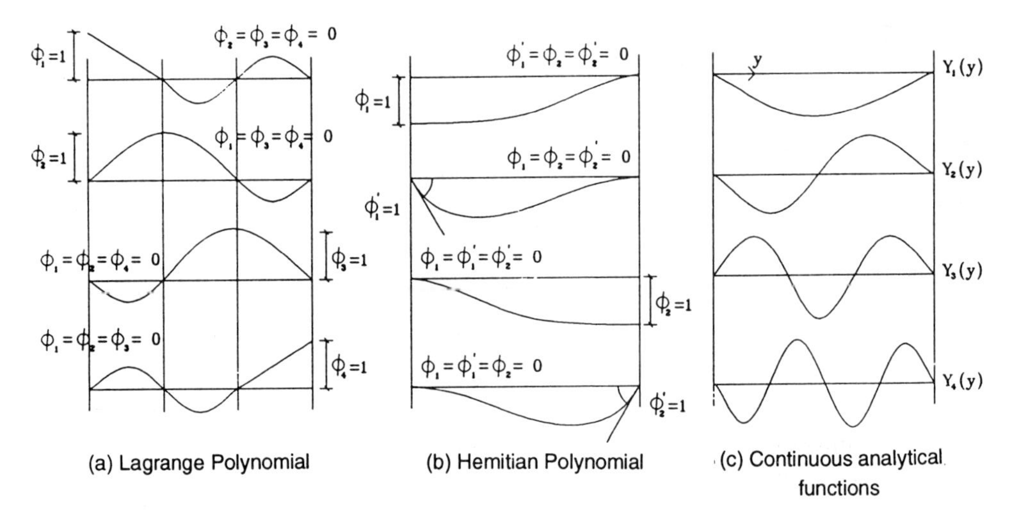

Fig. 1.7 Interpolation functions

The two types of commonly used osculatory polynomials are:
(i) Lagrange polynomials, which are often used for the construction of interpolation functions of elements in which only the functional values are specified at the nodes, as shown in Fig. 1.7(a). Examples are lower order plane stress elements [7] involving only nodal displacements u and v, and the Mindlin plate bending elements [27] involving w, θ_x and θ_y.
(ii) Hermitian polynomials, which are used for the construction of interpolation functions of elements in which not only the functional values but also the partial derivatives at the nodes are required, as shown in Fig. 1.7(b). Examples are the beam type 4-node plane stress element [28] with displacements u, v and slope $\partial u/\partial y$ as nodal parameters, and the Kirchhoff plate bending element [29] involving w, $\partial w/\partial x$, $\partial w/\partial y$ at each nodal point.
(c) Displacement function given in terms of analytical functions
The displacement functions of a plate bending finite strip [30] with two nodal lines can be written as

$$w = \sum_{k=1}^{n} \left[H_1(x)w_{1k} + H_2(x)\theta_{1k} + H_3(x)w_{2k} + H_4(x)\theta_{2k} \right] Y_k(y) \qquad (1.3)$$

in which $H_i(x)$ are cubic interpolation functions of a simple beam element, w_{ik}, θ_{ik} are the nodal parameters, and $Y_k(y)$ are continuous functions in the y direction, as shown in Fig. 1.7(c). A commonly used $Y_k(y)$ is the eigenfunction series which are derived from the solution of the beam vibration equation.

1.7 A SIMPLE EXAMPLE

A simple example concerning a bar with stepped sections under axial forces, as shown in Fig. 1.8, is used to illustrate the essential steps in a finite element analysis.
The system shown in Fig. 1.8(a) can be discretized into three subsystems, each of which consists of just one single bar element as depicted in Fig. 1.8(b). The

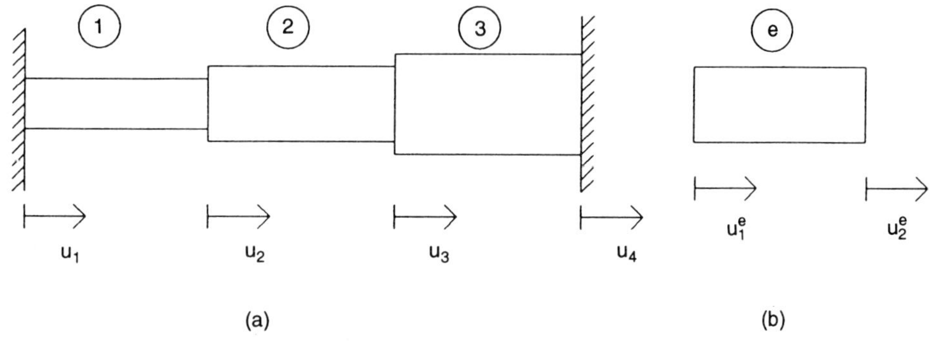

Fig. 1.8 A bar with stepped sections under axial loads

stiffness equation of a bar element relating the forces and displacements at the two nodal points is given by

$$\begin{bmatrix} k_e & -k_e \\ -k_e & k_e \end{bmatrix} \begin{bmatrix} u_1^e \\ u_2^e \end{bmatrix} = \begin{bmatrix} f_1^e \\ f_2^e \end{bmatrix} \quad \text{or} \quad \mathbf{K}_e \cdot \mathbf{u}_e = \mathbf{f}_e \qquad e = 1, 2, 3 \tag{1.4}$$

where $k_e = \mathrm{E}_e A_e / L_e$, with E_e, A_e, L_e being respectively the Young's modulus, cross-sectional area and length of the bar element e; $\mathbf{u}_e$ and $\mathbf{f}_e$ are respectively the displacements and forces at the nodal points.

From Fig. 1.8(a), it can be seen that the global parameters of the system are

$$\mathbf{u} = [u_1 u_2 u_3 u_4]^T \tag{1.5}$$

and that it is possible to relate the element displacement vector $\mathbf{u}_e$ and the global parameters $\mathbf{u}$ through a transformation matrix

$$\mathbf{u}_e = \mathbf{T}_e \cdot \mathbf{u} \qquad e = 1, 2, 3 \tag{1.6}$$

For the present case,

$$\mathbf{u}_1 = \begin{bmatrix} 1 & 0 & 0 & 0 \\ 0 & 1 & 0 & 0 \end{bmatrix} [u_1 u_2 u_3 u_4]^T = \mathbf{T}_1 \cdot \mathbf{u}$$

$$\mathbf{u}_2 = \begin{bmatrix} 0 & 1 & 0 & 0 \\ 0 & 0 & 1 & 0 \end{bmatrix} [u_1 u_2 u_3 u_4]^T = \mathbf{T}_2 \cdot \mathbf{u}$$

$$\mathbf{u}_3 = \begin{bmatrix} 0 & 0 & 1 & 0 \\ 0 & 0 & 0 & 1 \end{bmatrix} [u_1 u_2 u_3 u_4]^T = \mathbf{T}_3 \cdot \mathbf{u}$$

$$\mathbf{K}_e \cdot \mathbf{u}_e = \mathbf{f}_e \Rightarrow \mathbf{K}_e \cdot \mathbf{T}_e \cdot \mathbf{u} = \mathbf{f}_e \Rightarrow \mathbf{T}_e^{\mathrm{T}} \cdot \mathbf{K}_e \cdot \mathbf{T}_e \cdot \mathbf{u} = \mathbf{T}_e^{\mathrm{T}} \cdot \mathbf{f}_e \qquad e = 1, 2, 3$$

By simple direct multiplication, the element stiffness equation can be expressed in terms of the global displacements and forces.

For element 1,

$$\begin{bmatrix} k_1 & -k_1 & 0 & 0 \\ -k_1 & k_1 & 0 & 0 \\ 0 & 0 & 0 & 0 \\ 0 & 0 & 0 & 0 \end{bmatrix} \begin{bmatrix} u_1 \\ u_2 \\ u_3 \\ u_4 \end{bmatrix} = \begin{bmatrix} f_1^1 \\ f_2^1 \\ 0 \\ 0 \end{bmatrix} \tag{1.7a}$$

For element 2,

$$\begin{bmatrix} 0 & 0 & 0 & 0 \\ 0 & k_2 & -k_2 & 0 \\ 0 & -k_2 & k_2 & 0 \\ 0 & 0 & 0 & 0 \end{bmatrix} \begin{bmatrix} u_1 \\ u_2 \\ u_3 \\ u_4 \end{bmatrix} = \begin{bmatrix} 0 \\ f_1^2 \\ f_2^2 \\ 0 \end{bmatrix} \tag{1.7b}$$

For element 3,

$$\begin{bmatrix} 0 & 0 & 0 & 0 \\ 0 & 0 & 0 & 0 \\ 0 & 0 & k_3 & -k_3 \\ 0 & 0 & -k_3 & k_3 \end{bmatrix} \begin{bmatrix} u_1 \\ u_2 \\ u_3 \\ u_4 \end{bmatrix} = \begin{bmatrix} 0 \\ 0 \\ f_1^3 \\ f_2^3 \end{bmatrix} \tag{1.7c}$$

The global stiffness matrix equation of the system can now be obtained by adding up the equilibrium equations (1.7a–c) of individual elements.

$$\begin{bmatrix} k_1 & -k_1 & 0 & 0 \\ -k_1 & k_1+k_2 & -k_2 & 0 \\ 0 & -k_2 & k_2+k_3 & -k_3 \\ 0 & 0 & -k_3 & k_3 \end{bmatrix} \begin{bmatrix} u_1 \\ u_2 \\ u_3 \\ u_4 \end{bmatrix} = \begin{bmatrix} f_1^1 \\ f_2^1 + f_1^2 \\ f_2^2 + f_1^3 \\ f_2^3 \end{bmatrix} = \begin{bmatrix} f_1 \\ f_2 \\ f_3 \\ f_4 \end{bmatrix} \tag{1.8}$$

As stated earlier, the system stiffness matrix equation has to be modified in accordance with the specified boundary conditions. From Fig. 1.8(a), it is seen that nodes 1 and 4 are fixed and therefore $u_1 = u_4 = 0$. With the introduction of the boundary conditions, the final system can be reduced to

$$\begin{bmatrix} k_1+k_2 & -k_2 \\ -k_2 & k_2+k_3 \end{bmatrix} \begin{bmatrix} u_2 \\ u_3 \end{bmatrix} = \begin{bmatrix} f_2 \\ f_3 \end{bmatrix} \tag{1.9}$$

It is obvious that u_2 and u_3 can be solved for given values of f_2 and f_3. All the internal forces can then be computed from eq. 1.4.

1.8 Energy approach

The energy concept is one of the most common approaches in the finite element method. It uses the stationary principle of total potential energy which states that among all the permissible configurations, the actual configuration (u) will make the total potential energy $V(u)$ stationary, or, mathematically

$$\delta V(u) = 0 \tag{1.10}$$

In a static analysis,

$$V(u) = U(u) - W(u) \tag{1.11}$$

where U is the strain energy given by

$$U = \tfrac{1}{2} \int \boldsymbol{\varepsilon}^{\mathrm{T}} \cdot \mathbf{C} \cdot \boldsymbol{\varepsilon} \, \mathrm{d}\Omega \tag{1.12}$$

and W is the work done due to external load q

$$W = \int q u \, \mathrm{d}\Omega \tag{1.13}$$

in which $\boldsymbol{\varepsilon}$ is the strain vector, $\mathbf{C}$ is the constitutive matrix and $\mathrm{d}\Omega$ is the infinitesimal volume. If the system is discretized into elements, then

$$U = \sum_e U_e \qquad W = \sum_e W_e \tag{1.14}$$

where

$$U_e = \tfrac{1}{2}\int \boldsymbol{\varepsilon}_e^{\mathrm{T}} \cdot \mathbf{C}_e \cdot \boldsymbol{\varepsilon}_e \, \mathrm{d}\Omega_e \quad \text{and} \quad W_e = \int q_e u_e \, \mathrm{d}\Omega_e \tag{1.15}$$

By means of interpolation functions (eq. 1.2),

$$u_e = H \cdot \mathbf{u}_e \tag{1.16}$$

where $\mathbf{u}_e$ is the nodal displacement vector. The strain vector can then be obtained by direct differentiation

$$\boldsymbol{\varepsilon}_e = \partial u_e = (\partial H) \cdot \mathbf{u}_e = \mathbf{B} \cdot \mathbf{u}_e \tag{1.17}$$

where ∂ is an appropriate differential operator and $\mathbf{B}$ is the strain–displacement matrix. Hence, we can write

$$U_e = \tfrac{1}{2}\mathbf{u}_e^{\mathrm{T}} \cdot \left(\int \mathbf{B}^{\mathrm{T}} \cdot \mathbf{C} \cdot \mathbf{B} \, \mathrm{d}\Omega_e \right) \cdot \mathbf{u}_e = \tfrac{1}{2}\mathbf{u}_e^{\mathrm{T}} \cdot \mathbf{K}_e \cdot \mathbf{u}_e \tag{1.18}$$

and

$$W_e = \int q_e H \cdot \mathbf{u}_e \, \mathrm{d}\Omega_e = \mathbf{f}_e \cdot \mathbf{u}_e \tag{1.19}$$

where element stiffness matrix $\mathbf{K}_e = \int \mathbf{B}^{\mathrm{T}} \cdot \mathbf{C} \cdot \mathbf{B} \, \mathrm{d}\Omega_e$,
and element equivalent force $\mathbf{f}_e = \int q_e H \, \mathrm{d}\Omega_e$. (1.20)

Suppose that the element displacement vector $\mathbf{u}_e$ is related to the system nodal displacement vector $\mathbf{u}$ through the transformation $\mathbf{T}_e$, then

$$\mathbf{u}_e = \mathbf{T}_e \cdot \mathbf{u} \tag{1.21}$$

Equations 1.11, 1.14, 1.18 and 1.19 give

$$\begin{aligned} V(u) &= \sum_e (U_e - W_e) \\ &= \sum_e \tfrac{1}{2}\mathbf{u}_e^{\mathrm{T}} \cdot \mathbf{K}_e \cdot \mathbf{u}_e - \sum_e \mathbf{f}_e \cdot \mathbf{u}_e \\ &= \tfrac{1}{2}\mathbf{u}^{\mathrm{T}} \cdot \left(\sum_e \mathbf{T}_e^{\mathrm{T}} \mathbf{K}_e T_e \right) \cdot \mathbf{u} - \left(\sum_e \mathbf{f}_e \cdot \mathbf{T}_e \right) \cdot \mathbf{u} \end{aligned}$$

The variational principle (1.10) gives

$$\delta V(u) = \delta \mathbf{u}^{\mathrm{T}} \cdot \left[\left(\sum_e \mathbf{T}_e^{\mathrm{T}} \mathbf{K}_e \mathbf{T}_e \right) \cdot \mathbf{u} - \left(\sum_e \mathbf{f}_e \cdot \mathbf{T}_e \right) \right] = 0$$

or

$$\mathbf{K} \cdot \mathbf{u} = \mathbf{f} \tag{1.22}$$

where $\mathbf{K} = \sum_e \mathbf{T}_e^{\mathrm{T}} \mathbf{K}_e \mathbf{T}_e$ and $\mathbf{f} = \sum_e \mathbf{f}_e \cdot \mathbf{T}_e$
are the system stiffness and the force vector respectively.

Equation 1.22 is to be solved for system nodal displacement vector **u** when **f** is given. In the example of Section 1.7, if $\xi = x/L_e$, then

$$u_e = [1-\xi \quad \xi] \cdot \mathbf{u}_e = H \cdot u_e, \qquad \varepsilon_e = \frac{\partial}{\partial x} u_e = \frac{1}{L_e}[-1 \; 1] \cdot \mathbf{u}_e = \mathbf{B} \cdot \mathbf{u}_e$$

$$\mathbf{K}_e = \int \mathbf{B}^{\mathrm{T}} \cdot \mathbf{C} \cdot \mathbf{B} \, \mathrm{d}\Omega_e = \int \mathrm{E}_e (\mathbf{B}^{\mathrm{T}} \cdot \mathbf{B}) \, \mathrm{d}A \, \mathrm{d}x = \begin{bmatrix} k & -k \\ -k & k \end{bmatrix} \qquad \text{with } k = \frac{E_e A_e}{L_e}.$$

1.9 A BRIEF REVIEW OF THE CHAPTERS

The book is divided into eight chapters, including this introductory chapter. Two-dimensional and axisymmetric problems are discussed in Chapter 2. An explicit derivation of the element matrices of two practical 6-node and 8-node quadratic elements T6 and Q8 are presented. The use of infinite elements for problems involving infinite or semi-infinite domains is briefly discussed at the end of the chapter. Three-dimensional problems are dealt with in Chapter 3. Three quadratic isoparametric elements T10, P15 and H20 are first introduced at the beginning of the chapter. The numerical integration in three dimensions is then discussed. The steady-state problem governed by the quasi-harmonic equation is studied in Section 3.4, whereas the three-dimensional elasticity problem is examined in Section 3.5. The plate bending and shell finite elements are addressed in Chapter 4, and a comprehensive discussion on general Reissner–Mindlin plate elements is given. Plate bending elements with straight edges is also presented. This is followed by a thorough discussion on the degenerated shell element. At the end of the chapter, shell structures made up of an assembly of flat elements are investigated. The detailed formulation of a plate bending element L9P and that of a degenerated shell element L9S are given in Sections 4.5 and 4.9 respectively.

A substructure technique is discussed in Chapter 5, and it is demonstrated that the structural analysis can be much simplified if proper use is made on the symmetry and periodicity of the structure. The implication on finite element analysis due to structural symmetry and periodicity is explored in Sections 5.3 and 5.4. The two-level finite element method, which is a relatively new technique designed to reduce the number of unknowns without sacrificing the accuracy of the solution, is introduced in Chapter 6. Its application to building frames, plate bending, plane cracks, and laminated plates is discussed in respective sections of the chapter. From experience, it has been found that it is almost impossible, or at least impractical, to discretize a complex engineering system into finite elements by hand, and that it would be much better to leave the job of domain discretization to the computer. Accordingly, in Chapter 7, an exposé on the various approaches of automatic mesh generation is given. The mesh generation problem for different types of domains – planar regions, curved surfaces and solid volumes – is discussed separately in three sections. Chapter 8 is the last and longest chapter in which various aspects concerning finite element implementation are addressed. Optimization of matrix profile, solution of a system of linear equations, assembly of a system stiffness matrix, and introduction of boundary and loading conditions are the main topics discussed. A compact program named FACILE, for solving linear elastic and field problems based on the theory presented, is given in Section 8.6. A number of illustrative example runs of the program are also included.

1.10 INDICIAL NOTATION

Sometimes it is more convenient to refer to similar physical quantities by using indices. For instance, the axes of a Cartesian coordinate system in three orthogonal directions can be named x_1, x_2 and x_3 instead of x, y and z. An index itself presents no new ideas, it is just another way of expressing the same thing. However, by using indices instead of classical symbols, equations and formulae can now be written in a much more compact form. Familiarity with indicial notation will mean not only the saving of a lot of valuable space but also the streamlining of derivations and the generalizing of concepts.

Let us consider the Cauchy formula relating the internal stresses and the surface traction. Using the classical notation, the coordinate axes are named x, y and z. Hence, stresses will be written as σ_{xx}, σ_{xy}, σ_{xz}, etc., and the traction forces p, q, r on a plane defined by direction cosines l, m, n are given by

$$
\begin{aligned}
p &= \sigma_{xx}l + \sigma_{xy}m + \sigma_{xz}n \\
q &= \sigma_{yx}l + \sigma_{yy}m + \sigma_{yz}n \\
r &= \sigma_{zx}l + \sigma_{zy}m + \sigma_{zz}n
\end{aligned}
\tag{1.23}
$$

If indicial notation is employed, the coordinate axes will be named x_1, x_2 and x_3, and stresses will be written as σ_{11}, σ_{12}, σ_{13}, etc. The traction forces t_1, t_2, t_3 on a plane defined by direction cosines n_1, n_2, n_3 are given by

$$
\begin{aligned}
t_1 &= \sigma_{11}n_1 + \sigma_{12}n_2 + \sigma_{13}n_3 \\
t_2 &= \sigma_{21}n_1 + \sigma_{22}n_2 + \sigma_{23}n_3 \\
t_3 &= \sigma_{31}n_1 + \sigma_{32}n_2 + \sigma_{33}n_3
\end{aligned}
\tag{1.24}
$$

Putting it in a more compact form, we can write

$$
t_i = \sigma_{i1}n_1 + \sigma_{i2}n_2 + \sigma_{i3}n_3 \qquad i = 1, 2, 3 \tag{1.25}
$$

If summation convention is enforced on repeated indices (summing over the known range), a further simplification can be made, and we have

$$
t_i = \sum_{j=1}^{3} \sigma_{ij}n_j = \sigma_{ij}n_j \qquad i = 1, 2, 3 \tag{1.26}
$$

Compared to eq. 1.23, it is seen that eq. 1.26 is more compact and is easier to interpret than the system of equations expressed in terms of classical symbols.

Instead of working with individual components, traction force, stress and unit normal can each be considered as a single physical quantity, and the Cauchy formula can be expressed in an even more elegant way as

$$
\mathbf{t} = \boldsymbol{\sigma} \cdot \mathbf{n} \tag{1.27}
$$

in which $\mathbf{t} = (t_1, t_2, t_3)$ is the traction force vector, $\boldsymbol{\sigma} = [\sigma_{ij}]_{3\times3}$ is the stress tensor, and $\mathbf{n} = (n_1, n_2, n_3)$ is the unit normal to the surface. An additional advantage in using vectors and tensors is that the Cauchy formula in two dimensions will take exactly the same form.

Indicial notation will be used extensively in this book. However, for the more important end results, the matrix equation will also be given for readers who are not yet accustomed to using indices. Some of the basic operations on indices will be explained in the following paragraphs. For further details readers can refer to a standard mathematical text, e.g. Sokolnikoff [31].

When describing Cartesian coordinates or vector quantities, it is convenient to use Latin subscripts to represent its components. For example,

$$x_i \qquad i = 1, 2, 3$$

are the coordinates of position vector $\mathbf{x} = (x_1, x_2, x_3)$ or (x, y, z) of a three-dimensional Cartesian coordinate system; and

$$x_i \qquad i = 1, 2$$

will represent the coordinates (x_1, x_2) or (x, y) of a two-dimensional Cartesian coordinate system.

Similarly, the same notation can be used to describe displacement,

$$u_i \qquad i = 1, 2, 3$$

represent the components of the displacement vector $\mathbf{u} = (u_1, u_2, u_3)$ or (u, v, w).

To avoid confusion, the displacement at a nodal point will be identified by a Greek letter. For instance, the displacement vector at node α of an element is denoted by $\mathbf{u}_\alpha$, whereas its components are given by

$$u^i_\alpha \qquad i = 1, 2, 3$$

The submatrix $\mathbf{K}_{\alpha\beta}$ relating the force at node α due to a displacement at node β consists of 2×2 or 3×3 entries depending on whether a two- or three-dimensional domain is involved. In describing a stiffness submatrix coefficient, two subscripts have to be used. Hence, the components of submatrix $\mathbf{K}_{\alpha\beta}$ can be written as

$$\mathbf{K}^{ij}_{\alpha\beta} \qquad i, j = 1, 2, 3$$

In a two-dimensional stress analysis problem, the element stiffness matrix of a 6-node quadratic triangular element will have $6 \times 6 = 36$ submatrices, each of which contains $2 \times 2 = 4$ entries, making up a total of $36 \times 4 = 144$ coefficients in the element stiffness matrix. Each of these coefficients can be identified by four indices, α, β, i, j where $\alpha, \beta = 1, 2, 3, 4, 5, 6$ (or $\alpha, \beta = 1, 6$) and $i, j = 1, 2$.

In indicial notation, the derivative of any quantity with respect to a coordinate x_i is written compactly as

$$\frac{\partial}{\partial x_i} \equiv (\ )_{,i}$$

Thus we can write the gradient of the displacement vector as

$$\frac{\partial u_i}{\partial x_j} \equiv u_{i,j} \qquad i, j = 1, 2, 3$$

When i and j are allowed to vary from 1 to 3, we can see that $u_{i,j}$ describes the nine components of the displacement gradient.

Using the notion that strain is the symmetric part of the displacement gradient,

$$\varepsilon_{ij} = \tfrac{1}{2}(u_{i,j} + u_{j,i}) = \varepsilon_{ji} \qquad i, j = 1, 2, 3$$

and rotation is the antisymmetric part of the displacement gradient

$$\omega_{ij} = \tfrac{1}{2}(u_{i,j} - u_{j,i}) = -\omega_{ji} \qquad i, j = 1, 2, 3$$

The components ε_{ij} of the strain tensor $\boldsymbol{\varepsilon}$ can also be put in a matrix form:

$$\boldsymbol{\varepsilon} = [\varepsilon_{ij}] = \begin{bmatrix} \varepsilon_{11} & \varepsilon_{12} & \varepsilon_{13} \\ \varepsilon_{21} & \varepsilon_{22} & \varepsilon_{23} \\ \varepsilon_{31} & \varepsilon_{32} & \varepsilon_{33} \end{bmatrix}$$

or in vector form,

$$\vec{\varepsilon} = \begin{bmatrix} \varepsilon_{11} \\ \varepsilon_{22} \\ \varepsilon_{33} \\ \varepsilon_{12} \\ \varepsilon_{23} \\ \varepsilon_{31} \end{bmatrix}$$

Similarly, the stress components of the stress tensor $\boldsymbol{\sigma}$ may also be written as $\sigma_{ij}(i, j = 1, 2, 3)$ and arranged in matrix form,

$$\boldsymbol{\sigma} = [\sigma_{ij}] = \begin{bmatrix} \sigma_{11} & \sigma_{12} & \sigma_{13} \\ \sigma_{21} & \sigma_{22} & \sigma_{23} \\ \sigma_{31} & \sigma_{32} & \sigma_{33} \end{bmatrix}$$

or in vector form,

$$\vec{\sigma} = \begin{bmatrix} \sigma_{11} \\ \sigma_{22} \\ \sigma_{33} \\ \sigma_{12} \\ \sigma_{23} \\ \sigma_{31} \end{bmatrix}$$

Let $\vec{\gamma} = [\gamma_i] = (\gamma_1, \gamma_2, \gamma_3)$ be the body force vector. We may write the equilibrium equations (linear momentum balance) for an element of volume as

$$\sigma_{ij,j} + \gamma_i = 0 \qquad i = 1, 2, 3$$

where the repeated dummy index implies a summation over the known range of the index, i.e.

$$\sigma_{ij,j} \equiv \sum_{j=1}^{3} \sigma_{ij,j} \equiv \sigma_{i1,1} + \sigma_{i2,2} + \sigma_{i3,3}$$

The following is a second example of summation over repeated indices. The field variable ϕ over a 6-node triangular element T6 can be expressed in terms of the nodal values $\phi_1, \phi_2, \ldots, \phi_6$ and the interpolation functions $H_1, H_2, \ldots, H_6$ such that

$$\phi = H_1\phi_1 + H_2\phi_2 + H_3\phi_3 + H_4\phi_4 + H_5\phi_5 + H_6\phi_6 = H_\alpha\phi_\alpha$$

As a further example of the summation convention, consider a term $\sigma_{ij}\varepsilon_{ij}$ which has a unit of work per unit volume. This term contains two pairs of repeated indices representing a double summation; hence summing first over i gives

$$\sigma_{ij}\varepsilon_{ij} = \sigma_{1j}\varepsilon_{1j} + \sigma_{2j}\varepsilon_{2j} + \sigma_{3j}\varepsilon_{3j}$$

and then summing on j gives

$$\sigma_{ij}\varepsilon_{ij} = \sigma_{11}\varepsilon_{11} + \sigma_{21}\varepsilon_{21} + \sigma_{31}\varepsilon_{31} + \sigma_{12}\varepsilon_{12} + \sigma_{22}\varepsilon_{22} + \sigma_{32}\varepsilon_{32} + \sigma_{13}\varepsilon_{13} + \sigma_{23}\varepsilon_{23} + \sigma_{33}\varepsilon_{33}$$

The dot product between two vectors $\mathbf{a} = [a_i]$ and $\mathbf{b} = [b_i]$ is a scalar given by

$$\mathbf{a} \cdot \mathbf{b} = a_i b_i = a_1 b_1 + a_2 b_2 + a_3 b_3$$

The dot product between a second order tensor $\mathbf{T}$ and a vector $\mathbf{a}$ is again a vector which is given by

$$\mathbf{T} \cdot \mathbf{a} = [T_{ij}a_j] = (T_{1j}a_j, T_{2j}a_j, T_{3j}a_j)$$

$$\mathbf{a} \cdot \mathbf{T} = [a_i T_{ij}] = (a_i T_{i1}, a_i T_{i2}, a_i T_{i3})$$

It is noted that $\mathbf{T} \cdot \mathbf{a} = \mathbf{a} \cdot \mathbf{T}$ if $\mathbf{T}$ is symmetric.
The double dot product between two tensors $\mathbf{S}$ and $\mathbf{T}$ is a scalar given by

$$\mathbf{S} : \mathbf{T} = S_{ij} T_{ij}$$

For a linear elastic material, the most general linear relationship we can write to express the stress–strain characteristics is

$$\sigma_{ij} = \mathbf{C}_{ijkl}\varepsilon_{kl} \tag{1.28}$$

The elastic modulus appearing in eq. 1.28 is a fourth order tensor whose components are addressed by four subscripts. Hence $\mathbf{C}_{ijkl}$ characterizes the change in stress σ_{ij} due to a variation in strain ε_{kl}. The elastic modulus for an isotropic linear elastic material may be written as

$$\mathbf{C}_{ijkl} = \lambda\delta_{ij}\delta_{kl} + \mu(\delta_{ik}\delta_{jl} + \delta_{il}\delta_{jk}) \tag{1.29}$$

where λ and μ are the Lamé constants which can be deduced from the Young's modulus E and the Poisson's ratio ν as

$$\lambda = \frac{\nu E}{(1+\nu)(1-2\nu)} \quad \text{and} \quad \mu = \frac{E}{2(1+\nu)} \tag{1.30}$$

δ_{ij} is the Kronecker delta defined as

$$\delta_{ij} = \begin{cases} 1 & i = j \\ 0 & i \neq j \end{cases}$$

Hence for linear elastic material, the stress–strain relationship can be simplified to

$$\begin{aligned} \sigma_{ij} &= [\lambda\delta_{ij}\delta_{kl} + \mu(\delta_{ik}\delta_{jl} + \delta_{il}\delta_{jk})]\varepsilon_{kl} \\ &= \lambda\delta_{ij}\varepsilon_{kk} + \mu(\delta_{ik}\varepsilon_{kj} + \delta_{il}\varepsilon_{jl}) \\ &= \lambda\delta_{ij}\varepsilon_{kk} + \mu(\varepsilon_{ij} + \varepsilon_{ji}) \\ &= \lambda\delta_{ij}\varepsilon_{kk} + 2\mu\varepsilon_{ij} \end{aligned} \tag{1.31}$$

When the index equation is expanded, we get the individual stress components as

$$
\begin{aligned}
\sigma_{11} &= \lambda(\varepsilon_{11} + \varepsilon_{22} + \varepsilon_{33}) + 2\mu\varepsilon_{11} \\
\sigma_{22} &= \lambda(\varepsilon_{11} + \varepsilon_{22} + \varepsilon_{33}) + 2\mu\varepsilon_{22} \\
\sigma_{33} &= \lambda(\varepsilon_{11} + \varepsilon_{22} + \varepsilon_{33}) + 2\mu\varepsilon_{33} \\
\sigma_{12} &= 2\mu\varepsilon_{12} = \sigma_{21} \\
\sigma_{23} &= 2\mu\varepsilon_{23} = \sigma_{32} \\
\sigma_{31} &= 2\mu\varepsilon_{31} = \sigma_{13}
\end{aligned} \tag{1.32}
$$

This can be compared with the corresponding matrix equation between stress and strain

$$
\begin{bmatrix} \sigma_{11} \\ \sigma_{22} \\ \sigma_{33} \\ \sigma_{12} \\ \sigma_{23} \\ \sigma_{31} \end{bmatrix} = \begin{bmatrix} \lambda + 2\mu & \lambda & \lambda & 0 & 0 & 0 \\ \lambda & \lambda + 2\mu & \lambda & 0 & 0 & 0 \\ \lambda & \lambda & \lambda + 2\mu & 0 & 0 & 0 \\ 0 & 0 & 0 & 2\mu & 0 & 0 \\ 0 & 0 & 0 & 0 & 2\mu & 0 \\ 0 & 0 & 0 & 0 & 0 & 2\mu \end{bmatrix} \begin{bmatrix} \varepsilon_{11} \\ \varepsilon_{22} \\ \varepsilon_{33} \\ \varepsilon_{12} \\ \varepsilon_{23} \\ \varepsilon_{13} \end{bmatrix} \tag{1.33}
$$

It can be seen that the index equation (1.31) is more compact than eq. 1.33, and it is also more suitable for the programming purpose for two reasons. Firstly, the multiplication on zeros can be avoided; and secondly, the index equation can be used directly in a program loop with the loop counters serving as the index variables.

A spatial rate of change of a physical quantity can be obtained by the differential operator ∇ defined as follows:

$$
\nabla \equiv \left(\frac{\partial}{\partial x}, \frac{\partial}{\partial y}, \frac{\partial}{\partial z} \right) \equiv \left(\frac{\partial}{\partial x_1}, \frac{\partial}{\partial x_2}, \frac{\partial}{\partial x_3} \right)
$$

The gradient of a scalar field ψ is a vector which is given by

$$
\nabla\psi \equiv \left(\frac{\partial\psi}{\partial x}, \frac{\partial\psi}{\partial y}, \frac{\partial\psi}{\partial z} \right) \equiv \left(\frac{\partial\psi}{\partial x_1}, \frac{\partial\psi}{\partial x_2}, \frac{\partial\psi}{\partial x_3} \right)
$$

The gradient of a vector field **u** is a second order tensor given by

$$
\nabla\mathbf{u} = [u_{j,i}] = \begin{bmatrix} u_{1,1} & u_{2,1} & u_{3,1} \\ u_{1,2} & u_{2,2} & u_{3,2} \\ u_{1,3} & u_{2,3} & u_{3,3} \end{bmatrix}
$$

$$
\mathbf{u}\nabla = [u_{i,j}] = \begin{bmatrix} u_{1,1} & u_{1,2} & u_{1,3} \\ u_{2,1} & u_{2,2} & u_{2,3} \\ u_{3,1} & u_{3,2} & u_{3,3} \end{bmatrix}
$$

The divergence of a vector **u** is a scalar given by

$$
\nabla \cdot \mathbf{u} = \mathbf{u} \cdot \nabla = u_{1,1} + u_{2,2} + u_{3,3} = u_{i,i}
$$

Green–Gauss theorem

The Green–Gauss theorem describes the relationship between the surface integral for a vector field **v** in the direction normal to the bounding surface and the volume integral of the divergence of **v**.

$$\int_{\Gamma} \mathbf{v} \cdot \mathbf{n} \, \mathrm{d}\Gamma = \int_{\Omega} \mathbf{v} \cdot \nabla \, \mathrm{d}\Omega$$

where Γ and Ω denote the boundary surface of a region and the domain of the region respectively, and **n** is the unit normal vector to the surface.

1.11 References to Chapter 1

1. N. H. Hoff (1956) *Analysis of Structures*, Wiley, New York.
2. K. Wieghardt (1906) Über einen Grenzübergang der Elastizitätslehre und seine Anwendung auf die Statik hochgradig statisch uberstimmter Fachwerke, *Verhandlungen des Verein z. Beförderung des Gewerbefleisses. Abhandlungen*, **85**, 139–76.
3. W. Riedel (1927) Beiträge zur Lösung des ebenen Problems eines elastischen Körpers mittels der Airyschen Spannungsfunktion, *Zeitschrift für Angewandte Mathematik und Mechanik*, **7**, No. 3, 169–88.
4. D. McHenry (1943) A lattice analogy of the solution of plane stress problems, *J. Inst. Civil Engrs*, 59–82.
5. R. Courant (1943) Variational method for the solution of problems of equilibrium and vibrations, *Bulletin of the American Mathematical Society*, **49**, 1–23.
6. S. Levy (1953) Structural analysis and influence coefficients for Delta wings, *J. Aero. Sci.*, **20**, No. 7, 449–54.
7. M. J. Turner, R. W. Clough, H. C. Martin & L. J. Topp (1956) Stiffness and deflection analysis of complex structures, *J. Aero. Sci.*, **23**, No. 9, 805–23.
8. R. W. Clough (1960) The finite element method in plane stress analysis, *Proceedings of the Second Conference on Electronic Computation, ASCE*, 345–77.
9. R. J. Melosh (1963) Basis for derivation of matrices for the direct stiffness method, *J. Am. Inst. Aero. Astr.*, **1**, No. 7, 1631–7.
10. S. H. Crandall (1956) *Engineering Analysis*, McGraw-Hill, New York.
11. I. Holland & K. Bell (eds) (1969) *Finite Element Methods in Stress Analysis*, Tapir, Trondheim, Norway.
12. Symposium on Applied Finite Element Methods in Civil Engineering, Vanderbilt University, Nashville, USA (1969) ASCE.
13. R. H. Gallagher, Y. Yamada & J. T. Oden (eds) (1971) *Recent Advances in Matrix Methods of Structural Analysis and Design*, University of Alabama Press, Huntsville, USA.
14. B. F. de Veubeke (ed.) (1971) *High Speed Computing of Elastic Structures*, University of Liège, Belgium.
15. C. A. Brebbia & H. Tottenham (eds) (1973) *Variational Methods in Engineering*, Southampton University, UK.
16. S. J. Fenves, N. Perrone, J. Robinson & W. C. Schnobrich (eds) (1973) *Numerical and Computational Methods in Structural Mechanics*, Academic Press, New York.
17. R. H. Gallagher, J. T. Oden, C. Taylor & O. C. Zienkiewicz (eds) (1974) *International Symposium on Finite Element Method in Flow Problems*, Wiley.
18. K. J. Bathe, J. T. Oden & W. Wunderlich (eds) (1977) *Formulation and Computational Algorithms in Finite Element Analysis* (US–Germany Symposium), MIT Press.

19. W. G. Gray, G. F. Pinder & C. A. Brebbia (eds) (1977) *Finite Elements in Water Resources*, Pentech Press, London.
20. J. Robinson (ed.) (1978) *Finite Element Method in Commercial Environment*, Robinson and Associates, Dorset, UK.
21. R. Glowinski, E. Y. Rodin & O. C. Zienkiewicz (eds) (1979) *Energy Methods in Finite Element Analysis*, Wiley, Chichester, UK.
22. A. K. Aziz (ed.) (1972) *The Mathematical Foundations of the Finite Element Method with Applications to Partial Differential Equations*, Academic Press, New York.
23. J. R. Whiteman (ed.) (1973) *The Mathematics of Finite Elements and Applications*, Academic Press, London.
24. D. Norrie & G. de Vries (1976) *Finite Element Bibliography*, IFI/Plenum, New York.
25. O. C. Zienkiewicz (1991) *The Finite Element Method*, Volumes 1 & 2, 4th Edn, McGraw-Hill.
26. S. H. Lo & C. K. Lee (1992) On using meshes of mixed element types in adaptive finite element analysis, *J. Finite Elements in Analysis and Design.*
27. T. Rock & E. Hinton (1974) Free vibration and transient analysis of thick and thin plates using the finite element method, *Earthquake Engng Struct. Dym.*, **3**, 51–63.
28. R. G. Sisodiya & Y. K. Cheung (1971) A higher order in-plane parallelogram element and its applications to skewed curved box-girder bridges, *Developments in Bridge Design and Construction*, ed. by K. C. Rockey, J. L. Bannister & H. R. Evans, Crosby Lockwood, UK.
29. O. C. Zienkiewicz & Y. K. Cheung (1964) The finite element method for analysis of elastic isotropic and orthotropic slabs, *Proc. Inst. Civ. Engrs*, **28**, 471–80.
30. Y. K. Cheung (1976) *Finite Strip Method in Structural Analysis*, Pergamon Press.
31. I. S. Sokolnikoff (1964) *Tensor Analysis*, Wiley.

2 Two-dimensional and Axisymmetric Problems

2.1 Introduction

Although all structural engineering problems are three-dimensional in nature, some of them can be simplified to two-dimensional problems with little loss in accuracy. In this chapter, the plane stress/strain and axisymmetric problems will be discussed in detail. However, the more general three-dimensional formulation will be given in Chapter 3, from which the element stiffness matrix and element force vectors for the two-dimensional problems can be derived as special cases.

In plane strain problems, one deals with a situation in which the dimension of the structure in one direction, say the z-coordinate direction, is very large compared with the dimensions of the structure in the other two directions (x- and y-coordinate axes), and the applied forces which act in the $x-y$ plane do not vary in the z-direction. Practical applications of this representation occur in the analysis of dams, tunnels, retaining walls and other geotechnical works. Smaller scale problems include bars, rollers and tubes being acted upon by forces normal to their longitudinal axis. On the other hand, the plane stress situation will apply if a thin plate is loaded by forces applied at the boundary, parallel to the plane of the plate and distributed uniformly over the thickness, with the result that the stress components normal to the plate on both faces of the plate may be assumed to be zero. Examples of plane stress problems include plane frame analysis, shear wall analysis, and thin wall structures dominated by in-plane stresses.

Axisymmetric structures form another important class of three-dimensional problems. Numerous civil, mechanical, nuclear and aerospace engineering structures fall in the category of axisymmetric solids, including concrete tanks, nuclear pressure vessels, etc. The distinction with the general three-dimensional situation is that the material and geometric properties do not vary in the circumferential direction.

Steady-state field problems in two dimensions represent quite a large range of practical engineering problems such as torsion of a prismatic shaft, heat transfer, ground water flow, distribution of electric and magnetic potentials, etc. With appropriate interpretation for two dimensions, the finite element solution for the general quasi-harmonic equation developed in Chapter 3 can be applied directly without any special modification.

Although the 3-node triangular element has been used extensively in finite element analysis of two-dimensional problems, the more efficient quadratic isoparametric elements will however be developed in this chapter. In the following sections, the detailed formulation of element stiffness matrix and element force vectors of two popular isoparametric elements, namely the 6-node triangular element T6 and the 8-node quadrilateral element Q8 are given. The linear strain isoparametric elements T6 and Q8, apart from being very efficient in solving plane stress, plane strain and axisymmetric elasticity problems, have the further advantage that they can be fitted closely to curved boundaries without difficulty.

For regular domains, the Q8 element is the ideal choice both in terms of accuracy and domain discretization. However, when there is a need for mesh gradation, the T6 element could be regarded as more appropriate as the element properties of a Q8 element deteriorate rapidly as the element shape deviates from the regular square or rectangular shape [1,2]. A T6 element can also give better results than a Q8 element in the calculation of stresses around a crack tip by shifting a mid-side node to the 1/4 position towards the point of singularity [3]. Nevertheless, Q8 elements can still be very useful in various situations as rectangular or square elements can always be generated at the interior part of the domain, while the remaining parts may be filled up with the more flexible triangular elements by means of a general triangulation program [4]. Element stiffness matrices and element force vectors of subparametric triangular and rectangular elements can be obtained in explicit form by direct analytical integration, thereby reducing drastically the amount of computation at the element level [9,13]. The optimum use of T6 and Q8 elements can produce the most efficient finite element solution to two-dimensional boundary value continuum problems.

2.2 Plane stress and plane strain

The plane stress and plane strain conditions are the simplest form of behaviour for continuum structures and represent situations often encountered in practice. In fact, two-dimensional elasticity problems were the first successful examples of the application of the finite element method [5]. The strain and stress to be considered are the three components in the x_1–x_2 $(x-y)$ plane. In the case of plane stress, by definition, all other components of stress are zero and therefore do not contribute to the internal work. In plane strain, the stress in the direction perpendicular to the x_1–x_2 plane is not zero. However, by definition, the strain in that direction is zero, and therefore no contribution to the internal work is made by this stress, which can be evaluated from the three main stress components, if desired, at the end of the computation.

2.2.1 Linear elasticity

In plane stress and plane strain problems, the strains and stresses that have to be considered are the three components in the x_1–x_2 $(x-y)$ plane. Due to symmetry of the elasticity tensor, the elasticity matrix $\mathbf{C}_{ijkl}$ has only six independent components [6,7].

$$\mathbf{C} = \begin{bmatrix} C_{1111} & C_{1122} & C_{1112} \\ & C_{2222} & C_{2212} \\ \text{sym.} & & C_{1212} \end{bmatrix} \tag{2.1}$$

For practical engineering applications, however, we shall concentrate only on orthotropic materials for which, $C_{1112} = 0$ and $C_{2212} = 0$. The following table summarizes the elastic constants for isotropic and orthotropic materials:

	Orthotropic	Isotropic	
		Plane stress	Plane strain
C_{1111} C_{2222}	λ_1 λ_2	$\dfrac{E}{1-\nu^2}$	$\dfrac{E(1-\nu)}{(1+\nu)(1-2\nu)}$
C_{1212}	λ_3	$\dfrac{E}{2(1+\nu)}$	$\dfrac{E}{2(1+\nu)}$
C_{1122}	λ_4	$\dfrac{\nu E}{1-\nu^2}$	$\dfrac{\nu E}{(1+\nu)(1-2\nu)}$

where E = Young's modulus and ν = Poisson's ratio

(2.2)

2.2.2 Element stiffness matrix and element force vectors

The general formulations for the element stiffness matrix and the element force vectors are developed in Section 3.5.5 for three-dimensional problems. As mentioned previously, the two-dimensional formulations are also derived by limiting the Latin indices from 1, 2 and 3 to 1 and 2 only.

(a) Element stiffness matrix

Consider components $K_{ik}^{\alpha\beta} = C_{ijkl} I_{ji}^{\alpha\beta}$, where $i, j, k, l = 1, 2$, of the 2×2 element sub-matrix $\mathbf{K}^{\alpha\beta}$ associated with the element nodal force vector $\begin{bmatrix} f_1 \\ f_2 \end{bmatrix}_\alpha$ at node α and the displacement vector $\begin{bmatrix} u_1 \\ u_2 \end{bmatrix}_\beta$ at node β.

$$K_{11}^{\alpha\beta} = C_{1j1l} I_{jl}^{\alpha\beta} = C_{1111} I_{11}^{\alpha\beta} + C_{1112} I_{12}^{\alpha\beta} + C_{1211} I_{21}^{\alpha\beta} + C_{1212} I_{22}^{\alpha\beta}$$
$$= \lambda_1 I_{11}^{\alpha\beta} + \lambda_3 I_{22}^{\alpha\beta}$$

$$K_{12}^{\alpha\beta} = C_{1j2l} I_{jl}^{\alpha\beta} = C_{1121} I_{11}^{\alpha\beta} + C_{1122} I_{12}^{\alpha\beta} + C_{1221} I_{21}^{\alpha\beta} + C_{1222} I_{22}^{\alpha\beta}$$
$$= \lambda_4 I_{12}^{\alpha\beta} + \lambda_3 I_{21}^{\alpha\beta}$$

$$K_{21}^{\alpha\beta} = C_{2j1l} I_{jl}^{\alpha\beta} = C_{2111} I_{11}^{\alpha\beta} + C_{2112} I_{12}^{\alpha\beta} + C_{2211} I_{21}^{\alpha\beta} + C_{2212} I_{22}^{\alpha\beta}$$
$$= \lambda_3 I_{12}^{\alpha\beta} + \lambda_4 I_{21}^{\alpha\beta}$$

$$K_{22}^{\alpha\beta} = C_{2j2l} I_{jl}^{\alpha\beta} = C_{2121} I_{11}^{\alpha\beta} + C_{2122} I_{12}^{\alpha\beta} + C_{2221} I_{21}^{\alpha\beta} + C_{2222} I_{22}^{\alpha\beta}$$
$$= \lambda_3 I_{11}^{\alpha\beta} + \lambda_2 I_{22}^{\alpha\beta}$$

$$\underset{2\times 2}{K}^{\alpha\beta} = \begin{bmatrix} \lambda_1 I_{11}^{\alpha\beta} + \lambda_3 I_{22}^{\alpha\beta} & \lambda_4 I_{12}^{\alpha\beta} + \lambda_3 I_{21}^{\alpha\beta} \\ \lambda_3 I_{12}^{\alpha\beta} + \lambda_4 I_{21}^{\alpha\beta} & \lambda_3 I_{11}^{\alpha\beta} + \lambda_2 I_{22}^{\alpha\beta} \end{bmatrix} \tag{2.3}$$

Let $f_\alpha = \dfrac{\partial H_\alpha}{\partial x}$ and $g_\alpha = \dfrac{\partial H_\alpha}{\partial y}$, $\alpha = 1, N$, then

$$I_{11}^{\alpha\beta} = \int_{\Omega_e} f_\alpha f_\beta \, \mathrm{d}\Omega \quad I_{12}^{\alpha\beta} = \int_{\Omega_e} f_\alpha g_\beta \, \mathrm{d}\Omega \quad I_{21}^{\alpha\beta} = \int_{\Omega_e} g_\alpha f_\beta \, \mathrm{d}\Omega \quad I_{22}^{\alpha\beta} = \int_{\Omega_e} g_\alpha g_\beta \, \mathrm{d}\Omega$$

It has been noted by Zienkiewicz and Taylor [8] that the element sub-matrices $\mathbf{K}^{\alpha\beta}$ can also be obtained through matrix manipulations. The element stiffness matrix is given by

$$\underset{2N\times 2N}{\mathbf{K}_e} = \int_{\Omega_e} \underset{2N\times 3}{B^{\mathrm{T}}} \; \underset{3\times 3}{C} \; \underset{3\times 2N}{B} \, \mathrm{d}\Omega$$

where the strain matrix $\mathbf{B}^{\mathrm{T}} = \left[\underset{2\times 3}{B_1}, \ldots, \underset{2\times 3}{B_N}\right]$

and the elasticity matrix $\mathbf{C} = \begin{bmatrix} \lambda_1 & \lambda_4 & 0 \\ \lambda_4 & \lambda_2 & 0 \\ 0 & 0 & \lambda_3 \end{bmatrix}$

A typical 2×2 sub-matrix of $\mathbf{K}$ is given by

$$\underset{2\times 2}{\mathbf{K}^{\alpha\beta}} = \int_{\Omega_e} \mathbf{B}_\alpha^{\mathrm{T}} \mathbf{C} \, \mathbf{B}_\beta \, \mathrm{d}\Omega \qquad \text{with} \quad \mathbf{B}_\alpha^{\mathrm{T}} \begin{bmatrix} f_\alpha & 0 & g_\alpha \\ 0 & g_\alpha & f_\alpha \end{bmatrix} \tag{2.4}$$

$$= \int_{\Omega_e} \begin{bmatrix} f_\alpha & 0 & g_\alpha \\ 0 & g_\alpha & f_\alpha \end{bmatrix} \begin{bmatrix} \lambda_1 & \lambda_4 & 0 \\ \lambda_4 & \lambda_2 & 0 \\ 0 & 0 & \lambda_3 \end{bmatrix} \begin{bmatrix} f_\beta & 0 \\ 0 & g_\beta \\ g_\beta & f_\beta \end{bmatrix} \mathrm{d}\Omega$$

$$\underset{2\times 2}{\mathbf{K}^{\alpha\beta}} = \int_{A_e} t \begin{bmatrix} \lambda_1 f_\alpha f_\beta + \lambda_3 g_\alpha g_\beta & \lambda_4 f_\alpha g_\beta + \lambda_3 g_\alpha f_\beta \\ \lambda_3 f_\alpha g_\beta + \lambda_4 g_\alpha f_\beta & \lambda_3 f_\alpha f_\beta + \lambda_2 g_\alpha g_\beta \end{bmatrix} \mathrm{d}A \tag{2.5}$$

where A_e is the area and t is the thickness of the element respectively.

(b) Element force vectors

(i) Element initial stress(strain) force vector $\vec{r}_e$

$$r_i^\alpha = \int_{\Omega_e} H_{\alpha,j} \tau_{ij}^{\mathrm{o}} \, \mathrm{d}\Omega = \int_{A_e} t H_{\alpha,j} \tau_{ij}^{\mathrm{o}} \, \mathrm{d}A \tag{2.6}$$

For elements whose initial stress and strain are constant within the element, we have

$$r_i^\alpha = h_{\alpha j} \tau_{ij}^{\mathrm{o}} \qquad \text{where } h_{\alpha j} = \int_{A_e} t H_{\alpha,j} \, \mathrm{d}A$$

(ii) Element body force vector $\vec{\mathrm{b}}_e$

$$b_i^\alpha = \int_{\Omega_e} \gamma_i H_\alpha \, \mathrm{d}\Omega = \int_{\Omega_e} \gamma_i^\beta H_\beta H_\alpha \, \mathrm{d}\Omega = \gamma_i^\beta V_{\alpha\beta} \qquad \text{where } V_{\alpha\beta} = \int_{A_e} t H_\alpha H_\beta \, \mathrm{d}A$$

(iii) Element surface traction vector $\vec{\mathrm{s}}_e$

$$s_i^\alpha = \int_{\Gamma_e} q_i H_\alpha \, \mathrm{d}a = \int_{\mathrm{e}} t q_i H_\alpha \, \mathrm{d}\ell = q_i^\beta A_{\alpha\beta} \qquad \text{where } A_{\alpha\beta} = \int_{\mathrm{e}} t H_\alpha H_\beta \, \mathrm{d}\ell$$

(iv) Element surface traction vector $\vec{\mathrm{s}}_\alpha$ for tangential and normal forces

The equivalent force vector at node α due to surface traction $\vec{\mathrm{q}}$ is

$$\vec{\mathrm{s}}_\alpha = \int_{\mathrm{e}} t \vec{\mathrm{q}} H_\alpha \, \mathrm{d}\ell$$

where t is the thickness of the element along a curve $\mathcal{C}$, which is one of the edges of an isoparametric element as shown in Fig. 2.1. As traction forces on an edge can only contribute to nodes lying on that edge, α will take the values 1, 2 and 3.

The unit tangent vector $\hat{t}$ and unit normal vector $\hat{n}$ corresponding to infinitesimal element $d\ell$ are given by

$$\hat{t} = \left(\frac{dx}{d\ell}, \frac{dy}{d\ell}\right) \qquad \text{and} \qquad \hat{n} = \left(\frac{dy}{d\ell}, -\frac{dx}{d\ell}\right)$$

Let q_t and q_n be the tangent and normal components of the surface traction vector $\vec{q}$ (Fig. 2.2), then the tangential and normal forces on the surface element $d\ell$ are

$$\begin{aligned}\vec{q}\, d\ell &= \left[q_t\left(\frac{dx}{d\ell}, \frac{dy}{d\ell}\right) + q_n\left(\frac{dy}{d\ell}, -\frac{dx}{d\ell}\right)\right] d\ell \\ &= [q_t\, dx + q_n\, dy, q_t\, dy - q_n\, dx] \\ &= \begin{bmatrix} q_t & q_n \\ -q_n & q_t \end{bmatrix} \begin{bmatrix} dx \\ dy \end{bmatrix} \\ \vec{s}_\alpha &= \int_{\mathcal{C}} tH_\alpha \begin{bmatrix} q_t & q_n \\ -q_n & q_t \end{bmatrix} \begin{bmatrix} dx \\ dy \end{bmatrix} \end{aligned} \tag{2.7}$$

Since $\mathcal{C}$ is a parabolic curve for a quadratic element, any point $P(x, y)$ on the curve can be represented by means of quadratic interpolation functions defined on the

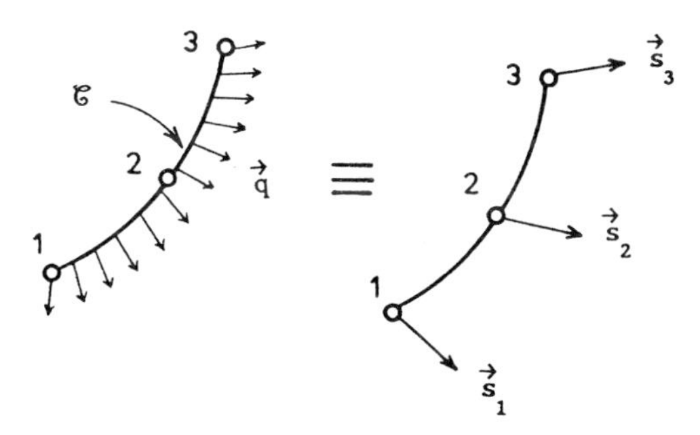

Fig. 2.1 Equivalent nodal traction forces

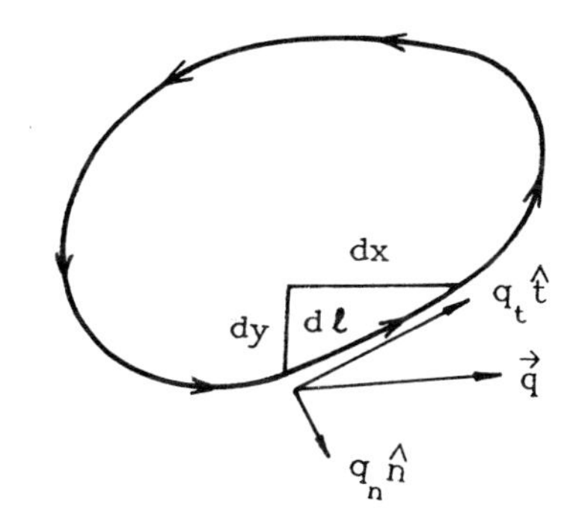

Fig. 2.2 Traction $\vec{q}$ on element $d\ell$

interval $[-1, 1]$ such that

$$x = H_1 x_1 + H_2 x_2 + H_3 x_3 \qquad\qquad y = H_1 y_1 + H_2 y_2 + H_3 y_3$$

where (x_1, y_1), (x_2, y_2), (x_3, y_3), are the coordinates of nodes 1, 2 and 3 on $\mathcal{C}$, and the interpolation functions $H_1 = \frac{1}{2}\varphi(\varphi - 1)$, $H_2 = 1 - \varphi^2$, $H_3 \times \frac{1}{2}\varphi(\varphi + 1)$, $\varphi \in [-1, 1]$.

$$\mathrm{d}x = (H'_1 x_1 + H'_2 x_2 + H'_3 x_3)\,\mathrm{d}\varphi \qquad \mathrm{d}y = (H'_1 y_1 + H'_2 y_2 + H'_3 y_3)\,\mathrm{d}\varphi$$

with $H'_1 = \frac{1}{2}(2\varphi - 1)$, $H'_2 = \quad 2\varphi$, $H'_3 = \frac{1}{2}(2\varphi + 1)$.

Hence,
$$\vec{s}_\alpha = \int_{-1}^{1} tH_\alpha \begin{bmatrix} q_t & q_n \\ -q_n & q_t \end{bmatrix} \begin{bmatrix} H'_1 x_1 + H'_2 x_2 + H'_3 x_3 \\ H'_1 y_1 + H'_2 y_2 + H'_3 y_3 \end{bmatrix} \mathrm{d}\varphi \tag{2.8}$$

The integral can be evaluated numerically using a two-point Gauss–Legendre integration formula such that

$$\int_{-1}^{1} f(\varphi)\,\mathrm{d}\varphi \approx f(\varphi = -1/\sqrt{3}) + f(\varphi = +1/\sqrt{3})$$

However, if the thickness is constant along curve $\mathcal{C}$ and q_t and q_n are constant as in the case of pressure load ($q_t = 0$, $q_n = -p$), an explicit expression for $\vec{s}_\alpha$ can be obtained,

$$[\vec{s}_1, \vec{s}_2, \vec{s}_3] = \frac{t}{6} \begin{bmatrix} q_t & q_n \\ -q_n & q_t \end{bmatrix} \begin{bmatrix} 4x_2 - 3x_1 - x_3 & 4(x_3 - x_1) & x_1 - 4x_2 + 3x_3 \\ 4y_2 - 3y_1 - y_3 & 4(y_3 - y_1) & y_1 - 4y_2 + 3y_3 \end{bmatrix} \tag{2.9}$$

2.2.3 6-node isoparametric triangular element T6

For isoparametric elements, the shape functions G_α, $\alpha = 1, N$, for element geometry definition are the same as the interpolation functions H_α used for the approximation of the field variables. The element reference domain for the 6-node isoparametric element T6 is a right-angle triangle as shown in Fig. 2.3, in which ξ and η will be

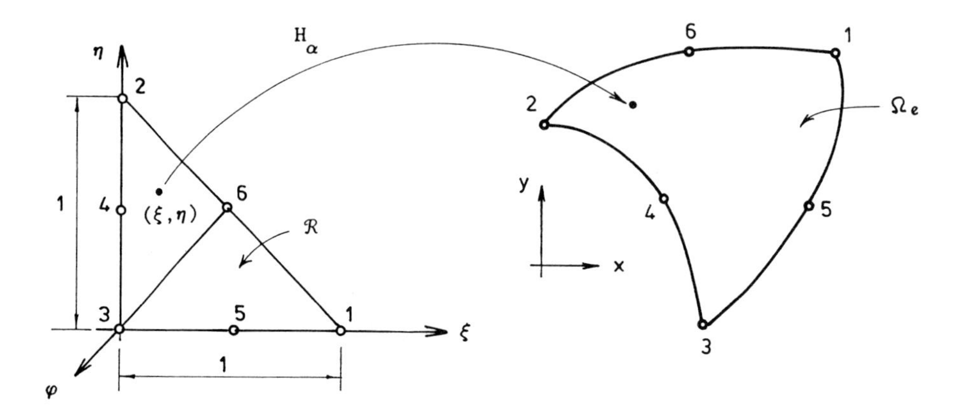

Fig. 2.3 Reference domain and mapping of isoparametric T6 element

Table 2.1 Interpolation functions H_α and their derivatives with respect to ξ and η of a T6 isoparametric element

α	H_α	$\dfrac{\partial H_\alpha}{\partial \xi}$	$\dfrac{\partial H_\alpha}{\partial \eta}$
1	$\xi(2\xi - 1)$	$4\xi - 1$	0
2	$\eta(2\eta - 1)$	0	$4\eta - 1$
3	$\varphi(2\varphi - 1)$	$1 - 4\varphi$	$1 - 4\varphi$
4	$4\eta\varphi$	-4η	$4(\varphi - \eta)$
5	$4\xi\varphi$	$4(\varphi - \xi)$	-4ξ
6	$4\xi\eta$	4η	4ξ

taken as independent variables. However, a point of the triangular element is always referred to by the triple (ξ, η, φ), with $\varphi = 1 - \xi - \eta$.

2.2.4 6-node subparametric triangular element

In the event that the isoparametric T6 element has only straight edges (subparametric), the element geometry is completely defined by the three corner nodes, and the element stiffness and the element force vectors can all be obtained in explicit forms by direct integration. In fact, the (ξ, η, φ) coordinates of any point P(x, y) in the subparametric triangle, Fig. 2.4, can be expressed by

$$\xi = \frac{2}{\Delta}(a_1 + b_1 x + c_1 y)$$

$$\eta = \frac{2}{\Delta}(a_2 + b_2 x + c_2 y)$$

$$\varphi = \frac{2}{\Delta}(a_3 + b_3 x + c_3 y)$$

where $2\Delta = a_1 + a_2 + a_3$

$$\begin{aligned} a_1 &= x_2 y_3 - x_3 y_2 & b_1 &= y_2 - y_3 & c_1 &= x_3 - x_2 \\ a_2 &= x_3 y_1 - x_1 y_3 & b_2 &= y_3 - y_1 & c_2 &= x_1 - x_3 \\ a_3 &= x_1 y_2 - x_2 y_1 & b_3 &= y_1 - y_2 & c_3 &= x_2 - x_1 \end{aligned}$$

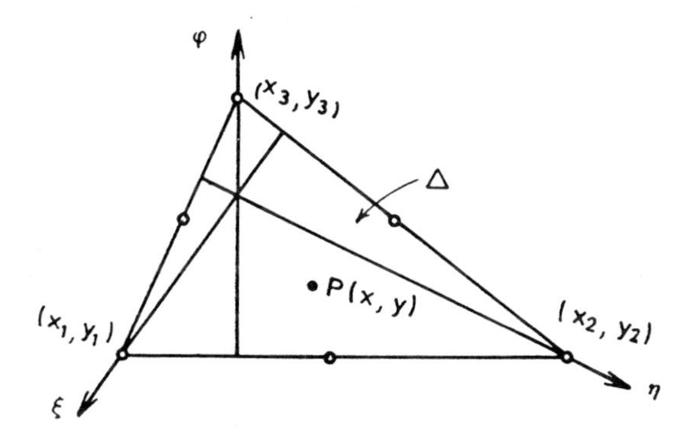

Fig. 2.4 A 6-node subparametric triangle

Using the chain rule, the derivatives of the interpolation functions H_α with respect to x and y can be found,

$$f_\alpha = \frac{\partial H_\alpha}{\partial x} = \frac{\partial H_\alpha}{\partial \xi}\frac{\partial \xi}{\partial x} + \frac{\partial H_\alpha}{\partial \eta}\frac{\partial \eta}{\partial x} + \frac{\partial H_\alpha}{\partial \varphi}\frac{\partial \varphi}{\partial x}$$

$$g_\alpha = \frac{\partial H_\alpha}{\partial y} = \frac{\partial H_\alpha}{\partial \xi}\frac{\partial \xi}{\partial y} + \frac{\partial H_\alpha}{\partial \eta}\frac{\partial \eta}{\partial y} + \frac{\partial H_\alpha}{\partial \varphi}\frac{\partial \varphi}{\partial y}$$

$$\begin{aligned}
f_1 &= \frac{b_1}{2\Delta}(4\xi - 1) & g_1 &= \frac{c_1}{2\Delta}(4\xi - 1) & f_4 &= \frac{2}{\Delta}(b_3\eta + b_2\varphi) & g_4 &= \frac{2}{\Delta}(c_3\eta + c_2\varphi) \\
f_2 &= \frac{b_2}{2\Delta}(4\eta - 1) & g_2 &= \frac{c_2}{2\Delta}(4\eta - 1) & f_5 &= \frac{2}{\Delta}(b_1\varphi + b_3\xi) & g_5 &= \frac{2}{\Delta}(c_1\varphi + c_3\xi) \\
f_3 &= \frac{b_3}{2\Delta}(4\varphi - 1) & g_3 &= \frac{c_3}{2\Delta}(4\varphi - 1) & f_6 &= \frac{2}{\Delta}(b_2\xi + b_1\eta) & g_6 &= \frac{2}{\Delta}(c_2\xi + c_1\eta)
\end{aligned} \qquad (2.10)$$

(a) Element stiffness matrix

The integration formula over the subparametric triangle Δ is given by

$$\int_\Delta \xi^i \eta^j \varphi^k \, dA = \frac{i!j!k!}{(i+j+k+2)!} 2\Delta \qquad (2.11)$$

The integrals $I_{11}^{\alpha\beta}$, $I_{12}^{\alpha\beta}$, $I_{21}^{\alpha\beta}$ and $I_{22}^{\alpha\beta}$ required for the sub-matrix $\mathbf{K}^{\alpha\beta}$ can be evaluated analytically by direct integration. The resulting element stiffness matrix of the 6-node subparametric element of constant thickness t is given in Table 2.2.

Table 2.2 Element stiffness matrix for a 6-node subparametric triangular element

$$\underset{12\times12}{\mathbf{K}^e} = \frac{t}{12\Delta}\left[\begin{array}{cc|cc|cc|cc|cc|cc}
u_1 & w_1 & -p_{12} & -r_{12} & -p_{13} & -r_{13} & 0 & 0 & 4p_{13} & 4r_{13} & 4p_{12} & 4r_{12} \\
 & v_1 & -s_{12} & -q_{12} & -s_{13} & -q_{13} & 0 & 0 & 4s_{13} & 4q_{13} & 4s_{12} & 4q_{12} \\
 & & u_2 & w_2 & -p_{23} & -r_{23} & 4p_{23} & 4r_{23} & 0 & 0 & 4p_{12} & 4s_{12} \\
 & & & v_2 & -s_{23} & -q_{23} & 4s_{23} & 4q_{23} & 0 & 0 & 4r_{12} & 4q_{12} \\
 & & & & u_3 & w_3 & 4p_{23} & 4s_{23} & 4p_{13} & 4s_{13} & 0 & 0 \\
 & & & & & v_3 & 4r_{23} & 4q_{23} & 4r_{13} & 4q_{13} & 0 & 0 \\
 & & & & & & t_1 & t_3 & 8p_{12} & 4h_{12} & 8p_{13} & 4h_{13} \\
 & & & & & & & t_2 & 4h_{12} & 8q_{12} & 4h_{13} & 8q_{13} \\
 & & & & & & & & t_1 & t_3 & 8p_{23} & 4h_{23} \\
 & \text{sym.} & & & & & & & & t_2 & 4h_{23} & 8q_{23} \\
 & & & & & & & & & & t_1 & t_3 \\
 & & & & & & & & & & & t_2
\end{array}\right]$$

where

$$\left.\begin{aligned}
u_i &= 3\lambda_1 b_i^2 + 3\lambda_3 c_i^2 \\
v_i &= 3\lambda_3 b_i^2 + 3\lambda_2 c_i^2 \\
w_i &= 3(\lambda_3 + \lambda_4) b_i c_i
\end{aligned}\right\} i = 1, 2, 3 \qquad
\left.\begin{aligned}
p_{ij} &= \lambda_1 b_i b_j + \lambda_3 c_i c_j \\
q_{ij} &= \lambda_3 b_i b_j + \lambda_2 c_i c_j \\
r_{ij} &= \lambda_4 b_i c_j + \lambda_3 c_i b_j \\
s_{ij} &= \lambda_3 b_i c_j + \lambda_4 c_i b_j \\
h_{ij} &= r_{ij} + s_{ij}
\end{aligned}\right] ij = 12, 13, 23$$

$$\begin{aligned}
t_1 &= 8\lambda_1(b_1^2 - b_2 b_3) + 8\lambda_3(c_1^2 - c_2 c_3) \\
t_2 &= 8\lambda_3(b_1^2 - b_2 b_3) + 8\lambda_2(c_1^2 - c_2 c_3) \\
t_3 &= 4(\lambda_3 + \lambda_4)(b_1 c_1 + b_2 c_2 + b_3 c_3)
\end{aligned}$$

(b) Element force vectors

(i) Element initial stress(strain) force vector

Components of nodal force due to initial stress and strain are

$$r_i^\alpha = \overset{\circ}{\tau}_{ij} h_{\alpha,j} \qquad i, j = 1, 2, \qquad \alpha = 1, 6$$

or

$$r_1^\alpha - \overset{\circ}{\tau}_{11} h_{\alpha 1} + \overset{\circ}{\tau}_{12} h_{\alpha 2} \qquad r_2^\alpha = \overset{\circ}{\tau}_{21} h_{\alpha 1} + \overset{\circ}{\tau}_{22} h_{\alpha 2}$$

For elements with constant thickness t,

$$h_{\alpha 1} = t \int_\Delta f_\alpha \, \mathrm{d}A \quad \text{and}$$

$$h_{\alpha 2} = t \int_\Delta g_\alpha \, \mathrm{d}A \qquad \alpha = 1, 6.$$

By direct integration, we have

$$\begin{aligned} [h_{11}, h_{21}, h_{31}, h_{41}, h_{51}, h_{61}] &= \frac{t}{6}[b_1, b_2, b_3, -4b_1, -4b_2, -4b_3] \\ [h_{12}, h_{22}, h_{32}, h_{42}, h_{52}, h_{62}] &= \frac{t}{6}[c_1, c_2, c_3, -4c_1, -4c_2, -4c_3] \end{aligned} \tag{2.12}$$

(ii) Element body force vector

In the case where the body force per unit volume $\vec{\gamma}$ over the element is constant, as in the case of gravity load, $\frac{1}{3}\vec{W}$ will be assigned to each of the mid-side nodes as shown in Fig. 2.5, where $\vec{W} = \Delta t \vec{\gamma}$ is the total body force of the element.

$$b_i^\alpha = t \int_\Delta \gamma_i H_\alpha \, \mathrm{d}A = t\gamma_i \int_\Delta H_\alpha \, \mathrm{d}A$$

$$\Rightarrow \quad b_i^1 = b_i^2 = b_i^3 = 0$$

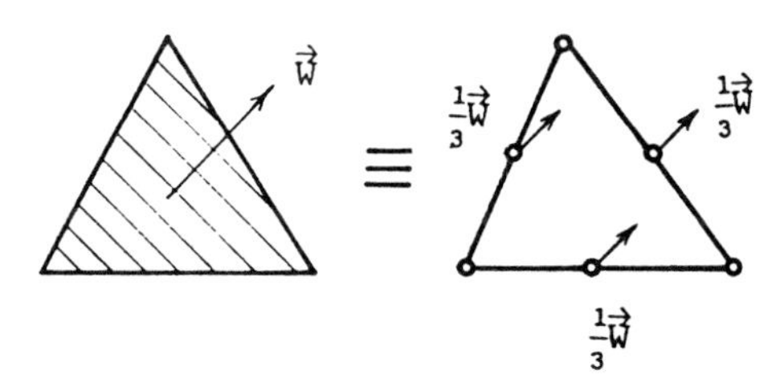

Fig. 2.5 Equivalent nodal forces for the case of constant body force

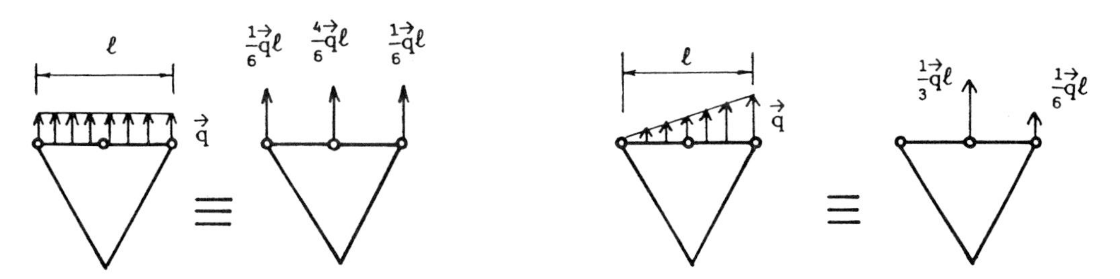

Fig. 2.6 Equivalent nodal forces due to surface traction

and $$b_i^4 = b_i^5 = b_i^6 = \tfrac{1}{3}\Delta t\gamma_i = \tfrac{1}{3}W_i$$

(iii) Element surface traction vector

The equivalent nodal forces due to a uniformly distributed load and a triangular load acting along a straight edge can be obtained by direct integration and are given in Fig. 2.6.

2.2.5 8-node isoparametric quadrilateral element Q8

The reference domain of an 8-node isoparametric element Q8 is a 2×2 square as shown in Fig. 2.7.

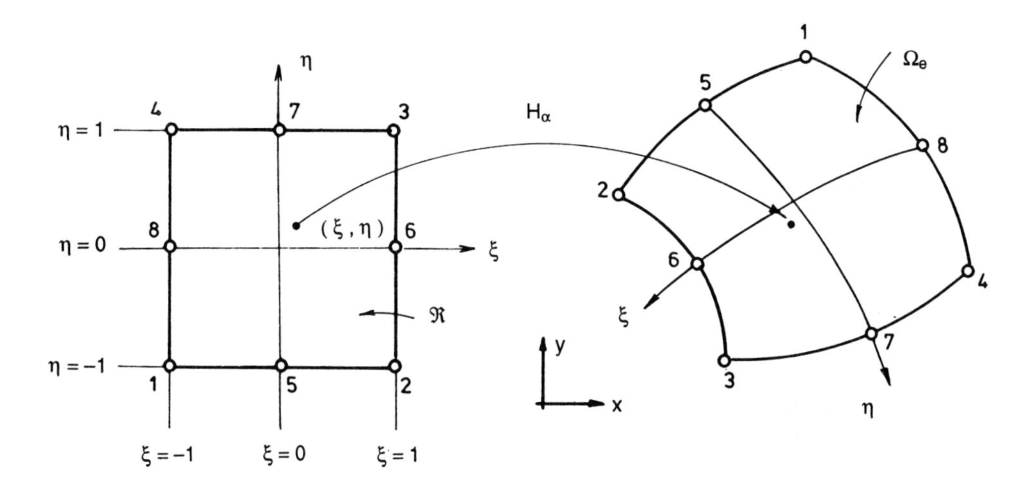

Fig. 2.7 Reference domain and mapping of Q8 isoparametric element

Table 2.3 The interpolation functions H_α and their derivatives with respect to ξ and η of the Q8 isoparametric element

α	H_α	$\dfrac{\partial H_\alpha}{\partial \xi}$	$\dfrac{\partial H_\alpha}{\partial \eta}$
1	$-\frac{1}{4}(1-\xi)(1-\eta)(1+\xi+\eta)$	$\frac{1}{4}(1-\eta)(2\xi+\eta)$	$\frac{1}{4}(1-\xi)(\xi+2\eta)$
2	$-\frac{1}{4}(1+\xi)(1-\eta)(1-\xi+\eta)$	$\frac{1}{4}(1-\eta)(2\xi-\eta)$	$-\frac{1}{4}(1+\xi)(\xi-2\eta)$
3	$-\frac{1}{4}(1+\xi)(1+\eta)(1-\xi-\eta)$	$\frac{1}{4}(1+\eta)(2\xi+\eta)$	$\frac{1}{4}(1+\xi)(\xi+2\eta)$
4	$-\frac{1}{4}(1-\xi)(1+\eta)(1+\xi-\eta)$	$\frac{1}{4}(1+\eta)(2\xi-\eta)$	$-\frac{1}{4}(1-\xi)(\xi-2\eta)$
5	$\frac{1}{2}(1-\xi^2)(1-\eta)$	$-\xi(1-\eta)$	$-\frac{1}{2}(1-\xi^2)$
6	$\frac{1}{2}(1+\xi)(1-\eta^2)$	$\frac{1}{2}(1-\eta^2)$	$-(1+\xi)\eta$
7	$\frac{1}{2}(1-\xi^2)(1+\eta)$	$-\xi(1+\eta)$	$\frac{1}{2}(1-\xi^2)$
8	$\frac{1}{2}(1-\xi)(1-\eta^2)$	$-\frac{1}{2}(1-\eta^2)$	$-(1-\xi)\eta$

2.2.6 8-node rectangular element

A Q8 isoparametric element is most effective for problems with a regular or nearly regular domain, in which a natural subdivision into rectangular elements is allowed. Similar to the 6-node subparametric element, the element stiffness matrix and element force vectors of a Q8 element in rectangular form with constant thickness t can also be obtained explicitly by direct integration. Over the rectangular element, the (x, y) coordinates are given by $x = x_o + \xi a$ and $y = y_o + \eta b$, where (x_o, y_o) are the coordinates of the centroid of the rectangular element, as shown in Fig. 2.8. The Jacobian matrix (Section 2.2.7(a)) in this case is constant, and is given by

$$\mathbf{J} = \begin{bmatrix} a & 0 \\ 0 & b \end{bmatrix}$$

Thus we can write

$$\frac{\partial H_\alpha}{\partial x} = \frac{1}{a}\frac{\partial H_\alpha}{\partial \xi} \quad \text{and} \quad \frac{\partial H_\alpha}{\partial y} = \frac{1}{b}\frac{\partial H_\alpha}{\partial \eta}$$

(a) Element stiffness matrix

The integrals $I_{11}^{\alpha\beta}, I_{12}^{\alpha\beta}, I_{21}^{\alpha\beta}$ and $I_{22}^{\alpha\beta}$ for sub-matrix $\mathbf{K}^{\alpha\beta}$ can be expressed as

$$\begin{aligned} I_{11}^{\alpha\beta} &= t\frac{b}{a}\int_\square \frac{\partial H_\alpha}{\partial \xi}\frac{\partial H_\beta}{\partial \xi}\,\mathrm{d}A \qquad & I_{12}^{\alpha\beta} &= t\int_\square \frac{\partial H_\alpha}{\partial \xi}\frac{\partial H_\beta}{\partial \eta}\,\mathrm{d}A \\ I_{21}^{\alpha\beta} &= t\int_\square \frac{\partial H_\alpha}{\partial \eta}\frac{\partial H_\beta}{\partial \xi}\,\mathrm{d}A \qquad & I_{22}^{\alpha\beta} &= t\frac{a}{b}\int_\square \frac{\partial H_\alpha}{\partial \eta}\frac{\partial H_\beta}{\partial \eta}\,\mathrm{d}A \end{aligned} \tag{2.13}$$

Integration of polynomials over a standard 2×2 square domain can be easily done using the following formula,

$$\int_\square \xi^i \eta^j \,\mathrm{d}A = \begin{cases} 0 & \text{if } i \text{ or } j \text{ is odd} \\ \dfrac{4}{(i+1)(j+1)} & \text{if } i \text{ and } j \text{ are even} \end{cases} \tag{2.14}$$

The resulting element stiffness matrix is given in Table 2.4.

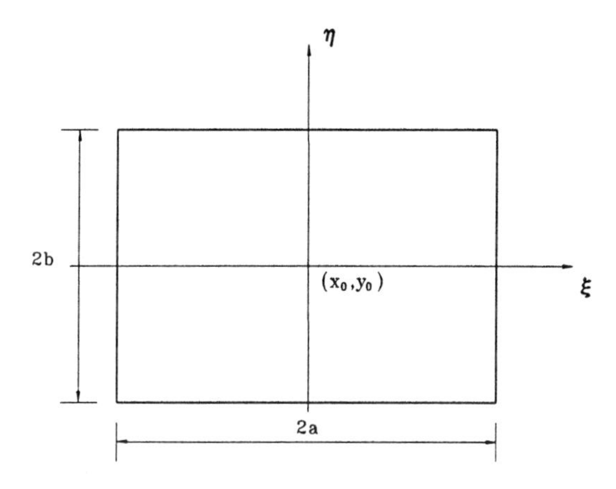

Fig. 2.8 An 8-node rectangular element

Table 2.4 Stiffness matrix for 8-node rectangular element in explicit form

$$
\underset{16\times 16}{\mathbf{K}_e} = t\begin{bmatrix}
f_{12} & 17S & s_{12} & d & t_{12} & 7S & s_{21} & -d & p_{12} & q_{43} & r_{21} & -4S & r_{12} & -4S & p_{21} & q_{34} \\
 & f_{34} & -d & s_{34} & 7S & t_{34} & d & s_{43} & q_{34} & p_{34} & -4S & r_{43} & -4S & r_{34} & q_{43} & p_{43} \\
 & & f_{12} & -17S & s_{21} & d & t_{12} & -7S & p_{12} & -q_{43} & p_{21} & -q_{34} & r_{12} & 4S & r_{21} & 4S \\
 & & & f_{34} & -d & s_{43} & -7S & t_{34} & -q_{34} & p_{34} & -q_{43} & p_{43} & 4S & r_{34} & 4S & r_{43} \\
 & & & & f_{12} & 17S & s_{12} & d & r_{12} & -4S & p_{21} & q_{34} & p_{12} & q_{43} & r_{21} & -4S \\
 & & & & & f_{34} & -d & s_{34} & -4S & r_{34} & q_{43} & p_{43} & q_{34} & p_{34} & -4S & r_{43} \\
 & & & & & & f_{12} & -17S & r_{12} & 4S & r_{21} & 4S & p_{12} & -q_{43} & p_{21} & -q_{34} \\
 & & & & & & & f_{34} & 4S & r_{34} & 4S & r_{43} & -q_{34} & p_{34} & -q_{43} & p_{43} \\
 & & & & & & & & g_{12} & 0 & 0 & -16S & h_{12} & 0 & 0 & 16S \\
 & & & & & & & & & g_{34} & -16S & 0 & 0 & h_{34} & 16S & 0 \\
 & & & & & & & & & & g_{21} & 0 & 0 & 16S & h_{21} & 0 \\
 & & & & & & & & & & & g_{43} & 16S & 0 & 0 & h_{43} \\
 & & & & & & & & & & & & g_{12} & 0 & 0 & -16S \\
 & & & & & & & & & & & & & g_{34} & -16S & 0 \\
 & \text{Sym.} & & & & & & & & & & & & & g_{21} & 0 \\
 & & & & & & & & & & & & & & & g_{43}
\end{bmatrix}
$$

where

$$\mu_1 = \frac{b\lambda_1}{45a} \qquad \mu_2 = \frac{a\lambda_3}{45b} \qquad \mu_3 = \frac{b\lambda_3}{45a} \qquad \mu_4 = \frac{a\lambda_2}{45b}$$

$$S = \frac{1}{36}(\lambda_3 + \lambda_4) \qquad d = \frac{1}{12}(\lambda_3 - \lambda_4) \qquad q_{ij} = \frac{1}{9}(\lambda_i - 5\lambda_j)$$

$$f_{ij} = 26\mu_i + 26\mu_j \qquad g_{ij} = 80\mu_i + 24\mu_j$$

$$h_{ij} = 40\mu_i - 24\mu_j \qquad p_{ij} = -40\mu_i + 3\mu_j$$

$$r_{ij} = -20\mu_i - 3\mu_j \qquad s_{ij} = 14\mu_i + \frac{17}{2}\mu_j \qquad t_{ij} = \frac{23}{2}\mu_i + \frac{23}{2}\mu_j$$

(b) Element force vectors

(i) Element initial stress(strain) force vector

The $h_{\alpha 1}$, $h_{\alpha 2}$ values required for the calculation of the initial stress/strain nodal forces can be evaluated by direct integration.

$$
\begin{aligned}
[h_{11}, h_{21}, h_{31}, h_{41}, h_{51}, h_{61}, h_{71}, h_{81}] &= \frac{bt}{3}[-1, 1, 1, -1, 0, 4, 0, -4] \\
[h_{12}, h_{22}, h_{32}, h_{42}, h_{52}, h_{62}, h_{72}, h_{82}] &= \frac{at}{3}[-1, -1, 1, 1, -4, 0, 4, 0]
\end{aligned}
\tag{2.15}
$$

(ii) Element body force vector

If the body force per unit volume $\vec{\gamma}$ over the element is constant, $\frac{1}{3}\vec{W}$ will be acting at the four mid-side nodes and $-\frac{1}{12}\vec{W}$ will be acting on the four corner nodes, as shown in Fig. 2.9, where $\vec{W} = 4abt\vec{\gamma}$ is the total element body force.

(iii) Element surface traction vector

The equivalent node traction forces for uniformly distributed load and triangular load are shown in Fig. 2.10.

2.2.7 Numerical integration for isoparametric T6 and Q8 elements

For general isoparametric T6 and Q8 elements with curved boundaries, the functions to be integrated cannot be expressed in terms of simple polynomials. In such cases,

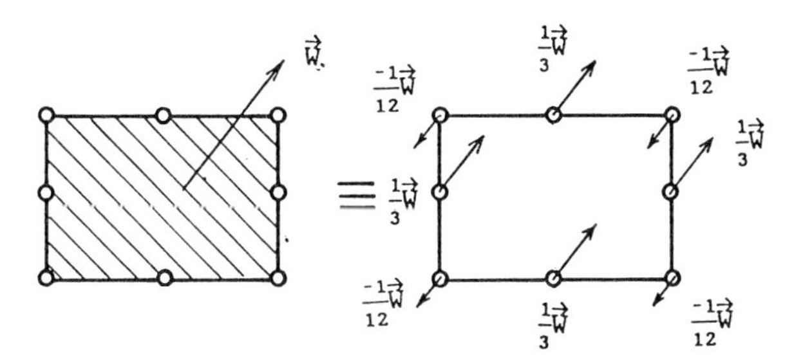

Fig. 2.9 Equivalent nodal forces for the case of constant body force

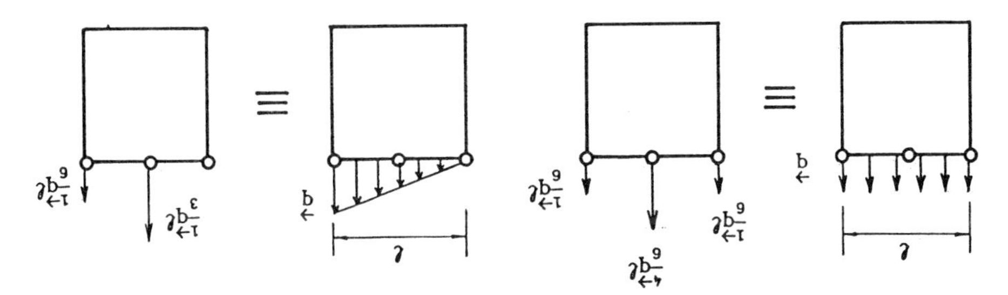

Uniformly distributed load Triangular load

Fig. 2.10 Equivalent nodal forces for surface traction

the numerical integration formulae will give fairly accurate values to the integrals to be evaluated [14]. In mechanics of solids, it has been observed that numerical integration errors very often compensate for the geometrical discretization error with overall beneficial results.

(a) Transformation of differential operator (2D)

The governing field equations of a physical problem involve unknowns u, v and their derivatives $\partial u/\partial x$, $\partial u/\partial y$, $\partial v/\partial x$, $\partial v/\partial y$, etc. Since u and v are defined in terms of interpolation functions H_α which are functions of ξ and η, it is necessary to devise some means of expressing global derivatives, $\partial u/\partial x$, $\partial u/\partial y$, etc., in terms of local derivatives, $\partial H_\alpha/\partial\xi$, $\partial H_\alpha/\partial\eta$.

$$u = H_\alpha u_\alpha \Rightarrow \frac{\partial u}{\partial x} = \frac{\partial H_\alpha}{\partial x}u_\alpha \qquad \frac{\partial u}{\partial y} = \frac{\partial H_\alpha}{\partial y}u_\alpha$$

$$v = H_\alpha v_\alpha \Rightarrow \frac{\partial v}{\partial x} = \frac{\partial H_\alpha}{\partial x}v_\alpha \qquad \frac{\partial v}{\partial y} = \frac{\partial H_\alpha}{\partial y}v_\alpha$$

Hence, in the determination of $\partial u/\partial x$, $\partial u/\partial y$, etc., $\partial H_\alpha/\partial x$ and $\partial H_\alpha/\partial y$ are required.

$$\begin{bmatrix} \dfrac{\partial H_\alpha}{\partial\xi} \\ \dfrac{\partial H_\alpha}{\partial\eta} \end{bmatrix} = \begin{bmatrix} \dfrac{\partial x}{\partial\xi} & \dfrac{\partial y}{\partial\xi} \\ \dfrac{\partial x}{\partial\eta} & \dfrac{\partial y}{\partial\eta} \end{bmatrix} \begin{bmatrix} \dfrac{\partial H_\alpha}{\partial x} \\ \dfrac{\partial H_\alpha}{\partial y} \end{bmatrix} = \mathbf{J} \begin{bmatrix} \dfrac{\partial H_\alpha}{\partial x} \\ \dfrac{\partial H_\alpha}{\partial y} \end{bmatrix} \tag{2.16}$$

$$\begin{bmatrix} \dfrac{\partial H_\alpha}{\partial x} \\ \dfrac{\partial H_\alpha}{\partial y} \end{bmatrix} = \mathbf{J}^{-1} \begin{bmatrix} \dfrac{\partial H_\alpha}{\partial\xi} \\ \dfrac{\partial H_\alpha}{\partial\eta} \end{bmatrix} = \frac{1}{\det(\mathbf{J})} \begin{bmatrix} J_{22} & -J_{12} \\ -J_{21} & J_{11} \end{bmatrix} \begin{bmatrix} \dfrac{\partial H_\alpha}{\partial\xi} \\ \dfrac{\partial H_\alpha}{\partial\eta} \end{bmatrix} \tag{2.17}$$

Table 2.5 Numerical integration formulae for T6 and Q8 elements

Reference domain	Order n	Number of points, m	Coordinates ξ_k	η_k	Weight, W_k
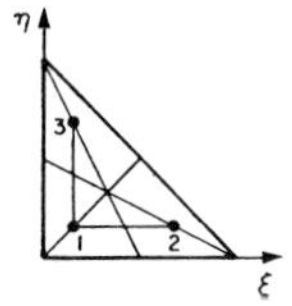	2	3	$\frac{1}{6}$	$\frac{1}{6}$	
			$\frac{2}{3}$	$\frac{1}{6}$	$\frac{1}{6}$
			$\frac{1}{6}$	$\frac{2}{3}$	
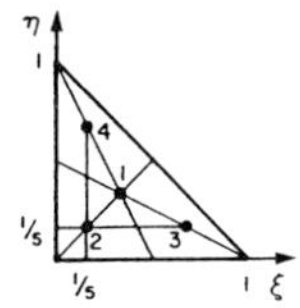	3	4	$\frac{1}{3}$	$\frac{1}{3}$	$-\frac{27}{96}$
			$\frac{1}{5}$	$\frac{1}{5}$	
			$\frac{3}{5}$	$\frac{1}{5}$	$\frac{25}{96}$
			$\frac{1}{5}$	$\frac{3}{5}$	
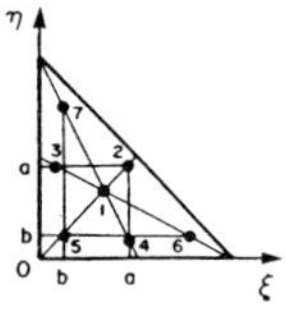	5	7	$\frac{1}{3}$	$\frac{1}{3}$	$\frac{9}{80}$
			a	a	
			$1-2a$	a	$A = \frac{155+\sqrt{15}}{2400}$
	$a = \frac{6+\sqrt{15}}{21}$		a	$1-2a$	
			b	b	
	$b = \frac{4}{7} - a$		$1-2b$	b	$\frac{31}{240} - A$
			b	$1-2b$	
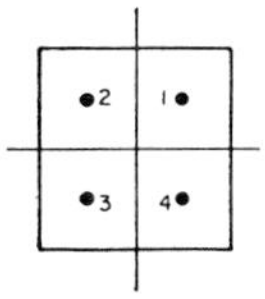	3	4	$\pm\sqrt{\frac{1}{3}}$	$\pm\sqrt{\frac{1}{3}}$	1
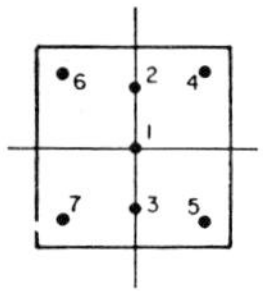			0	0	$\frac{8}{7}$
	5	7	0	$\pm\sqrt{\frac{14}{15}}$	$\frac{20}{63}$
			$\pm\sqrt{\frac{3}{5}}$	$\pm\sqrt{\frac{1}{3}}$	$\frac{20}{36}$
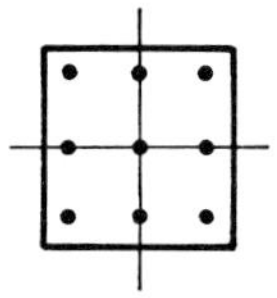			0	0	$\frac{64}{81}$
	5	9	$\left(0, \pm\sqrt{\frac{3}{5}}\right)$	$\left(\pm\sqrt{\frac{3}{5}}, 0\right)$	$\frac{40}{81}$
			$\left(\pm\sqrt{\frac{3}{5}},\right.$	$\left.\pm\sqrt{\frac{3}{5}}\right)$	$\frac{25}{81}$

J is known as the Jacobian matrix, whose components are given by

$$
\begin{aligned}
J_{11} &= \frac{\partial x}{\partial \xi} = \frac{\partial G_\beta}{\partial \xi} x_\beta \qquad J_{12} = \frac{\partial y}{\partial \xi} = \frac{\partial G_\beta}{\partial \xi} y_\beta \\
J_{21} &= \frac{\partial x}{\partial \eta} = \frac{\partial G_\beta}{\partial \eta} x_\beta \qquad J_{22} = \frac{\partial y}{\partial \eta} = \frac{\partial G_\beta}{\partial \eta} y_\beta \\
\det(\mathbf{J}) &= J_{11} J_{22} - J_{12} J_{21}
\end{aligned}
\tag{2.18}
$$

where G_α are geometric shape functions. For isoparametric elements, $G_\alpha = H_\alpha$.

(b) Transformation of integral (2D)

The change of variables allows us to change the integration of a function $f(x, y)$ defined over the element domain A_e into a simple integration referring to the natural coordinates (ξ, η) of the reference element, such that

$$\int_{A_e} f(x, y)\, \mathrm{d}\Omega = \int_{\mathcal{R}} \hat{\mathrm{f}}(\xi, \eta)\, \mathrm{d}\mathcal{R} \tag{2.19}$$

where $\hat{\mathrm{f}}(\xi, \eta) = f(x(\xi, \eta), y(\xi, \eta)) \cdot \det(\mathbf{J}(\xi, \eta))$, and region $\mathcal{R}$ is the element reference domain; for a T6 element, it is a right-angle triangle, whereas for a Q8 element, it is a 2×2 square.

(c) Numerical integration (2D)

Exact integration of complicated functions $\hat{\mathrm{f}}(\xi, \eta)$ arising from complex continuum problems and distorted element geometries is not practical, and numerical integration techniques have to be used [10,11]. The integration is usually replaced by a summation as follows:

$$\int_{\mathcal{R}} \hat{\mathrm{f}}(\xi, \eta)\, \mathrm{d}\mathcal{R} \approx \sum_{k=1}^{m} W_k \hat{\mathrm{f}}(\xi_k, \eta_k) \tag{2.20}$$

where m is the number of integration points used, and W_k and (ξ_k, η_k) are the weights and positions of the sampling points respectively. The common numerical integration schemes for T6 and Q8 elements are summarized in Table 2.5, where order n is the degree of polynomial that can be integrated exactly.

2.2.8 Examples

1 Bending of a cantilever beam

The finite element analysis of the bending of a cantilever beam is quoted as an example to study the convergence characteristics of the Q8 element. The calculation was done on a micro PC-AT with 80286 processor running at 8 MHz. The dimensions, boundary conditions and loading of the beam structure are shown in Fig. 2.11.

The results of analysis for the beam divided progressively into $M \times N$ 8-node quadrilateral elements are summarized in Table 2.6, in which NE = number of elements, NN = number of node points, NEQ = number of equations, NA = number of off-diagonal terms, CPU time = time required for matrix decomposition, δ = deflection at the tip (value from the elementary strength of material including shear effects = 42.92 mm for Young's modulus E = 1200 N/mm^2 and shear modulus

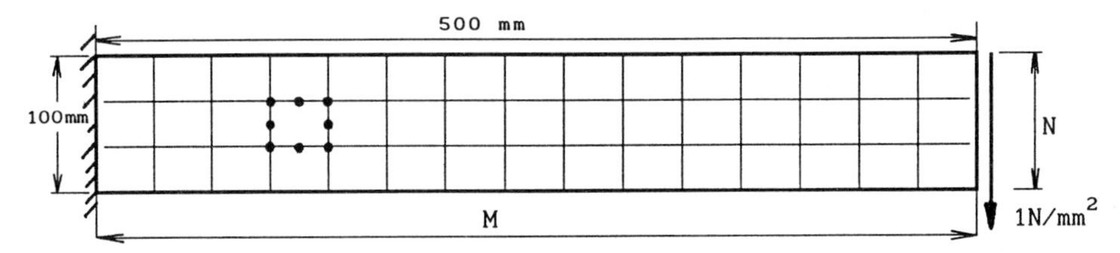

Fig. 2.11 Finite element analysis of cantilever beam

Table 2.6 Finite element analysis of the cantilever beam

	M	*N*	*NE*	*NN*	*NEQ*	*NA*	CPU time (s)	δ (mm)
1	5	1	5	28	56	540	1.8	42.21
2	10	2	20	85	170	2589	11.4	42.69
3	15	3	45	172	344	7048	38.9	42.77
4	20	4	80	289	578	14817	110.2	42.79
5	25	5	125	436	872	26796	228.0	42.80
6	30	6	180	613	1226	43885	417.6	42.81
7	35	7	245	820	1640	66984	717.4	42.82

$G = 480$ N/mm^2). From (M, N) equal to (20,4) onwards, direct access-files had to be used in the matrix reduction process, and the CPU time required was much more than the last three cases in which no backing storage was needed.

A comparison is made on the performance of the T6 element and the Q8 element using a simple example of the bending of a cantilever beam. Five different meshes having more or less the same number of DOF were used in the finite element analysis as shown in Fig. 2.12. The accuracy of the finite element solution is to be assessed by an error norm $\| \mathbf{e} \|$, which is defined as

$$\| \mathbf{e} \| = \left[\int_{\Omega} (\sigma - \hat{\sigma})^{\mathrm{T}} \mathbf{C}^{-1} (\sigma - \hat{\sigma}) \, \mathrm{d}\Omega \right]^{1/2}$$

where σ and $\hat{\sigma}$ are respectively the exact solution and the finite element solution of stress, and $\mathbf{C}$ is the constitutive matrix. The relative error in energy norm ε may also be defined as

$$\varepsilon = \frac{\| \mathbf{e} \|}{\| \mathbf{u} \|}$$

where $\| \mathbf{u} \|$ is the energy norm of the exact solution, which can be calculated by

$$\| \mathbf{u} \| = \left[\int_{\Omega} \sigma^{\mathrm{T}} \mathbf{C}^{-1} \sigma \, \mathrm{d}\Omega \right]^{1/2}$$

From Fig. 2.12, the mesh (a) of the four rectangular Q8 elements gave a better result than the mesh (b) of eight T6 elements obtained by subdividing the Q8 elements. However, the result could be much improved if nearly equilateral triangles are used

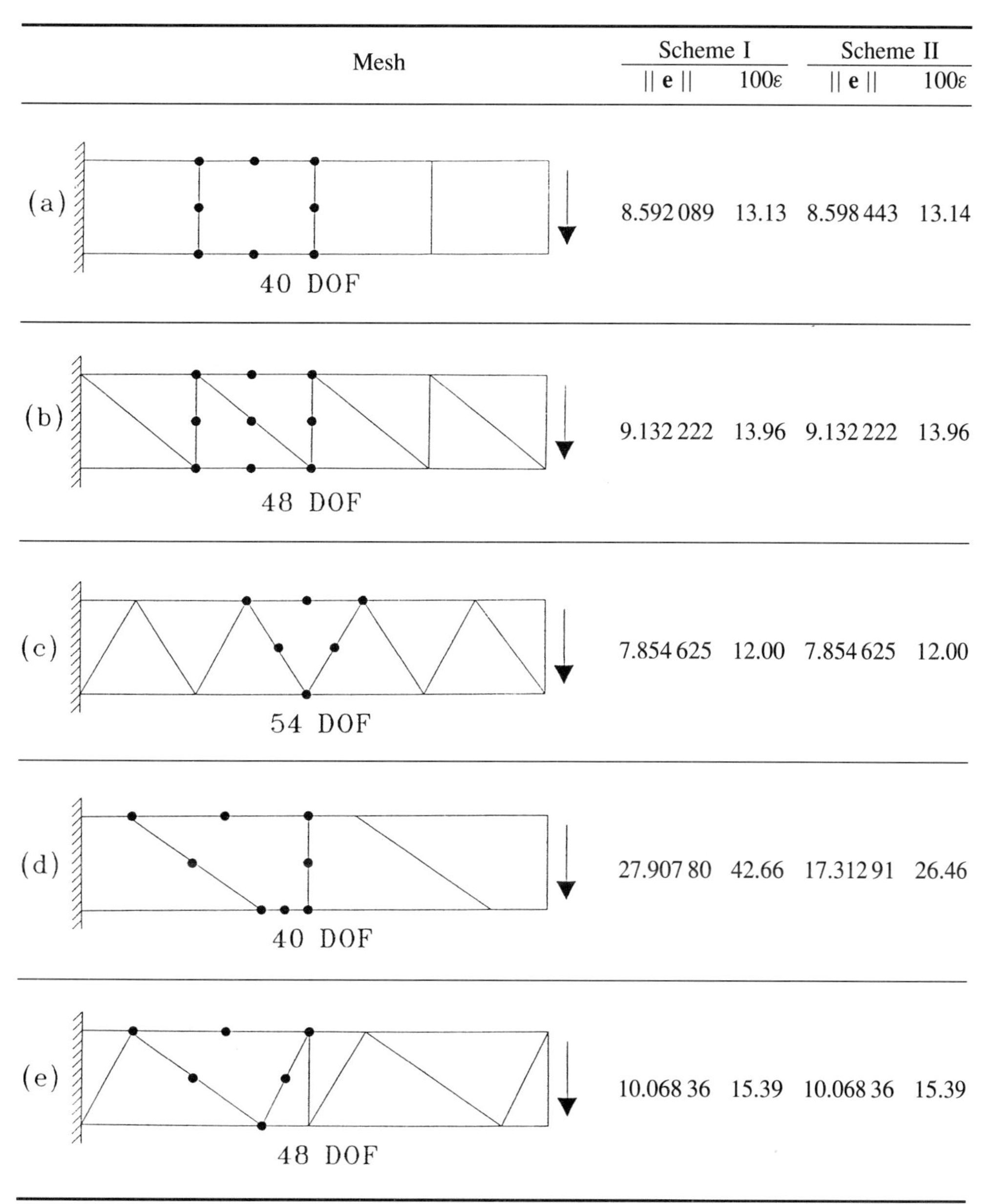

Mesh	Scheme I ‖**e**‖	Scheme I 100ε	Scheme II ‖**e**‖	Scheme II 100ε
(a) 40 DOF	8.592 089	13.13	8.598 443	13.14
(b) 48 DOF	9.132 222	13.96	9.132 222	13.96
(c) 54 DOF	7.854 625	12.00	7.854 625	12.00
(d) 40 DOF	27.907 80	42.66	17.312 91	26.46
(e) 48 DOF	10.068 36	15.39	10.068 36	15.39

Scheme I: 2 × 2 (reduced) integration scheme for Q8 elements
3-point integration scheme for T6 elements
Scheme II: 3 × 3 integration scheme for Q8 elements
4-point integration scheme for T6 elements

Fig. 2.12 A cantilever beam is modelled using various mesh patterns

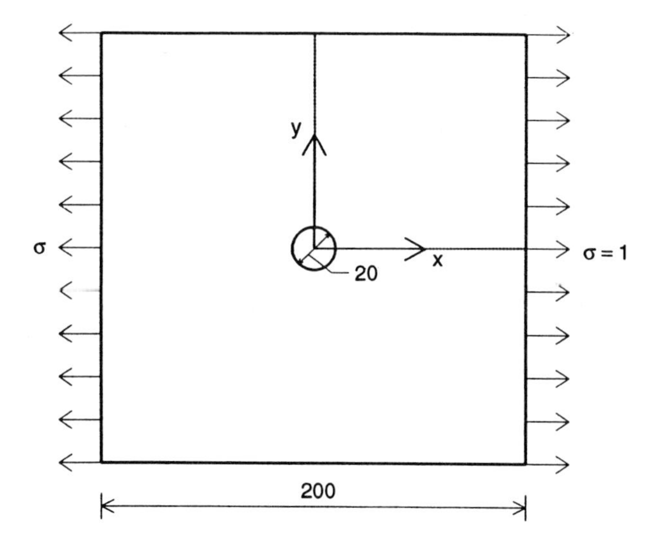

Fig. 2.13 Stress flow around a circular hole

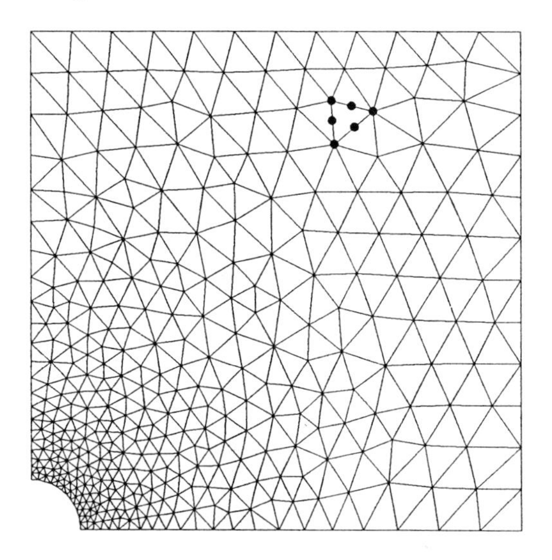

Fig. 2.14 The finite element mesh for one quarter of the plate

as shown in the mesh (c). Q8 elements tend to stiffen and lose accuracy if their form becomes distorted from a rectangle. Hence, using the mesh (d) of distorted Q8 elements resulted in a much higher error, which can, however, be reduced by employing a higher integration rule as indicated in the column of Scheme II of Fig. 2.12. This also suggests that distorted Q8 elements have to be integrated by a more exact integration rule. On the other hand, T6 triangular elements are much less sensitive to distortions. The error of the distorted mesh (e) is only slightly higher than that of the mesh (b). Finally, for the integration of the subparametric (straight-edged) T6 elements, the 3-point integration scheme is just as good as the 4-point integration scheme.

2 A circular hole under a uniform stress field

The stress flow around a circular hole under a uniform stress field is modelled by a square plate with a small hole at the centre. The plate is assumed to be isotropic and its dimensions are shown in Fig. 2.13. By symmetry, only one quarter of the plate needs to be considered, which is divided into 1473 T6 elements as shown in Fig. 2.14. The stress concentration factor $S_c = \sigma_{\max}/\sigma$ obtained is 3.10, which is only 3% higher than the analytical value [12]. The principal stresses along the x- and y-axes are plotted respectively in Fig. 2.15(a) and 2.15(b).

3 Double edge notched strip under tension

The collapsed quarter-point isoparametric element (CQPE) is very effective in producing a singular stress field at the crack tip [35,36]. Barsoum [34] showed that the accuracy of the solutions produced by CQPE compares favourably with those obtained by the hybrid principle [37]. The problem of double edge notched strip under tension, shown in Fig. 2.16, is analyzed using the CQPE. The numerical

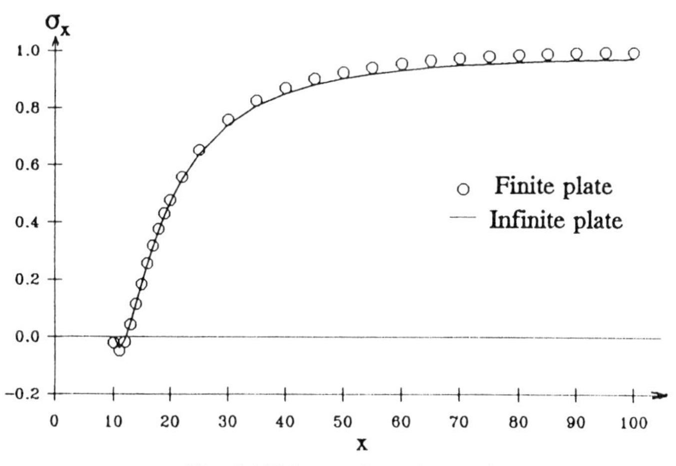

Fig. 2.15(a) σ_x along the x-axis

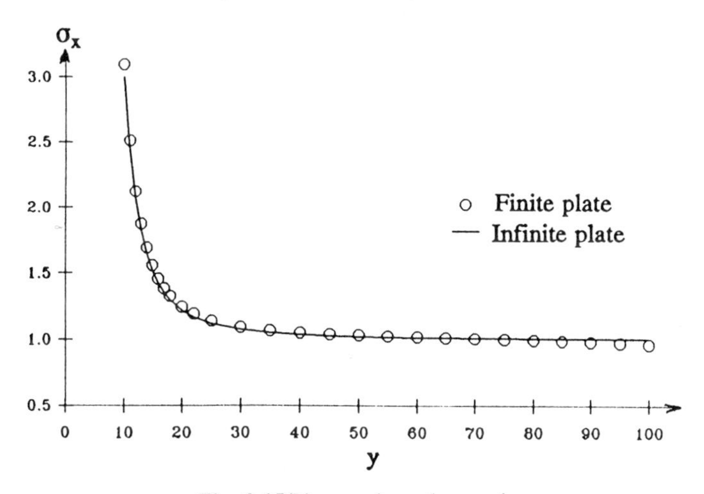

Fig. 2.15(b) σ_x along the y-axis

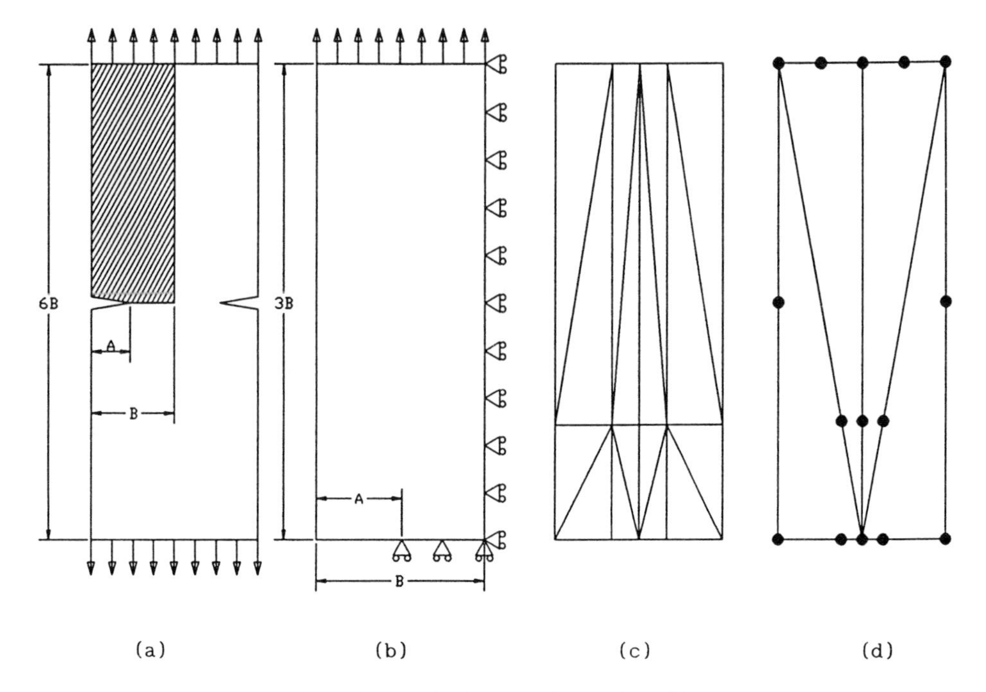

Fig. 2.16 (a) The double notched strip under tension. (b) By symmetry, one quarter of the strip is considered. (c) A finite element mesh of 45 nodes and 16 elements. (d) A magnified view of the mesh around the crack tip, mid-side nodes are shifted to quarter positions

constants used are $A = 0.5$, $B = 1.0$, Young's modulus $E = 1.0$ and Poisson's ratio $\nu = 0.3$. With a coarse mesh of 80 degrees of freedom for one quarter of the strip (Fig. 2.16(c)), the error in strain energy is 8.8%; without using CQPE at the crack tip, the error would have been 16.9%.

2.3 AXISYMMETRIC STRESS ANALYSIS

The problem of stress distribution in bodies of revolution (axisymmetric solids) under axisymmetric loading is of considerable practical interest. The mathematical problems presented are very similar to those of plane stress and plane strain as, once again, the situation is two-dimensional [15,16]. By axisymmetry, the two components of displacements in any plane section of the body through the axis of symmetry define completely the state of strain and stress. Such a typical section is shown in Fig. 2.17, where r and z denote respectively the radial and axial coordinates of a point, with u and v being the corresponding displacements. In the axisymmetric situation, any radial displacement automatically induces a strain in the circumferential direction, and as the stresses in this direction are certainly non-zero, this fourth component of strain and the associated stress has to be considered.

2.3.1 Transversely isotropic material

A general case of anisotropy need not be considered since axial symmetry would be impossible to achieve under such circumstances. A case of practical interest is that of

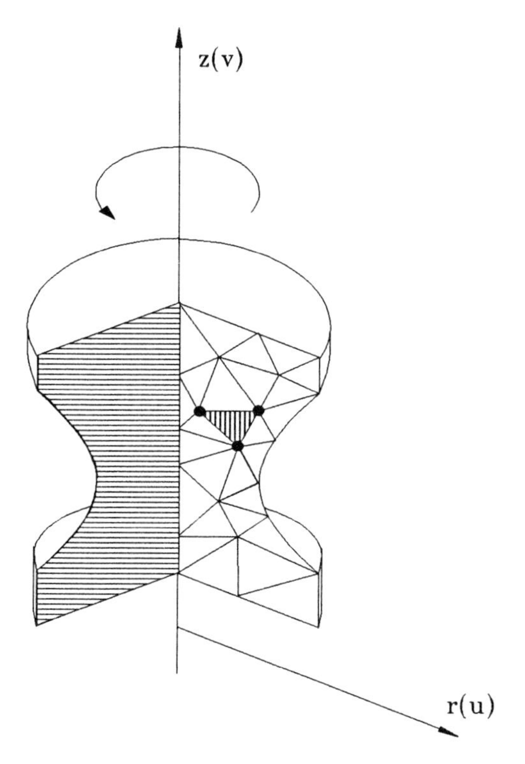

Fig. 2.17 Axisymmetric stress analysis

a stratified material, in which the plane of isotropy is normal to the axis of symmetry [6–8]. With the z-axis representing the normal to the plane of stratification, we can write

$$\varepsilon_r = \frac{\sigma_r}{E_1} - \frac{\nu_2 \sigma_z}{E_2} - \frac{\nu_1 \sigma_\theta}{E_1} \qquad \varepsilon_z = \frac{\sigma_z}{E_2} - \frac{\nu_2 \sigma_r}{E_2} - \frac{\nu_2 \sigma_\theta}{E_2}$$

$$\varepsilon_\theta = \frac{\sigma_\theta}{E_1} - \frac{\nu_2 \sigma_z}{E_2} - \frac{\nu_1 \sigma_r}{E_1} \qquad \gamma_{rz} = \frac{\tau_{rz}}{G_2} \qquad \gamma_{z\theta} = \gamma_{\theta r} = 0 \tag{2.21}$$

in which the constants E_1, ν_1 are associated with the behaviour in the plane of strata and E_2, G_2, ν_2 with a direction normal to the strata plane.

***Elasticity matrix,* C**

The elastic constants C_{ijkl} for stratified and isotropic material can now be summarized as:

$$\begin{bmatrix} \sigma_r \\ \sigma_z \\ \sigma_\theta \\ \sigma_{rz} \end{bmatrix}_{4\times1} = \begin{bmatrix} \lambda_2 & \lambda_4 & \lambda_5 & 0 \\ \lambda_4 & \lambda_1 & \lambda_4 & 0 \\ \lambda_5 & \lambda_4 & \lambda_2 & 0 \\ 0 & 0 & 0 & \lambda_3 \end{bmatrix}_{4\times4} \begin{bmatrix} \varepsilon_r \\ \varepsilon_z \\ \varepsilon_\theta \\ \gamma_{rz} \end{bmatrix}_{4\times1} \tag{2.22}$$

where $\lambda_1, \lambda_2, \lambda_3, \lambda_4, \lambda_5$, are given in Table 2.7.

Table 2.7 Elastic constants for axisymmetric stress analysis

	Stratified material	Isotropic material
λ_1	$\dfrac{E_2(1-\nu_1)}{\mu}$	$\dfrac{E(1-\nu)}{(1+\nu)(1-2\nu)}$
λ_2	$\dfrac{E_1(1-\kappa\nu_2^2)}{(1+\nu_1)\mu}$	
λ_3	G_2	$\dfrac{E}{2(1+\nu)}$
λ_4	$\dfrac{\nu_2 E_1}{\mu}$	$\dfrac{E\nu}{(1+\nu)(1-2\nu)}$
λ_5	$\dfrac{E_1(\nu_1+\kappa\nu_2^2)}{(1+\nu_1)\mu}$	

where $\kappa = \dfrac{E_1}{E_2}$ and $\mu = \dfrac{1}{(1-\nu_1-2\kappa\nu_2^2)}$

2.3.2 Element stiffness matrix

Using matrix notations as given in Zienkiewicz and Taylor [8], the element nodal force vector $\begin{bmatrix} f_1 \\ f_2 \end{bmatrix}_\alpha$ at node α and the displacement vector $\begin{bmatrix} u_1 \\ u_2 \end{bmatrix}_\beta$ at node β is related by 2×2 element submatrix $\mathbf{K}^{\alpha\beta}$ such that

$$\underset{2\times 2}{\mathbf{K}^{\alpha\beta}} = 2\pi \int_{A_e} \mathbf{B}_\alpha^{\mathrm{T}} \mathbf{C} \mathbf{B}_\beta r \, \mathrm{d}A \tag{2.23}$$

where
$$\underset{2\times 4}{\mathbf{B}_\alpha^{\mathrm{T}}} = \begin{bmatrix} f_\alpha & 0 & h_\alpha & g_\alpha \\ 0 & g_\alpha & 0 & f_\alpha \end{bmatrix} \qquad \underset{4\times 2}{\mathbf{B}_\beta} = \begin{bmatrix} f_\beta & 0 \\ 0 & g_\beta \\ h_\beta & 0 \\ g_\beta & f_\beta \end{bmatrix}$$

$$f_\alpha = \frac{\partial H_\alpha}{\partial r} \qquad g_\alpha = \frac{\partial H_\alpha}{\partial z} \qquad h_\alpha = \frac{H_\alpha}{r}$$

$$\underset{2\times 2}{\mathbf{B}_\alpha^T \mathbf{C} \mathbf{B}_\beta} = \left[\begin{array}{l|l} \lambda_2 f_\alpha f_\beta + \lambda_3 g_\alpha g_\beta + & \lambda_4 f_\alpha g_\beta + \lambda_3 g_\alpha f_\beta + \\ \lambda_5 (f_\alpha h_\beta + h_\alpha f_\beta) + \lambda_2 h_\alpha h_\beta & \lambda_4 h_\alpha g_\beta \\ \hline \lambda_3 f_\alpha g_\beta + \lambda_4 g_\alpha f_\beta + & \lambda_3 f_\alpha f_\beta + \lambda_1 g_\alpha g_\beta \\ \lambda_4 g_\alpha h_\beta & \end{array}\right] \tag{2.24}$$

It is noted that the plane strain case is recovered if h_α are set to zero. Using numerical integration, we have

$$\underset{2\times 2}{\mathbf{K}^{\alpha\beta}} = 2\pi \int_{\mathcal{R}} \mathbf{B}_\alpha^T \mathbf{C} \mathbf{B}_\beta r \det(\mathbf{J}) \, \mathrm{d}\mathcal{R} \approx 2\pi \sum_{k=1}^{m} W_k [\mathbf{B}_\alpha^T \mathbf{C} \mathbf{B}_\beta r \det(\mathbf{J})]_{(\xi_k, \eta_k)} \tag{2.25}$$

2.3.3 Element nodal force vectors

(i) Nodal force vector due to initial stress(strain)

The nodal force vector at node α due to initial stress(strain) is given by

$$\vec{r}_\alpha = 2\pi \int_{\Omega_e} \mathbf{B}_\alpha^{\mathrm{T}} \vec{\tau}_o \, d\Omega \qquad \text{where } \vec{\tau}_o = \begin{bmatrix} \tau_{or} \\ \tau_{oz} \\ \tau_{o\theta} \\ \tau_{orz} \end{bmatrix} = \mathbf{C} \begin{bmatrix} \varepsilon_{or} \\ \varepsilon_{oz} \\ \varepsilon_{o\theta} \\ \varepsilon_{orz} \end{bmatrix} - \begin{bmatrix} \sigma_{or} \\ \sigma_{oz} \\ \sigma_{o\theta} \\ \sigma_{orz} \end{bmatrix}$$

$$\vec{r}_\alpha = 2\pi \int_{A_e} \mathbf{B}_\alpha^{\mathrm{T}} \vec{\tau}_o r \, dA = 2\pi \int_{A_e} \begin{bmatrix} f_\alpha & 0 & h_\alpha & g_\alpha \\ 0 & g_\alpha & 0 & f_\alpha \end{bmatrix} \begin{bmatrix} \tau_{or} \\ \tau_{oz} \\ \tau_{o\theta} \\ \tau_{orz} \end{bmatrix} r \, dA \tag{2.26}$$

For elements with constant initial stress(strain),

$$\vec{r}_\alpha = 2\pi \begin{bmatrix} \tau_{or} & \tau_{orz} & \tau_{o\theta} \\ \tau_{orz} & \tau_{oz} & 0 \end{bmatrix} \int_{A_e} \begin{bmatrix} f_\alpha \\ g_\alpha \\ h_\alpha \end{bmatrix} r \, dA \tag{2.27}$$

$$\int_{A_e} \begin{bmatrix} f_\alpha \\ g_\alpha \\ h_\alpha \end{bmatrix} r \, dA = \int_{\mathcal{R}} \begin{bmatrix} r f_\alpha \\ r g_\alpha \\ H_\alpha \end{bmatrix} \det(\mathbf{J}) \, d\mathcal{R} \approx \sum_{k=1}^{m} W_k \begin{bmatrix} r f_\alpha \det(\mathbf{J}) \\ r g_\alpha \det(\mathbf{J}) \\ H_\alpha \det(\mathbf{J}) \end{bmatrix}_{(\xi_k, \eta_k)} \tag{2.28}$$

(ii) Element body force vector

$$\vec{b}_\alpha = \int_{\Omega_e} \vec{\gamma} H_\alpha \, d\Omega = 2\pi \int_{A_e} \vec{\gamma} H_\alpha r \, dA = 2\pi \int_{\mathcal{R}} \vec{\gamma} H_\alpha r \det(\mathbf{J}) \, d\mathcal{R}$$

$$\vec{b}_\alpha \approx 2\pi \sum_{k=1}^{m} W_k \left[\vec{\gamma} H_\alpha r \det(\mathbf{J}) \right]_{(\xi_k, \eta_k)} \tag{2.29}$$

If the body forces vary with r, as for the case of rotating machinery, we have $\vec{\gamma} = \begin{bmatrix} \rho r \omega^2 \\ 0 \end{bmatrix}$, where ω is the angular velocity and ρ the density of the material.

(iii) Element surface traction vector

The nodal force vector is given by

$$\vec{s}_\alpha = \int_{\Gamma_e} \vec{q} H_\alpha \, dA = 2\pi \int_{\mathcal{C}} \vec{q} H_\alpha r \, dl$$

or
$$\vec{s}_\alpha = 2\pi \int_{-1}^{1} r H_\alpha \begin{bmatrix} q_t & q_n \\ -q_n & q_t \end{bmatrix} \begin{bmatrix} H'_1 x_1 & + & H'_2 x_2 & + & H'_3 x_3 \\ H'_1 y_1 & + & H'_2 y_2 & + & H'_3 y_3 \end{bmatrix} d\varphi \tag{2.30}$$

where $H'_1 = \frac{1}{2}(2\varphi - 1)$, $H'_2 = -2\varphi$, $H'_3 = \frac{1}{2}(2\varphi + 1)$

Readers can refer to Section 2.2.2 for details.

2.3.4 Evaluation of element stresses

The stresses within an element can be calculated by

$$\underset{4\times1}{\vec{\sigma}} = \underset{4\times4}{\mathbf{C}} \cdot \underset{4\times2N}{\mathbf{B}} \cdot \underset{2N\times1}{\vec{\mathrm{u}}_e} - \underset{4\times1}{\vec{\tau}_o} \tag{2.31}$$

where $\underset{4\times2N}{\mathbf{B}} = [\underset{4\times2}{\mathbf{B}_1} \ldots, \underset{4\times2}{\mathbf{B}_N}]$ and $\vec{\mathrm{u}}_e$ is the element displacement vector.

2.3.5 Examples

1 Sphere under internal pressure

The problem can be taken as an axisymmetric analysis, for which the analytical solution of the radial displacement at a point r from the centre of the sphere is given by Reference 38

$$u_r = \frac{pa^3}{E(b^3 - a^3)}\left[(1 - 2\nu)r + \frac{b^3}{2r^2}(1 + \nu)\right]$$

where a and b are respectively the internal and external radii of the sphere, p is the applied internal pressure, E is the Young's modulus and ν is the Poisson's ratio.

In the finite element analysis, one quarter of a typical section is divided into quadrilateral mesh and triangular mesh as shown in Fig. 2.18. The results of analysis for different Poisson's ratios and integration rules are given in Table 2.8. It should be pointed out that the computed displacements can be dramatically improved by using reduced integration as the Poisson's ratio approaches 0.5.

2 Stress in a rotating disc

This is another example of axisymmetric analysis, in which the stress of a thin disc of radius b rotating at a constant angular velocity ω is examined. The analytical

Table 2.8 Results of using different numerical integration rules

	Radial displacement							
	$\nu = 0.3$		$\nu = 0.4$		$\nu = 0.49$		$\nu = 0.499$	
	Point A	Point B	Point A	Point B	Point A	Point B	Point A	Point B
Q8 (2 × 2)	0.399 17	0.150 04	0.413 35	0.128 55	0.426 22	0.109 27	0.427 58	0.107 37
Error (%)	0.21	−0.02	0.23	0.02	0.22	0.02	0.20	−0.01
Q8 (3 × 3)	0.398 51	0.149 58	0.411 23	0.127 77	0.398 23	0.102 17	0.248 87	0.063 11
Error (%)	0.37	0.28	0.74	0.62	6.77	6.51	41.91	41.22
T6 (3pts)	0.394 51	0.149 03	0.407 24	0.127 17	0.417 97	0.107 26	0.419 02	0.105 26
Error (%)	1.37	0.65	1.70	1.09	2.15	1.86	2.20	1.96
T6 (4pts)	0.394 92	0.148 37	0.407 46	0.128 04	0.416 82	0.106 91	0.397 75	0.098 39
Error (%)	1.27	1.09	1.65	0.41	2.42	2.17	7.16	8.35
Exact	0.400 000	0.150 000	0.414 286	0.128 571	0.427 143	0.109 286	0.428 429	0.107 357

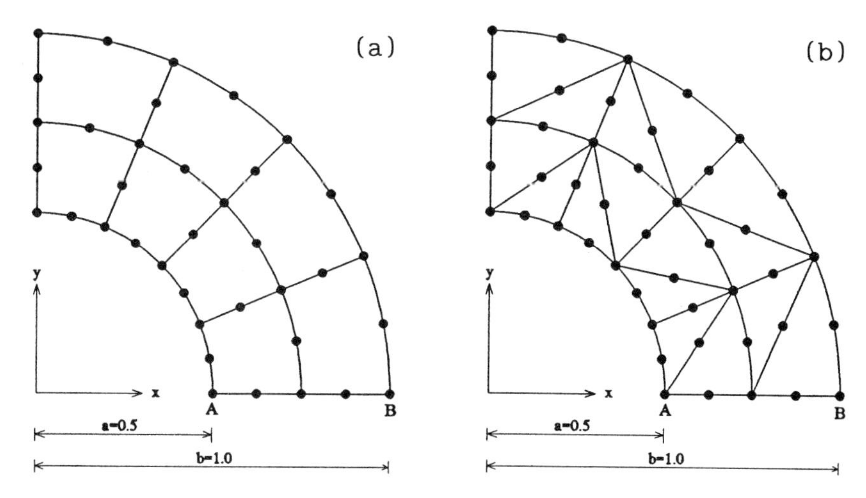

Fig. 2.18 (a) Quadrilateral mesh, (b) triangular mesh

solution for the radial displacement and the principal stresses under the plane stress condition are given by [12]

$$u_r = \frac{1-\nu}{8E}\rho\omega^2[(3+\nu)b^2r - (1+\nu)r^3]$$

$$\sigma_r = \frac{3+\nu}{8}\rho\omega^2(b^2 - r^2) \qquad \sigma_\theta = \rho\omega^2\left(\frac{3+\nu}{8}b^2 - \frac{1+3\nu}{8}r^2\right)$$

where E is the Young's modulus, ν is the Poisson's ratio, ρ is the mass per unit volume, and Ω is the angular velocity.

Quadrilateral and triangular meshes are used in the finite element analysis as shown in Fig. 2.19. The maximum errors in the radial displacement and major principal stress for a thin disc with $b = 10$ and $E = 1000$ are shown in Table 2.9. The results are very accurate – the maximum error in radial displacement is 0.26% and that in major principal stress is 0.4%.

Table 2.9 Maximum errors in radial displacement and major principal stress

Max. error (%) in:	T6 (3 pts)	T6 (4 pts)	Q8 (4 pts)	Q8 (9 pts)
u_r	0.40	0.33	0.39	0.38
σ_θ	0.26	0.26	0.25	0.25

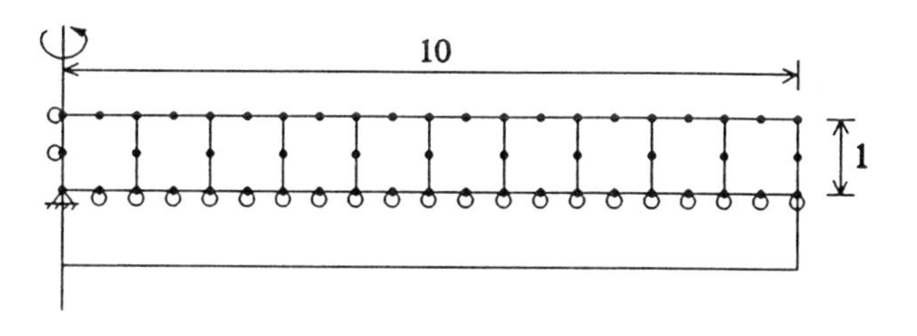

Fig. 2.19(a) Q8 element mesh

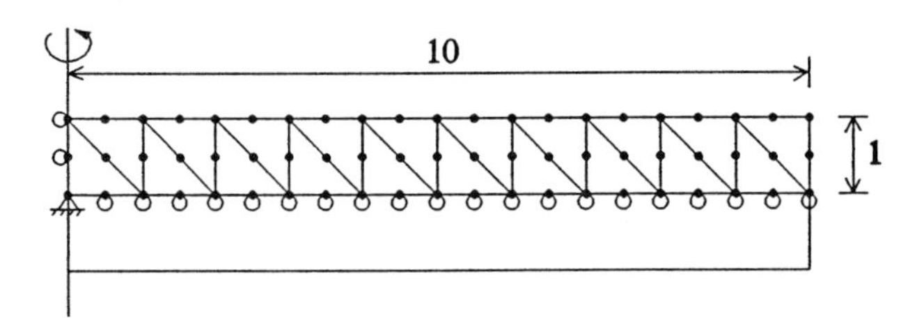

Fig. 2.19(b) T6 element mesh

2.4 STEADY-STATE FIELD PROBLEMS (2D)

Almost without any modification, the finite element solution developed in Section 3.4 of the next chapter for general quasi-harmonic equation can be directly applied to two-dimensional situations [17–30]. Although a simple 3-node triangular element has been used extensively, the more efficient quadratic isoparametric elements T6 and Q8 will be developed in this chapter for the solution of the quasi-harmonic equation in two dimensions.

By applying direct analytical integration, explicit expressions of element stiffness matrix and element force vectors for subparametric 6-node triangular and 8-node rectangular elements are derived. Once again numerical integration has to be used for the general curved isoparametric T6 and Q8 elements. Practical numerical integration formulae for isoparametric elements are already given in Table 2.5. The performance of the quadratic elements is assessed by comparing the numerical results of some practical examples with available analytical solutions.

2.4.1 Element stiffness matrix and element force vectors

Expressions for 3D element stiffness matrix and element force vectors are developed in Section 3.4.6. Assuming that the conductivity coefficients k_{ij} and the radiation(convection) coefficient c are constant within the element, and using notations

$$I_{ij}^{\alpha\beta} = \int_{\Omega_e} H_{\alpha,i} H_{\beta,j}\, \mathrm{d}\Omega \qquad V_{\alpha\beta} = \int_{\Omega_e} H_\alpha H_\beta\, \mathrm{d}\Omega \qquad A_{\alpha\beta} = \int_{\Gamma_e} H_\alpha H_\beta\, \mathrm{d}\Gamma$$

then,

1. *Element stiffness matrix*, $\mathbf{K}_e = [k_{ij} I_{ij}^{\alpha\beta} + cA_{\alpha\beta}]$ (2.32)
2. *Element body force*, $\vec{\mathrm{b}}_e = [Q_\beta V_{\alpha\beta}]$ (2.33)
3. *Element surface force*, $\vec{\mathrm{s}}_e = [g_\beta A_{\alpha\beta}]$ (2.34)

where Q_β is the relevant quantity generated(removed) at node β, and g_β is the specified flux at node β.

The Latin indices i, j take the values 1 and 2, and the Greek indices α, β range from 1 to N (number of nodes in the element).

If the local element coordinate axes coincide with the principal directions of material stratification, the expressions for element stiffness matrix $\mathbf{K}_e$ can be simplified to

$$\mathbf{K}_e = [k_1 I_{11}^{\alpha\beta} + k_2 I_{22}^{\alpha\beta} + cA_{\alpha\beta}] \tag{2.35}$$

where k_1 and k_2 are conductivity coefficients along the principal directions.

2.4.2 6-node subparametric triangular element

Within the interior part of the domain, triangular elements with straight edges can be used. Subparametric triangular elements are considered to be computationally more efficient as the element stiffness matrix and element force vectors can be obtained in explicit form by direct integration (Section 2.2.4).

(a) Element stiffness matrix

$$\mathbf{K}_e = [k_1 I_{11}^{\alpha\beta} + k_2 I_{22}^{\alpha\beta} + cA_{\alpha\beta}] \qquad \alpha, \beta = 1, 6 \tag{2.36}$$

For an element of constant thickness t, the integrals $I_{11}^{\alpha\beta}$ and $I_{22}^{\alpha\beta}$, $\alpha, \beta = 1, 6$, required for the element stiffness matrix $\mathbf{K}_e$ can be evaluated analytically as given below:

$$\underset{6\times6}{[I_{11}^{\alpha\beta}]} = \frac{t}{12\Delta}\begin{bmatrix} 3b_1^2 & -b_1b_2 & -b_1b_3 & 0 & 4b_1b_3 & 4b_1b_2 \\ & 3b_2^2 & -b_2b_3 & 4b_2b_3 & 0 & 4b_1b_2 \\ & & 3b_3^2 & 4b_2b_3 & 4b_1b_3 & 0 \\ & & & 8(b_1^2 - b_2b_3) & 8b_1b_2 & 8b_1b_3 \\ & & & & 8(b_1^2 - b_2b_3) & 8b_2b_3 \\ \text{sym.} & & & & & 8(b_1^2 - b_2b_3) \end{bmatrix} \tag{2.37}$$

$$\underset{6\times6}{[I_{22}^{\alpha\beta}]} = \frac{t}{12\Delta}\begin{bmatrix} 3c_1^2 & -c_1c_2 & -c_1c_3 & 0 & 4c_1c_3 & 4c_1c_2 \\ & 3c_2^2 & -c_2c_3 & 4c_2c_3 & 0 & 4c_1c_2 \\ & & 3c_3^2 & 4c_2c_3 & 4c_1c_3 & 0 \\ & & & 8(c_1^2 - c_2c_3) & 8c_1c_2 & 8c_1c_3 \\ & & & & 8(c_1^2 - c_2c_3) & 8c_2c_3 \\ \text{sym.} & & & & & 8(c_1^2 - c_2c_3) \end{bmatrix} \tag{2.38}$$

For the evaluation of the surface integral $A_{\alpha\beta} = \int_{\Gamma_e} H_\alpha H_\beta \, d\Gamma$, surfaces with contribution to the integral have to be considered in turn.

(1) Over the top(bottom) surface of the triangular element

$$[A_{\alpha\beta}]_o = \int_{A_e} H_\alpha H_\beta \, dA \tag{2.39}$$

$$\underset{6\times6}{[A_{\alpha\beta}]_o} = \frac{\Delta}{180}\begin{bmatrix} 6 & -1 & -1 & -4 & 0 & 0 \\ -1 & 6 & -1 & 0 & -4 & 0 \\ -1 & -1 & 6 & 0 & 0 & -4 \\ -4 & 0 & 0 & 32 & 16 & 16 \\ 0 & -4 & 0 & 16 & 32 & 16 \\ 0 & 0 & -4 & 16 & 16 & 32 \end{bmatrix} \tag{2.40}$$

where Δ is the area of the triangular element.

(2) Along an edge of the triangular element

Only those components associated with the three nodes on that edge will have non-zero values (Fig. 2.20), and are given by

$$\underset{3\times 3}{[A_{\alpha\beta}]} = \frac{tl}{30}\begin{bmatrix} 4 & 2 & -1 \\ 2 & 16 & 2 \\ -1 & 2 & 4 \end{bmatrix} \tag{2.41}$$

where t is the thickness of the element and l is the length of the edge under consideration.

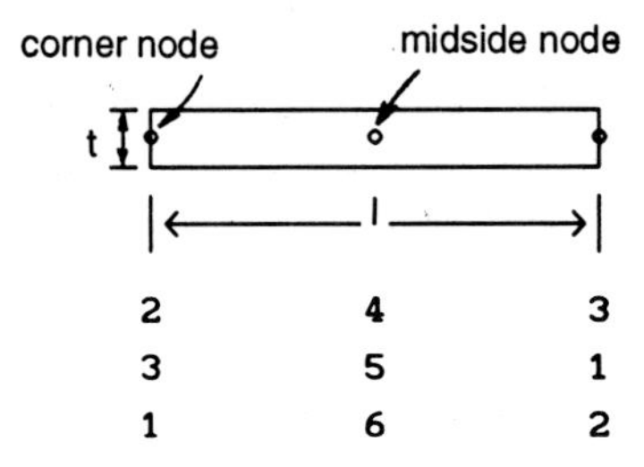

Fig. 2.20 An edge of a quadratic element

(i) Along edge 2–4–3 ($\xi = 0$)

$$\underset{6\times 6}{[A_{\alpha\beta}]_1} = \frac{tl_1}{30}\begin{bmatrix} & & & & & \\ & 4 & -1 & 2 & & \\ & -1 & 4 & 2 & & \\ & 2 & 2 & 16 & & \\ & & & & & \\ & & & & & \end{bmatrix} \tag{2.42}$$

Zero elements have been omitted.

(ii) Along edge 3–5–1 ($\eta = 0$)

$$\underset{6\times 6}{[A_{\alpha\beta}]_2} = \frac{tl_2}{30}\begin{bmatrix} 4 & & -1 & & 2 & \\ & & & & & \\ -1 & & 4 & & 2 & \\ & & & & & \\ 2 & & 2 & & 16 & \\ & & & & & \end{bmatrix} \tag{2.43}$$

(iii) Along edge 1–6–2 ($\varphi = 0$)

$$\underset{6\times 6}{[A_{\alpha\beta}]_3} = \frac{tl_3}{30}\begin{bmatrix} 4 & -1 & & & & 2 \\ -1 & 4 & & & & 2 \\ & & & & & \\ & & & & & \\ & & & & & \\ 2 & 2 & & & & 16 \end{bmatrix} \tag{2.44}$$

(b) Element force vectors

(i) Element body force

$$V_{\alpha\beta} = \int_{\Omega_e} H_\alpha H_\beta \, d\Omega = t\int_{A_e} H_\alpha H_\beta \, dA = t[A_{\alpha\beta}]_o \tag{2.45}$$

$$\vec{b}_e = [Q_\beta V_{\alpha\beta}] = tQ_\beta [A_{\alpha\beta}]_o \tag{2.46}$$

In the case where the heat generation per unit volume is constant within the element, $Q_\beta = Q,\ \beta = 1, 6,$ we have

$$\vec{b}_e^T = \tfrac{1}{3} Qt\Delta(0, 0, 0, 1, 1, 1) \tag{2.47}$$

(ii) Element surface force vector

$$\vec{s}_e = [g_\beta A_{\alpha\beta}] \qquad \alpha, \beta = 1, 6 \tag{2.48}$$

where the $A_{\alpha\beta}$ values over the top(bottom) surface and along the edges have already been considered and derived in Section 2.4.2.

In the case where the flux g is constant over the top(bottom) surface, we have

$$\vec{s}_e^T = \tfrac{1}{3} \Delta g(0, 0, 0, 1, 1, 1) \tag{2.49}$$

Similarly, if the flux g is constant along an edge, we have

(i) along edge 2–4–3

$$\vec{s}_e^T = \tfrac{1}{6} tl_1 g(0, 1, 1, 4, 0, 0)$$

(ii) along edge 3–5–1

$$\vec{s}_e^T = \tfrac{1}{6} tl_2 g(1, 0, 1, 0, 4, 0)$$

(iii) along edge 1–6–2

$$\vec{s}_e^T = \tfrac{1}{6} tl_3 g(1, 1, 0, 0, 0, 4)$$

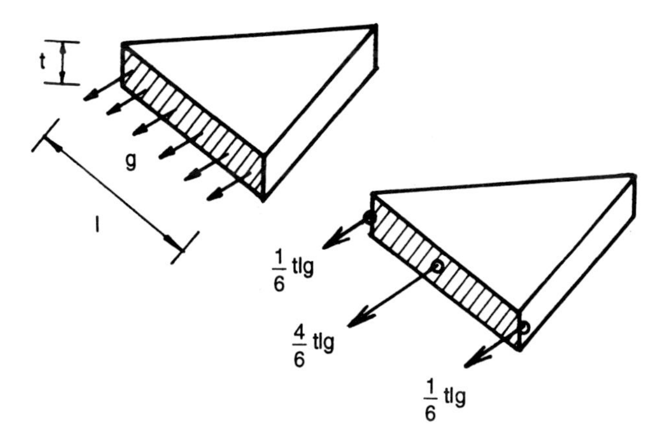

Fig. 2.21 Equivalent nodal forces for constant flux along an edge

2.4.3 8-node subparametric rectangular element

For the discretization of rectangular domains, and at the interior part of an irregular domain, square or rectangular elements are used more often because of its high efficiency.

(a) ***Element stiffness matrix***

$$\mathbf{K}_e = \left[k_1 I_{11}^{\alpha\beta} + k_2 I_{22}^{\alpha\beta} + cA_{\alpha\beta}\right] \qquad \alpha, \beta = 1, 8 \tag{2.50}$$

The integrals $I_{11}^{\alpha\beta}$ and $I_{22}^{\alpha\beta}$, $\alpha, \beta = 1, 8$, of a subparametric rectangular element of constant thickness t can be evaluated analytically, and the resulting expressions are given below:

$$\underset{8\times 8}{\left[I_{11}^{\alpha\beta}\right]} = \frac{bt}{45a}\begin{bmatrix} 26 & 14 & 23/2 & 17/2 & -40 & -3 & -20 & 3 \\ & 26 & 17/2 & 23/2 & -40 & 3 & -20 & -3 \\ & & 26 & 14 & -20 & 3 & -40 & -3 \\ & & & 26 & -20 & -3 & -40 & 3 \\ & & & & 80 & 0 & 40 & 0 \\ & \text{sym.} & & & & 24 & 0 & -24 \\ & & & & & & 80 & 0 \\ & & & & & & & 24 \end{bmatrix} \tag{2.51}$$

$$\underset{8\times 8}{\left[I_{22}^{\alpha\beta}\right]} = \frac{at}{45b}\begin{bmatrix} 26 & 17/2 & 23/2 & 14 & 3 & -20 & -3 & -40 \\ & 26 & 14 & 23/2 & 3 & -40 & -3 & -20 \\ & & 26 & 17/2 & -3 & -40 & 3 & -20 \\ & & & 26 & -3 & -20 & 3 & -40 \\ & & & & 24 & 0 & -24 & 0 \\ & \text{sym.} & & & & 80 & 0 & 40 \\ & & & & & & 24 & 0 \\ & & & & & & & 80 \end{bmatrix} \tag{2.52}$$

Evaluation of $A_{\alpha\beta} = \int_{\Gamma_e} H_\alpha H_\beta \, d\Gamma$:

(i) Over the top(bottom) surface of the rectangular element

$$[A_{\alpha\beta}]_o = \frac{ab}{45}\begin{bmatrix} 6 & 2 & 3 & 2 & -6 & -8 & -8 & -6 \\ & 6 & 2 & 3 & -6 & -6 & -8 & -8 \\ & & 6 & 2 & -8 & -6 & -6 & -8 \\ & & & 6 & -8 & -8 & -6 & -6 \\ & & & & 32 & 20 & 16 & 20 \\ & & & & & 32 & 20 & 16 \\ & \text{sym.} & & & & & 32 & 20 \\ & & & & & & & 32 \end{bmatrix} \tag{2.53}$$

(ii) Along an edge of the rectangular element

As an edge is identified only by the three relevant nodes, there is no difference between an edge of a triangular element and that of a rectangular element. As a consequence, the results obtained in Section 2.4.2(a) can be directly applied here.

(b) ***Element force vector***

(i) Element body force vector, $\vec{b}_e = tQ_\beta[A_{\alpha\beta}]_o$

In the case where the heat generation per unit volume is constant within the element, $Q_\beta = Q$, $\beta = 1, 8$, we have

$$\vec{b}_e^{\,T} = \tfrac{1}{3}Qtab(-1, -1, -1, -1, 4, 4, 4, 4) \tag{2.54}$$

(ii) Element surface force vector

$$\vec{s}_e = [g_\beta A_{\alpha\beta}] \qquad \alpha, \beta = 1, 8 \tag{2.55}$$

$A_{\alpha\beta}$ values over the top(bottom) surface and along edges have already been considered and derived in Section 2.4.3(a).

In the case where the flux g is constant over the top(bottom) surface, we have

$$\vec{s}_e^{\,\mathrm{T}} = \tfrac{1}{3} gab(-1, -1, -1, -1, 4, 4, 4, 4) \tag{2.56}$$

2.4.4 Numerical integration for T6 and Q8 elements

While the element stiffness matrix and element force vectors of subparametric triangular and rectangular elements can be obtained in explicit form by direct integration, for isoparametric elements T6 and Q8 with curved boundaries, numerical integration has to be used. The numerical integration formulae given in Table 2.5 can once again be applied to the integrals. The order of integration with no loss of convergence is the most advantageous in the evaluation of element matrices [8]. For quadratic isoparametric elements, the optimum integration scheme is the one which integrates exactly polynomials up to degree 3.

2.4.5 Examples

1 Temperature distribution over a square plate

The temperature distribution over a square plate is studied by the finite element method. The plate is insulated at the top and bottom surfaces, and is subjected to a linearly varying temperature profile along its four edges, as shown in Fig. 2.22. Analysis is carried out using a 5 × 5 Q8 element mesh and a 50 T6 element mesh obtained by dividing each square element along the diagonal as depicted in Fig. 2.23. The plate is assumed to be orthotropic with $k_x = 1$ and $k_y = 2$. Both meshes give the same result as the theoretical solution which is a bilinear variation over the plate. The temperature profiles along lines parallel to the x- and y-axis are shown in Fig. 2.24.

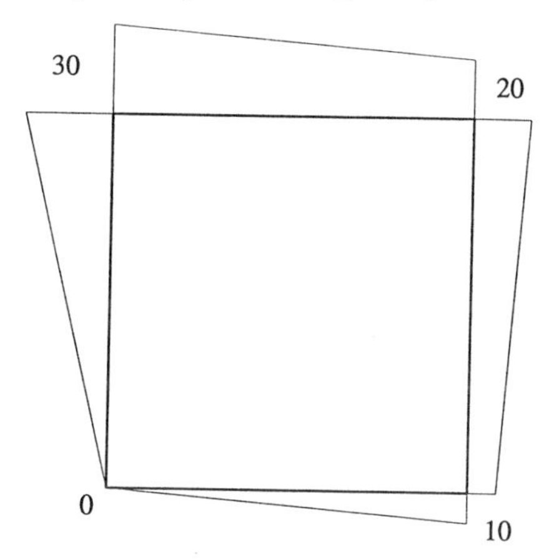

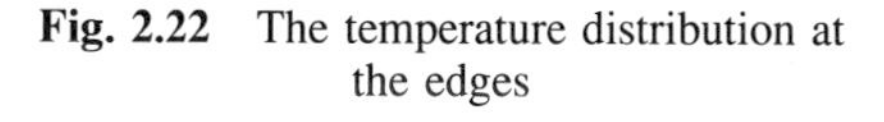
Fig. 2.22 The temperature distribution at the edges

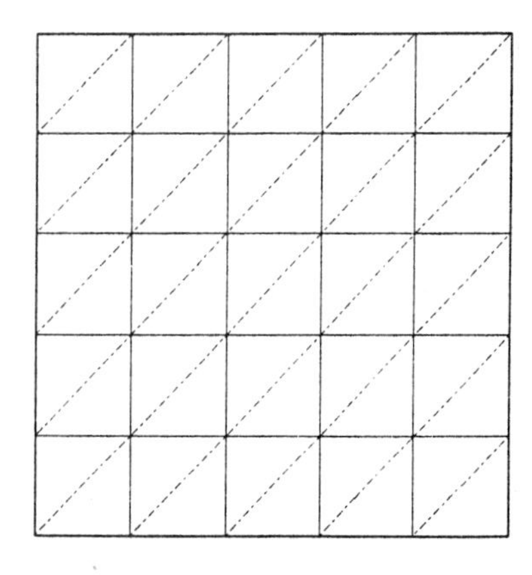

Fig. 2.23 The finite element mesh. — · — lines of division for T6 elements

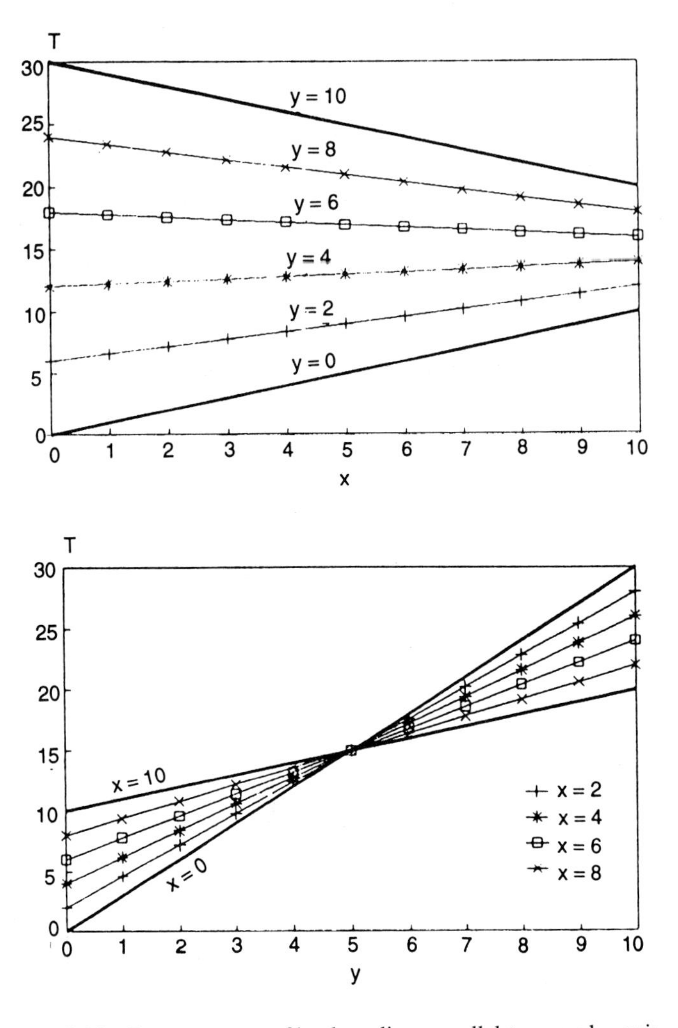

Fig. 2.24 Temperature profile along lines parallel to x- and y-axis

2 Torsion of a prismatic cross-section

The torsion of a prismatic cross-section can be solved easily by the finite element method as a steady-state potential problem similar to the heat conduction problem. The governing equation of a two-dimensional steady-state field problem of potential ϕ is in the form:

$$\frac{\partial}{\partial x}\left(k_x \frac{\partial \phi}{\partial x}\right) + \frac{\partial}{\partial y}\left(k_y \frac{\partial \phi}{\partial y}\right) + Q = 0 \tag{2.57}$$

where Q is the rate of generation(removal) of the relevant quantity per unit volume, and k_x and k_y are the permeability coefficients.

For the torsion problem, the governing equation for the Prandtl stress function ϕ

is given by

$$\frac{\partial}{\partial x}\left(\frac{1}{G_x}\frac{\partial \phi}{\partial x}\right)+\frac{\partial}{\partial y}\left(\frac{1}{G_y}\frac{\partial \phi}{\partial y}\right)+2\theta=0 \tag{2.58}$$

where θ is the angle of twist per unit length, and G_x, G_y are the shear moduli.

The boundary condition, ϕ = constant, along the section boundary, can be arbitrarily chosen to be zero for a simply-connected domain. Comparing eqs 2.57 and 2.58, the torsion problem can be recast into a field problem with

$$k_x=\frac{1}{G_x} \qquad k_y=\frac{1}{G_y} \qquad \text{and } Q=2\theta$$

The applied torque T can then be evaluated using

$$T=-\int_A \frac{\partial \phi}{\partial x}x+\frac{\partial \phi}{\partial y}y\,\mathrm{d}A \tag{2.59}$$

Integrating by parts and noting that $\phi=0$ on the boundary, we have

$$T=2\int_A \phi\,\mathrm{d}A \tag{2.60}$$

Hence the torsional rigidity can be evaluated by

$$D=\frac{T}{\theta}=\frac{2}{\theta}\int_A \phi\,\mathrm{d}A \tag{2.61}$$

Two sections having analytical expressions for ϕ are tested –an ellipse $(x/a)^2+(y/b)^2=1$, $a=1.0$, $b=0.6$, and an equilateral triangle with unit sides [12]. The material is assumed to be isotropic such that $G_x=G_y=1$ and θ is taken equal to 1. Two meshes were generated for each section – a triangular T6 mesh and a mixed mesh of Q8 and T6 elements as shown in Fig. 2.25. The nodal ϕ values and the torsional rigidity for the meshes are shown in Table 2.10. It is observed that all the meshes give good approximations to both the values of the stress function and the

Table 2.10 Numerical results for the two cross-sections

	Number of nodes	Number of elements	Max. error in ϕ (%)	Torsional rigidity, D
		Elliptic cross-section		
T6 mesh	157	66	0.09	0.498 956 8
Q8 + T6 mesh	129	38	0.18	0.498 945 6
			Exact =	0.498 958 8
		Equilateral triangular cross-section		
T6 mesh	66	25	0.00	0.021 615 98
Q8 + T6 mesh	56	15	0.72	0.021 663 92
			Exact =	0.021 650 63

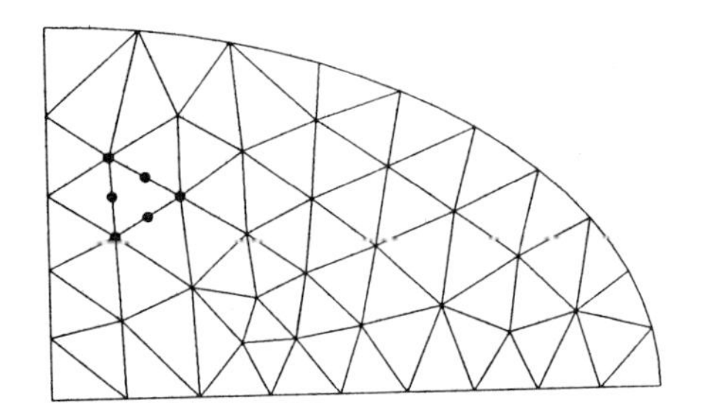

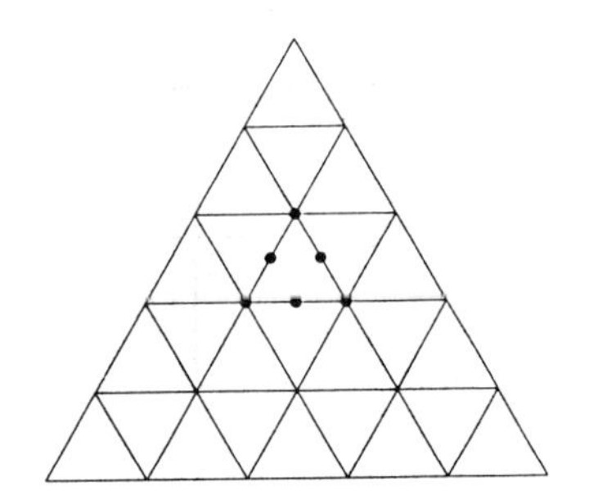

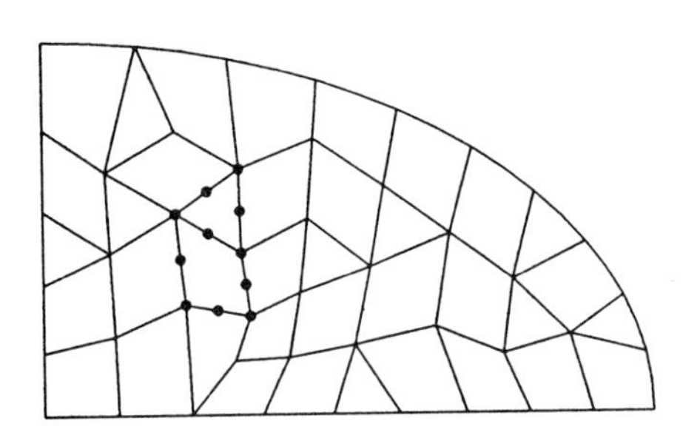

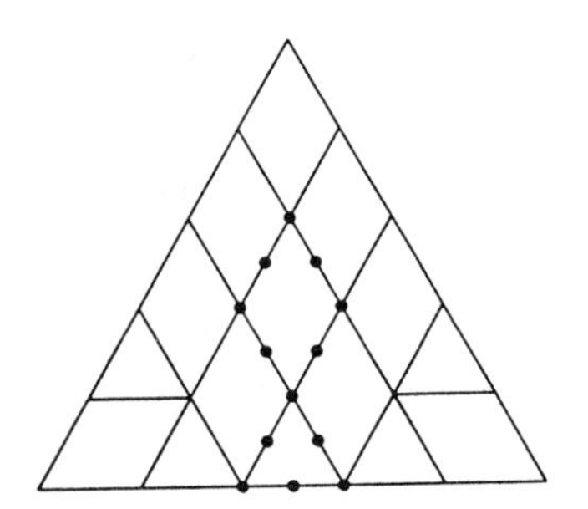

Fig. 2.25 Finite element meshes for the test cross-sections

torsional rigidity. The $Q8 + T6$ meshes give a slightly less accurate result, probably due to the fewer number of elements and nodes employed.

2.5 INFINITE DOMAIN AND INFINITE ELEMENT

A difficulty often encountered in finite element analysis is in the treatment of an unbounded domain, and one of the solutions to this problem is in the use of the 'infinite element' [31–33]. In this process, the conventional finite elements are coupled to elements of the type shown in Fig. 2.26 which model the domain stretching to infinity.

The (x, y) coordinates of any point P are given by

$$x = \sum_{\alpha=1}^{6} G_\alpha x_\alpha \quad \text{and} \quad y = \sum_{\alpha=1}^{6} G_\alpha y_\alpha \tag{2.62}$$

where

$$\begin{aligned} G_1 &= \frac{\xi\eta(1-\eta)}{1-\xi} & G_2 &= \frac{-2\xi(1-\eta^2)}{1-\xi} & G_3 &= \frac{-\xi\eta(1+\eta)}{1-\xi} \\ G_4 &= \frac{-(1+\xi)\eta(1-\eta)}{2(1-\xi)} & G_5 &= \frac{(1+\xi)(1-\eta^2)}{1-\xi} & G_6 &= \frac{(1+\xi)\eta(1+\eta)}{2(1-\xi)} \end{aligned} \tag{2.63}$$

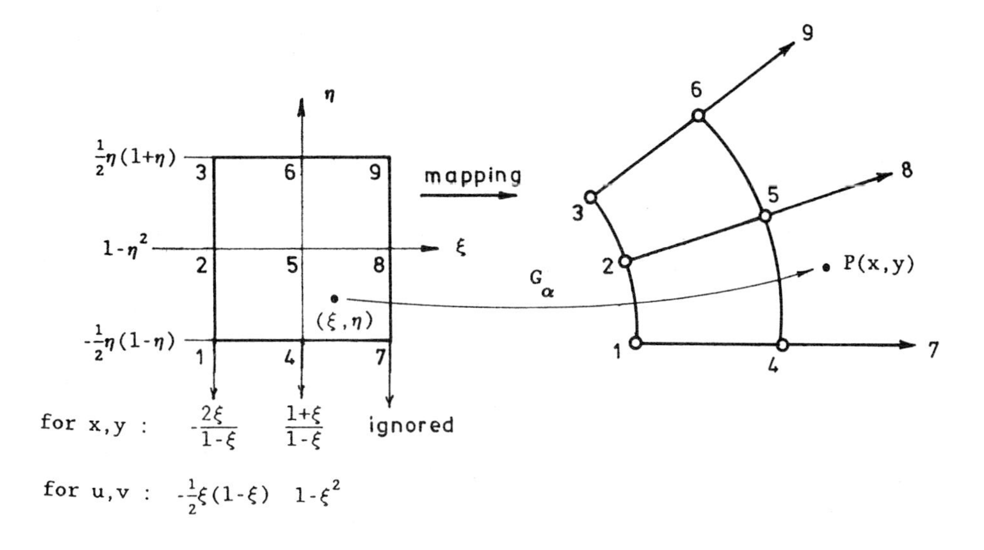

Fig. 2.26 Mapping a Lagrangian 9-node element to an infinite domain

And the displacement (u, v) at point P is given by the standard Lagrangian interpolation functions such that

$$u = \sum_{\alpha=1}^{6} H_\alpha u_\alpha \quad \text{and} \quad v = \sum_{\alpha=1}^{6} H_\alpha v_\alpha \tag{2.64}$$

where

$$\begin{aligned} &H_1 = \tfrac{1}{4}\xi(1-\xi)\eta(1-\eta) \quad H_2 = -\tfrac{1}{2}\xi(1-\xi)(1-\eta^2) \quad H_3 = -\tfrac{1}{4}\xi(1-\xi)\eta(1+\eta) \\ &H_4 = -\tfrac{1}{2}(1-\xi^2)\eta(1-\eta) \quad H_5 = (1-\xi^2)(1-\eta^2) \quad H_6 = \tfrac{1}{2}(1-\xi^2)\eta(1+\eta) \end{aligned} \tag{2.65}$$

The terms involving nodes 7, 8 and 9 are simply ignored since displacements at infinity are assumed to be zero. The only change needed to be made to a finite element program is in the computation of derivatives with respect to x and y and the Jacobian matrix.

2.5.1 Example

1 Boussinesq problem – a point load on an elastic half-space

In the finite element analysis of the Boussinesq problem, a mesh consisting of 16 standard Q8 elements is used as shown in Fig. 2.27. Besides the boundary condition $u_r = 0$ along the axis of symmetry, an additional condition of zero displacement along the truncated boundary is imposed. A second mesh produced by replacing the four outermost Q8 elements with the 6-node infinite elements I6 is shown in Fig. 2.28. No far field boundary condition is required because it has already been implicitly imposed in the infinite elements. Finally, a third mesh consisting of eight T6, eight Q8 and four I6 elements is generated by dividing the four quadrilateral elements near the applied load into T6 elements as shown in Fig. 2.29. The material is assumed to be isotropic with Young's modulus $E = 1$ and Poisson's ratio $\nu = 0.1$.

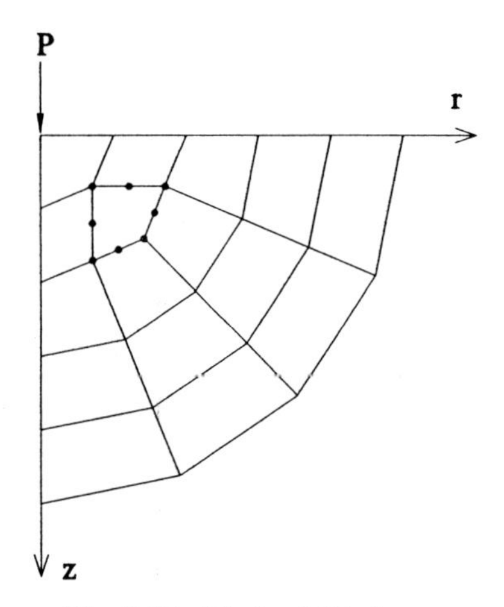

Fig. 2.27 Mesh of Q8 elements

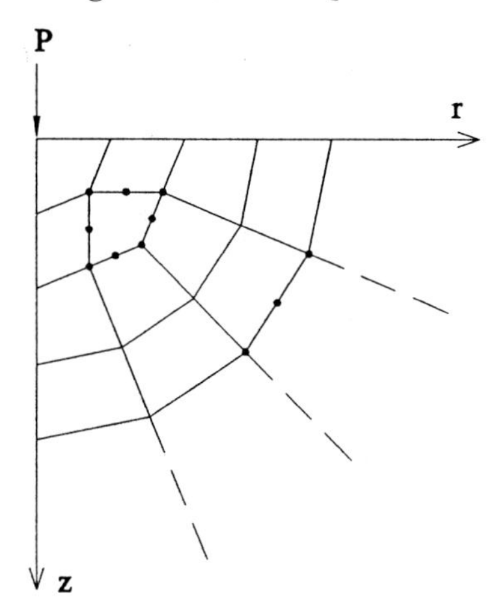

Fig. 2.28 Mesh of Q8 and I6 elements

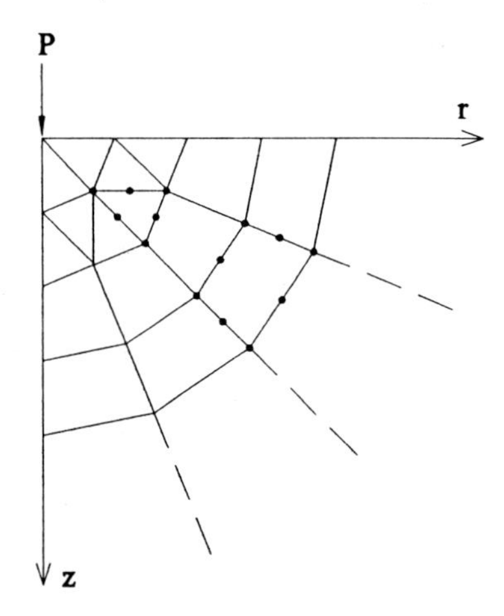

Fig. 2.29 Mesh of Q8, I6 and T6 elements

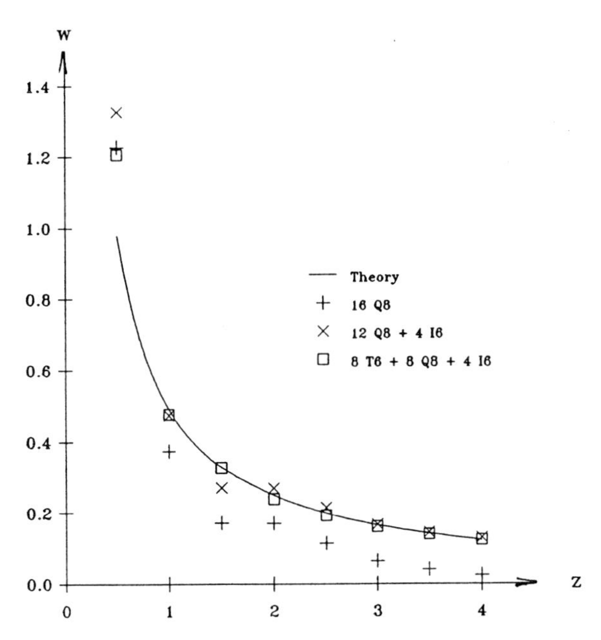

Fig. 2.30 Vertical displacement along $r = 0$ axis

Table 2.11 Vertical displacement along $r = 0$ axis for different meshes

	Vertical displacement			
z	Theory	Q8	Q8 + I6	T6 + Q8 + I6
0.5	0.980 39	1.227 50	1.325 80	1.206 70
1.0	0.490 20	0.374 24	0.473 12	0.475 76
1.5	0.326 80	0.171 95	0.271 50	0.327 78
2.0	0.245 10	0.170 36	0.270 70	0.239 12
2.5	0.196 08	0.113 70	0.214 96	0.193 20
3.0	0.163 40	0.063 62	0.166 15	0.161 29
3.5	0.140 06	0.040 27	0.143 99	0.138 95
4.0	0.122 55	0.024 39	0.129 16	0.123 15
8.0	0.061 27		0.067 40	0.064 79

The vertical displacement under a unit point load along the $r = 0$ axis is tabulated in Table 2.11, and is graphically presented in Fig. 2.30.

The analytical solution for the vertical displacement along the axis of symmetry ($r = 0$) is given by Reference 12

$$w = \frac{P(1+\nu)}{2\pi E\sqrt{(r^2+z^2)}}\left[\frac{z^2}{r^2+z^2} + 2(1-\nu)\right]$$

where P is the applied point load, and r and z are respectively horizontal and vertical distances from the applied load P.

All three finite element solutions overestimate the vertical displacement near the applied load ($z = 0.5$) due to the presence of singularity at the origin. At positions away from the load ($z \geq 1$), the standard Q8 element mesh always underestimates the vertical displacements. The situation is greatly improved by the introduction of the infinite elements. Better results still can be achieved from the slightly refined T6 + Q8 + I6 mesh within the region $1 \leq z \leq 2$.

2.6 References to Chapter 2

1. R. D. Cook (1981) *Concepts and Application of Finite Element Analysis*, Wiley.
2. J. A. Stricklin, W. S. Ho, E. Richardson & W. E. Haisler (1977) On isoparametric vs linear strain triangular elements, *Int. J. Num. Methods Engng.*, **11**, 1041–1043.
3. R. S. Barsoum (1977) Triangular quarter point elements as elastic and perfectly elastic crack tip elements, *Int. J. Num. Methods Engng.*, **11**, 85–98.
4. S. H. Lo (1985) A new mesh generation scheme for arbitrary planar domains, *Int. J. Num. Methods Engng.*, **21**, 1403–26.
5. R. W. Clough (1960) The finite element in plane stress analysis, *Proc. 2nd ASCE Conf. on Electronic Computation*, Pittsburgh, Pa., USA. Sept.
6. T. J. Chung (1988) *Continuum Mechanics*, Prentice-Hall International Editions.
7. L. E. Malvern (1969) *Introduction to the Mechanics of a Continuous Medium*, Prentice-Hall Inc.
8. O. C. Zienkiewicz & R. L. Taylor (1989) The Finite Element Method, Volume 1 – *Basic Formulation and Linear Problems*, 4th edn, McGraw-Hill International Editions.
9. S. W. Sloan (1981) A fast stiffness formulation for finite element analysis of two-dimensional solids, *Int. J. Num. Methods Engng.*, **17**, 1313–23.
10. A. H. Stroud, *Approximate Calculation of Multiple Integrals*, Prentice-Hall series in automatic computation.
11. B. M. Irons (1966) Engineering application of numerical integration in stiffness methods, *JAIAA*, **4**, 2035–7.
12. S. P. Timoshenko & J. N. Goodier, *Theory of Elasticity*, McGraw-Hill Kogakusha Ltd.
13. H. C. Martin & G. F. Carey, *Introduction to Finite Element Analysis*, Tata McGraw-Hill.
14. G. Dhatt & G. Touzot, *Une Présentation de la méthode des éléments finis*, Maloine S. A. Editeur, 27 rue de l'Ecole de Médecine, 75006 Paris.
15. R. W. Clough & Y. R. Rashid (1965) Finite element analysis of axisymmetric solids, *Proc. ASCE*, **91**, EM1, 71.
16. E. L. Wilson (1965) Structural analysis of axisymmetric solids, *JAIAA*, **3**, 2269–74.
17. O. C. Zienkiewicz & Y. K. Cheung (1965) Finite elements in the solution of field problems, *The Engineer*, 507–10, Sept.
18. O. C. Zienkiewicz, P. Mayer & Y. K. Cheung (1966) Solution of anisotropic seepage problems by finite elements, *Proc. Am. Soc. Civ. Eng.*, **92**, EM1, 111–20.
19. J. F. Ely & O. C. Zienkiewicz (1960) Torsion of compound bars – a relaxation solution, *Int. J. Num. Mech. Sci.*, **1**, 356–65.
20. P. Silvester & M. S. Hsieh (1971) Finite element solution of two-dimensional exterior field problems, *Proc. IEEE*, **118**.

21. B. H. McDonald & A. Wexler (1972) Finite element solution of unbounded field problems, *Proc. IEEE*, MTT-20, No. 12.
22. G. de Vries & D. H. Norrie (1969) *Application of the finite element technique to potential flow problems*, Report 7 and 8, Dept. Mech. Engng, Univ. of Calgary, Alberta, Canada.
23. L. J. Doctors (1970) An application of finite element technique to boundary value problems of potential flow, *Int. J. Num. Methods Engng.*, **2**, 243-52.
24. B. E. Larock (1969) Jets from two dimensional symmetric nozzles of arbitrary shape, *J. Fluid Mech.*, **37**, 479-83.
25. R. L. Taylor & C. B. Brown (1967) Darcy flow solution with a free surface, *Proc. Am. Soc. Civ. Eng.*, **93**, HY2, 25-33.
26. J. C. Luke (1957) A variational principle for a fluid with a free surface, *J. Fluid Mech.*, **27**, 395-7.
27. J. M. Sloss & J. C. Bruch (1978) Free surface seepage problem, *Proc. ASCE*, **108**, EM5, 1099-1111.
28. N. Kikuchi (1977) Seepage flow problems by variational inequalities, *Int. J. Num. Anal. Meth. Geomech.*, **1**, 283-90.
29. C. S. Desai (1976) Finite element residual schemes for unconfined flow, *Int. J. Num. Methods Engng.*, **10**, 1415-18.
30. K. J. Bathe & M. Koshgoftar (1979) Finite element from surface seepage analysis without mesh iteration, *Int. J. Num. Anal. Meth. Geomech.*, **3**, 13-22.
31. P. Bettess (1977) Infinite elements, *Int. J. Num. Methods Engng.*, **11**, 53-64.
32. B. Beer & J. L. Meek (1981) Infinite domain elements, *Int. J. Num. Methods Engng.*, **17**, 43-52.
33. O. C. Zienkiewicz, C. Emson & P. Bettess (1983) A novel boundary infinite element, *Int. J. Num. Methods Engng.*, **19**, 293-404.
34. R. S. Barsoum (1976) On the use of isoparametric finite elements in linear fracture mechanics, *Int. J. Num. Methods Engng.*, **10**, 25-37.
35. R. D. Henshell & K. G. Shaw (1975) Crack tip finite elements are unnecessary, *Int. J. Num. Methods Engng.*, **9**, 495-507.
36. H. D. Hibbitt (1977) Some properties of singular isoparametric elements, *Int. J. Num. Methods Engng.*, **11**, 180-84.
37. T. H. H. Pian, P. Tong & C. H. Luk (1971) Elastic crack analysis by a finite element hybrid method, *Proc. Air Force 3rd Conf. Matrix Meth. in Struct. Mech.*, AFFDL-TR-71-160, Wright-Patterson Air Force Base, Dayton, OH, USA.
38. Little, Robert William (1973) *Elasticity*, Prentice-Hall, Englewood Cliffs, New Jersey, USA.

3 Three-dimensional Problems

3.1 Introduction

Although in many practical situations, two-dimensional approximations can provide very economical solutions, there are also a large number of physical engineering problems that can only be adequately described in terms of a three-dimensional model. Such three-dimensional problems can be solved numerically using solid finite elements. It has already been pointed out that the simplest two-dimensional element is a triangle, and its three-dimensional counterpart is therefore a tetrahedron. Other common solid elements which enjoy widespread use include the hexahedron and pentahedron elements.

As far as finite element analysis is concerned, there is very little difference in the formulation and solution procedures between a two-dimensional problem and a three-dimensional one. The main differences lie in (a) the choice of shape function, (b) the vastly increased amount of input and output data, and (c) the much higher computation cost involved. A simple example of a cantilever beam can be used to illustrate the above points. For a two-dimensional analysis using a 2×12 mesh of Q8 elements, there are 24 elements, 101 nodes involving 202 equations with a half-band width of 22. For a three-dimensional analysis of the same cantilever beam using a $2 \times 12 \times 1$ mesh of H20 elements, there are 24 elements, 241 nodes involving 723 equations with a half-band width of 81. While the number of elements remains the same, the number of equations has increased by almost four times, and the computation cost would have gone up more than 48 times, since it is well-known that the solution time for a system of simultaneous equations is proportional to the total number of equations and the square of the half-band width. This is the primary reason why three-dimensional models were seldom used in the earlier numerical analyses of common engineering problems.

Nevertheless, the availability of high-speed powerful computers at much reduced cost and the introduction of better geometrical modelling and modern numerical solution techniques have made drastic changes in the general attitude and feasibility of using three-dimensional models in practical engineering problems. More realistic three-dimensional analysis with fewer engineering assumptions is expected to become increasingly popular as automatic mesh generation techniques become further and further refined such that most three-dimensional objects can be discretized in some easy robust manner.

In this chapter, the finite element procedures for both steady-state field problems and the three-dimensional linear elasticity analysis, which embrace a majority of common engineering problems, will be discussed in detail. Before entering into the general formulation of these problems, several popular isoparametric elements will be introduced first. Their interpolation functions and the corresponding derivatives with respect to local element coordinates are given explicitly in tabular form. The formulae for the transformation of differential operators and integration limits in three dimensions are also derived.

3.2 THREE-DIMENSIONAL ISOPARAMETRIC ELEMENTS

Although the 4-node tetrahedron, the 6-node pentahedron and the 8-node hexahedron are the simplest solid elements, they are not the preferred elements in engineering analysis because of their low accuracy. It is now generally accepted that the higher order isoparametric elements, in particular the quadratic elements, are better elements because of the higher computational efficiency and the ability to adapt to curved surfaces. The three-dimensional counterparts of the two-dimensional T6 and Q8 elements are the 10-node tetrahedron element T10 and the 20-node hexahedron element H20. Similar to the two-dimensional situation, the tetrahedron elements are better suited to the filling up of arbitrarily shaped solids, while the hexahedron elements are best applied to the discretization of nearly regular domains. The interpolation functions and their derivatives which are required in the calculation of the element stiffness matrix and element force vectors are given explicitly in the later sections. A useful 15-node pentahedron element which can be connected to the T10 and H20 elements is also presented.

3.2.1 10-node isoparametric tetrahedron element T10

The complete polynomial basis function for T10 is given by $[1\ \xi\ \eta\ \zeta\ \xi^2\ \xi\eta\ \eta^2\ \eta\zeta\ \zeta^2\ \xi\zeta]$, from which the interpolation(shape) functions can be derived through the standard transformation. The mapping of the T10 element is shown in Fig. 3.1, while the interpolation(shape) functions and the relevant derivatives are listed in Table 3.1.

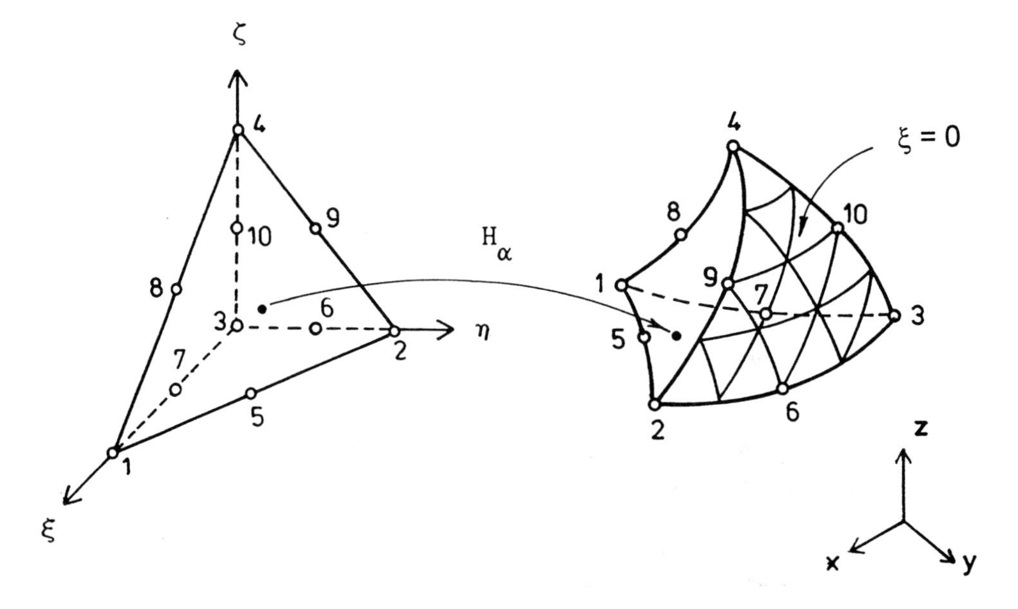

Fig. 3.1 Reference domain and mapping of tetrahedron T10 element

Table 3.1 Interpolation functions H_α and their derivatives with respect to ξ, η and ζ of a tetrahedron T10 element

	H	$\frac{\partial H}{\partial \xi}$	$\frac{\partial H}{\partial \eta}$	$\frac{\partial H}{\partial \zeta}$
1	$\xi(2\xi - 1)$	$4\xi - 1$	0	0
2	$\eta(2\eta - 1)$	0	$4\eta - 1$	0
3	$\varphi(2\varphi - 1)$	$1 - 4\varphi$	$1 - 4\varphi$	$1 - 4\varphi$
4	$\zeta(2\zeta - 1)$	0	0	$4\zeta - 1$
5	$4\xi\eta$	4η	4ξ	0
6	$4\eta\varphi$	-4η	$4(\varphi - \eta)$	-4η
7	$4\xi\varphi$	$4(\varphi - \xi)$	-4ξ	-4ξ
8	$4\xi\zeta$	4ζ	0	4ξ
9	$4\eta\zeta$	0	4ζ	4η
10	$4\varphi\zeta$	-4ζ	-4ζ	$4(\varphi - \zeta)$

where $\varphi = 1 - \xi - \eta - \zeta$

3.2.2 20-node isoparametric hexahedron element H20

Similar to T10, the complete polynomial basis function for H20 is given by $[1\ \xi\ \eta\ \zeta\ \xi^2\ \xi\eta\ \eta^2\ \eta\zeta\ \zeta^2\ \zeta\xi\ \eta\xi^2\ \xi\eta^2\ \zeta\eta^2\ \eta\zeta^2\ \xi\zeta^2\ \zeta\xi^2\ \xi\eta\zeta\ \eta\zeta\xi^2\ \zeta\xi\eta^2\ \xi\eta\zeta^2]$, from which once again the interpolation(shape) functions can be derived through the standard transformation. The mapping of the H20 element is shown in Fig. 3.2, while the interpolation(shape) functions and the relevant derivatives are listed in (i) to (iv) of Table 3.2 for the corner nodes, the mid-side nodes when $\xi = 0$, the mid-side nodes when $\eta = 0$, and the mid-side nodes when $\zeta = 0$ respectively.

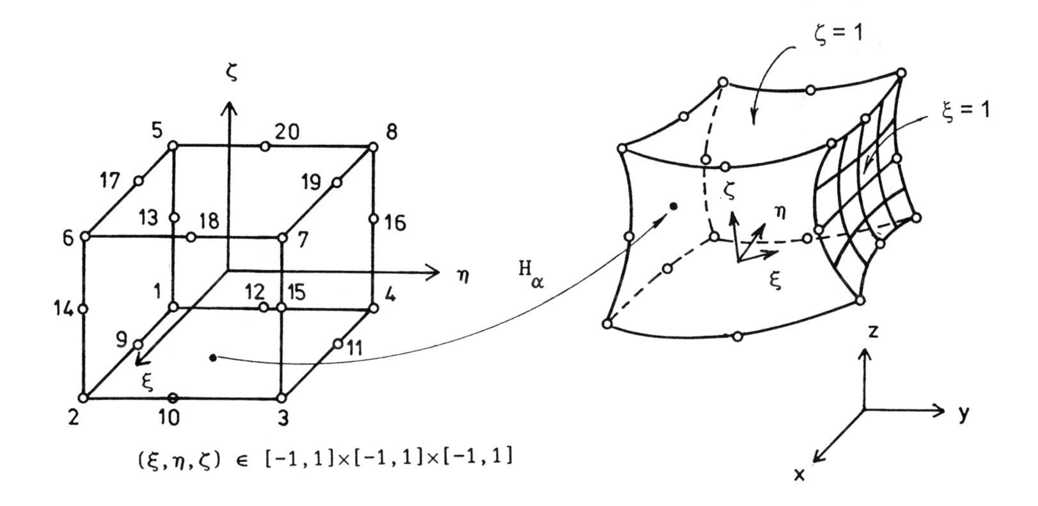

Fig. 3.2 Reference domain and mapping of hexahedron H20 element

Table 3.2 Interpolation functions H_α and their derivatives with respect to ξ, η and ζ of a hexahedron H20 element

(i) Corner nodes

Node	1	2	3	4	5	6	7	8
ξ_α	−1	1	1	−1	−1	1	1	−1
η_α	−1	−1	1	1	−1	−1	1	1
ζ_α	−1	−1	−1	−1	1	1	1	1

$$H_\alpha = \tfrac{1}{8}(1+\xi_o)(1+\eta_o)(1+\zeta_o)(\xi_o+\eta_o+\zeta_o-2)$$

$$\frac{\partial H_\alpha}{\partial \xi} = \tfrac{1}{8}\xi_\alpha(1+\eta_o)(1+\zeta_o)(2\xi_o+\eta_o+\zeta_o-1)$$

$$\frac{\partial H_\alpha}{\partial \eta} = \tfrac{1}{8}\eta_\alpha(1+\zeta_o)(1+\xi_o)(\xi_o+2\eta_o+\zeta_o-1)$$

$$\frac{\partial H_\alpha}{\partial \zeta} = \tfrac{1}{8}\zeta_\alpha(1+\xi_o)(1+\eta_o)(\xi_o+\eta_o+2\zeta_o-1)$$

where $\xi_o = \xi_\alpha\xi,\ \eta_o = \eta_\alpha\eta,\ \zeta_o = \zeta_\alpha\zeta$.

(ii) Mid-side nodes ($\xi = 0$)

Node	9	11	17	19
ξ_α	0	0	0	0
η_α	−1	1	−1	1
ζ_α	−1	−1	1	1

$$H_\alpha = \tfrac{1}{4}(1+\eta_o)(1+\zeta_o)(1-\xi^2)$$

$$\frac{\partial H_\alpha}{\partial \eta} = \tfrac{1}{4}\eta_\alpha(1+\zeta_o)(1-\xi^2)$$

$$\frac{\partial H_\alpha}{\partial \zeta} = \tfrac{1}{4}\zeta_\alpha(1+\eta_o)(1-\xi^2)$$

$$\frac{\partial H_\alpha}{\partial \xi} = -\tfrac{1}{2}\xi(1+\eta_o)(1+\zeta_o)$$

(iii) Mid-side nodes ($\eta = 0$)

Node	10	12	18	20
ξ_α	1	−1	1	−1
η_α	0	0	0	0
ζ_α	−1	−1	1	1

$$H_\alpha = \tfrac{1}{4}(1+\zeta_o)(1+\xi_o)(1-\eta^2)$$

$$\frac{\partial H_\alpha}{\partial \zeta} = \tfrac{1}{4}\zeta_\alpha(1+\xi_o)(1-\eta^2)$$

$$\frac{\partial H_\alpha}{\partial \xi} = \tfrac{1}{4}\xi_\alpha(1+\zeta_o)(1-\eta^2)$$

$$\frac{\partial H_\alpha}{\partial \eta} = -\tfrac{1}{2}\eta(1+\zeta_o)(1+\xi_o)$$

(iv) Mid-side nodes ($\zeta = 0$)

Node	13	14	15	16
ξ_α	−1	1	1	−1
η_α	−1	−1	1	1
ζ_α	0	0	0	0

$$H_\alpha = \tfrac{1}{4}(1+\xi_o)(1+\eta_o)(1-\zeta^2)$$

$$\frac{\partial H_\alpha}{\partial \xi} = \tfrac{1}{4}\xi_\alpha(1+\eta_o)(1-\zeta^2)$$

$$\frac{\partial H_\alpha}{\partial \eta} = \tfrac{1}{4}\eta_\alpha(1+\xi_o)(1-\zeta^2)$$

$$\frac{\partial H_\alpha}{\partial \zeta} = -\tfrac{1}{2}\zeta(1+\xi_o)(1+\eta_o)$$

3.2.3 15-node isoparametric pentahedron element P15

The mapping of the P15 element is shown in Fig. 3.3, whereas the interpolation(shape) functions and their derivatives are listed in Table 3.3.

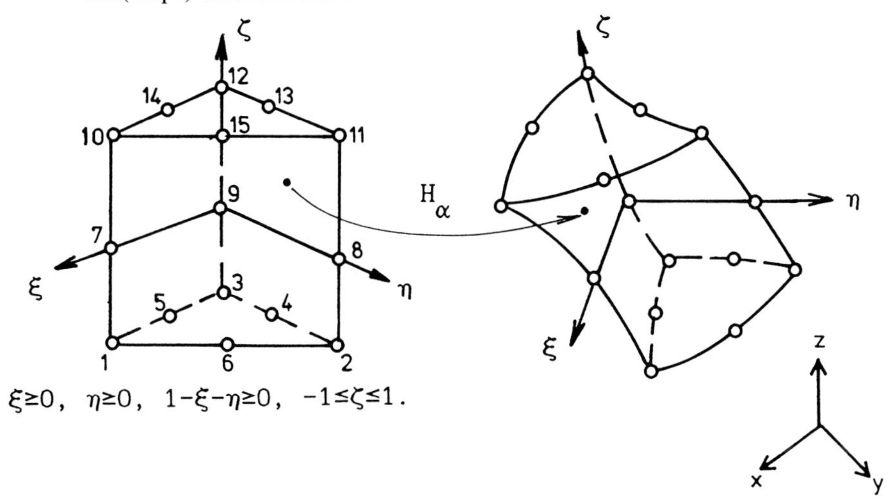

Fig. 3.3 Reference domain and mapping of pentahedron P15 element

Table 3.3 Interpolation functions H_α and their derivatives with respect to ξ, η and ζ of a pentahedron P15 element

	H	$\dfrac{\partial H}{\partial \xi}$	$\dfrac{\partial H}{\partial \eta}$	$\dfrac{\partial H}{\partial \zeta}$
1	$\xi(1-\zeta)(\xi-1-\frac{1}{2}\zeta)$	$(1-\zeta)(2\xi-1-\frac{1}{2}\zeta)$	0	$\xi(\zeta-\xi+\frac{1}{2})$
2	$\eta(1-\zeta)(\eta-1-\frac{1}{2}\zeta)$	0	$(1-\zeta)(2\eta-1-\frac{1}{2}\zeta)$	$\eta(\zeta-\eta+\frac{1}{2})$
3	$\varphi(1-\zeta)(\varphi-1-\frac{1}{2}\zeta)$	$(1-\zeta)(1+\frac{1}{2}\zeta-2\varphi)$	$(1-\zeta)(1+\frac{1}{2}\zeta-2\varphi)$	$\varphi(\zeta-\varphi+\frac{1}{2})$
4	$2\eta\varphi(1-\zeta)$	$-2\eta(1-\zeta)$	$2(1-\zeta)(\varphi-\eta)$	$-2\eta\varphi$
5	$2\varphi\xi(1-\zeta)$	$2(1-\zeta)(\varphi-\xi)$	$-2\xi(1-\zeta)$	$-2\varphi\xi$
6	$2\xi\eta(1-\zeta)$	$2\eta(1-\zeta)$	$2\xi(1-\zeta)$	$-2\xi\eta$
7	$\xi(1-\zeta^2)$	$1-\zeta^2$	0	$-2\xi\zeta$
8	$\eta(1-\zeta^2)$	0	$1-\zeta^2$	$-2\eta\zeta$
9	$\varphi(1-\zeta^2)$	ζ^2-1	ζ^2-1	$-2\varphi\zeta$
10	$\xi(1+\zeta)(\xi-1+\frac{1}{2}\zeta)$	$(1+\zeta)(2\xi-1+\frac{1}{2}\zeta)$	0	$\xi(\zeta+\xi-\frac{1}{2})$
11	$\eta(1+\zeta)(\eta-1+\frac{1}{2}\zeta)$	0	$(1+\zeta)(2\eta-1+\frac{1}{2}\zeta)$	$\eta(\zeta+\eta-\frac{1}{2})$
12	$\varphi(1+\zeta)(\varphi-1+\frac{1}{2}\zeta)$	$(1+\zeta)(1-\frac{1}{2}\zeta-2\varphi)$	$(1+\zeta)(1-\frac{1}{2}\zeta-2\varphi)$	$\varphi(\zeta+\varphi-\frac{1}{2})$
13	$2\eta\varphi(1+\zeta)$	$-2\eta(1+\zeta)$	$2(1+\zeta)(\varphi-\eta)$	$2\eta\varphi$
14	$2\varphi\xi(1+\zeta)$	$2(1+\zeta)(\varphi-\xi)$	$-2\xi(1+\zeta)$	$2\varphi\xi$
15	$2\xi\eta(1+\zeta)$	$2\eta(1+\zeta)$	$2\xi(1+\zeta)$	$2\xi\eta$

where $\varphi = 1-\xi-\eta$

3.3 NUMERICAL INTEGRATION IN THREE DIMENSIONS

Since analytical integration for three-dimensional curved elements cannot be regarded as a practical proposition, all the expressions for the element stiffness matrix and the element force vectors have to be evaluated by numerical integration techniques. Two transformations are necessary if standard numerical integration formulae are to be applied to evaluate integrals over general curved isoparametric elements. As the interpolation functions, H_α, are defined in terms of local element coordinates, it is necessary to devise some means of expressing the global derivatives in terms of the local derivatives. The element volume over which the integration has to be carried out needs to be expressed in terms of the local coordinates with an appropriate change of limits of integration. These two transformations of the differential operator and the integral in three-dimensional situations will be discussed in detail in Sections 3.3.1 and 3.3.2 respectively. It should be stressed that an efficient numerical integration scheme is essential in the implementation of a general-purpose three-dimensional finite element program. Practical numerical integration formulae for both tetrahedron and hexahedron elements are given in Section 3.3.3.

3.3.1 Transformation of differential operator (3D)

A set of local coordinates (ξ, η, ζ) and a corresponding set of global coordinates (x, y, z) are related through shape functions, G_α, such that

$$x = G_\alpha(\xi, \eta, \zeta)x_\alpha \qquad y = G_\alpha(\xi, \eta, \zeta)y_\alpha \qquad z = G_\alpha(\xi, \eta, \zeta)z_\alpha \qquad \alpha = 1, N$$

By the chain rule of partial differentiation, we can write

$$\frac{\partial H}{\partial \xi} = \frac{\partial H}{\partial x}\frac{\partial x}{\partial \xi} + \frac{\partial H}{\partial y}\frac{\partial y}{\partial \xi} + \frac{\partial H}{\partial z}\frac{\partial z}{\partial \xi}$$

$$\frac{\partial H}{\partial \eta} = \frac{\partial H}{\partial x}\frac{\partial x}{\partial \eta} + \frac{\partial H}{\partial y}\frac{\partial y}{\partial \eta} + \frac{\partial H}{\partial z}\frac{\partial z}{\partial \eta}$$

$$\frac{\partial H}{\partial \zeta} = \frac{\partial H}{\partial x}\frac{\partial x}{\partial \zeta} + \frac{\partial H}{\partial y}\frac{\partial y}{\partial \zeta} + \frac{\partial H}{\partial z}\frac{\partial z}{\partial \zeta}$$

or

$$\begin{bmatrix} \dfrac{\partial H}{\partial \xi} \\ \dfrac{\partial H}{\partial \eta} \\ \dfrac{\partial H}{\partial \zeta} \end{bmatrix} = \begin{bmatrix} \dfrac{\partial x}{\partial \xi} & \dfrac{\partial y}{\partial \xi} & \dfrac{\partial z}{\partial \xi} \\ \dfrac{\partial x}{\partial \eta} & \dfrac{\partial y}{\partial \eta} & \dfrac{\partial z}{\partial \eta} \\ \dfrac{\partial x}{\partial \zeta} & \dfrac{\partial y}{\partial \zeta} & \dfrac{\partial z}{\partial \zeta} \end{bmatrix} \begin{bmatrix} \dfrac{\partial H}{\partial x} \\ \dfrac{\partial H}{\partial y} \\ \dfrac{\partial H}{\partial z} \end{bmatrix} = \mathbf{J} \begin{bmatrix} \dfrac{\partial H}{\partial x} \\ \dfrac{\partial H}{\partial y} \\ \dfrac{\partial H}{\partial z} \end{bmatrix} \tag{3.1}$$

from which

$$\begin{bmatrix} \dfrac{\partial H}{\partial x} \\ \dfrac{\partial H}{\partial y} \\ \dfrac{\partial H}{\partial z} \end{bmatrix} = \mathbf{J}^{-1} \begin{bmatrix} \dfrac{\partial H}{\partial \xi} \\ \dfrac{\partial H}{\partial \eta} \\ \dfrac{\partial H}{\partial \zeta} \end{bmatrix} = \frac{1}{\det(\mathbf{J})} \begin{bmatrix} C_{11} & C_{12} & C_{13} \\ C_{21} & C_{22} & C_{23} \\ C_{31} & C_{32} & C_{33} \end{bmatrix} \begin{bmatrix} \dfrac{\partial H}{\partial \xi} \\ \dfrac{\partial H}{\partial \eta} \\ \dfrac{\partial H}{\partial \zeta} \end{bmatrix} \tag{3.2}$$

where

$$\det(\mathbf{J}) = J_{11}(J_{22}J_{33} - J_{23}J_{32}) + J_{12}(J_{23}J_{31} - J_{21}J_{33}) + J_{13}(J_{21}J_{32} - J_{22}J_{31})$$

$$C_{11} = J_{22}J_{33} - J_{32}J_{23} \qquad C_{12} = J_{32}J_{13} - J_{12}J_{33} \qquad C_{13} = J_{12}J_{23} - J_{22}J_{13}$$

$$C_{21} = J_{23}J_{31} - J_{33}J_{21} \qquad C_{22} = J_{33}J_{11} - J_{13}J_{31} \qquad C_{23} = J_{13}J_{21} - J_{23}J_{11}$$

$$C_{31} = J_{21}J_{32} - J_{31}J_{22} \qquad C_{32} = J_{31}J_{12} - J_{11}J_{32} \qquad C_{33} = J_{11}J_{22} - J_{21}J_{12}$$

and

$$J_{11} = \frac{\partial x}{\partial \xi} = \frac{\partial G_\alpha}{\partial \xi} x_\alpha \qquad J_{12} = \frac{\partial y}{\partial \xi} = \frac{\partial G_\alpha}{\partial \xi} y_\alpha \qquad J_{13} = \frac{\partial z}{\partial \xi} = \frac{\partial G_\alpha}{\partial \xi} z_\alpha$$

$$J_{21} = \frac{\partial x}{\partial \eta} \qquad J_{22} = \frac{\partial y}{\partial \eta} \qquad J_{23} = \frac{\partial z}{\partial \eta}$$

$$J_{31} = \frac{\partial x}{\partial \zeta} \qquad J_{32} = \frac{\partial y}{\partial \zeta} \qquad J_{33} = \frac{\partial z}{\partial \zeta}$$

3.3.2 Transformation of integral (3D)

The change of variables allows us to change the integration of a function $f(x, y, z)$ defined over the element domain Ω_e into another integration referring to the natural coordinates (ξ, η, ζ) of the reference element such that

$$\int_{\Omega_e} f(x, y, z)\, \mathrm{d}\Omega = \int_{\mathcal{R}} \hat{\mathrm{f}}(\xi, \eta, \zeta)\, \mathrm{d}\mathcal{R} \tag{3.3}$$

where $\hat{\mathrm{f}}(\xi, \eta, \zeta) = f(x(\xi, \eta, \zeta), y(\xi, \eta, \zeta), z(\xi, \eta, \zeta)) \cdot \det(\mathbf{J}(\xi, \eta, \zeta))$, and the region $\mathcal{R}$ is the element reference domain (for hexahedron elements, it is a $2 \times 2 \times 2$ cube, whereas for tetrahedron elements, it is a right-angled tetrahedron).

3.3.3 Numerical integration (3D)

It has already been indicated that the exact integration of a complicated function $\hat{\mathrm{f}}(\xi, \eta, \zeta)$ arising from complex field problems and distorted element geometries is not a practical proposition, and that numerical integration techniques will have to be used, in which an integral is replaced by a summation,

$$\int_{\mathcal{R}} \hat{\mathrm{f}}(\xi, \eta, \zeta)\, \mathrm{d}\mathcal{R} \approx \sum_{k=1}^{m} W_k \hat{\mathrm{f}}(\xi_k, \eta_k, \zeta_k) \tag{3.4}$$

where m is the number of integration points used, and W_k and (ξ_k, η_k, ζ_k) are the weights and positions of sampling points respectively. Some useful integration formulae for H20 and T10 elements are given in Tables 3.4 and 3.5 respectively. However, for the integration over a P15 pentahedron element, the following eight-point integration scheme, which allows for polynomials up to degree 3 to be integrated exactly, can be used.

	1, 5	2, 6	3, 7	4, 8
ξ_k	1/3	1/5	3/5	1/5
η_k	1/3	1/5	1/5	3/5

$$\zeta_1 = \zeta_2 = \zeta_3 = \zeta_4 = -1/\sqrt{3}$$
$$\zeta_5 = \zeta_6 = \zeta_7 = \zeta_8 = 1/\sqrt{3}$$
$$W_1 = W_5 = -27/96$$
$$W_2 = W_3 = W_4 = W_6 = W_7 = W_8 = 25/96$$

Table 3.4 Numerical integration formulae over a 2 × 2 × 2 cube

Order n	Number of points, m	Coordinates ξ_i	η_i	ζ_i	Weight W_i
2	4	0	$\pm\sqrt{\frac{2}{3}}$	$-\frac{1}{\sqrt{3}}$	2
		$\pm\sqrt{\frac{2}{3}}$	0	$\frac{1}{\sqrt{3}}$	
3	6	$\frac{1}{\sqrt{6}}$	$\pm\frac{1}{\sqrt{2}}$	$-\frac{1}{\sqrt{3}}$	4/3
		$-\frac{1}{\sqrt{6}}$	$\pm\frac{1}{\sqrt{2}}$	$\frac{1}{\sqrt{3}}$	
		$-\sqrt{\frac{2}{3}}$	0	$-\frac{1}{\sqrt{3}}$	
		$\sqrt{\frac{2}{3}}$	0	$\frac{1}{\sqrt{3}}$	
3	6	± 1	0	0	4/3
		0	± 1	0	
		0	0	± 1	
5	14	$\pm a$	0	0	320/361
		0	$\pm a$	0	
		0	0	$\pm a$	
		$\pm b$	$\pm b$	$\pm b$	121/361
		$a=\sqrt{\frac{19}{30}}$		$b=\sqrt{\frac{19}{33}}$	
7	34	$\pm a$	0	0	0.295 747 599 451 303
		0	$\pm a$	0	
		0	0	$\pm a$	
$a = 0.925\,820\,099\,772\,552$		$\pm a$	$\pm a$	0	0.094 101 508 916 324
$b = 0.406\,703\,186\,426\,716$		0	$\pm a$	$\pm a$	
		$\pm a$	0	$\pm a$	
$c = 0.734\,112\,528\,752\,115$		$\pm b$	$\pm b$	$\pm b$	0.412 333 862 271 436
		$\pm c$	$\pm c$	$\pm c$	0.224 703 174 765 601

$\int_{-1}^{1}\int_{-1}^{1}\int_{-1}^{1} f(\xi,\eta,\zeta)\,d\xi\,d\eta\,d\zeta \approx \sum_{k=1}^{m} W_k f(\xi_k,\eta_k,\zeta_k)$

3.3.4 Surface integral (3D)

In many physical engineering problems, integration over curved surfaces is frequently required. This is particularly true for the evaluation of equivalent nodal forces due to surface tractions $\vec{q}$. A typical expression for evaluating the contribution of such surface traction is

$$s_i^\alpha = q_i^\beta A_{\alpha\beta}$$

where

$$A_{\alpha\beta} = \int_{\Gamma_e} H_\alpha H_\beta \, da \tag{3.5}$$

Table 3.5 Numerical integration formulae over a tetrahedron

Order, n	Number of points, m	Coordinates ξ_i	η_i	ζ_i		Weight W_i
1	1	$\frac{1}{4}$	$\frac{1}{4}$	$\frac{1}{4}$		$\frac{1}{6}$
2	4	a	a	a		$\frac{1}{24}$
	$a=\frac{5-\sqrt{5}}{20}$	a	a	b		
	$b=\frac{5+3\sqrt{5}}{20}$	a	b	a		
		b	a	a		
3	5	a	a	a		$-\frac{2}{15}$
	$a=\frac{1}{4}$	b	b	b		$\frac{3}{40}$
	$b=\frac{1}{6}$	b	b	c		
	$c=\frac{1}{2}$	b	c	b		
		c	b	b		
5	15	a	a	a		$\frac{112}{5670}=\frac{8}{405}$
	$a=\frac{1}{4}$	b_i	b_i	b_i	$i=1,2$	$\left.\begin{matrix} w_1 \\ w_2 \end{matrix}\right\}=\frac{2665 \mp 14\sqrt{15}}{226\,800}$
	$\left.\begin{matrix} b_1 \\ b_2 \end{matrix}\right\}=\frac{7\pm\sqrt{15}}{34}$	b_i	b_i	c_i		
	$\left.\begin{matrix} c_1 \\ c_2 \end{matrix}\right\}=\frac{13\mp 3\sqrt{15}}{34}$	b_i	c_i	b_i		
		c_i	b_i	b_i		
	$d=\frac{5-\sqrt{15}}{20}$	d	d	e		$\frac{5}{567}$
	$e=\frac{5+\sqrt{15}}{20}$	d	e	d		
		e	d	d		
		d	e	e		
		e	d	e		
		e	e	d		

$$\int_0^1 \int_0^{1-\xi} \int_0^{1-\xi-\eta} f(\xi, \eta, \zeta)\, d\xi\, d\eta\, d\zeta \approx \sum_{k=1}^{m} W_k f(\xi_k, \eta_k, \zeta_k)$$

The element da will generally lie on a surface where one of the coordinates is constant. Furthermore, if $\overrightarrow{da}$ is considered as a vector oriented in the direction normal to the surface, as shown in Fig. 3.4, then $\overrightarrow{da}$ can be expressed as a cross product of the base vectors $\overrightarrow{dx_1}$ and $\overrightarrow{dx_2}$ such that

$$\begin{aligned} \overrightarrow{da} &= d\vec{x}_1 \times d\vec{x}_2 \\ &= \frac{\partial \vec{x}}{\partial \xi} d\xi \times \frac{\partial \vec{x}}{\partial \eta} d\eta \\ &= \left(\frac{\partial \vec{x}}{\partial \xi} \times \frac{\partial \vec{x}}{\partial \eta} \right) d\xi\, d\eta \end{aligned}$$

or

$$da \| \overrightarrow{da} \| = \left\| \frac{\partial \vec{x}}{\partial \xi} \times \frac{\partial \vec{x}}{\partial \eta} \right\| d\xi\, d\eta \tag{3.6}$$

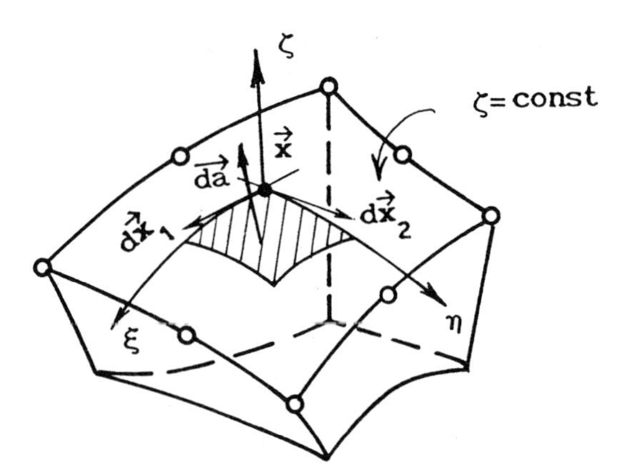

Fig. 3.4 Integration over curved surfaces

3.3.5 Element nodal forces due to a pressure load

Considerable simplifications can be made in the case of uniform pressure load which always acts normal to the curved surface. Let $\vec{s}_\alpha, \alpha = 1, N$, be the equivalent nodal forces due to uniform pressure p acting on the curved surface Γ defined by N nodal points as shown in Fig. 3.5.

$$\vec{s}_\alpha = \int_\Gamma \vec{q} H_\alpha \, d\Gamma \qquad \alpha = 1, N$$

where $\vec{q}$ is the traction force on Γ, and is equal to $-p\hat{n}$, with $\hat{n}$ being the unit normal vector on Γ.

$$\vec{s}_\alpha = \int_\Gamma -p\hat{n} H_\alpha \, d\Gamma = -p \int_\Gamma H_\alpha \, d\vec{a} = -p \int_{\mathcal{R}} H_\alpha \left(\frac{\partial \vec{x}}{\partial \xi} \times \frac{\partial \vec{x}}{\partial \eta} \right) d\xi \, d\eta \quad (3.7)$$

$$\vec{x} = (x, y, z) \Rightarrow \frac{\partial \vec{x}}{\partial \xi} = \left(\frac{\partial x}{\partial \xi}, \frac{\partial y}{\partial \xi}, \frac{\partial z}{\partial \xi} \right) \quad \text{and} \quad \frac{\partial \vec{x}}{\partial \eta} = \left(\frac{\partial x}{\partial \eta}, \frac{\partial y}{\partial \eta}, \frac{\partial z}{\partial \eta} \right)$$

$$\frac{\partial \vec{x}}{\partial \xi} \times \frac{\partial \vec{x}}{\partial \eta} = \left(\frac{\partial y \partial z}{\partial \xi \partial \eta} - \frac{\partial z \partial y}{\partial \xi \partial \eta}, \frac{\partial z \partial x}{\partial \xi \partial \eta} - \frac{\partial x \partial z}{\partial \xi \partial \eta}, \frac{\partial x \partial y}{\partial \xi \partial \eta} - \frac{\partial y \partial x}{\partial \xi \partial \eta} \right) \quad (3.8)$$

with $\qquad \frac{\partial x}{\partial \xi} = \frac{\partial H_\beta}{\partial \xi} x_\beta \qquad \frac{\partial y}{\partial \xi} = \frac{\partial H_\beta}{\partial \xi} y_\beta \qquad \frac{\partial z}{\partial \xi} = \frac{\partial H_\beta}{\partial \xi} z_\beta \qquad$ etc.

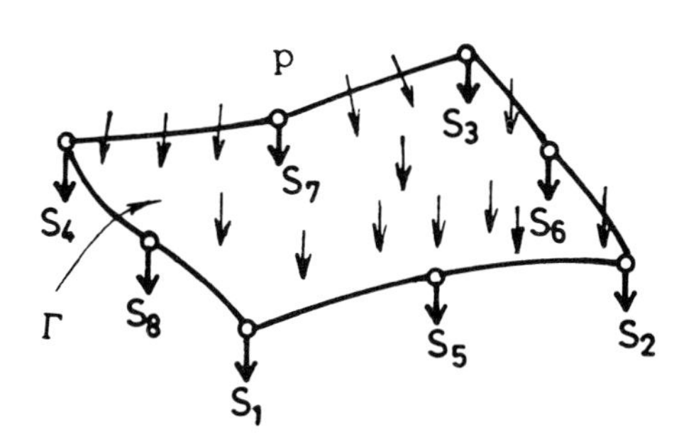

Fig. 3.5 Surface Γ under pressure load p

For triangular isoparametric curved surfaces defined by six nodal points ($N = 6$), the reference domain of integration $\mathcal{R} = \Delta$ (a right-angled isosceles triangle) and H_β, $\partial H_\beta/\partial\xi$, $\partial H_\beta/\partial\eta$, $\beta = 1, 6$ are given in Table 2.1; whereas for rectangular isoparametric curved surfaces defined by eight nodal points ($N = 8$), $\mathcal{R} = \square$ (a2 × 2 square) and H_β, $\partial H_\beta/\partial\xi$, $\partial H_\beta/\partial\eta$, $\beta = 1, 8$ are given in Table 2.3.

3.4 STEADY-STATE FIELD PROBLEMS

Steady-state field problems are governed by the general 'quasi-harmonic' equation [13], the particular cases of which are the well-known Laplace and Poisson equations [1-12]. The range of physical problems falling into this category is large. To name but just a few frequently encountered in engineering practice, we have

- heat conduction and convection.
- seepage through porous media,
- irrotational flow of ideal fluid,
- distribution of electrical or magnetic potential,
- torsion of prismatic shaft, etc.

A general formulation without reference to the actual physical quantities (such as the temperature, T, or the pizometric head, H, etc.) and the corresponding numerical procedures will be developed, so that the results can be applied equally well to all physical situations governed by the quasi-harmonic equation.

3.4.1 The general quasi-harmonic equation

Governing field equation: $\nabla \cdot \vec{q} - Q = 0$ in Ω (3.9)

(i) Forced boundary condition, $\phi = \phi_o$ on Γ_1 (3.10)

(ii) Natural boundary condition, $q_n = \vec{q} \cdot \hat{n} = c\phi + g$ on Γ_2 (3.11)

Furthermore, the rate of flow $\vec{q}$ is related to the gradient of the potential ϕ by the generalized Fourier law: $\vec{q} = -\mathbf{K}(\nabla\phi)$

where Ω = problem domain
Γ = boundary(surface) of Ω
$\vec{q}$ = rate of flow

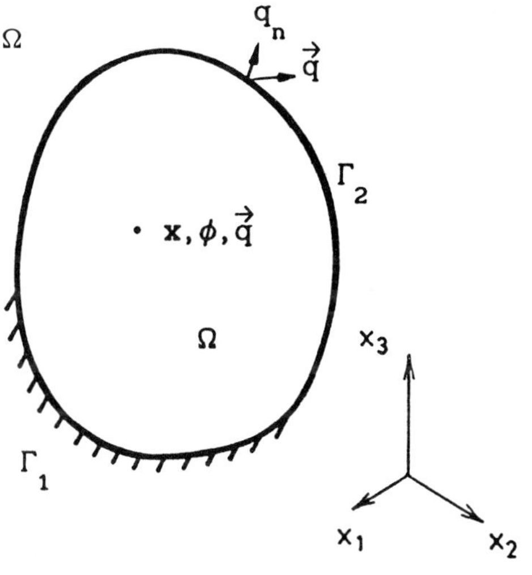

Fig. 3.6

Q = rate at which the relevant quantity is generated(removed) per unit volume
ϕ = related potential
$\mathbf{K} = [k_{ij}]$ = conductivity matrix
ϕ_o = potential specified on boundary Γ_1
q_n = the normal component of flow on Γ_2
$\hat{n}$ = unit normal vector on Γ
c = radiation(convection) coefficient
g = specified normal flow on boundary Γ_2

3.4.2 Weak form of general quasi-harmonic equation

A functional (potential) is an integral expression that yields the governing differential equations and the non-essential boundary conditions of a problem when given the standard treatment of calculus of variation. In areas other than the mechanics of solids, it is more likely that a potential may not be known or may not exist at all. This happens in fluid mechanics problems where, for some types of flow, all that is available are the differential equations and boundary conditions. Under these circumstances, the finite element method can still be applied by transforming the governing differential equations and the non-essential boundary conditions into an 'equivalent' integral expression or the so-called 'weak form' using the weighted residual technique [13–15].

Multiplying the governing field equation by an arbitrary function ψ and integrating the resulting expression over the problem domain Ω, we can obtain

$$\int_\Omega \psi(\nabla\cdot\vec{q} - Q)\,d\Omega = 0 \qquad \forall\psi \tag{3.12}$$

Integrating by parts will result in

$$\int_\Omega \nabla\cdot(\psi\vec{q})\,d\Omega - \int_\Omega (\nabla\psi)\cdot\vec{q}\,d\Omega = \int_\Omega \psi Q\,d\Omega \tag{3.13}$$

Using the Gaussian theorem, we can transform the first term on the left-hand side into a surface integral,

$$\int_\Gamma \hat{n}\cdot(\psi\vec{q})\,d\Gamma - \int_\Omega (\nabla\psi)\cdot\vec{q}\,d\Omega = \int_\Omega \psi Q\,d\Omega \tag{3.14}$$

Choosing $\psi = 0$ on Γ_1, we have,

$$\int_\Omega (\nabla\psi)\cdot\vec{q}\,d\Omega = \int_{\Gamma_2} (\psi q_n)\,d\Gamma - \int_\Omega \psi Q\,d\Omega \tag{3.15}$$

Using boundary condition $q_n = c\phi + g$, we obtain

$$\int_\Omega (\nabla\psi)\cdot\vec{q}\,d\Omega = \int_{\Gamma_2} \psi(g + c\phi)\,d\Gamma - \int_\Omega \psi Q\,d\Omega \qquad \forall\psi \tag{3.16}$$

3.4.3 Galerkin formulation

By the Galerkin approximation, the weighting function ψ is defined with the aid of the element interpolation function H_α such that $\psi = H_\alpha\psi_\alpha$, where arbitrary values

ψ_α are the values of the function ψ at the nodes. Substituting ψ into eq. 3.16,

$$\int_\Omega H_{\alpha,i}\psi_\alpha q_i \,\mathrm{d}\Omega = \int_{\Gamma_2} H_\alpha \psi_\alpha (g + c\phi)\,\mathrm{d}\Gamma - \int_\Omega H_\alpha \psi_\alpha Q \,\mathrm{d}\Omega \qquad \forall \psi_\alpha \quad (3.17)$$

$$\Rightarrow \int_\Omega H_{\alpha,i} q_i \,\mathrm{d}\Omega = \int_{\Gamma_2} H_\alpha (g + c\phi)\,\mathrm{d}\Gamma - \int_\Omega H_\alpha Q \,\mathrm{d}\Omega \qquad (3.18)$$

with $q_i = -k_{ij}\dfrac{\partial \phi}{\partial x_j}$, we can write

$$\int_\Omega H_{\alpha,i} k_{ij} \frac{\partial \phi}{\partial x_j}\,\mathrm{d}\Omega = \int_\Omega H_\alpha Q \,\mathrm{d}\Omega - \int_{\Gamma_2} H_\alpha (g + c\phi)\,\mathrm{d}\Gamma \qquad (3.19)$$

3.4.4 Finite element discretization

By the finite element approximation, $\phi = H_\beta \phi_\beta$, $\dfrac{\partial \phi}{\partial x_j} = H_{\beta,j}\phi_\beta$, we have

$$\left[\int_\Omega k_{ij} H_{\alpha,i} H_{\beta,j}\,\mathrm{d}\Omega\right]\phi_\beta + \left[\int_{\Gamma_2} c H_\alpha H_\beta \,\mathrm{d}\Gamma\right]\phi_\beta = \int_\Omega H_\alpha Q \,\mathrm{d}\Omega - \int_{\Gamma_2} H_\alpha g \,\mathrm{d}\Gamma \qquad (3.20)$$

In matrix notation, we have

$$\mathbf{K}\vec{\phi} = \vec{\mathrm{f}} \qquad K_{\alpha\beta}\phi_\beta = f_\alpha \qquad (3.21)$$

where it should be noted that:

1. $\mathbf{K}$ is the global stiffness matrix, with components

$$K_{\alpha\beta} = \int_\Omega k_{ij} H_{\alpha,i} H_{\beta,j}\,\mathrm{d}\Omega + \int_{\Gamma_2} c H_\alpha H_\beta \,\mathrm{d}\Gamma \qquad (3.22)$$

2. $\vec{\phi} = [\phi_\beta]$ are the potential values at the nodes. (3.23)
3. $\vec{f} = \vec{b} - \vec{s}$, and $[f_\alpha]$ is the global force vector. (3.24)

The components of the force vector are given by
(i) Body force vector due to source Q

$$\vec{b} = \left[\int_\Omega H_\alpha Q \,\mathrm{d}\Omega\right] \qquad (3.25)$$

(ii) Surface force vector due to flux g

$$\vec{s} = \left[\int_{\Gamma_2} H_\alpha g \,\mathrm{d}\Gamma\right] \qquad (3.26)$$

3.4.5 Conductivity matrix K

By the energy arguments, matrix $\mathbf{K}$ can be shown to be symmetric and positive-definite. If the local element coordinate axes coincide with the principal direction of stratification, three coefficients k_1, k_2 and k_3 need to be specified, and $\mathbf{K}$ is then a diagonal matrix given by

$$\mathbf{K} = \begin{bmatrix} k_1 & 0 & 0 \\ 0 & k_2 & 0 \\ 0 & 0 & k_3 \end{bmatrix} \tag{3.27}$$

A common situation in practice is that the material is isotropic, in such a case $k_1 = k_2 = k_3 = k$, or $k_{ij} = k\delta_{ij}$.

***Transformation of conductivity matrix* K**

Let $\mathbf{K} = [k_{ij}]$ and $\overline{\mathbf{K}} = [\overline{k}_{ij}]$ be the conductivity matrices associated with coordinate basis vectors $\{\mathbf{e}_1, \mathbf{e}_2, \mathbf{e}_3\}$ and $\{\overline{\mathbf{e}}_1, \overline{\mathbf{e}}_2, \overline{\mathbf{e}}_3\}$ respectively. Then $\overline{\mathbf{K}} = [\overline{k}_{ij}]$ is related to $\mathbf{K} = [k_{ij}]$ by the following transformation:

$$\overline{\mathbf{K}} = \mathbf{R}^{\mathrm{T}} \cdot \mathbf{K} \cdot \mathbf{R} \qquad \text{or} \qquad \overline{k}_{ij} = R_{mi} R_{nj} k_{mn} \tag{3.28}$$

where rotation matrix $\mathbf{R}\,(3 \times 3)$ has the property that $\overline{\mathbf{e}}_i = \mathbf{R} \cdot \mathbf{e}_i$, which in component form is given by

$$R_{ij} = \mathbf{e}_i \cdot \overline{\mathbf{e}}_j \tag{3.29}$$

3.4.6 Element stiffness matrix and element force vectors

In line with eqs 3.22–3.26, the element stiffness matrix and element force vectors can be similarly defined as

$$\underset{N\times N}{\mathbf{K}_e} = \left[\int_{\Omega_e} k_{ij} H_{\alpha,i} H_{\beta,j} \, \mathrm{d}\Omega + \int_{\Gamma_e} c H_\alpha H_\beta \, \mathrm{d}\Gamma \right] \tag{3.30}$$

$$\underset{N\times 1}{\vec{\mathrm{b}}_e} = \left[\int_{\Omega_e} H_\alpha Q \, \mathrm{d}\Omega \right] \qquad \underset{N\times 1}{\vec{\mathrm{s}}_e} = \left[\int_{\Gamma_e} H_\alpha g \, \mathrm{d}\Gamma \right] \tag{3.31}$$

Assuming that k_{ij} and c are constant within the element, and using notations

$$I_{ij}^{\alpha\beta} = \int_{\Omega_e} H_{\alpha,i} H_{\beta,j} \, \mathrm{d}\Omega \qquad V_{\alpha\beta} = \int_{\Omega_e} H_\alpha H_\beta \, \mathrm{d}\Omega \qquad A_{\alpha\beta} = \int_{\Gamma_e} H_\alpha H_\beta \, \mathrm{d}\Gamma \tag{3.32}$$

then we have the following concise expressions:

1. Element stiffness matrix, $\mathbf{K}_e = \left[k_{ij} I_{ij}^{\alpha\beta} + c A_{\alpha\beta} \right]$ (3.33)
2. Element body force, $\vec{\mathrm{b}}_e = \left[Q_\beta V_{\alpha\beta} \right]$ (3.34)
3. Element surface force, $\vec{\mathrm{s}}_e = \left[g_\beta A_{\alpha\beta} \right]$ (3.35)

where Q_β is the relevant quantity generated(removed) at node β, and g_β is the specified flux at node β.

It should be noted that if isotropic material is considered, $k_{ij} = k\delta_{ij}$, and the expression for the element stiffness matrix $\mathbf{K}_e$ can be further simplified to

$$\mathbf{K}_e = \left[k(I_{11}^{\alpha\beta} + I_{22}^{\alpha\beta} + I_{33}^{\alpha\beta}) + c A_{\alpha\beta} \right] \tag{3.36}$$

Obviously, since for any definite integral it is required that the sum of parts should be equal to the total, we can write the global system stiffness equations in terms of the sum of all element stiffness equations, i.e.

$$\int_{\Omega} (.)\,\mathrm{d}\Omega = \sum_{e=1}^{N_e} \int_{\Omega_e} (.)\,\mathrm{d}\Omega \qquad \int_{\Gamma} (.)\,\mathrm{d}\Gamma = \sum_{e=1}^{N_e} \int_{\Gamma_e} (.)\,\mathrm{d}\Gamma \tag{3.37}$$

3.4.7 Examples

1 Temperature distribution in a hemisphere

For a hemisphere with radius R subjected to a temperature T_o over its convex surface and a zero temperature at its base, the temperature distribution inside the hemisphere expressed in in terms of spherical coordinates (r, ϕ) is given by [19]

$$T(r,\phi) = T_o \sum_{k=0}^{\infty} (-1)^k \left(\frac{4k+3}{2k+2}\right) \left(\frac{(2k)!}{2^{2k}(k!)^2}\right) \left(\frac{r}{R}\right)^{2k+1} P_{2k+1}(\cos\phi)$$

where r = radius (distance from the origin),
ϕ = angle measured from the z-axis towards the x–y plane,
$P_{2k+1}(x)$ = Legendre polynomial of degree $2k+1$ which is defined as

$$P_n(x) = \frac{1}{2^n n!} \frac{\mathrm{d}^n}{\mathrm{d}x^n} (x^2 - 1)^n$$

The problem of $R = 1$ and $T_o = 100$ was analyzed by the finite element method using four H20 solid elements as shown in Fig. 3.7. The numerical results along with the theoretical series solution are tabulated in Table 3.6. Although only four H20 elements were used in the finite element analysis, the solution is fairly accurate. It must be pointed out that the series solution is not exact either; at the point (0, 0, 1),

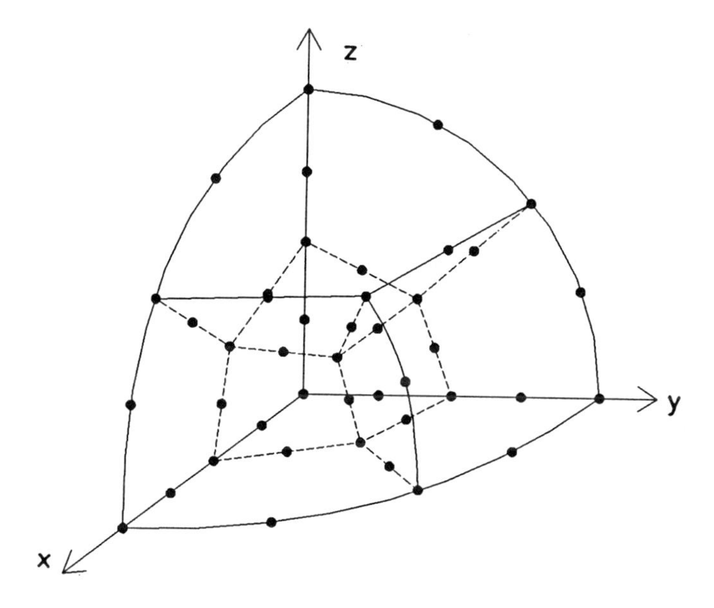

Fig. 3.7 One quarter of the hemisphere divided into four H20 elements

Table 3.6 Temperature at nodal points

Nodal coordinates			Nodal temperature (°C)		
x	y	z	FE	Analytical	Error (%)
0.0000	0.0000	0.2500	33.81	36.20	−6.59
0.0000	0.4268	0.1768	31.02	30.69	1.07
0.4268	0.0000	0.1768	31.02	30.69	1.07
0.3211	0.3211	0.1443	24.75	25.95	−4.63
0.0000	0.0000	0.5000	68.80	65.84	4.50
0.0000	0.1768	0.4268	58.97	59.20	−0.39
0.0000	0.3536	0.3536	51.56	54.13	−4.75
0.0000	0.5303	0.5303	83.92	80.04	4.85
0.1768	0.0000	0.4268	58.97	59.20	−0.39
0.1443	0.3211	0.3211	47.96	49.90	−3.88
0.3536	0.0000	0.3536	51.56	54.13	−4.75
0.3211	0.1443	0.3211	47.96	49.90	−3.88
0.2887	0.2887	0.2887	45.52	47.15	−3.46
0.4330	0.4330	0.4330	77.95	75.14	3.75
0.5303	0.0000	0.5303	83.92	80.04	4.85
0.0000	0.0000	0.7500	86.54	86.67	−0.14

an error of 1.26% still exists in the series approximation calculated using 2000 terms. Therefore, although a strict comparison is difficult, as far as computational effort is concerned, the finite element method does appear to be more economical, as the CPU time required for the finite element solution is less than one second, compared with about 4 minutes for the series solution.

2 Temperature distribution in a rectangular parallelepiped

The temperature of a rectangular parallelepiped of size $\ell_1 \times \ell_2 \times d$ is kept at zero at all faces except the top surface which is maintained at temperature $f(x, y)$, as shown in Fig. 3.8. The analytical solution of the steady-state temperature distribution in the solid is given by the series [19]

$$T(x, y, z) = \sum_{m=1}^{\infty} \sum_{n=1}^{\infty} c_{mn} \sin \frac{m\pi x}{\ell_1} \sin \frac{n\pi y}{\ell_2} \frac{\sinh(k_{mn} z)}{\sinh(k_{mn} d)}$$

where the coefficients c_{mn} and k_{mn} are given by

$$c_{mn} = \frac{4}{\ell_1 \ell_2} \int_0^{\ell_1} \int_0^{\ell_2} f(x, y) \sin \frac{m\pi x}{\ell_1} \sin \frac{n\pi y}{\ell_2} \, dx \, dy$$

$$k_{mn} = \pi \sqrt{\left(\frac{m^2}{\ell_1^2} + \frac{n^2}{\ell_2^2} \right)}$$

A 3D finite element analysis has been carried out to find the temperature distribution inside a unit cube ($\ell_1 = \ell_2 = d = 1$), due to a temperature variation $f(x, y) = xy(1 - x)(1 - y)$ at the top surface. The cube is divided into 64 H20

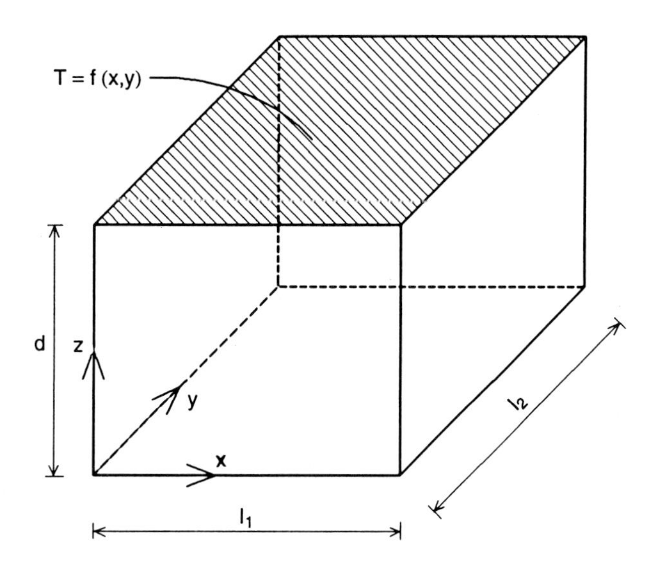

Fig. 3.8 Temperature distribution in a rectangular prism

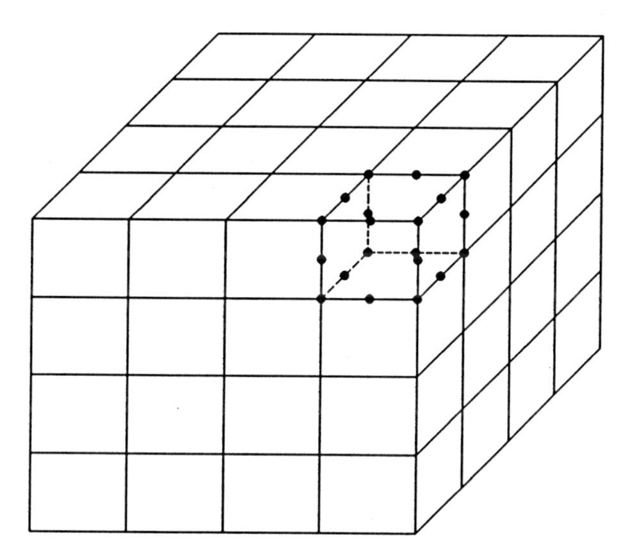

Fig. 3.9 A unit cube divided into H20 solid finite elements

elements ($4 \times 4 \times 4 \times$ mesh) as shown in Fig. 3.9 which gives a total of 425 equations. The temperature profiles at the middle section ($y = 0.5$) at various z-levels are plotted in Fig. 3.10; the solid lines represent the analytical series solution and the crosses are the nodal temperatures of the finite element solution.

3.5 THREE-DIMENSIONAL ELASTICITY PROBLEMS

The general formulation of three-dimensional stress–strain analysis based on the virtual work principle will be developed in Section 3.5.1. The special cases of

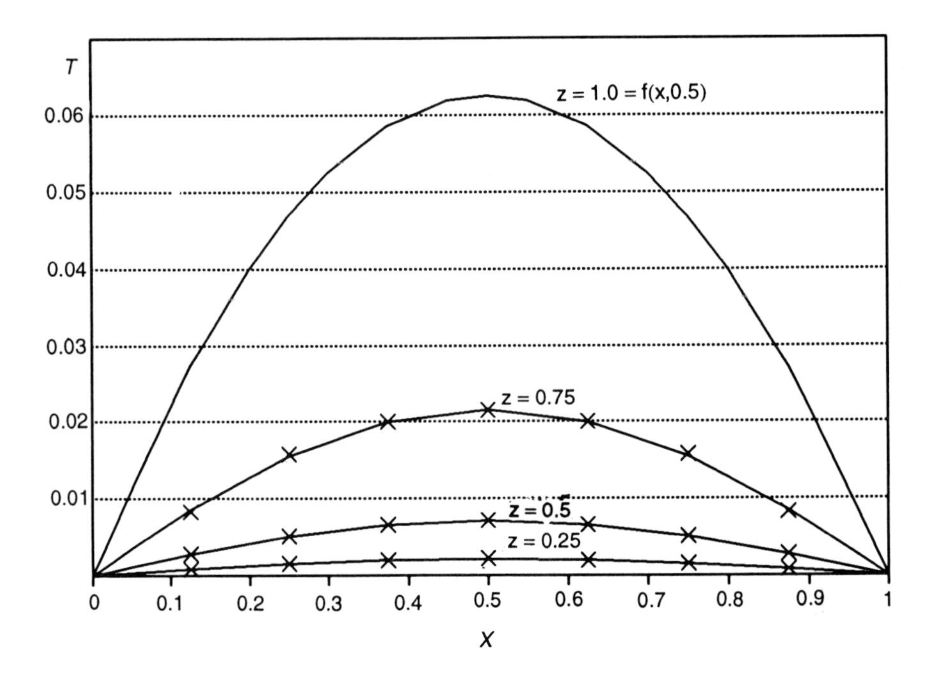

Fig. 3.10 Temperature profile at section $y = 0.5$

plane stress, plane strain and axisymmetric problems have already been dealt with in Chapter 2. Following exactly the same numerical procedures as in the two-dimensional cases, the finite element matrices for three-dimensional elasticity problems are obtained when suitable interpolation functions and their derivatives are substituted in the general expressions as given in Section 3.5.5.

3.5.1 Principle of virtual work

The virtual work equation for quasi-static problems is given by [16]:

$$\int_{\Omega} \sigma : (\vec{\psi}\nabla)\,\mathrm{d}\Omega = \int_{\Omega} \vec{\gamma}\cdot\vec{\psi}\,\mathrm{d}\Omega + \int_{\Gamma} \vec{\mathrm{q}}\cdot\vec{\psi}\,\mathrm{d}a \qquad \forall\vec{\psi} \tag{3.38}$$

where Ω = problem domain
Γ = boundary(surface) of Ω
σ = Cauchy stress tensor
$\vec{\psi}$ = arbitrary C^1 functions defined on Ω
$\vec{\gamma}$ = body force per unit volume
$\vec{\mathrm{q}}$ = surface traction per unit area
and ∇ = gradient differential operator.

In index notations, we have

$$\sigma : \vec{\psi}\nabla = \sigma_{ij}\psi_{i,j} \text{ with } \psi_{i,j} = \frac{\partial \psi_i}{\partial x_j} \quad \vec{\gamma}\cdot\vec{\psi} = \gamma_i\psi_i \text{ and } \vec{\mathrm{q}}\cdot\vec{\psi} = q_i\psi_i \quad i, j = 1, N_{\mathrm{d}} \tag{3.39}$$

where N_{d} is the number of dimensions of the problem domain.

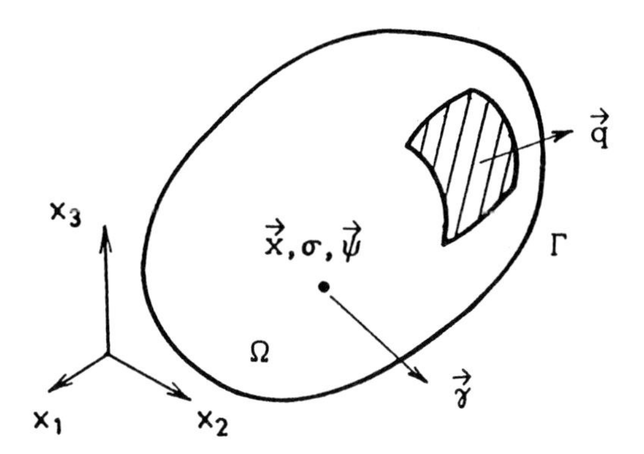

Fig. 3.11 Body in equilibrium

3.5.2 Galerkin formulation

By the Galerkin approximation, the weighting function $\vec{\psi}$ is defined with the aid of the element interpolation functions H_α such that $\psi_i = H_\alpha \psi_i^\alpha$, where ψ_i^α is the value of function ψ_i at node α. Substituting ψ_i into the virtual work equation (3.38), we have

$$\sigma : (\psi\nabla) = \sigma_{ij} H_{\alpha,j} \psi_i^\alpha \quad \text{with } H_{\alpha,j} = \frac{\partial H_\alpha}{\partial x_j} \qquad \vec{\gamma} \cdot \vec{\psi} = \gamma_i H_\alpha \psi_i^\alpha \qquad \vec{q} \cdot \psi = q_i H_\alpha \psi_i^\alpha$$

$$\text{Hence} \quad \psi_i^\alpha \left[\int_\Omega \sigma_{ij} H_{\alpha,j} \, d\Omega \right] = \psi_i^\alpha \left[\int_\Omega \gamma_i H_\alpha \, d\Omega + \int_\Gamma q_i H_\alpha \, da \right] \qquad \forall \psi_i^\alpha \tag{3.40}$$

$$\Rightarrow \int_\Omega \sigma_{ij} H_{\alpha,j} \, d\Omega = \int_\Omega \gamma_i H_\alpha \, d\Omega + \int_\Gamma q_i H_\alpha \, da \tag{3.41}$$

3.5.3 Linear elasticity

For an elastic material, the most general linear relationship to express the stress–strain characterization is

$$\sigma = \mathbf{C} : (\varepsilon - \varepsilon_o) + \sigma_o = \mathbf{C} : \varepsilon - \tau_o \qquad \text{with} \quad \tau_o = \mathbf{C} : \varepsilon_o - \sigma_o \tag{3.42}$$

$$\sigma_{ij} = C_{ijkl}(\varepsilon_{kl} - \varepsilon_{kl}^o) + \sigma_{ij}^o = C_{ijkl}\varepsilon_{kl} - \tau_{ij}^o \qquad \tau_{ij}^o = C_{ijkl}\varepsilon_{kl}^o - \sigma_{ij}^o \tag{3.43}$$

where C_{ijkl} are material elastic coefficients, $\varepsilon_{kl} = \frac{1}{2}(u_{k,l} + u_{l,k})$ are the strain components, and ε_{kl}^o and σ_{ij}^o are the components of initial strain and stress respectively.

Since $\mathbf{C}$ is symmetric in its last two indices, $C_{ijkl} = C_{ijlk}$,

$$C_{ijkl}\varepsilon_{kl} = C_{ijkl}\tfrac{1}{2}(u_{k,l} + u_{l,k}) = C_{ijkl}u_{k,l} \tag{3.44}$$

3.5.4 Stiffness matrix, displacement vector and force vectors

In terms of interpolation functions, the displacement u_k can be written as $u_k = H_\beta u_k^\beta$, hence

$$C_{ijkl}u_{k,l} = C_{ijkl}H_{\beta,l}u_k^\beta$$

Substituting into eq. 3.41, we get

$$\left[\int_\Omega H_{\alpha,j} C_{ijkl} H_{\beta,l}\, \mathrm{d}\Omega\right] u_k^\beta = \int_\Omega H_{\alpha,j} \tau_{ij}^{\mathrm{o}}\, \mathrm{d}\Omega + \int_\Omega \gamma_i H_\alpha\, \mathrm{d}\Omega + \int_\Gamma q_i H_\alpha\, \mathrm{d}a \tag{3.45}$$

In matrix notation, we have

$$\mathbf{K}\vec{\mathrm{u}} = \vec{\mathrm{f}} \qquad K_{\alpha\beta}^{ik} u_k^\beta = f_i^\alpha \tag{3.46}$$

where it should be noted that:

1. $\mathbf{K}$ is the global stiffness matrix, with components

$$K_{\alpha\beta}^{ik} = \int_\Omega H_{\alpha,j} C_{ijkl} H_{\beta,l}\, \mathrm{d}\Omega \tag{3.47}$$

2. $\vec{\mathrm{u}}$ is the displacement vector with components u_k^β (3.48)
3. $\vec{\mathrm{f}} = \vec{\mathrm{r}} + \vec{\mathrm{b}} + \vec{\mathrm{s}} + \vec{\mathrm{p}}$, where $[f_i^\alpha]$ is the global force vector, (3.49)

with the component force vectors being given by
(i) Initial stress and strain force vector, $\vec{r}$

$$\vec{\mathrm{r}} = \left[\int_\Omega H_{\alpha,j} \tau_{ij}^{\mathrm{o}}\, \mathrm{d}\Omega\right] \tag{3.50}$$

(ii) Body force vector, $\vec{b}$

$$\vec{\mathrm{b}} = \left[\int_\Omega \gamma_i H_\alpha\, \mathrm{d}\Omega\right] \tag{3.51}$$

(iii) Surface force vector $\vec{s}$

$$\vec{\mathrm{s}} = \left[\int_\Gamma q_i H_\alpha\, \mathrm{d}a\right] \tag{3.52}$$

(iv) Concentrated nodal force vector
(this has been isolated from the body force and the surface integrals)

$$\vec{\mathrm{p}} = [p_i^\alpha] \tag{3.53}$$

where p_i^α is the concentrated point force applied at node α in the direction i.

3.5.5 Element stiffness matrix and element force vectors

Based on eqs 3.47–3.52, the element stiffness matrix and the element force vectors are defined by

$$\mathbf{K}_e = \left[\int_{\Omega_e} H_{\alpha,j} C_{ijkl} H_{\beta,l}\, \mathrm{d}\Omega\right] \tag{3.54}$$

$$\vec{r}_e = \left[\int_{\Omega_e} H_{\alpha,j} \tau_{ij}^{o}\, \mathrm{d}\Omega\right] \qquad \vec{\mathrm{b}}_e = \left[\int_{\Omega_e} \gamma_i H_\alpha\, \mathrm{d}\Omega\right] \qquad \vec{\mathrm{s}}_e = \left[\int_{\Gamma_e} q_i H_\alpha\, \mathrm{d}a\right] \tag{3.55}$$

Similar to Section 3.4.6, we can also obtain the system stiffness and force relationships as a sum of all the element contributions, i.e.

$$\mathbf{K} = \sum_{e=1}^{N_e} \mathbf{K}_e \qquad \vec{r} = \sum_{e=1}^{N_e} \vec{r}_e \qquad \vec{\mathrm{b}} = \sum_{e=1}^{N_e} \vec{\mathrm{b}}_e \qquad \vec{\mathrm{s}} = \sum_{e=1}^{N_e} \vec{\mathrm{s}}_e \tag{3.56}$$

where N_e is the number of finite elements in the discretized domain.

(i) Element stiffness matrix, $\mathbf{K}_e$

$$K_{ik}^{\alpha\beta} = \int_{\Omega_e} H_{\alpha,j} C_{ijkl} H_{\beta,l} \, \mathrm{d}\Omega \tag{3.57}$$

For elements with homogeneous elastic properties,

$$K_{ik}^{\alpha\beta} = C_{ijkl} I_{jl}^{\alpha\beta} \tag{3.58}$$

with

$$I_{jl}^{\alpha\beta} = \int_{\Omega_e} H_{\alpha,j} H_{\beta,l} \, \mathrm{d}\Omega \tag{3.59}$$

(ii) Element initial stress and strain force vector, $\vec{r}_e$

$$r_i^{\alpha} = \int_{\Omega_e} H_{\alpha,j} \tau_{ij}^{\mathrm{o}} \, \mathrm{d}\Omega \tag{3.60}$$

For elements whose initial stress and strain are constant within the element, we have

$$r_i^{\alpha} = \tau_{ij}^{o} h_{\alpha j} \tag{3.61}$$

where

$$h_{\alpha j} = \int_{\Omega_e} H_{\alpha,j} \, \mathrm{d}\Omega \tag{3.62}$$

(iii) Element body force vector, $\vec{\mathrm{b}}_e$

$$b_i^{\alpha} = \int_{\Omega_e} \gamma_i H_{\alpha} \, \mathrm{d}\Omega \tag{3.63}$$

If the variation of body force within the element can be adequately represented by means of the element interpolation functions, $\gamma_i = \gamma_i^{\beta} H_{\beta}$, where γ_i^{β} is the body force per unit volume at node β in the direction i. Then

$$b_i^{\alpha} = \gamma_i^{\beta} V_{\alpha\beta} \qquad \text{where} \quad V_{\alpha\beta} = \int_{\Omega_e} H_{\alpha} H_{\beta} \, \mathrm{d}\Omega \tag{3.64}$$

(iv) Element surface traction vector, $\vec{\mathrm{s}}_e$

$$s_i^{\alpha} = \int_{\Gamma_e} q_i H_{\alpha} \, \mathrm{d}a = \int_{\Gamma_e} q_i^{\beta} H_{\beta} H_{\alpha} \, \mathrm{d}a = q_i^{\beta} A_{\alpha\beta} \tag{3.65}$$

where q_i^{β} is the surface traction at node β in the direction i.

3.5.6 Isotropic material

For a general anisotropic material, 21 independent elastic constants are necessary to define completely the three-dimensional stress–strain relationship [17]. However, the number of independent elastic constants can be reduced to only two when isotropic materials are considered in a linear elastic analysis. For such a problem, a great deal of simplification can be made in the expressions of the element stiffness matrix, and, as a result the computational effort at element level is significantly reduced. The elastic coefficients C_{ijkl} characterizing the stress–strain relationship for isotropic materials can be written as [17]

$$C_{ijkl} = \lambda\delta_{ij}\delta_{kl} + \mu(\delta_{ik}\delta_{jl} + \delta_{il}\delta_{jk}) \tag{3.66}$$

where λ and μ are the Lamé constants which can be deduced from the Young's modulus E and the Poisson's ratio ν as

$$\lambda = \frac{\nu E}{(1+\nu)(1-2\nu)} \qquad \mu = \frac{E}{2(1+\nu)} \tag{3.67}$$

From Section 3.5.5, the element stiffness matrix $\mathbf{K}_e$ is given by

$$\begin{aligned} K_{ik}^{\alpha\beta} &= \int_{\Omega_e} H_{\alpha,j} C_{ijkl} H_{\beta,l}\, \mathrm{d}\Omega = C_{ijkl} I_{jl}^{\alpha\beta} \\ &= \lambda\delta_{ij}\delta_{kl} I_{jl}^{\alpha\beta} + \mu(\delta_{ik}\delta_{jl} + \delta_{il}\delta_{jk}) I_{jl}^{\alpha\beta} \\ &= \lambda I_{ik}^{\alpha\beta} + \mu(\delta_{ik} I_{jj}^{\alpha\beta} + I_{ki}^{\alpha\beta}) \end{aligned} \tag{3.68}$$

By direct index substitutions, the sub-matrix $\mathbf{K}^{\alpha\beta}$ can be expressed as

$$\underset{3\times3}{\mathbf{K}^{\alpha\beta}} = \left[\begin{array}{c|c|c} (\lambda+\mu)I_{11}^{\alpha\beta} + \mu I^{\alpha\beta} & \lambda I_{12}^{\alpha\beta} + \mu I_{21}^{\alpha\beta} & \lambda I_{13}^{\alpha\beta} + \mu I_{31}^{\alpha\beta} \\ \hline \lambda I_{21}^{\alpha\beta} + \mu I_{12}^{\alpha\beta} & (\lambda+\mu)I_{22}^{\alpha\beta} + \mu I^{\alpha\beta} & \lambda I_{23}^{\alpha\beta} + \mu I_{32}^{\alpha\beta} \\ \hline \lambda I_{31}^{\alpha\beta} + \mu I_{13}^{\alpha\beta} & \lambda I_{32}^{\alpha\beta} + \mu I_{23}^{\alpha\beta} & (\lambda+\mu)I_{33}^{\alpha\beta} + \mu I^{\alpha\beta} \end{array}\right] \alpha, \beta = 1, N \tag{3.69}$$

$$\text{where } I^{\alpha\beta} = I_{jj}^{\alpha\beta} = I_{11}^{\alpha\beta} + I_{22}^{\alpha\beta} + I_{33}^{\alpha\beta} \tag{3.70}$$

3.5.7 Examples

1 Bending of a cantilever beam

The bending of a cantilever beam is studied by the finite element method using H20 solid elements. The dimensions and the finite element discretization of the cantilever beam are shown in Fig. 3.12. A unit shear force is applied uniformly over the free

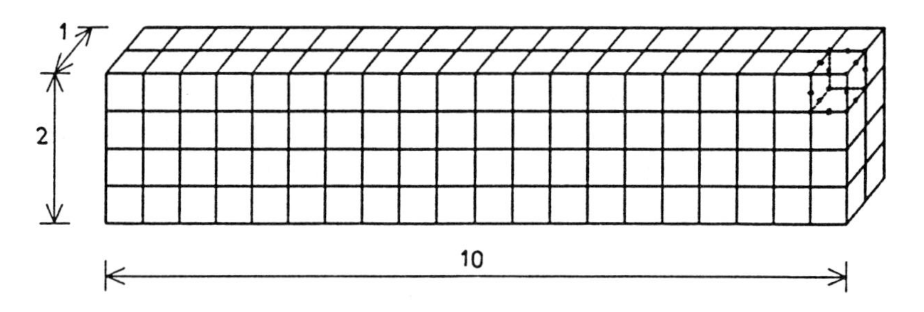

Fig. 3.12 A cantilever beam discretized into 160 H20 elements

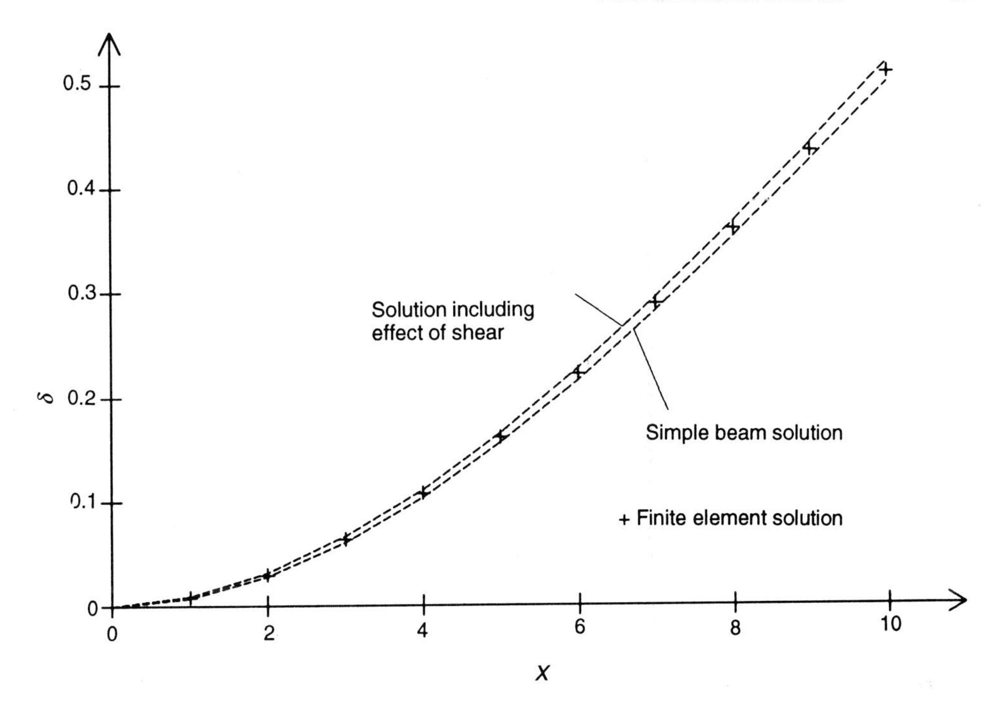

Fig. 3.13 Vertical displacement along the centroidal axis of the beam

end section. The material of the beam is assumed to be isotropic, with Young's modulus $E = 1000$ and Poisson's ratio $\nu = 0.25$.

A mesh of 160 ($2 \times 4 \times 20$) H20 elements and 1077 nodes was used in the analysis, and each node has three DOF making up a total of 3231 equations. The half-bandwidth and the profile of the global stiffness matrix were equal to 151 and 488 637 respectively. The calculation was done on an IBM 320 Powerstation, and the problem was solved in 15 seconds.

The vertical displacement δ along the centroidal axis of the beam is given in Fig. 3.13. The analytical values of the vertical displacement from the simple beam theory with and without considering the shear effects are also plotted in the same figure for comparison. As expected, the simple beam solution neglecting the shear effects underestimates the vertical displacement, and the vertical displacement is slightly overestimated when shear effects are included, demonstrating the inconsistency of elementary beam theory when considering bending and shearing effects.

The basic assumption of 'plane sections remain plane' made in the simple beam theory is checked by the three-dimensional analysis. It is found that the assumption is practically valid at all cross-sections of the beam. A typical plot of the vertical position of a point after deformation $y' = y+\delta$ against the longitudinal displacement u at the section $x = 5.0$ along the y-axis and along a vertical edge, is depicted in Fig. 3.14.

2 Twisting of a prismatic shaft

The twisting of a triangular cross-section prismatic shaft is studied by the three-dimensional finite element analysis. The dimensions and the loading conditions of

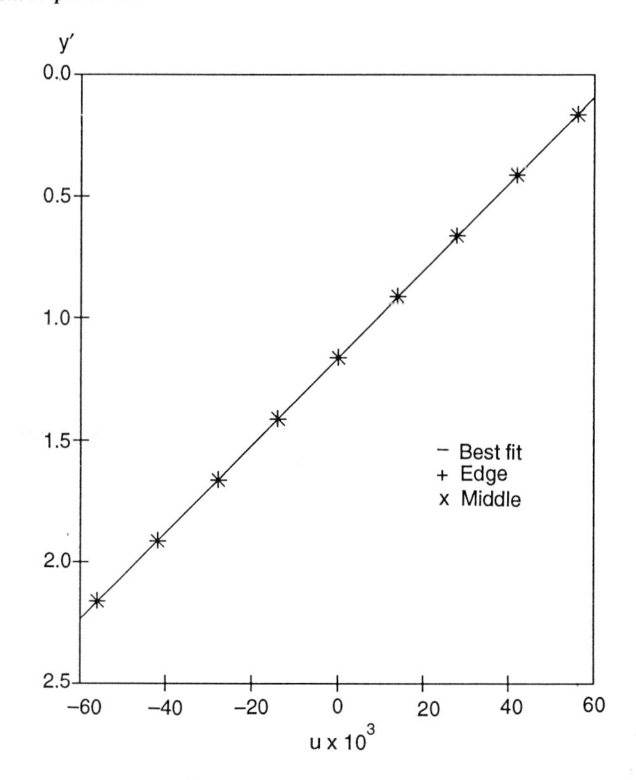

Fig. 3.14 Graph of y' against u at section $x = 5.0$

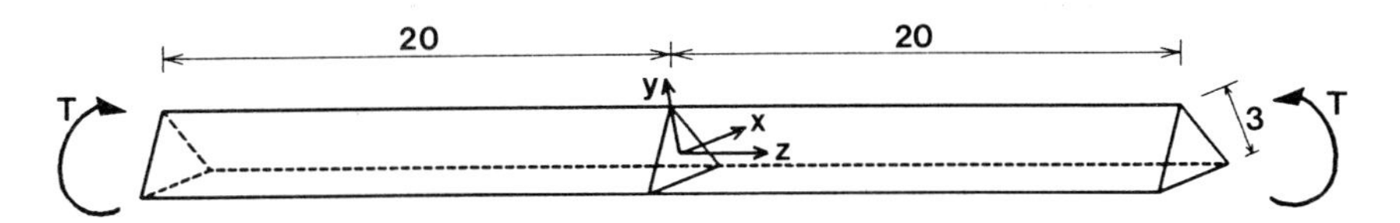

Fig. 3.15 Dimensions and loading conditions of shaft

the shaft are shown in Fig. 3.15. By symmetry, only half of the shaft is modelled. It is divided into 90 P15 solid elements involving 1224 DOF, as shown in Fig. 3.16. The magnitude of the applied torque T is equal to 3 units, which is generated by three unit shear forces acting along the edges of the triangular cross-section as depicted in Fig. 3.17.

From the displacement (u, v) of the finite element solution at a point $P(x, y)$, the angle of twist θ is given by

$$\cos\theta = \frac{\mathrm{OP} \cdot \mathrm{OP}'}{\|\mathrm{OP}\| \; \|\mathrm{OP}'\|} \tag{3.71}$$

where origin O is at the centroid of the triangular section and $P' = (x + u, y + v)$ is the position of point P after deformation. A number of sampling points are taken at each cross-section and the average is calculated as the value of θ for that section. It is found that at each point P, the distances $\|\mathrm{OP}\|$ and $\|\mathrm{OP}'\|$ are nearly identical,

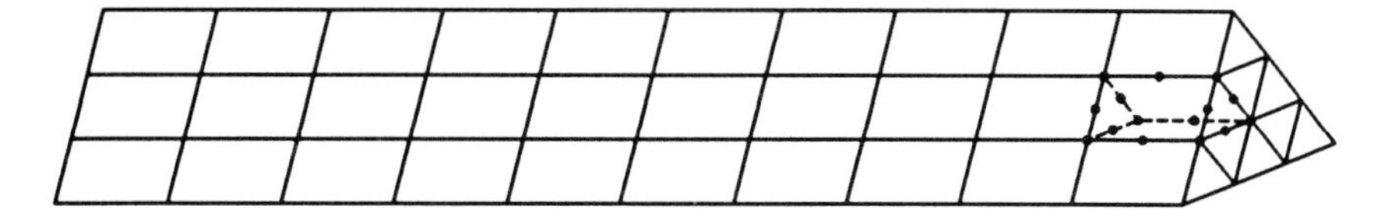

Fig. 3.16 Half of shaft discretized into finite elements

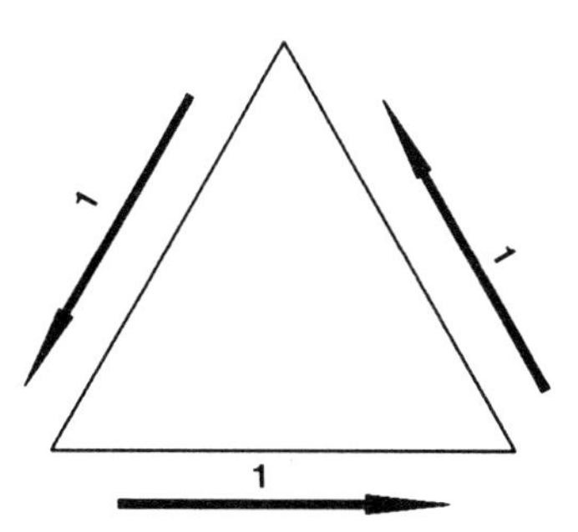

Fig. 3.17 Shear forces acting at end of shaft ($z = 20$)

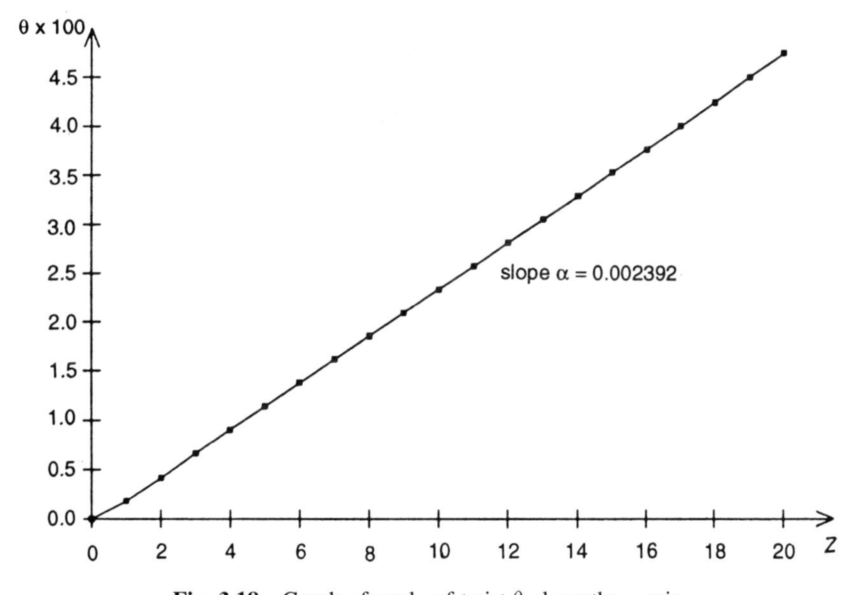

Fig. 3.18 Graph of angle of twist θ along the z-axis

and the θs computed by eq. 3.71 at various sampling points are almost equal. This demonstrates that the deformation of the shaft is mainly governed by a rotation, and that an angle of twist θ can be unambiguously defined for each cross-section.

When θ is plotted against z, as shown in Fig. 3.18, a linear relationship is obtained, except at sections close to the fixed end. This agrees well with the assumption made in Saint Venant torsion [18] that $\theta = \alpha z$, where α is the angle of twist per unit length.

Table 3.7 Comparison of numerical and analytical values of warping

Nodal coordinates		$w \times 10^3$		
x	y	FE	Analytical	Error (%)
0.0000	2.0000	0.000 000	0	0
−0.5774	1.0000	−0.612 227	−0.613 557	−0.22
0.5774	1.0000	0.612 227	0.613 557	−0.22
−1.1547	0.0000	0.612 229	0.613 556	−0.22
0.0000	0.0000	0.000 000	0	0
1.1547	0.0000	−0.612 229	−0.613 556	−0.22
−1.7321	−1.0000	−0.000 011	0	0
−0.5774	−1.0000	−0.612 226	−0.613 557	−0.22
0.5774	−1.0000	0.612 226	0.613 557	−0.22
1.7321	−1.0000	0.000 011	0	0

From the Saint Venant torsion of a prismatic bar, the value of α is given by

$$\alpha = \frac{T}{J}$$

where the value of the torsional rigidity, J, of an equilateral triangular cross-section can be computed using the following formula

$$J = \frac{Gh^4}{15\sqrt{3}}$$

In the finite element calculation, a value of 400 is used for the shear modulus G, and the height of the section, h, is taken as 3 units. From the numerical constants chosen, J is found equal to 1247 and α is equal to 0.002 406, which is in excellent agreement with the slope of the graph θ against z in Fig. 3.17.

The warping of a section along the triangular prismatic shaft is given by

$$w = \alpha \frac{3xy^2 - x^3}{2h}$$

The numerical and analytical values of w at the mid-span section ($z = 10$) are listed in Table 3.7, and again are found to be in a very good agreement. However, due to the end disturbances, the errors at the two ends are larger than those at the mid-span section. An error of up to 2% is found at the cross-section near the end at level $z = 3$.

*3 **Boussinesq problem – A point load on an elastic half-space***

The Boussinesq problem of a point load on an elastic half-space analyzed as an axisymmetric stress problem has been described in Section 2.5.1. In this section, the same problem is re-analyzed as a three-dimensional problem using solid H20 elements. The original problem involves the solution of an infinite half-space, but for the finite element analysis, the domain is truncated at a distance of 10 units from the point load P along the three coordinate directions. The truncated domain is divided into 125 ($5 \times 5 \times 5$) H20 elements, as shown in Fig. 3.19. Along each coordinate axis, an element size propagation ratio of 1.5 is applied to give a higher element density near the point load. The boundary conditions are (i) by symmetry

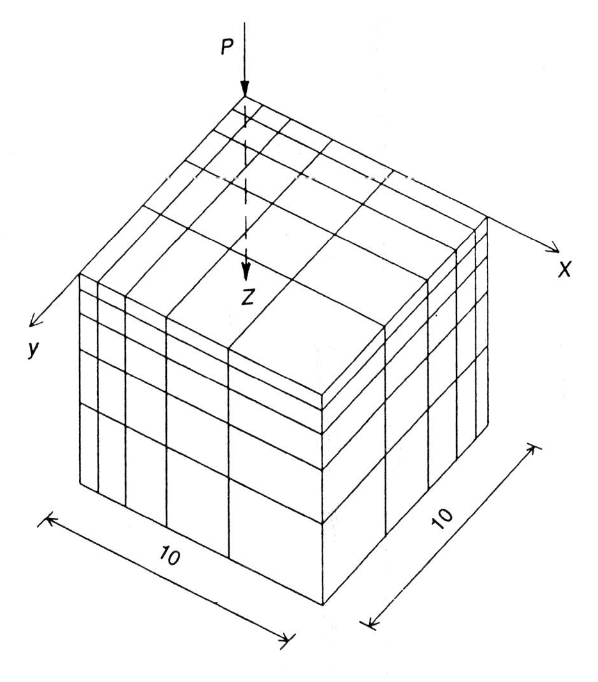

Fig. 3.19 Finite element model for the Boussinesq problem

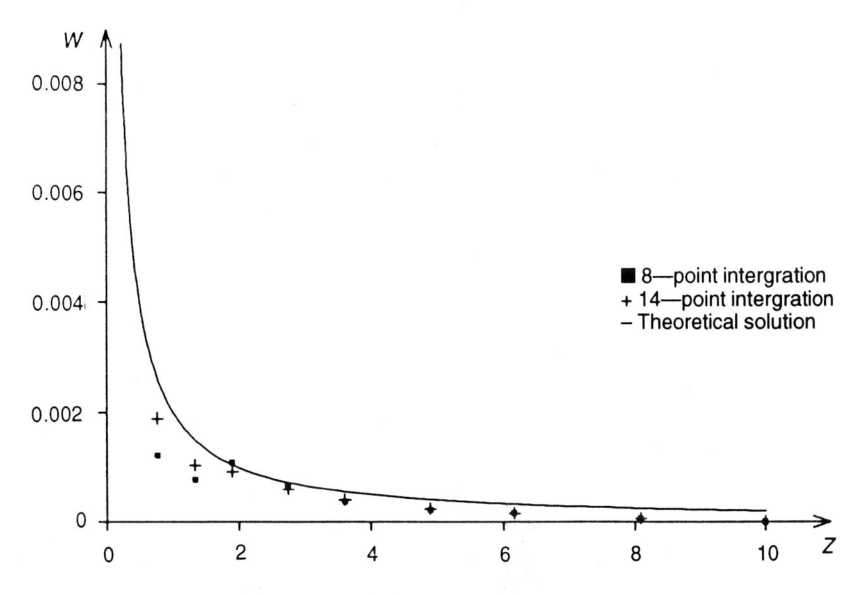

Fig. 3.20 Vertical displacement under the point load

the normal component of displacements at xy- and yz-planes are zero and (ii) the displacements at the bottom face ($z = 10$) are set equal to zero to approximate the far field boundary condition of zero displacement at infinity.

In the numerical analysis, Young's modulus and Poisson's ratio were taken as 1000 and 0.25 respectively. Two different integration rules, namely the 8-point and

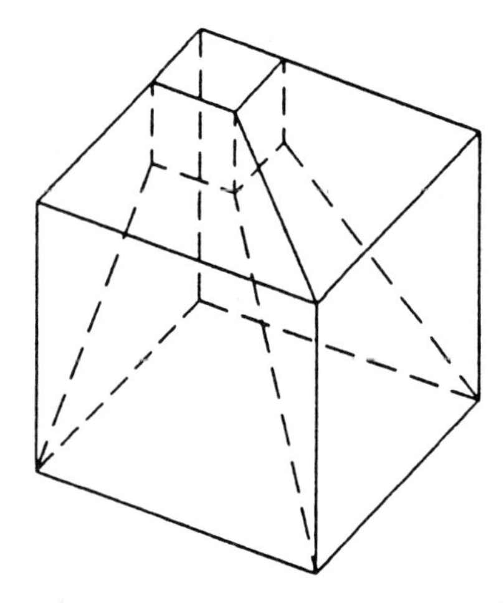

Fig. 3.21 Local refinement under the point load

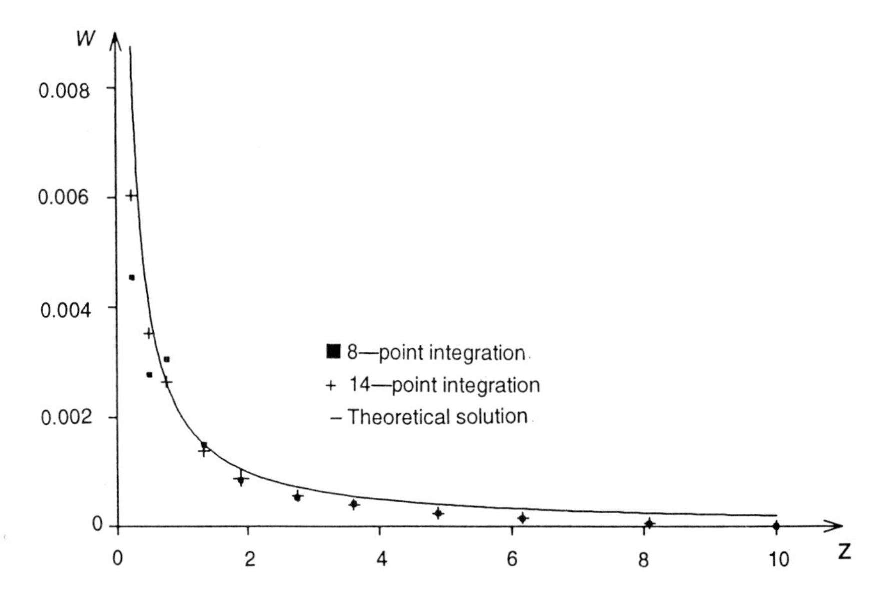

Fig. 3.22 Vertical displacement calculated using an FE mesh with local refinement

the 14-point rules, were used in the evaluation of the element stiffness matrix. The vertical downward displacements under the point load at various depths are shown in Fig. 3.20. A rather large discrepancy is observed between the numerical and the analytical solutions, with more fluctuations associated with the results obtained using 8-point integration.

A local refinement was made to the element mesh directly under the point load, as shown in Fig. 3.21. The results were much improved by this local refinement, as shown in Fig. 3.22. The vertical displacements obtained using 14-point integration almost coincide with the analytical values except at the far field, which is somewhat expected as the boundary condition was not modelled exactly in the finite element analysis. Although the results obtained using 8-point integration are also improved, fluctuation still exists in the vertical displacement.

3.6 REFERENCES TO CHAPTER 3

1. O. C. Zienkiewicz, P. L. Arlett & Y. K. Cheung (1967) Solution of three-dimensional field problems by the finite element method, *The Engineer*, **27**, October.
2. L. R. Herrmann (1965) Elastic torsion analysis of irregular shapes, *Proc. Am. Soc. Civ. Eng.*, **91**, EM6, 11-19.
3. J. Simkin & C. W. Trowbridge (1979) On the use of the total scalar potential in the numerical solution of field problems in electromagnets, *Int. J. Num. Methods Engng.*, **14**, 423-40.
4. J. Simkin & C. W. Trowbridge (1980) Three-dimensional non-linear electromagnetic field computations using scalar potentials, *Proc. Inst. Elec. Eng.*, **127**, B(6).
5. M. M. Sussman (1988) Remarks on computational magnetostatics, *Int. J. Num. Methods Engng.*, **26**, 987-1000.
6. S. T. K. Chan, B. E. Larock & L. R. Herrmann (1973) Free surface ideal fluid flows by finite elements, *Proc. Am. J. Civ. Eng.*, **99**, HY6.
7. J. H. Argyris, G. Mareczek & D. W. Scharpf (1969) Two and three dimensional flow using finite elements, *J. Roy. Aero. Soc.*, **73**, 961-4.
8. G. de Vries & D. H. Norrie (1971). The application of the finite element technique to potential flow problems, *J. Appl. Mech., ASME*, **38**, 978-82.
9. C. S. Desai (1975) Finite element methods for flow in porous media, in *Finite Elements in Fluids*, ed. by R. H. Gallagher, **1**, 157-82, Wiley.
10. I. Javandel & P. A. Witherspoon (1968) Applications of the finite element method to transient flow in porous media, *Trans. Soc. Petrol. Eng.*, **243**, 241-51.
11. J. C. Bruch (1980) A survey of free-boundary value problems in the theory of fluid flow through porous media, *Advances in Water Resources*, **3**, 65-80.
12. C. Baiocchi, V. Comincioli & V. Maione (1975) Unconfined flow through porous media, *Meccanice, Ital. Ass. Theor. Appl. Mech.*, **10**, 51-60.
13. O. C. Zienkiewicz & R. L. Taylor (1989) The Finite Element Method, 4th edn, Vol. 1 – *Basic Formulation and Linear Problems*, McGraw-Hill International Editions.
14. R. Courant (1943) Variational methods for the solution of problems of equilibrium and vibration, *Bull. Am. Math. Soc.*, **49**, 1-23.
15. K. Washizu (1975) *Variational Methods in Elasticity and Plasticity, 2nd edn*, Pergamon Press.
16. T. J. R. Hughes (1987) *The Finite Element Method (Linear Static and Dynamic Finite Element Analysis)*, Prentice-Hall International Edition.
17. T. J. Chung (1988) *Continuum Mechanics*, Prentice-Hall International Edition.
18. S. P. Timoshenko & J. N. Goodier (1970) *Theory of Elasticity*, McGraw-Hill Kogakusha.
19. F. B. Hildebrand (1977) *Advanced Calculus for Applications*, Prentice-Hall, India.

4 Plate and Shell Elements

4.1 Introduction to plate elements

Plate bending elements are of considerable importance for modelling engineering structures. Over the last three decades, a great deal of effort has been devoted to the formulation of efficient and reliable bending elements. The reviews compiled in references [1-5] provide a clear indication of the widespread interest in the subject. They also illustrate the degree of difficulty which is inherent in the formulation of successful plate bending elements. The main obstacle is due to the fact that the classical Kirchhoff equations of thin plate require C^1-continuity of deflection for the strain energy to be bounded and the finite element approximations to be convergent. While C^0 finite elements (based on the Reissner-Mindlin theory for moderately thick plates in which shear deformation has been included) of various shapes and with different number of nodes are readily formulated, the higher order C^1 elements prove to be far more challenging and difficult to develop.

Several fully conforming, convergent triangular elements were proposed [6-9], in the early days of research on the subject. However, these elements were found to be rather complicated in their formulations and excessively stiff in distorted configurations. Some of these deficiencies were overcome by Razzaque [10] by means of a technique which was subsequently shown to be equivalent to a stress-hybrid formulation of the type introduced by Pian [11] and Pian and Tong [12].

An alternative approach was introduced in the late sixties [13-16] in which plates are treated as a degenerated case of a three-dimensional solid with a simplified strain distribution in the thickness direction. Plate bending elements based upon the Reissner-Mindlin theory include transverse shear strains and require only C^0-continuity. This approach opens the way to a greater variety of interpolatory schemes, but is not without its own inherent difficulties [17-36].

A large number of Kirchhoff plate elements have been presented in various finite element texts. In this chapter, several Reissner-Mindlin plate elements and a special 'discrete' Kirchhoff element will be discussed in detail.

4.2 Reissner-Mindlin plate element

4.2.1 Theory

The domain Ω of a plate element is given by

$$\Omega = \left\{ (x, y, z) \in \mathbb{R}^3,\ z \in \left[-\frac{h}{2}, \frac{h}{2} \right],\ (x, y) \in A \subset \mathbb{R}^2 \right\}$$

where h (or $h(x, y)$) is the thickness of the plate and A is its area. The boundary of A is denoted by ∂A. In the subsequent discussions, the capital letter indices, I, J, K take on values 1, 2, and the small letter indices i, j, k take on values 1, 2, 3.

The basic assumptions of the theory are:

1. The vertical direct stress $\sigma_{33} = 0$. This implies that the lateral load is resisted by bending action.
2. $u_I(x, y, z) = -z\theta_I(x, y)$. This implies that the normal remains straight after deformation. However, because $w_{,I} = \gamma_I + \theta_I$ (Fig. 4.1 (b)), the normal is no longer necessarily perpendicular to the deflected mid-surface.
3. $u_3(x, y, z) = w(x, y)$. This means that the transverse displacement w does not vary with the thickness, and is a function of x and y only.
4. A shear correction factor $\mathfrak{K}$ is introduced in order to account for the true shear stress distribution across the plate thickness. Reissner [22] gave a value of 5/6 to account for the parabolic shear stress distribution in static problems, while Mindlin [23] proposed a value of $\pi^2/12$.

It should be noted that the Poisson–Kirchhoff theory for thin plates has only one assumption which is different from the Reissner–Mindlin plate – the normal to the mid-surface remains normal after deformation.

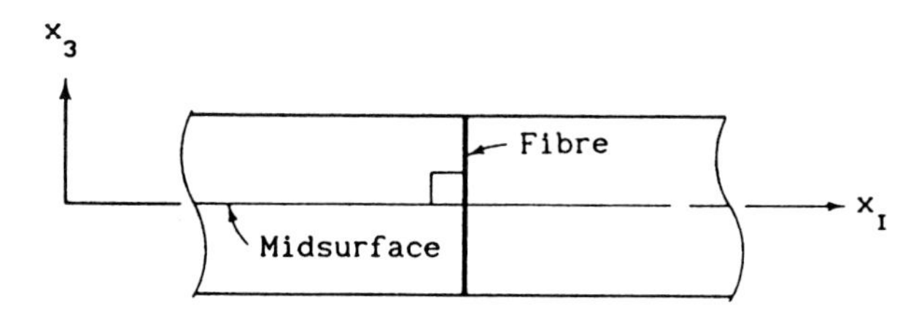

(a) Undeformed geometry

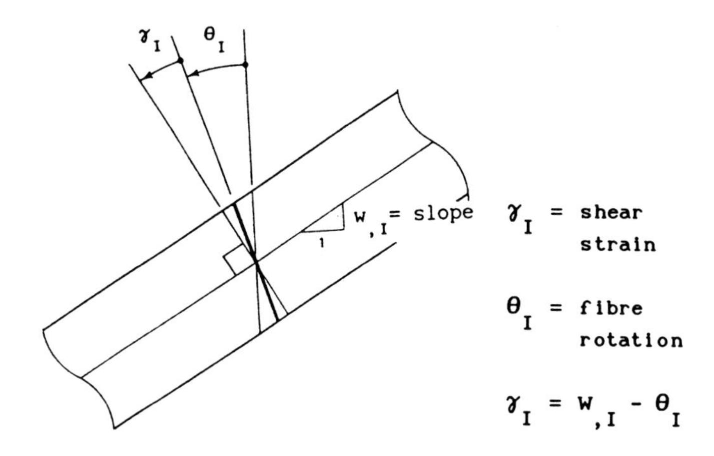

(b) Deformed geometry

Fig. 4.1 Sign conventions for fibre rotation angles, θ_1, θ_2

4.2.2 Strain-displacement relationship

Strain-displacement equations can be derived from assumptions 2 and 3 above.

$$\varepsilon_{IJ} = \tfrac{1}{2}(u_{I,J} + u_{J,I}) \equiv u_{(I,J)} = -z\theta_{(I,J)} \tag{4.1}$$

$$\varepsilon_{I3} = \tfrac{1}{2}(u_{I,3} + u_{3,I}) \equiv u_{(I,3)} = \tfrac{1}{2}(w_{,I} - \theta_I) \tag{4.2}$$

For a Reissner–Mindlin plate, the normal fibre rotation θ_I and slope $w_{,I}$ are not necessarily the same, and thus transverse shear strains can be accommodated. This is in contrast with the Poisson–Kirchhoff thin plate theory in which $\theta_I = w_{,I}$ and consequently $\gamma_I = 0$. Nevertheless, in the thin plate limit, the transverse shear strains are expected to be negligible.

4.2.3 Constitutive equation

The reduced form of the constitutive equation used in the plate theory can be obtained by eliminating ε_{33} in the three-dimensional constitutive equation using assumption 1 ($\sigma_{33} = 0$). For simplicity, the isotropic case is considered.

$$\sigma_{ij} = \lambda\delta_{ij}\varepsilon_{kk} + 2\mu\varepsilon_{ij} \tag{4.3}$$

where λ and μ are the Lamé coefficients and δ_{ij} is the Kronecker delta. Assumption 1, $\sigma_{33} = 0$, implies

$$\varepsilon_{33} = \frac{-\lambda}{\lambda + 2\mu}(\varepsilon_{11} + \varepsilon_{22}) \tag{4.4}$$

Hence,

$$\sigma_{IJ} = \bar{\lambda}\delta_{IJ}\varepsilon_{KK} + 2\mu\varepsilon_{IJ} \qquad \sigma_{I3} = 2\mu\varepsilon_{I3} \tag{4.5}$$

with

$$\bar{\lambda} = \frac{2\lambda\mu}{\lambda + 2\mu} \tag{4.6}$$

$\bar{\lambda}$ and μ may be expressed in terms of Young's modulus, E, and Poisson's ratio, ν

$$\bar{\lambda} = \frac{\nu E}{1 - \nu^2} \qquad \mu = \frac{E}{2(1 + \nu)} \tag{4.7}$$

4.2.4 Virtual work equation

From Section 3.5.1, the virtual work equation for quasi-static three-dimensional elasticity problems is given by

$$\int_\Omega \sigma : (\vec{\psi}\nabla)\,\mathrm{d}\Omega = \int_\Omega \vec{\varphi}\cdot\vec{\psi}\,\mathrm{d}\Omega + \int_{\partial\Omega} \vec{t}\cdot\vec{\psi}\,\mathrm{d}\Gamma \qquad \forall\vec{\psi} \tag{4.8}$$

where $\vec{\varphi}$ is the body force per unit volume and $\vec{t}$ is the surface traction per unit area. Taking into consideration $\sigma_{33} = 0$, and replacing arbitrary function $\vec{\psi}$ by virtual displacement $\hat{u}$, we have

$$\int_\Omega \hat{u}_{(I,J)}\sigma_{IJ} + 2\hat{u}_{(I,3)}\sigma_{I3}\,\mathrm{d}\Omega = \int_\Omega \hat{u}_i\varphi_i\,\mathrm{d}\Omega + \int_{\partial\Omega} \hat{u}_i t_i\,\mathrm{d}\Gamma \qquad \forall\hat{u}_i \tag{4.9}$$

Using $\hat{u}_I = -z\hat{\theta}_I$ and $\hat{u}_3 = \hat{w}$, we have

$$\overbrace{\int_A \int_h (-\hat{\kappa}_{IJ}\sigma_{IJ}z + \hat{\gamma}_I\sigma_{I3})\,dz\,dA}^{\text{internal stress}} - \overbrace{\int_A \int_h (-\hat{\theta}_I\varphi_I z + \hat{w}\varphi_3)\,dz\,dA}^{\text{body force}}$$

$$\overbrace{- \int_{A^-} -\left(\frac{-h}{2}\hat{\theta}_I\right) t_I + \hat{w}t_3\,dA - \int_{A^+} -\left(\frac{h}{2}\hat{\theta}_I\right) t_I + \hat{w}t_3\,dA}^{\text{external force on lower and upper surfaces}}$$

$$\overbrace{- \int_{\partial A} \int_h (-z\hat{\theta}_I t_I + \hat{w}t_3)\,dz\,ds}^{\text{external force along edge}} = 0 \tag{4.10}$$

where $\int_h = \int_{-h/2}^{+h/2}$ $\quad \hat{\kappa}_{IJ} = \hat{\theta}_{(I,J)}$ $\quad \hat{\gamma}_I = \hat{w}_{,I} - \hat{\theta}_I$ and A^- and A^+ are the lower and upper surfaces of the plate.

Using sign conventions as depicted in Fig. 4.2, and expressing the stresses in terms of force resultants, we have

$$\int_A \hat{\gamma}_I q_I - \hat{\kappa}_{IJ} m_{IJ}\,dA - \int_A \hat{w}F - \hat{\theta}_I C_I\,dA - \int_{\partial A} \hat{w}\overline{q} - \hat{\theta}_I \overline{m}_I\,ds = 0 \tag{4.11}$$

where

(i) moment, $m_{IJ} = \int_h \sigma_{IJ} z\,dz$ (4.12)

(ii) shear force, $q_I = \int_h \sigma_{I3}\,dz$ (4.13)

(iii) total transverse force per unit area,

$$F = t_3^+ + t_3^- + \int_h \varphi_3\,dz \tag{4.14}$$

where t_i^+ and t_i^- are the surface traction forces on the upper and lower surfaces respectively.

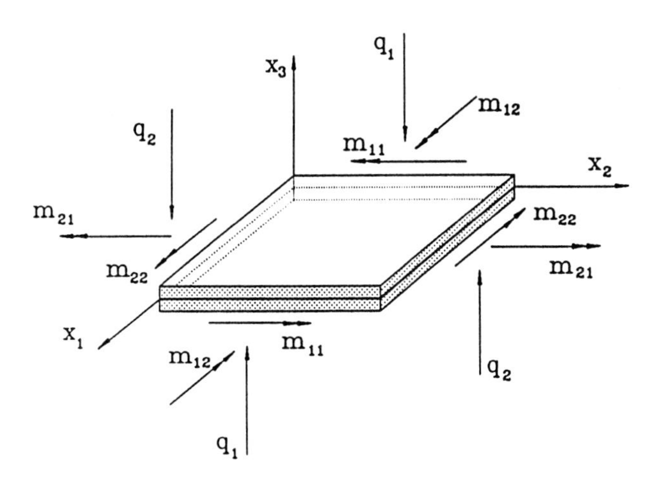

Fig. 4.2 Sign conventions for stress resultants

(iv) total applied couple per unit area,

$$C_I = \frac{h}{2}t_I^+ - \frac{h}{2}t_I^- + \int_h \varphi_I z\,dz \tag{4.15}$$

(v) Prescribed boundary moments, $\overline{m}_I = \int_h t_I z\,dz$ (4.16)

(vi) Prescribed boundary shear force, $\overline{q} = \int_h t_3\,dz$ (4.17)

Equations 4.12 and 4.13 can be expanded to give the following explicit expressions for the moment and the shear force:

(i) moment, $m_{IJ} = \int_h \sigma_{IJ} z\,dz = \int_{-h/2}^{+h/2} (\overline{\lambda}\delta_{IJ}\varepsilon_{KK} + 2\mu\varepsilon_{IJ})z\,dz$

$$m_{IJ} = \frac{-h^3}{12}[\overline{\lambda}\delta_{IJ}\theta_{K,K} + 2\mu\theta_{(I,J)}] = -D_{IJKL}\theta_{(K,L)} = -D_{IJKL}\kappa_{KL} \tag{4.18}$$

$$\text{where } D_{IJKL} = \frac{h^3}{12}[\overline{\lambda}\delta_{IJ}\delta_{KL} + \mu(\delta_{IK}\delta_{JL} + \delta_{IL}\delta_{JK})] \tag{4.19}$$

(ii) shear force, $q_I = \int_h \sigma_{I3}\,dz = \int_{-h/2}^{+h/2} 2\mu\varepsilon_{I3}\,dz$

$$q_I = \mu h(w_{,I} - \theta_I) = C_{IJ}\gamma_J \tag{4.20}$$

$$\text{where } C_{IJ} = \mu h \delta_{IJ} \tag{4.21}$$

At this stage it would be appropriate to include the following remarks:

1. Orthotropic and general anisotropic material behaviour can be incorporated by appropriately redefining the elastic constants D_{IJKL} and C_{IJ}.
2. The shear correction factor, $\mathfrak{K}$, introduced in assumption 4 can be accounted for simply by replacing C_{IJ} with $C_{IJ}^* = \mathfrak{K}C_{IJ}$ throughout.

4.2.5 Finite element stiffness matrix and load vector

By finite element approximation, we write for each element

$$w = H_\alpha w^\alpha \qquad \theta_I = H'_\alpha\theta_I^\alpha \qquad \hat{w} = H_\alpha\hat{w}^\alpha \qquad \hat{\theta}_I = H'_\alpha\hat{\theta}_I^\alpha \qquad \alpha = 1, N \tag{4.22}$$

where H_α and H'_α are respectively the interpolation functions for displacement w and rotations θ_I associated with node α; and w^α, θ_I^α, $\hat{w}^\alpha$ and $\hat{\theta}_I^\alpha$ are respectively the values of w, θ_I (real), $\hat{w}$ and $\hat{\theta}_I$ (virtual) at node α, and N is the number of nodes in the element.

The element stiffness matrix can now be derived from the virtual work equation (eq. 4.11), in which the first integral can be expanded as

$$\int_A \hat{\gamma}_I q_I - \hat{\kappa}_{IJ} m_{IJ}\,dA$$

$$= \int_A (\hat{\theta}_I^\alpha H'_{\alpha,J})D_{IJKL}(\theta_K^\beta H'_{\beta,L}) + (\hat{w}^\alpha H_{\alpha,I} - H'_\alpha\hat{\theta}_I^\alpha)C_{IK}(w^\beta H_{\beta,K} - H'_\beta\theta_K^\beta)\,dA$$

$$= \int_A \hat{\theta}_I^\alpha \left[(D_{IJKL} H'_{\alpha,J} H'_{\beta,L} + C_{IK} H'_\alpha H'_\beta) \theta_K^\beta - (C_{IK} H'_\alpha H_{\beta,K}) w^\beta \right]$$

$$+ \hat{w}^\alpha \left[(C_{IK} H_{\alpha,I} H_{\beta,K}) w^\beta - (C_{IK} H_{\alpha,I} H'_\beta) \theta_K^\beta \right] \mathrm{d}A$$

$$= \hat{\theta}_I^\alpha \left[A_{\alpha\beta}^{IK} \theta_K^\beta - A_{\alpha\beta}^I w^\beta \right] + \hat{w}^\alpha \left[B_{\alpha\beta} w^\beta - B_{\alpha\beta}^K \theta_K^\beta \right] \tag{4.23}$$

where $$A_{\alpha\beta}^{IK} = \int_A D_{IJKL} H'_{\alpha,J} H'_{\beta,L} + C_{IK} H'_\alpha H'_\beta \,\mathrm{d}A \tag{4.24}$$

$$A_{\alpha\beta}^I = \int_A C_{IK} H'_\alpha H_{\beta,K} \,\mathrm{d}A \tag{4.25}$$

$$B_{\alpha\beta} = \int_A C_{IK} H_{\alpha,I} H_{\beta,K} \,\mathrm{d}A \tag{4.26}$$

$$B_{\alpha\beta}^K = \int_A C_{IK} H_{\alpha,I} H'_\beta \,\mathrm{d}A \tag{4.27}$$

and the second and third integrals can be rewritten respectively as

1. $$\int_A \hat{w} F \,\mathrm{d}A + \int_{\partial A} \hat{w} \overline{q} \,\mathrm{d}s = \int_A H_\alpha \hat{w}^\alpha F \,\mathrm{d}A + \int_{\partial A} H_\alpha \hat{w}^\alpha \overline{q} \,\mathrm{d}s = Q_\alpha \hat{w}^\alpha \tag{4.28}$$

   where the shear at node α, $$Q_\alpha = \int_A H_\alpha F \,\mathrm{d}A + \int_{\partial A} H_\alpha \overline{q} \,\mathrm{d}s \tag{4.29}$$

2. $$\int_A \hat{\theta}_I C_I \,\mathrm{d}A + \int_{\partial A} \hat{\theta}_I \overline{m}_I \,\mathrm{d}s = \int_A H'_\alpha \hat{\theta}_I^\alpha C_I \,\mathrm{d}A + \int_{\partial A} H'_\alpha \hat{\theta}_I^\alpha \overline{m}_I \,\mathrm{d}s = M_\alpha^I \hat{\theta}_I^\alpha \tag{4.30}$$

   where the moment at node α, $$M_\alpha^I = -\left(\int_A H'_\alpha C_I \,\mathrm{d}A + \int_{\partial A} H'_\alpha \overline{m}_I \,\mathrm{d}s \right) \tag{4.31}$$

Using the results of eqs 4.23 to 4.31, eq. 4.11 can be written as:

$$\hat{\theta}_I^\alpha \left[A_{\alpha\beta}^{IK} \theta_K^\beta - A_{\alpha\beta}^I w^\beta \right] + \hat{w}^\alpha \left[B_{\alpha\beta} w^\beta - B_{\alpha\beta}^K \theta_K^\beta \right] = Q_\alpha \hat{w}^\alpha + M_\alpha^I \hat{\theta}_I^\alpha \tag{4.32}$$

Since $\hat{\theta}_I^\alpha$ and $\hat{w}^\alpha$ may take on arbitrary values, we must have

$$A_{\alpha\beta}^{IK} \theta_K^\beta - A_{\alpha\beta}^I w^\beta = M_\alpha^I \qquad \text{and} \qquad B_{\alpha\beta} w^\beta - B_{\alpha\beta}^K \theta_K^\beta = Q_\alpha \tag{4.33}$$

In matrix form, eq. 4.33 can be written as

$$\begin{bmatrix} B_{\alpha\beta} & -B_{\alpha\beta}^K \\ -A_{\alpha\beta}^I & A_{\alpha\beta}^{IK} \end{bmatrix} \begin{bmatrix} w^\beta \\ \theta_K^\beta \end{bmatrix} = \begin{bmatrix} Q_\alpha \\ M_\alpha^I \end{bmatrix} \tag{4.34}$$

$$\Rightarrow \quad \underset{3\times 3}{\mathbf{K}_{\alpha\beta}} \begin{bmatrix} w \\ \theta_1 \\ \theta_2 \end{bmatrix}_\beta = \begin{bmatrix} Q \\ M_1 \\ M_2 \end{bmatrix}_\alpha \tag{4.35}$$

where sub-matrix $\mathbf{K}_{\alpha\beta}$ relates the displacements at node β with the forces at node α of an element. The bending part and shear part of element sub-matrix $\mathbf{K}_{\alpha\beta}$ can be separated such that:

$$\mathbf{K}_{\alpha\beta} = \mathbf{K}_{\alpha\beta}^b + \mathbf{K}_{\alpha\beta}^s \tag{4.36}$$

where bending stiffness, $$\underset{3\times 3}{\mathbf{K}^{b}_{\alpha\beta}} = \begin{bmatrix} 0 & 0 \\ 0 & \underset{2\times 2}{\displaystyle\int_A D_{IJKL} H'_{\alpha,J} H'_{\beta,L}\, \mathrm{d}A} \end{bmatrix} \tag{4.37}$$

shear stiffness, $$\underset{3\times 3}{\mathbf{K}^{s}_{\alpha\beta}} = \begin{bmatrix} B_{\alpha\beta} & -B^{K}_{\alpha\beta} \\ -A^{I}_{\alpha\beta} & \underset{2\times 2}{\displaystyle\int_A C_{IK} H'_{\alpha} H'_{\beta}\, \mathrm{d}A} \end{bmatrix} \tag{4.38}$$

By direct index substitution, we obtain

$$\underset{3\times 3}{\mathbf{K}^{s}_{\alpha\beta}} = \int_A \left[\begin{array}{c|c|c} 0 & 0 & 0 \\ \hline 0 & \begin{array}{c} D_{11}H'_{\alpha,1}H'_{\beta,1} + D_{31}H'_{\alpha,2}H'_{\beta,1} \\ + D_{13}H'_{\alpha,1}H'_{\beta,2} + D_{33}H'_{\alpha,2}H'_{\beta,2} \end{array} & \begin{array}{c} D_{12}H'_{\alpha,1}H'_{\beta,2} + D_{32}H'_{\alpha,2}H'_{\beta,2} \\ D_{13}H'_{\alpha,1}H'_{\beta,1} + D_{33}H'_{\alpha,2}H'_{\beta,1} \end{array} \\ \hline 0 & \begin{array}{c} D_{21}H'_{\alpha,2}H'_{\beta,1} + D_{31}H'_{\alpha,1}H'_{\beta,1} \\ + D_{23}H'_{\alpha,2}H'_{\beta,2} + D_{33}H'_{\alpha,1}H'_{\beta,2} \end{array} & \begin{array}{c} D_{22}H'_{\alpha,2}H'_{\beta,2} + D_{32}H'_{\alpha,1}H'_{\beta,2} \\ D_{23}H'_{\alpha,2}H'_{\beta,1} + D_{33}H'_{\alpha,1}H'_{\beta,1} \end{array} \end{array}\right] \mathrm{d}A \tag{4.39}$$

$$\underset{3\times 3}{\mathbf{K}^{s}_{\alpha\beta}} = \int_A \left[\begin{array}{c|c|c} \begin{array}{c} C_{11}H_{\alpha,1}H_{\beta,1} + C_{21}H_{\alpha,2}H_{\beta,1} \\ + C_{12}H_{\alpha,1}H_{\beta,2} + C_{22}H_{\alpha,2}H_{\beta,2} \end{array} & \begin{array}{c} -C_{11}H_{\alpha,1}H'_{\beta} \\ -C_{21}H_{\alpha,2}H'_{\beta} \end{array} & \begin{array}{c} -C_{12}H_{\alpha,1}H'_{\beta} \\ -C_{22}H_{\alpha,2}H'_{\beta} \end{array} \\ \hline -C_{11}H_{\beta,1}H'_{\alpha} - C_{12}H_{\beta,2}H'_{\alpha} & C_{11}H'_{\alpha}H'_{\beta} & C_{12}H'_{\alpha}H'_{\beta} \\ \hline -C_{21}H_{\beta,1}H'_{\alpha} - C_{22}H_{\beta,2}H'_{\alpha} & C_{21}H'_{\alpha}H'_{\beta} & C_{22}H'_{\alpha}H'_{\beta} \end{array}\right] \mathrm{d}A \tag{4.40}$$

4.2.6 Isotropic plate

The constitutive matrices for isotropic plates are given by:

$$\underset{3\times 3}{\mathbf{D}} = \frac{Eh^3}{12(1-\nu^2)} \begin{bmatrix} 1 & \nu & 0 \\ \nu & 1 & 0 \\ 0 & 0 & \dfrac{1-\nu}{2} \end{bmatrix} = \begin{bmatrix} \lambda_1 & \lambda_2 & 0 \\ \lambda_2 & \lambda_1 & 0 \\ 0 & 0 & \lambda_3 \end{bmatrix} \qquad \mathbf{C} = \begin{bmatrix} C & 0 \\ 0 & C \end{bmatrix} \tag{4.41}$$

where $\lambda_1 = \dfrac{Eh^3}{12(1-\nu^2)}$ $\quad \lambda_2 = \nu\lambda_1 \quad \lambda_3 = \frac{1}{2}(1-\nu)\lambda_1$

$C = \mathfrak{K}\mu h \quad \mu = \dfrac{E}{2(1+\nu)}$ $\quad$ shear factor, $\mathfrak{K} = \frac{5}{6}$

$$\underset{3\times 3}{\mathbf{K}^{b}_{\alpha\beta}} = \int_A \left[\begin{array}{c|c|c} 0 & 0 & 0 \\ \hline 0 & \lambda_1 H'_{\alpha,1}H'_{\beta,1} + \lambda_3 H'_{\alpha,2}H'_{\beta,2} & \lambda_2 H'_{\alpha,1}H'_{\beta,2} + \lambda_3 H'_{\alpha,2}H'_{\beta,1} \\ 0 & \lambda_2 H'_{\alpha,2}H'_{\beta,1} + \lambda_3 H'_{\alpha,1}H'_{\beta,2} & \lambda_1 H'_{\alpha,2}H'_{\beta,2} + \lambda_3 H'_{\alpha,1}H'_{\beta,1} \end{array}\right] \mathrm{d}A \tag{4.42}$$

$$\underset{3\times 3}{\mathbf{K}^{s}_{\alpha\beta}} = \int_A \left[\begin{array}{c|c|c} CH_{\alpha,1}H_{\beta,1} + CH_{\alpha,2}H_{\beta,2} & -CH_{\alpha,1}H'_{\beta} & -CH_{\alpha,2}H'_{\beta} \\ \hline -CH_{\beta,1}H'_{\alpha} & CH'_{\alpha}H'_{\beta} & 0 \\ \hline -CH_{\beta,2}H'_{\alpha} & 0 & CH'_{\alpha}H'_{\beta} \end{array}\right] \mathrm{d}A \tag{4.43}$$

4.2.7 Convergence criterion for Reissner–Mindlin plate element

As the element size decreases, the solution will converge only to the 'exact' solution of the approximate model implied by the formulation as constrained by the basic assumption of the plate theory. For this to happen, the necessary conditions are:

1. All three rigid body modes must be exactly representable.
2. The following constant strain states must be exactly reproduced.
   Curvatures: $\theta_{1,1}, \theta_{2,2}, \theta_{1,2} + \theta_{2,1}$
   Transverse shear strains: $\gamma_1 = w_{,1} - \theta_1, \gamma_2 = w_{,2} - \theta_2$

4.2.8 Boundary conditions

The boundary conditions for Reissner–Mindlin plates may not always be imposed in the same way as those for the classical thin plate theory [24]. The differences occur in the specification of the 'simply supported' case. For Reissner–Mindlin plates, there are two ways of specifying the simply supported boundary condition, depending on the actual physical situation. Rather than being an additional complication, this freedom turns out to be a considerable advantage, for it enables the solution of problems in which thin plate finite elements have failed, such as those reported in Rossow [25] and Scott [26]. Consider a smooth portion of the plate boundary with a local (n, s) coordinate system, in which s denotes the tangential direction and n the outward normal direction, as shown in Fig. 4.3.

The common boundary conditions encountered in practice are:

1. *Clamped edge*
   $w = 0 \qquad \theta_s = 0 \qquad \theta_n = 0$
2. *Free end*
   $q = 0 \qquad m_s = 0 \qquad m_n = 0$
3. *Simply supported*
   $SS_1 : w = 0 \qquad m_s = 0 \qquad m_n = 0$
   $SS_2 : w = 0 \qquad \theta_s = 0 \qquad m_n = 0$

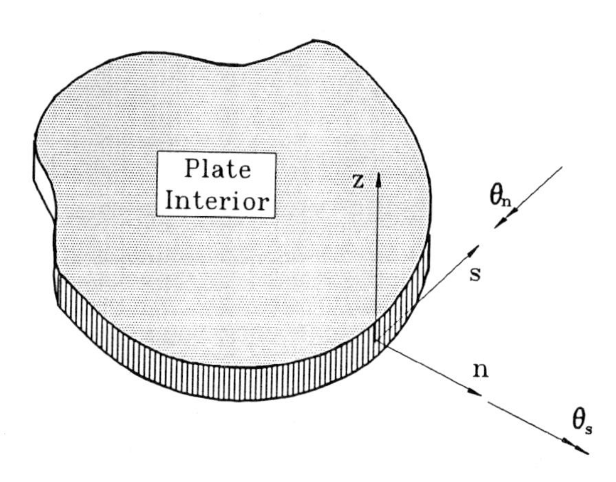

Fig. 4.3 Local tangent–normal coordinate system at plate boundary

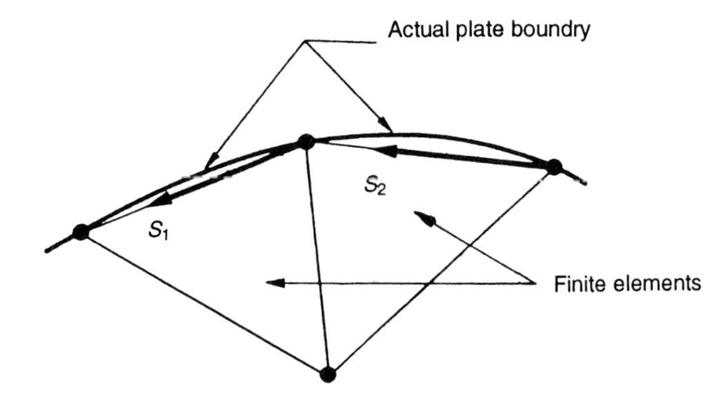

Fig. 4.4 Approximation of curved boundary by straight-edge finite elements

4. *Symmetric*
   $q = 0 \qquad m_s = 0 \qquad \theta_n = 0$
5. *Skew-symmetric (same as SS_2)*
   $w = 0 \qquad \theta_s = 0 \qquad m_n = 0$

In thin plate theory, SS_2 is the appropriate simply supported boundary condition, since $w = 0$ along the boundary necessitates $\partial w/\partial s = 0$, and the absence of shear strain implies $\theta_s = \partial w/\partial s$. Along a curved boundary, where simply supported plates are approximated by straight edges as shown in Fig. 4.4, SS_2 leads to difficulties. In this case, specifying $\partial w/\partial s = 0$ at an interelement boundary, for which s is not colinear implies $\partial w/\partial n = 0$ as well. As the mesh is refined, the clamped boundary condition is achieved. In other words, the correct solution to the wrong problem is attained. Strategies for circumventing this paradox tend to be cumbersome from the implementation point of view.

Another difficulty concerns the solution of a uniformly loaded, simply supported, rhombic plate [25, 89]. The analytical solution is singular in the vicinity of the obtuse vertices, with moments of opposite sign. However, Sander [27] has reported solutions in which thin plate elements yield moments of the same sign. This phenomenon can be attributed to an over-constraining of thin plate elements caused by SS_2. A pleasant feature of the Reissner–Mindlin theory is that the alternative simply supported boundary condition SS_1 is available, which completely obviates these paradoxes. However, in cases where there is no danger of over-constraining, such as simply supported rectangular plates, SS_2 may be employed to eliminate degrees of freedom for greater economy of solution.

4.2.9 Shear locking

Despite its mathematical elegance, application of the Reissner–Mindlin theory to the formulation of plate elements is not free from troubles. A common problem of the standard isoparametric C^0 displacement element is the well-known shear locking problem, especially when low-order interpolation functions are employed for thin plates [17]. Such a phenomenon is primarily due to the elements' inability to model Kirchhoff-type bending modes in the thin plate limits. Hence, it results in a magnification of errors in the shear energy term by a factor proportional to the square of the plate element aspect ratio (length/thickness). Shear locking is then manifested in displacement predictions that are grossly in error, and the resulting stresses show a marked oscillatory behaviour [5, 28].

4.2.10 Uniform/selective reduced integration

An approach to solving the locking problem in thin plates is the use of reduced integration [16, 29]. The two obvious possibilities are 'uniform reduced integration' and 'selective reduced integration'. In uniform reduced integration, both the bending and shear terms are integrated with the same rule, which is of lower order than the *normal* one. However, in the selective reduced integration, a low order quadrature formula is used for evaluation of the over-constrained or troublesome transverse shear components of the element stiffness matrix, whereas a normal quadrature formula is employed for the bending terms.

One of the early examples of a uniform reduced integration plate element was the 8-node Serendipity element proposed by Zienkiewicz *et al.* [15], in which 2×2 Gauss quadrature was employed. Although this element has been widely used, it has been shown to give poor results in the thin plate limit [30, 31]. Nevertheless, reduced integration still represents a considerable improvement over normal integration. Common integration rules for Lagrangian plate bending elements are depicted

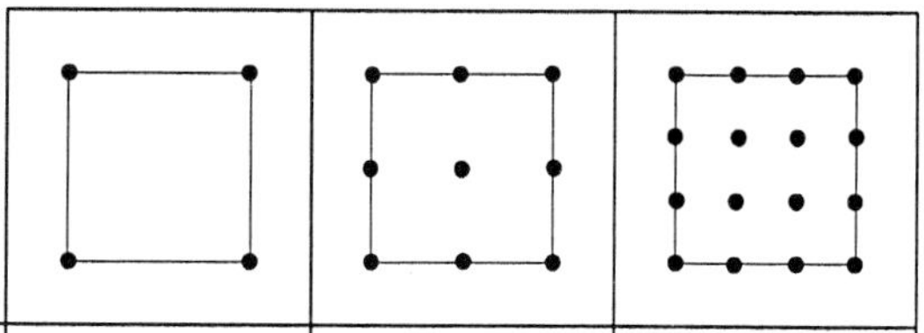

	(4-node element)	(9-node element)	(16-node element)
w, θ_1, θ_2 interpolation functions	Bilinear	Biquadratic	Bicubic
Uniform reduced integration	1×1 $U1$	2×2 $U2$	3×3 $U3$
Selective reduced integration	1×1 shear 2×2 bending $S1$	2×2 shear 3×3 bending $S2$	3×3 shear 4×4 bending $S3$

Fig. 4.5 Lagrangian plate elements; 3 DOF per node – w, θ_1, θ_2

in Fig. 4.5. These elements have been studied by, among others, Parisch [5], Hughes and Cohen [32], and Belytschko and Tsay [33].

4.2.11 Equivalence of reduced integration and mixed formulation

The selective reduced integration Lagrangian plate elements have been shown to be identical to elements derived from a mixed formulation (Malkus and Hughes [34]), in which in addition to w and θ_I, shear forces q_I are also included as the primary variables. The shear variables, interpolated from nodal values at Gauss points of shear integration schemes, are discontinuous between elements. However, to date no thorough robust elements have been developed based on the mixed formulation alone, and various schemes based on intuitive ideas with a physical rather than mathematical basis have been used to produce more reliable elements [35–37].

4.2.12 Rank deficiency

Although improved numerical results are obtained for thin plates, the uniform/selective reduced integration approach is not trouble-free. In particular, application of reduced integration schemes will engender spurious singular modes as reflected by a rank deficiency in the element stiffness matrix [18, 28, 29].

In Table 4.1, the number of zero-energy modes in excess of the three rigid body modes is shown for Lagrangian plate elements evaluated by reduced integration.

Table 4.1 Zero-energy modes in excess of rigid body modes for reduced integrated Lagrangian elements

Integration scheme	U_1	U_2	U_3	S_1	S_2	S_3
Number of zero-energy modes	4	4	4	2	1	1

In Table 4.1, the number of spurious singular modes is smaller for selective reduced integration than for the corresponding uniform reduced integration elements, and is therefore an advantage of the former. The patch test with a count on the number of parameters, introduced by Zienkiewicz and Taylor [17], provides the necessary condition that an element is singular or not. Elements with zero-energy modes may yield unreliable results, oscillatory errors in some cases, and even a singular global stiffness matrix under certain boundary conditions.

4.3 Simple plate bending elements with straight edges

4.3.1 A correct rank 4-node quadrilateral element, Q_1

It has been stated in Section 4.2.12 that some element stiffness matrices suffer from a rank deficiency. In this section we will introduce a correct rank Hughes and Tezduyar [38] element, Q_1, in which a special interpolation for the transverse shear strain energy is used. The geometric and kinematic descriptions of the element are shown in Fig. 4.6, where $\mathbf{a}_\alpha$ and $\mathbf{b}_\alpha$ are unit vectors. The transverse displacement

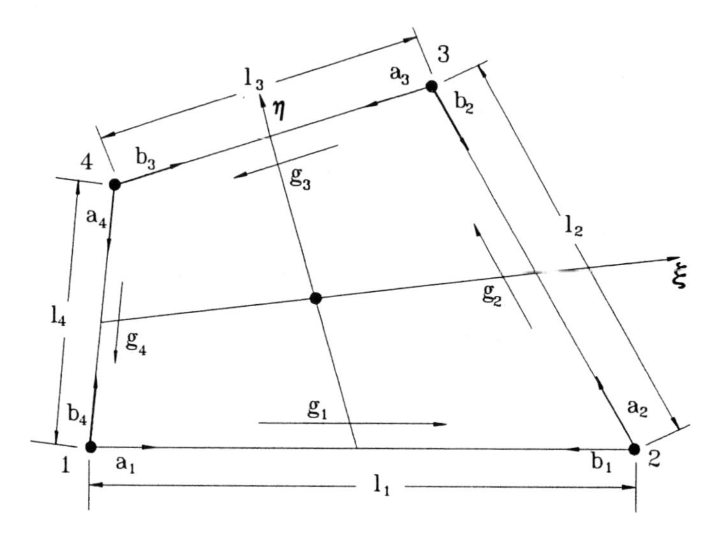

Fig. 4.6 Geometric and kinematic data for quadrilateral element Q_1

and rotation vector at node α are denoted by w_α and θ_α respectively. Throughout, subscript β equals $\alpha + 1$ modulus 4, i.e.

α	1	2	3	4
β	2	3	4	1

A more detailed description of the definition for the element shear strain would be appropriate, and the relevant points are as follows:

1. On each element side, a shear strain component g_α, located at the midpoint and in a direction parallel to the side, is given by

$$g_\alpha = \frac{w_\beta - w_\alpha}{l_\alpha} - \mathbf{a}_\alpha \cdot \frac{\theta_\alpha + \theta_\beta}{2} \qquad \alpha = 1, 4 \tag{4.44}$$

2. A shear strain vector, γ_β, is defined at each node β as

$$\left.\begin{aligned} \boldsymbol{\gamma}_\beta &= \lambda_\beta \mathbf{a}_\beta + \mu_\beta \mathbf{b}_\beta \\ \text{where} \quad \mu_\beta &= (g_\beta + g_\alpha c_\beta)/(1 - c_\beta^2) \\ \lambda_\beta &= -(g_\alpha + g_\beta c_\beta)/(1 - c_\beta^2) \\ c_\beta &= \mathbf{a}_\beta \cdot \mathbf{b}_\beta \end{aligned}\right\} \quad \text{no sum on } \beta \tag{4.45}$$

3. The shear strain can finally be expressed in terms of the nodal values using the bilinear interpolation functions, H_α, such that

$$\gamma = H_\alpha \gamma_\alpha \tag{4.46}$$

It should be noted that constant transverse shear deformation modes are represented exactly for general quadrilateral geometry. In other words, if the nodal transverse displacements and rotations are specified to give a constant transverse shear strain field, $\gamma = \gamma_0$, then by going through the steps specified above, a constant strain field, $\gamma = \gamma_0$, will indeed be obtained.

Now we come to the stiffness matrix of the element Q_1. All aspects of the stiffness formulation are identical to those of the standard 4-node bilinear quadrilateral element except for the shear stiffness matrix $\mathbf{K}^{s}_{\alpha\beta}$, which unlike that given in eq. 4.38, is now represented by the standard form

$$\mathbf{K}^{s}_{\alpha\beta} = \int_A \mathbf{B}_\alpha^T \mathbf{C} \mathbf{B}_\beta \, dA \tag{4.47}$$

where the strain matrix is given by

$$\underset{2\times 3}{\mathbf{B}_\alpha} = [B_\alpha^1 B_\alpha^2 B_\alpha^3]$$

and the property matrix is given by

$$\underset{2\times 2}{\mathbf{C}} = \begin{bmatrix} C_{11} & C_{12} \\ C_{21} & C_{22} \end{bmatrix}$$

with

$$\begin{aligned} B_\alpha^1 &= \frac{G_\alpha}{l_\alpha} - \frac{G_\beta}{l_\beta} \\ B_\alpha^2 &= \tfrac{1}{2}(b_{\beta 1} G_\alpha - a_{\beta 1} G_\beta) \\ B_\alpha^3 &= \tfrac{1}{2}(b_{\beta 2} G_\alpha - a_{\beta 2} G_\beta) \end{aligned} \tag{4.48}$$

and

$$G_\alpha = \frac{H_\alpha(\mathbf{a}_\alpha - c_\alpha \mathbf{b}_\alpha)}{1 - c_\alpha^2} - \frac{H_\beta(\mathbf{b}_\beta - c_\beta \mathbf{a}_\beta)}{1 - c_\beta^2} \quad \text{(no sum on } \alpha \text{ or } \beta) \tag{4.49}$$

where $(a_{\beta 1}, a_{\beta 2}) = a_\beta$ and $(b_{\beta 1}, b_{\beta 2}) = b_\beta$.

Two by two Gaussian quadrature is used to integrate all element contributions.

4.3.2 The linear triangular element, T_1

Three-node triangular elements employing linear shape functions for w and θ_I exhibit severe locking under normal circumstances. The technique used in Section 4.3.1 to define the transverse shear strain field for Q_1 can be used to alleviate locking while maintaining the correct rank for a triangular element [39]. The formulation follows closely through eqs 4.44–4.49. The only changes involve the nodal indices which are defined, in this case, by the relation

α	1	2	3
β	2	3	1

Figure 4.7 shows the element geometric and kinematic quantities. Implementation follows along the same lines as for Q_1. Although one-point centroid quadrature results in rank deficiency for a single element, this is not a practical detriment since rank deficiency disappears in an assembly of two or more elements. Due to the greater economy of one-point quadrature, it seems to be the obvious choice for practical applications.

4.3.3 The discrete Kirchhoff approach

By relaxing the C^1 continuity requirement, thin plate elements are developed by insisting that the Kirchhoff hypothesis of zero transverse shear strains be satisfied

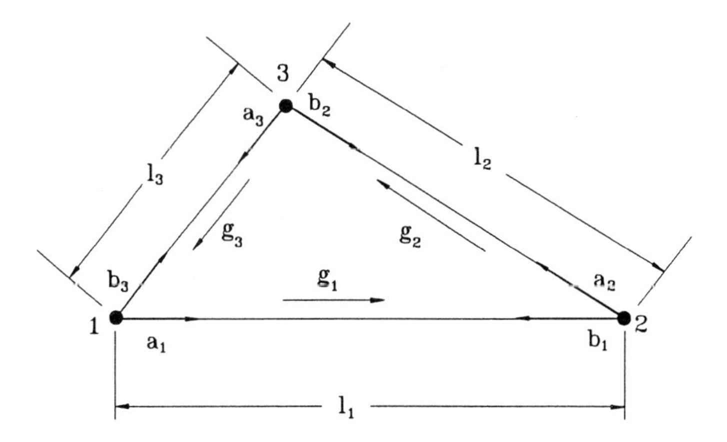

Fig. 4.7 Three-node triangular element T_1

only at a discrete number of points within the element [40, 41]. To illustrate the idea, let us consider the development of the 12 DOF, 4-node quadrilateral element DKQ (Discrete Kirchhoff Quadrilateral) [42,43] as shown in Fig. 4.8.

Some of the salient features are given below:

1. The domain of the element is a straight-edged quadrilateral. Rotations are defined by

$$\theta = H_\alpha \theta_\alpha \tag{4.50}$$

   where H_α are the 8-node Serendipity interpolation functions and $\theta_\alpha = (\theta_1, \theta_2)_\alpha$ is the rotation vector at node α. Hence we begin with 16 rotational degrees of freedom, in which the mid-side degrees of freedom are eventually eliminated.
2. Let **n** and **s** be the normal and tangent unit vectors along an edge. The normal component of θ is required to vary linearly along each side, i.e.

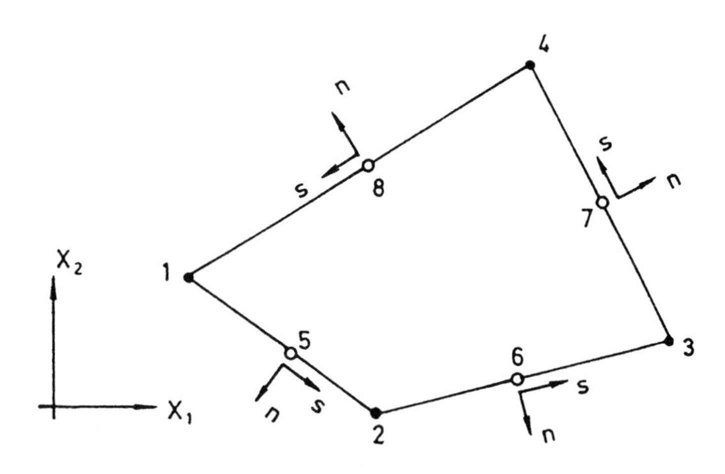

Fig. 4.8 Discrete Kirchhoff quadrilateral, DKQ

$$\mathbf{n} \cdot \theta_m = \tfrac{1}{2}\mathbf{n} \cdot (\theta_\alpha + \theta_\beta) \tag{4.51}$$

where $m = \alpha + 4$, $\alpha = 1, 2, 3, 4$, and $\beta = 2, 3, 4, 1$.

This linear variation requirement amounts to 4 constraints.

3. The transverse displacement is defined only at the element boundary, and is assumed to vary cubically along each edge according to the Hermite interpolation. This requires three nodal values of w and $w_{,I}(I = 1, 2)$, or a total of 12 more degrees of freedom, making a grand total of 28 unknowns.
4. The Kirchhoff constraints are imposed as follows:

   (i) At corner nodes: $w_{,I} = \theta_I$ (i.e. $\gamma_I = 0$) (4.52)

   (ii) At mid-side nodes: $w_{,s} = \mathbf{s} \cdot \theta_m$ (i.e. $\gamma_s = 0$) (4.53)

   Equation 4.52 represents 8 constraints and eq. 4.53 yields 4 more. These 12 Kirchhoff constraints and the 4 normal rotation constraints under (2) above enable a reduction of the unknowns to 12 DOF, namely the corner node displacements and rotations.
5. The transverse shear terms are simply ignored in the formulation. Thus the bending stiffness is completely defined by the derivatives of the rotation interpolations.

The following remarks can be made about the DKQ element:

1. As the tangential shear strain component is set to zero at three distinct points along each edge, it vanishes identically along the element boundary.
2. The 2×2 Gaussian quadrature is good enough to maintain the correct rank, but the 3 × 3 rule is required if the element stiffness is to be integrated exactly in the rectangular configuration. The former is recommended in practice because of its greater economy.
3. The transverse displacement is not defined at the interior of the element. This creates ambiguities in the correct definition of consistent transverse applied forces and element inertial properties in dynamics. Two *ad hoc* possibilities using a bilinear variation and an incomplete quartic scheme are studied. Both methods seem to perform adequately in practice.
4. The convergence to thin plate solution is to be expected because the transverse shear stiffness is neglected and the Kirchhoff hypothesis is satisfied tangentially along the element boundary.
5. A discrete Kirchhoff triangular element also exists. A triangular element DKT employing virtually identical concepts was the predecessor of DKQ [44, 45].

4.4 Some discussion on plate bending elements

In addition to the discrete Kirchhoff constraints, elements have been derived from the concept of *integral Kirchhoff constraints* (e.g. area or boundary mean shear strain may be required to vanish). Most prominent among these developments are perhaps Irons' Semiloof [46] and Lyons' Isoflex [47] elements. Crisfield [48] has developed a modified version of Lyons' 4-node element in which the Kirchhoff constraints are given in explicit algebraic form, thus avoiding cumbersome calculations at the element level. Plate bending elements based on the generalized concept of discrete collocation constraints can be found from the works of Hughes and Tezduyar [38], Bathe and Dvorkin [49], Hinton and Huang [31], and others [34, 41].

As an alternative to displacement models, multi-field variational principles can be employed for the development of both thin and thick plate elements [50–52]. The formulation involves assumed equilibrating stresses within each element and compatible displacements along interelement boundaries [53]. Since the pointwise equilibrium requirement is generally difficult to satisfy, the *hybrid/mixed method* was proposed by Pian *et al.* [54, 55], in which the equilibrium condition was initially relaxed, and subsequently enforced through constraint equations by using Lagrangian multipliers in the Hellinger–Reissner principle. One of the major drawbacks of the hybrid/mixed approach is the lack of specific rules in choosing stress/strain parameters. The selection was largely based on a trial-and-error procedure in conjunction with numerical tests to determine the *best* choice [56,57]. Unfortunately, the numerical performance of the element so formulated may be unreliable and problem-dependent. Discussions of the hybrid/mixed approach have also been given to separate specific situations – the conditions for suppressing singular modes [55], the suggestion of using *simple* assumed stress/strain fields to abate locking [58,59], frame-invariance conditions [60], investigation of the effects of various stress and strain assumptions on the behaviour of a particular element, etc. A need therefore exists for a critical assessment and synthesis of the concepts, procedures and criteria proposed in an attempt to develop a general framework for the selection of stress/strain parameters in the hybrid/mixed approaches.

4.5 PLATE BENDING ELEMENT L9P

In this section, the 9-node biquadratic Lagrangian plate bending element L9P will be presented in detail. The reference domain (a 2×2 square) and the mapping of the element L9P are shown in Fig. 4.9. The interpolation functions H_α and their derivatives with respect to ξ and η of the element L9P are given in Table 4.2.

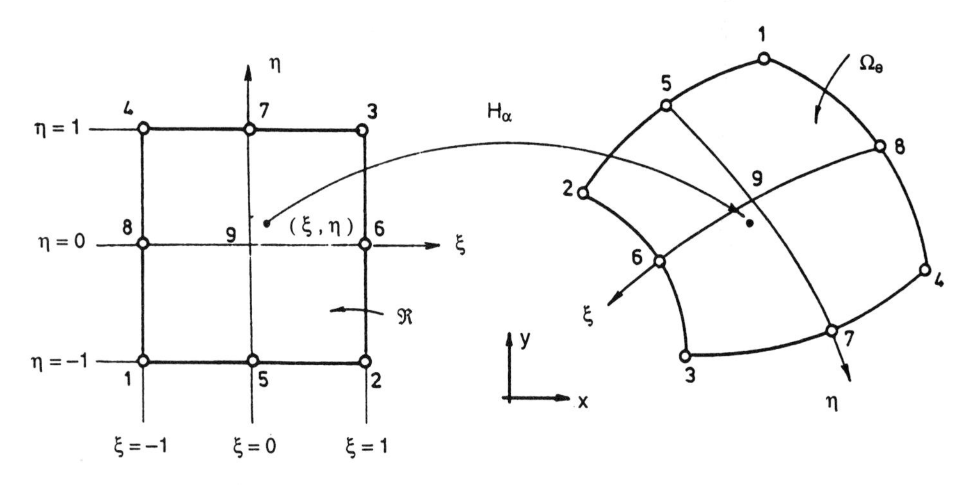

Fig. 4.9 Reference domain and mapping of the L9P element

Table 4.2 The interpolation functions H_α and their derivatives with respect to ξ and η of the L9P element

α	H_α	$\dfrac{\partial H_\alpha}{\partial \xi}$	$\dfrac{\partial H_\alpha}{\partial \eta}$
1	$f_1 g_1$	$f'_1 g_1$	$f_1 g'_1$
2	$f_3 g_1$	$f'_3 g_1$	$f_3 g'_1$
3	$f_3 g_3$	$f'_3 g_3$	$f_3 g'_3$
4	$f_1 g_3$	$f'_1 g_3$	$f_1 g'_3$
5	$f_2 g_1$	$f'_2 g_1$	$f_2 g'_1$
6	$f_3 g_2$	$f'_3 g_2$	$f_3 g'_2$
7	$f_2 g_3$	$f'_2 g_3$	$f_2 g'_3$
8	$f_1 g_2$	$f'_1 g_2$	$f_1 g'_2$
9	$f_2 g_2$	$f'_2 g_2$	$f_2 g'_2$

where

$f_1 = \frac{1}{2}\xi(\xi - 1)$ $\quad f'_1 = \xi - \frac{1}{2}$ $\quad g_1 = \frac{1}{2}\eta(\eta - 1)$ $\quad g'_1 = \eta - \frac{1}{2}$

$f_2 = 1 - \xi^2$ $\quad f'_2 = -2\xi$ $\quad g_2 = 1 - \eta^2$ $\quad g'_2 = -2\eta$

$f_3 = \frac{1}{2}\xi(\xi + 1)$ $\quad f'_3 = \xi + \frac{1}{2}$ $\quad g_3 = \frac{1}{2}\eta(\eta + 1)$ $\quad g'_3 = \eta + \frac{1}{2}$

The element stiffness matrix can be evaluated by substituting the interpolation functions of Table 4.2 into eqs 4.36–4.40 and integrated using the selective reduced integration scheme (2×2 for shear and 3×3 for bending). Such a plate bending element L9P for the case of isotropic material (eqs 4.41–4.43) has been developed, and the corresponding computer listing can be found in Section 8.6.

The accuracy and the convergence characteristics of the element L9P for the analysis of thin and thick plates have been studied. The bending of a simply-supported

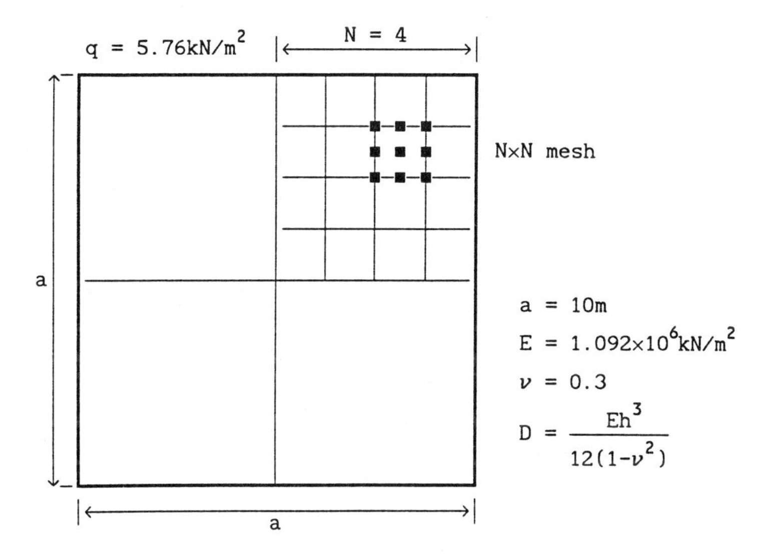

Fig. 4.10 A simply-supported square plate under uniformly distributed load

square plate under a uniformly distributed load is taken as an example. A uniformly distributed load of intensity 5.76 kN/m^2 is applied on a 10 m $\times$ 10 m square plate simply-supported along its boundary as shown in Fig. 4.10. In the finite element analysis, one quarter of the plate is divided into $N \times N$ L9P elements, where N is the number of elements along an edge. The plate is assumed to be isotropic with material constants $E = 1.092 \times 10^6$ kN/m^2 and $\nu = 0.3$. The central deflections δ for plates with thickness h ranging from 0.0001 m to 2.5 m are given in Table 4.3.

From Table 4.3, it is seen that the plate bending element L9P is able to give correct results for plates as thin as $h/a = 10^{-5}$ up to very thick plates with $h/a = 0.25$, where h is the thickness and a is the span of the plate respectively. The convergence is fairly rapid for all the plates with different thickness/span ratios; a very accurate deflection can already be obtained by using a 2 $\times$ 2 mesh. A typical plot of the deflection δ against the number of elements along an edge N for the plate with thickness/span ratio $h/a = 0.01$ is depicted in Fig. 4.11.

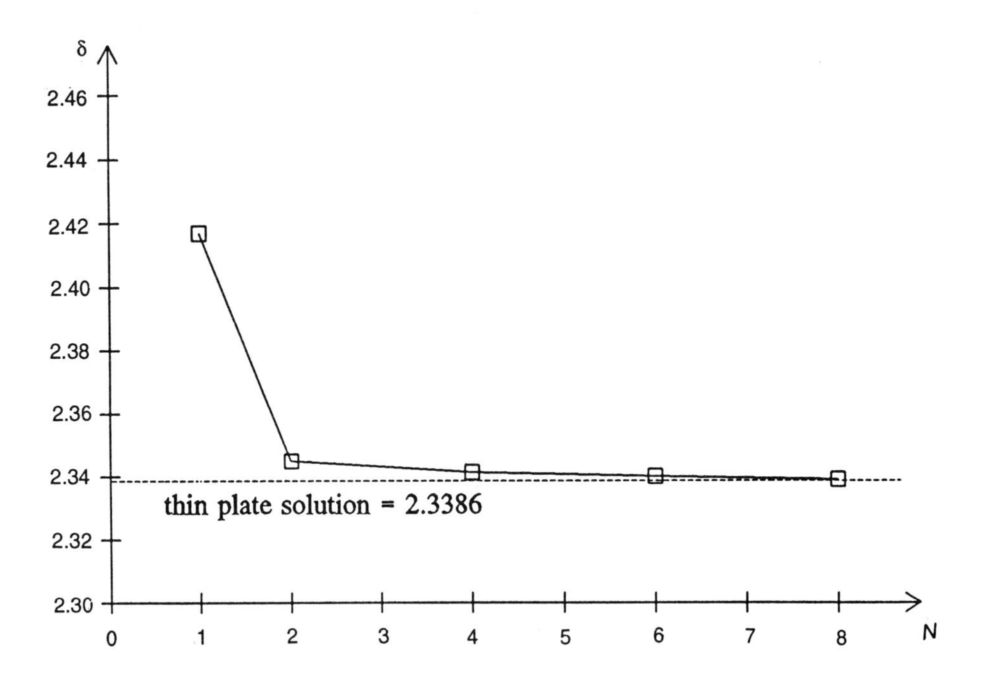

Fig. 4.11 Graph of central deflection δ plotted against N for plate with $h/a = 0.01$

Table 4.3 Central deflection ($10^{-5}qa^4/D$)

h/a	0.25	0.2	0.15	0.1	0.05	0.02	0.01	0.001	0.0001	0.00001
1 × 1	556.361	505.131	467.626	440.828	424.744	420.240	419.595	419.384	419.382	419.382
2 × 2	538.601	491.192	454.317	427.979	412.174	407.750	407.116	406.919	406.906	406.906
4 × 4	537.839	490.476	453.638	427.325	411.536	407.117	406.486	406.276	406.274	406.274
6 × 6	537.549	490.207	453.384	427.082	411.302	406.883	406.252	406.043	406.042	406.042
8 × 8	537.202	489.925	453.155	426.891	411.131	406.719	406.089	405.880	405.878	405.882
Reference 56		490		427		407	406	406	406	406

4.6 INTRODUCTION TO SHELL ELEMENTS

Many different approaches have been developed in finite element shell analysis, and an enormous amount of literature now exists [61–66]. These range from the earlier flat rectangular and triangular shell elements [67, 68], the cylindrical shell element [69] and the doubly curved element [70], to the doubly curved degenerated shell element of recent years. However, despite the intense interest focused on the formulation of shell finite elements for nearly thirty years, there are still dissatisfactions with the present available methodology [71, 72]. In the following sections, we will concentrate on a general formulation for curved shell elements using C^0 interpolations as for the plate bending elements, in which transverse shear deformations are taken into account [73–75]. Due to the space limitation, no attempt is made to review the various schemes.

The degenerated shell approach provides a number of distinct advantages. Firstly, it can treat shells of arbitrary geometries easily without invoking complicated assumptions for a specific shell theory. Secondly, the element can be readily extended to the analysis of non-linear problems [76–78]. This is considered to be an important attribute of this type of approach because practical shell analysis often involves consideration of non-linear effects, e.g. in problems of large amplitude vibration and post-buckling analysis. Thirdly, since the Reissner–Mindlin theory is employed to account for shear deformations, the approach can be applied to both thin and thick shells. Fourthly, only C^0 continuity is required in the approximations of kinematic variables. As a result, lower order interpolation functions can be used, as compared to the Kirchhoff-type elements for which higher order polynomials are necessary to satisfy C^1 continuity. Finally, elements of this type facilitate compatible modelling of shell–solid interfaces or shell-to-shell junctions.

The shell theory presented in this chapter is derived directly from three-dimensional elasticity with certain kinematic and mechanical assumptions built in. Based on a displacement approach, isoparametric shell elements are derived through the application of the virtual work equation. The approach is essentially the same as for the case of plate bending, where Reissner–Mindlin plate elements were also developed from three-dimensional theory.

4.7 A DEGENERATED SHELL ELEMENT

4.7.1 Geometrical definition of the element

Figure 4.12 shows a bi-unit cube being mapped into the physical shell element, in the same way as a square reference domain would be mapped into a distorted quadrilateral element in the two-dimensional analysis. For a fixed value of ζ, the surface defined by $\bar{\mathbf{x}}(\xi, \eta)$ is called a lamina; similarly, for fixed (ξ, η), the line $\tilde{\mathbf{x}}$ is called the fibre. However, the fibres are in general not perpendicular to the lamina.

The geometry of a typical quadrilateral shell element is defined by

$$\mathbf{x}(\xi, \eta, \zeta) = \bar{\mathbf{x}}(\xi, \eta) + \zeta\tilde{\mathbf{x}}(\xi, \eta) \tag{4.54}$$

$$\bar{\mathbf{x}}(\xi, \eta) = G_\alpha(\xi, \eta)\bar{\mathbf{x}}_\alpha \tag{4.55}$$

$$\tilde{\mathbf{x}}(\xi, \eta) = G_\alpha(\xi, \eta)\tilde{\mathbf{x}}_\alpha \tag{4.56}$$

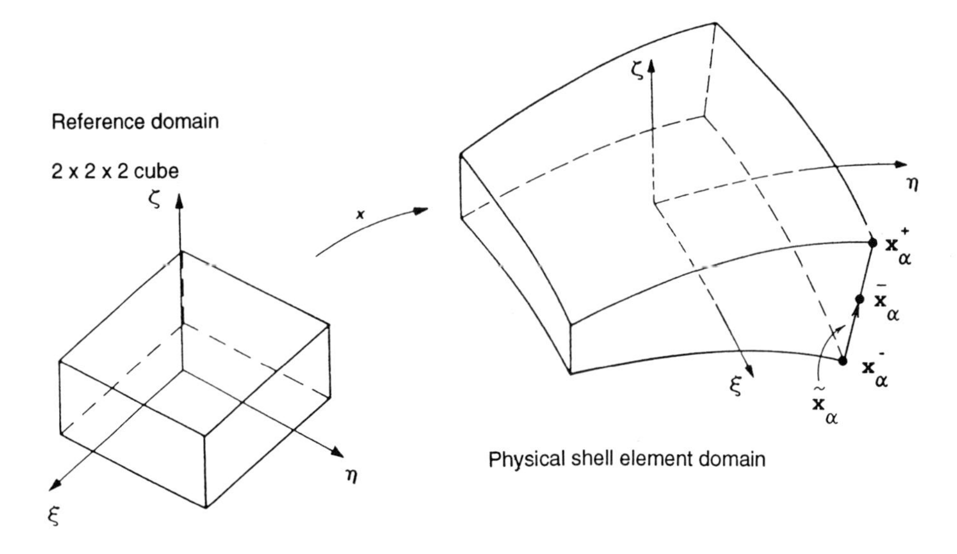

Fig. 4.12 Mapping a bi-unit cube to the physical shell domain

In eqs 4.54–4.56, $\mathbf{x}$ denotes the position vector of a generic point of the shell; $\bar{\mathbf{x}}$ is the position vector of a point on the reference surface; $\tilde{\mathbf{x}}$ is a vector directed from a point on the reference surface representing the *fibre direction*; $\bar{\mathbf{x}}_\alpha$ is the position vector of nodal point α; G_α denotes a two-dimensional shape function associated with node α; N is the number of element nodes; $\tilde{\mathbf{x}}_\alpha$ is a vector emanating from node α in the fibre direction.

For a particular choice of two-dimensional shape functions, the geometry of the shell element is precisely defined upon the specification of $\bar{\mathbf{x}}_\alpha$ and $\tilde{\mathbf{x}}_\alpha$, $\alpha = 1, N$. It is convenient in practice to take as input the coordinates of the top and bottom surface of the shell along each nodal fibre ($\mathbf{x}_\alpha^+$ and $\mathbf{x}_\alpha^-$, so that

$$\bar{\mathbf{x}}_\alpha = \tfrac{1}{2}(\mathbf{x}_\alpha^+ + \mathbf{x}_\alpha^-) \tag{4.57}$$

$$\tilde{\mathbf{x}}_\alpha = \tfrac{1}{2}(\mathbf{x}_\alpha^+ - \mathbf{x}_\alpha^-) \tag{4.58}$$

and the shell thickness at node α, $h_\alpha = 2\|\tilde{\mathbf{x}}_\alpha\|$ (4.59)

4.7.2 Laminar coordinate system

The components in the direction normal to the laminar surface (ξ = constant) are essential if the basic shell assumptions are to be taken into account. Thus, at each integration point of the element, a Cartesian reference frame is constructed in such a way that two axes are tangent to the lamina through the point. The frame is defined by its orthonormal basis vectors $\{\hat{\mathbf{e}}_1, \hat{\mathbf{e}}_2, \hat{\mathbf{e}}_3\}$, in which $\hat{\mathbf{e}}_3$ is perpendicular to the lamina as shown in Fig. 4.13.

Unit tangent vectors along ξ and η coordinate directions are given by

$$\mathbf{e}_\xi = \frac{\partial \mathbf{x}}{\partial \xi} \Big/ \left\| \frac{\partial \mathbf{x}}{\partial \xi} \right\| \qquad \mathbf{e}_\eta = \frac{\partial \mathbf{x}}{\partial \eta} \Big/ \left\| \frac{\partial \mathbf{x}}{\partial \eta} \right\| \tag{4.60}$$

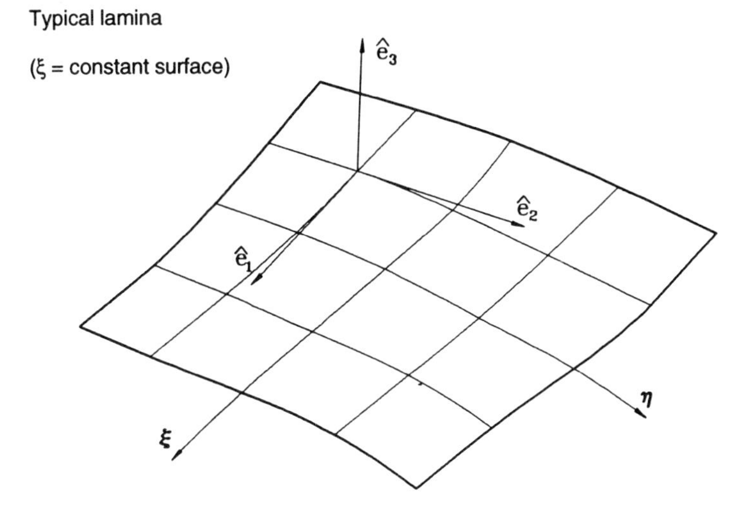

Fig. 4.13 Typical lamina coordinate system

From eq. 4.60, we can define

$$\hat{\mathbf{e}}_3 = \frac{\mathbf{e}_\xi \times \mathbf{e}_\eta}{\|\mathbf{e}_\xi \times \mathbf{e}_\eta\|} \tag{4.61}$$

The remaining orthonormal vectors tangent to the lamina can be defined as

$$\hat{\mathbf{e}}_1 = \frac{\sqrt{2}}{2}(\mathbf{a}_1 - \mathbf{a}_2) \qquad \hat{\mathbf{e}}_2 = \frac{\sqrt{2}}{2}(\mathbf{a}_1 + \mathbf{a}_2) \tag{4.62}$$

where

$$\mathbf{a}_1 = \frac{\mathbf{e}_\xi + \mathbf{e}_\eta}{\|\mathbf{e}_\xi + \mathbf{e}_\eta\|} \qquad \mathbf{a}_2 = \hat{\mathbf{e}}_3 \times \mathbf{a}_1 \tag{4.63}$$

It will be necessary to transform quantities from the global coordinate system to the laminar coordinate system. This is facilitated by the following transformation matrix:

$$\mathbf{R} = [R_{im}] = [\hat{\mathbf{e}}_1, \hat{\mathbf{e}}_2, \hat{\mathbf{e}}_3]^{\mathrm{T}} : \hat{\mathrm{u}}_i = R_{im} u_m \tag{4.64}$$

4.7.3 Fibre coordinate system

The displacement field over the shell element is based on the displacement parameters at the nodal points. Hence, at each nodal point, a unique coordinate system is erected, which is used as a reference frame for rotations. The sole requirement that the frame must satisfy is that one direction coincides with the fibre direction. This in itself is not sufficient to define the frame, since the other two base vectors can still be oriented arbitrarily. However, the following algorithm may be employed to select the basis at a node. Let $\{\tilde{\mathbf{e}}_1, \tilde{\mathbf{e}}_2, \tilde{\mathbf{e}}_3\}$ be the base vectors of the fibre coordinate system, and $\tilde{\mathbf{x}}$ denote the vector in the fibre direction. Then

1. $\tilde{\mathbf{e}}_3 = \tilde{\mathbf{x}}/\|\tilde{\mathbf{x}}\|$ (4.65)
2. Let $\mathbf{a} = \tilde{\mathbf{e}}_3 \times \mathbf{e}_1$ and $\mathbf{b} = \tilde{\mathbf{e}}_3 \times \mathbf{e}_2$ (4.66)
   where $\{\mathbf{e}_1, \mathbf{e}_2, \mathbf{e}_3\}$ is the global coordinate system.

3. Define $\mathbf{c} = \begin{cases} \mathbf{a} & \text{if } \|\mathbf{a}\| \geq \|\mathbf{b}\| \\ \mathbf{b} & \text{otherwise} \end{cases}$ (4.67)
4. $\tilde{\mathbf{e}}_1 = \mathbf{c}/\|\mathbf{c}\|$ and $\tilde{\mathbf{e}}_2 = \tilde{\mathbf{e}}_3 \times \tilde{\mathbf{e}}_1$ (4.68)

It should be noted that for nodes situated at the shell boundary, the fibre coordinate system can be uniquely defined by the vector in the fibre direction and a second vector tangent to the shell boundary.

4.7.4 Displacement field

The displacement of the shell will be defined by the displacement of the mid-surface and the displacement due to the rotations of the fibre. Hence, the displacement of the shell assumes the following form:

$$\mathbf{u}(\xi, \eta, \zeta) = \overline{\mathbf{u}}(\xi, \eta) + \zeta\tilde{\mathbf{u}}(\xi, \eta) \tag{4.69}$$

where

$$\overline{\mathbf{u}}(\xi, \eta) = H_\alpha(\xi, \eta)\overline{\mathbf{u}}_\alpha \tag{4.70}$$

$$\tilde{\mathbf{u}}(\xi, \eta) = H'_\alpha(\xi, \eta)\tilde{\mathbf{u}}_\alpha \tag{4.71}$$

where $\mathbf{u}$ is the displacement of a generic point, $\overline{\mathbf{u}}$ is the displacement of a point on the reference surface, and $\tilde{\mathbf{u}}$ is the fibre displacement. The vector $\tilde{\mathbf{u}}_\alpha$ is constructed such that the fibre may rotate, but may not stretch

$$\tilde{\mathbf{u}}_\alpha = \tfrac{1}{2}h_\alpha(\theta_2^\alpha\tilde{\mathbf{e}}_1^\alpha - \theta_1^\alpha\tilde{\mathbf{e}}_2^\alpha) \qquad \text{(no sum on } \alpha) \tag{4.72}$$

The quantities θ_1^α and θ_2^α represent the rotations of the fibre about the base vectors $\tilde{\mathbf{e}}_1^\alpha$ and $\tilde{\mathbf{e}}_2^\alpha$ respectively. The right-hand-rule sign convention for fibre rotations is illustrated in Fig. 4.14.

4.7.5 Constitutive equation

The stress–strain relationship for the shell element can be deduced from the constitutive equation of three-dimensional elasticity written with respect to the laminar

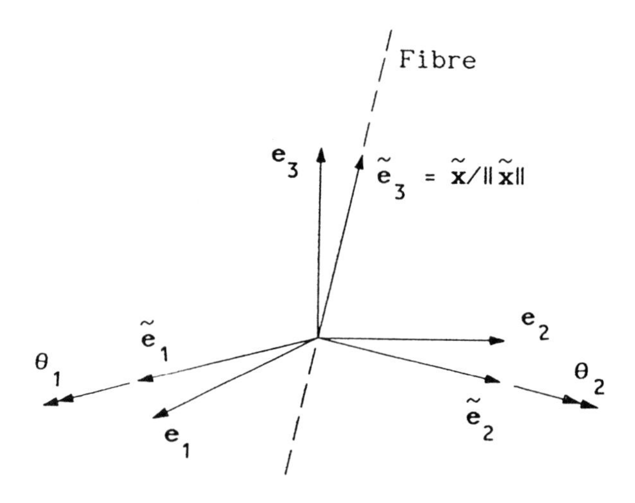

Fig. 4.14 Nodal fibre basis and sign convention for rotation

coordinate system.

$$\hat{\sigma} = \hat{\mathrm{C}}\,\hat{\varepsilon} \tag{4.73}$$

The symbol $\wedge$ is used to emphasize that the components are expressed in terms of the laminar coordinate system. Written in vector form, the stress and strain components are given by

$$\underset{6\times1}{\hat{\sigma}} = \begin{bmatrix} \hat{\sigma}_{11} \\ \hat{\sigma}_{22} \\ \hat{\sigma}_{12} \\ \hat{\sigma}_{23} \\ \hat{\sigma}_{31} \\ \hat{\sigma}_{33} \end{bmatrix} \qquad \underset{6\times1}{\hat{\varepsilon}} = \begin{bmatrix} \hat{\varepsilon}_{11} \\ \hat{\varepsilon}_{22} \\ \hat{\varepsilon}_{12} \\ \hat{\varepsilon}_{23} \\ \hat{\varepsilon}_{31} \\ \hat{\varepsilon}_{33} \end{bmatrix} = \begin{bmatrix} \frac{\partial \hat{\mathrm{u}}_1}{\partial \hat{\mathrm{x}}_1} \\ \frac{\partial \hat{\mathrm{u}}_2}{\partial \hat{\mathrm{x}}_2} \\ \frac{\partial \hat{\mathrm{u}}_1}{\partial \hat{\mathrm{x}}_2} + \frac{\partial \hat{\mathrm{u}}_2}{\partial \hat{\mathrm{x}}_1} \\ \frac{\partial \hat{\mathrm{u}}_2}{\partial \hat{\mathrm{x}}_3} + \frac{\partial \hat{\mathrm{u}}_3}{\partial \hat{\mathrm{x}}_2} \\ \frac{\partial \hat{\mathrm{u}}_3}{\partial \hat{\mathrm{x}}_1} + \frac{\partial \hat{\mathrm{u}}_1}{\partial \hat{\mathrm{x}}_3} \\ \frac{\partial \hat{\mathrm{u}}_3}{\partial \hat{\mathrm{x}}_3} \end{bmatrix} \qquad \underset{6\times6}{\hat{\mathrm{C}}} = [\hat{C}_{ij}] \tag{4.74}$$

The constitutive equation has to be modified in order to enforce the zero normal stress condition along the $\hat{\mathbf{e}}_3$ direction.

$$\hat{\sigma}_{33} = \hat{\sigma}_6 = 0 \quad \Rightarrow \quad \hat{C}_{61}\hat{\varepsilon}_1 + \hat{C}_{62}\hat{\varepsilon}_2 + \cdots + \hat{C}_{66}\hat{\varepsilon}_6 = 0$$

$$\Rightarrow \quad \hat{\varepsilon}_{33} = \hat{\varepsilon}_6 = \frac{-\left(\sum_{b=1}^{5} \hat{C}_{6b}\hat{\varepsilon}_b\right)}{\hat{C}_{66}} \tag{4.75}$$

Substituting eq. 4.75 into eq. 4.74 results in

$$\hat{\sigma}_a = \sum_{b=1}^{5} \hat{C}_{ab}\hat{\varepsilon}_b + \hat{C}_{a6}\hat{\varepsilon}_6 = \sum_{b=1}^{5} \left(\hat{C}_{ab} - \frac{\hat{C}_{a6}\hat{C}_{b6}}{\hat{C}_{66}}\right)\hat{\varepsilon}_b = \sum_{b=1}^{5} \hat{D}_{ab}\hat{\varepsilon}_b \tag{4.76}$$

where

$$\hat{D}_{ab} = \hat{C}_{ab} - \frac{\hat{C}_{a6}\hat{C}_{b6}}{\hat{C}_{66}} \tag{4.77}$$

Hence for $\hat{\sigma}_{33} = 0$, the stress–strain relationship can be expressed as

$$\underset{5\times1}{\hat{\sigma}} = \underset{5\times5}{\hat{\mathbb{D}}}\ \underset{5\times1}{\hat{\varepsilon}} \qquad \text{or} \qquad \hat{\sigma}_a = \sum_{b=1}^{5} \hat{D}_{ab}\hat{\varepsilon}_b \qquad a = 1,5 \tag{4.78}$$

For isotropic material, we have

$$\underset{5\times5}{\hat{\mathbb{D}}} = \frac{E}{1-\nu^2} \begin{bmatrix} 1 & \nu & 0 & 0 & 0 \\ & 1 & 0 & 0 & 0 \\ & & \frac{1-\nu}{2} & 0 & 0 \\ & & & \mathfrak{K}\left(\frac{1-\nu}{2}\right) & 0 \\ \text{sym.} & & & & \mathfrak{K}\left(\frac{1-\nu}{2}\right) \end{bmatrix} \tag{4.79}$$

where E is Young's modulus and ν is Poisson's ratio, and $\mathfrak{K} = 5/6$ is the shear correction factor to account for the non-uniform distribution of transverse shear stresses across the shell thickness.

4.7.6 Strain–displacement relationship

The strain–displacement matrix is given by

$$\underset{5\times 5N}{\mathbf{B}} = [B_1, B_2, \ldots, \underset{5\times 5}{B_N}] \tag{4.80}$$

Hence the strain–displacement transformation can be written as

$$\underset{5\times 1}{\hat{\varepsilon}} = \sum_{\alpha=1}^{N} \underset{5\times 5}{\mathbf{B}_\alpha} \underset{5\times 1}{\begin{bmatrix} \bar{\mathbf{u}} \\ \theta_1 \\ \theta_2 \end{bmatrix}_\alpha} \tag{4.81}$$

where each $\mathbf{B}_\alpha$ takes the form

$$\underset{5\times 5}{\mathbf{B}_\alpha} = \begin{bmatrix} \underset{5\times 3}{\mathbf{b}_a} & \underset{5\times 2}{\mathbf{c}_a} \end{bmatrix}_\alpha = \begin{bmatrix} \mathbf{b}_1 & \mathbf{c}_1 \\ \mathbf{b}_2 & \mathbf{c}_2 \\ \mathbf{b}_3 & \mathbf{c}_3 \\ \mathbf{b}_4 & \mathbf{c}_4 \\ \mathbf{b}_5 & \mathbf{c}_5 \end{bmatrix}_\alpha \tag{4.82}$$

in which $\underset{1\times 3}{\mathbf{b}_a} = (b_{a1}, b_{a2}, b_{a3}$ and $\underset{1\times 2}{\mathbf{c}_a} = (c_{a1}, c_{a2})$ are row vectors.

Explicit formulae can be obtained for the strains defined in eq. 4.74 by computing the displacement gradients:

$$\begin{aligned} \frac{\partial \hat{u}_i}{\partial \hat{x}_j} &= \frac{\partial}{\partial \hat{x}_j}(R_{im}u_m) = R_{im}\frac{\partial u_m}{\partial \hat{x}_j} = R_{im}\frac{\partial}{\partial \hat{x}_j}(\bar{u}_m + \zeta \tilde{u}_m) \\ &= R_{im}\frac{\partial}{\partial \hat{x}_j}(H_\alpha \bar{u}_m^\alpha + \zeta H'_\alpha \tilde{u}_m^\alpha) \\ &= R_{im}\left(\frac{\partial H_\alpha}{\partial \hat{x}_j}\bar{u}_m^\alpha + \left[\frac{1}{2}h_\alpha(\theta_2^\alpha \tilde{e}_{1m}^\alpha - \theta_1^\alpha \tilde{e}_{2m}^\alpha)\right]\frac{\partial}{\partial \hat{x}_j}(\zeta H'_\alpha)\right) \end{aligned} \tag{4.83}$$

where $\tilde{e}_{1m}^\alpha$ and $\tilde{e}_{2m}^\alpha (m = 1, 2, 3)$ are components of vectors $\tilde{\mathbf{e}}_1^\alpha$ and $\tilde{\mathbf{e}}_2^\alpha$.

From eq. 4.82, we can derive

membrane effects in plane stress due to displacements:

$$\left.\begin{aligned} & \left\{\begin{aligned} b_{1m} &= R_{1m}\frac{\partial H_\alpha}{\partial \hat{x}_1} \\ b_{2m} &= R_{2m}\frac{\partial H_\alpha}{\partial \hat{x}_2} \\ b_{3m} &= R_{1m}\frac{\partial H_\alpha}{\partial \hat{x}_2} + R_{2m}\frac{\partial H_\alpha}{\partial \hat{x}_1} \end{aligned}\right. \\ & \left\{\begin{aligned} b_{4m} &= R_{2m}\frac{\partial H_\alpha}{\partial \hat{x}_3} + R_{3m}\frac{\partial H_\alpha}{\partial \hat{x}_2} \\ b_{5m} &= R_{3m}\frac{\partial H_\alpha}{\partial \hat{x}_1} + R_{1m}\frac{\partial H_\alpha}{\partial \hat{x}_3} \end{aligned}\right. \end{aligned}\right\} m = 1, 2, 3 \tag{4.84}$$

transverse shear force due to displacements (rows b_{4m}, b_{5m})

$$\left.\begin{array}{ll}
\begin{array}{l}\text{membrane effects}\\ \text{in plane stress}\\ \text{due to rotations}\end{array} & \left\{\begin{array}{l}
c_{1I} = \omega_{1I}^{\alpha} \dfrac{\partial}{\partial \hat{x}_1}(\zeta H'_{\alpha})\\
c_{2I} = \omega_{2I}^{\alpha} \dfrac{\partial}{\partial \hat{x}_2}(\zeta H'_{\alpha})\\
c_{3I} = \omega_{1I}^{\alpha} \dfrac{\partial}{\partial \hat{x}_2}(\zeta H'_{\alpha}) + \omega_{2I}^{\alpha} \dfrac{\partial}{\partial \hat{x}_1}(\zeta H'_{\alpha})
\end{array}\right.\\
\begin{array}{l}\text{transverse shear}\\ \text{force due to}\\ \text{rotations}\end{array} & \left\{\begin{array}{l}
c_{4I} = \omega_{2I}^{\alpha} \dfrac{\partial}{\partial \hat{x}_3}(\zeta H'_{\alpha}) + \omega_{3I}^{\alpha} \dfrac{\partial}{\partial \hat{x}_2}(\zeta H'_{\alpha})\\
c_{5I} = \omega_{3I}^{\alpha} \dfrac{\partial}{\partial \hat{x}_1}(\zeta H'_{\alpha}) + \omega_{1I}^{\alpha} \dfrac{\partial}{\partial \hat{x}_3}(\zeta H'_{\alpha})
\end{array}\right.
\end{array}\right\}
\begin{array}{l} I = 1, 2\\ \text{no sum on } \alpha\end{array}
\tag{4.85}$$

where

$$\left.\begin{array}{l}
\omega_{i1}^{\alpha} = -\frac{1}{2} h_{\alpha} \sum_{m=1}^{3} R_{im} \tilde{e}_{2m}^{\alpha} = -\frac{1}{2} h_{\alpha} \hat{\mathbf{e}}_i \cdot \tilde{\mathbf{e}}_2^{\alpha}\\
\omega_{i2}^{\alpha} = \frac{1}{2} h_{\alpha} \sum_{m=1}^{3} R_{im} \tilde{e}_{1m}^{\alpha} = \frac{1}{2} h_{\alpha} \hat{\mathbf{e}}_i \cdot \tilde{\mathbf{e}}_1^{\alpha}
\end{array}\right\} \text{no sum on } \alpha \tag{4.86}$$

The differentiation indicated in eqs 4.84–4.85 is calculated by transforming corresponding global quantities:

$$\frac{\partial}{\partial \hat{x}_i} = \sum_{m=1}^{3} R_{im} \frac{\partial}{\partial x_m} \tag{4.87}$$

In shell elements, selective reduced integration is often used on membrane and transverse shear terms to avoid locking [79,80]. The terms in sub-matrix $\mathbf{B}_{\alpha}$ evaluated by reduced integration are indicated by the symbol *.

$$\underset{5\times 5}{\mathbf{B}_{\alpha}} = \begin{bmatrix} \mathbf{b}_1^* & \mathbf{c}_1\\ \mathbf{b}_2^* & \mathbf{c}_2\\ \mathbf{b}_3^* & \mathbf{c}_3\\ \mathbf{b}_4^* & \mathbf{c}_4^*\\ \mathbf{b}_5^* & \mathbf{c}_5^* \end{bmatrix}_{\alpha} \tag{4.88}$$

4.7.7 Element stiffness matrix

The element stiffness matrix is given by

$$\underset{5N\times 5N}{\mathbf{K}_e} = \underset{5\times 5}{[\mathbf{K}_{\alpha\beta}]} \tag{4.89}$$

where sub-matrix

$$\underset{5\times 5}{\mathbf{K}_{\alpha\beta}} = \int_{\Omega_e} \mathbf{B}_{\alpha}^{\mathrm{T}} \hat{\mathbb{D}} \mathbf{B}_{\beta}\, \mathrm{d}\Omega = \int_{\mathcal{R}} \mathbf{B}_{\alpha}^{\mathrm{T}} \hat{\mathbb{D}} \mathbf{B}_{\beta} \det(\mathbf{J}) \mathrm{d}\mathcal{R} \approx \sum_{k=1}^{m} W_k [\mathbf{B}_{\alpha}^{\mathrm{T}} \hat{\mathbb{D}} \mathbf{B}_{\beta} \det(\mathbf{J})]_k \tag{4.90}$$

where $\mathcal{R}$ is the reference domain (bi-unit cube) and $\mathbf{J}$ is the Jacobian matrix.

$$\underset{3\times 3}{\mathbf{J}(\xi, \eta, \zeta)} = \begin{bmatrix} \dfrac{\partial x_1}{\partial \xi} & \dfrac{\partial x_1}{\partial \eta} & \dfrac{\partial x_1}{\partial \zeta}\\ \dfrac{\partial x_2}{\partial \xi} & \dfrac{\partial x_2}{\partial \eta} & \dfrac{\partial x_2}{\partial \zeta}\\ \dfrac{\partial x_3}{\partial \xi} & \dfrac{\partial x_3}{\partial \eta} & \dfrac{\partial x_3}{\partial \zeta} \end{bmatrix} = \begin{bmatrix} \dfrac{\partial \mathbf{x}}{\partial \xi} & \dfrac{\partial \mathbf{x}}{\partial \eta} & \dfrac{\partial \mathbf{x}}{\partial \zeta} \end{bmatrix} \tag{4.91}$$

From the geometrical definition of the shell element (eq. 4.54), we have

$$\left.\begin{aligned} \frac{\partial \mathbf{x}}{\partial \xi} &= \frac{\partial G_\alpha}{\partial \xi}\bar{\mathbf{x}}_\alpha + \frac{\partial G_\alpha}{\partial \xi}\zeta\tilde{\mathbf{x}}_\alpha \\ \frac{\partial \mathbf{x}}{\partial \eta} &= \frac{\partial G_\alpha}{\partial \eta}\bar{\mathbf{x}}_\alpha + \frac{\partial G_\alpha}{\partial \eta}\zeta\tilde{\mathbf{x}}_\alpha \\ \frac{\partial \mathbf{x}}{\partial \zeta} &= G_\alpha\tilde{\mathbf{x}}_\alpha \end{aligned}\right] \tag{4.92}$$

4.7.8 Element force vectors

The element external force vector is made up of the body force, surface force and edge force vectors.

(i) Body force

The element body force vector is given by $\underset{5N\times 1}{\mathbf{b}_e} = \underset{5\times 1}{[\mathbf{b}_\alpha]}$, with

$$\mathbf{b}_\alpha = \int_{\Omega_e} \mathbf{H}_\alpha^{\mathrm{T}}\vec{\varphi}\,\mathrm{d}\Omega = \int_{\mathcal{R}} \mathbf{H}_\alpha^{\mathrm{T}}\vec{\varphi}\,\det(\mathbf{J})\mathrm{d}\mathcal{R} \approx \sum_{k=1}^{m} W_k[\mathbf{H}_\alpha^{\mathrm{T}}\vec{\varphi}\,\det(\mathbf{J})]_k \tag{4.93}$$

where

$$\underset{3\times 5}{\mathbf{H}_\alpha} = \begin{bmatrix} H_\alpha & 0 & 0 & -\frac{1}{2}H'_\alpha\zeta h_\alpha\tilde{e}^\alpha_{21} & \frac{1}{2}H'_\alpha\zeta h_\alpha\tilde{e}^\alpha_{11} \\ 0 & H_\alpha & 0 & -\frac{1}{2}H'_\alpha\zeta h_\alpha\tilde{e}^\alpha_{22} & \frac{1}{2}H'_\alpha\zeta h_\alpha\tilde{e}^\alpha_{12} \\ 0 & 0 & H_\alpha & -\frac{1}{2}H'_\alpha\zeta h_\alpha\tilde{e}^\alpha_{23} & \frac{1}{2}H'_\alpha\zeta h_\alpha\tilde{e}^\alpha_{13} \end{bmatrix} \text{ no sum on } \alpha \tag{4.94}$$

$$\underset{5\times 3}{\mathbf{H}_\alpha^{\mathrm{T}}}\underset{3\times 1}{\vec{\varphi}} = \begin{bmatrix} H_\alpha & 0 & 0 \\ 0 & H_\alpha & 0 \\ 0 & 0 & H_\alpha \\ & -\frac{1}{2}H'_\alpha\zeta h_\alpha\tilde{\mathbf{e}}^\alpha_2 & \\ & \frac{1}{2}H'_\alpha\zeta h_\alpha\tilde{\mathbf{e}}^\alpha_1 & \end{bmatrix} \begin{bmatrix} \varphi_1 \\ \varphi_2 \\ \varphi_3 \end{bmatrix} = \begin{bmatrix} H_\alpha\varphi_1 \\ H_\alpha\varphi_2 \\ H_\alpha\varphi_3 \\ -\frac{1}{2}H'_\alpha\zeta h_\alpha\tilde{\mathbf{e}}^\alpha_2\cdot\vec{\varphi} \\ \frac{1}{2}H'_\alpha\zeta h_\alpha\tilde{\mathbf{e}}^\alpha_1\cdot\vec{\varphi} \end{bmatrix} \text{ no sum on } \alpha \tag{4.95}$$

and $\vec{\varphi}$ is the body force per unit volume.

(ii) Surface force

The element surface force vector is given by $\underset{5N\times 1}{\mathbf{s}_e} = \underset{5\times 1}{[\mathbf{s}_\alpha]}$, with

$$\mathbf{s}_\alpha = \int_{A_e^\pm} \mathbf{H}_\alpha^{\mathrm{T}}\vec{\mathrm{t}}\,\mathrm{d}A = \int_{\mathcal{A}^\pm} \mathbf{H}_\alpha^{\mathrm{T}}\vec{\mathrm{t}}j\,\mathrm{d}\mathcal{A} \approx \sum_{k=1}^{m} W_k[\mathbf{H}_\alpha^{\mathrm{T}}\vec{\mathrm{t}}j]_k \quad \zeta = \begin{cases} +1 & \text{top surface} \\ -1 & \text{bottom surface} \end{cases} \tag{4.96}$$

where $\mathcal{A}^\pm$ is the top and bottom surfaces of the reference domain, surface Jacobian $j = \left\| \frac{\partial \mathbf{x}}{\partial \xi} \times \frac{\partial \mathbf{x}}{\partial \eta} \right\|$, and $\vec{\mathrm{t}}$ is the surface traction.

$$\underset{5\times 3}{\mathbf{H}_\alpha^{\mathrm{T}}}\underset{3\times 1}{\vec{\mathrm{t}}} = \begin{bmatrix} H_\alpha t_1 \\ H_\alpha t_2 \\ H_\alpha t_3 \\ -\frac{1}{2}H'_\alpha\zeta h_\alpha\tilde{\mathbf{e}}^\alpha_2\cdot\vec{\mathrm{t}} \\ \frac{1}{2}H'_\alpha\zeta h_\alpha\tilde{\mathbf{e}}^\alpha_1\cdot\vec{\mathrm{t}} \end{bmatrix} \text{ no sum on } \alpha \tag{4.97}$$

(a) Pressure load

Pressure load acts normal to the shell surface, and can be expressed as

$$\vec{t} = -\zeta p\mathbf{n} \qquad \zeta = \begin{cases} +1 & \text{top surface} \\ -1 & \text{bottom surface} \end{cases} \tag{4.98}$$

where p is the pressure and unit normal vector $\mathbf{n} = \dfrac{\mathbf{e}_\xi \times \mathbf{e}_\eta}{\|\mathbf{e}_\xi \times \mathbf{e}_\eta\|}$

(b) Shear load

It is assumed that the shear force is specified in the ξ and η directions on the surface under consideration.

$$\vec{t} = t_1\mathbf{e}_\xi + t_2\mathbf{e}_\eta \tag{4.99}$$

where t_1 and t_2 are the shears in the ξ and η directions respectively.

(iii) Edge force

Suppose we wish to apply a distributed loading on the edge surface ABCD ($\xi = -1$) as shown in Fig. 4.15.

Let $\vec{q}$ denote the distributed traction force on the edge surface, then the nodal force vector is given by

$$\underset{5N\times 1}{\mathbf{s}_e^*} = \underset{5\times 1}{[\mathbf{s}_\alpha^*]}$$

$$\underset{5\times 1}{\mathbf{s}_\alpha^*} = \int_{\text{ABCD}} \mathbf{H}_\alpha^{\mathrm{T}}\vec{q}\,\mathrm{d}A = \int_{-1}^{+1}\int_{-1}^{+1} [\mathbf{H}_\alpha^{\mathrm{T}}\vec{q}j]_{\xi=-1}\,\mathrm{d}\zeta\,\mathrm{d}\eta \approx \sum_{k=1}^{m} W_k[\mathbf{H}_\alpha^{\mathrm{T}}\vec{q}j]_k \tag{4.100}$$

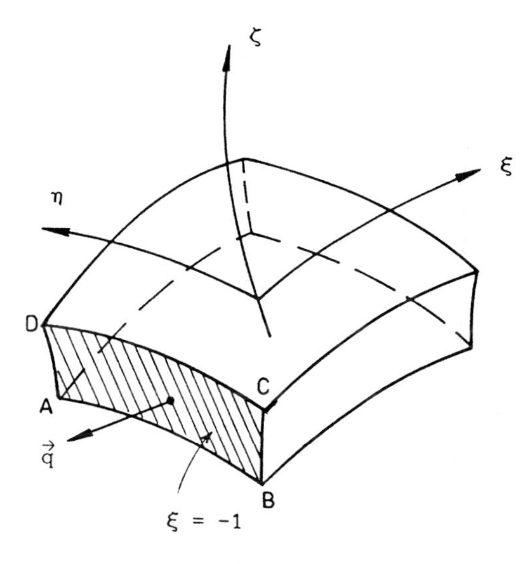

Fig. 4.15 Edge force on surface $\xi = -1$

where the surface Jacobian $j = \left\| \frac{\partial \mathbf{x}}{\partial \eta} \times \frac{\partial \mathbf{x}}{\partial \zeta} \right\|$.

$$\underset{5\times3}{\mathbf{H}_\alpha^{\mathrm{T}}}\underset{3\times1}{\vec{\mathrm{q}}} = \begin{bmatrix} H_\alpha q_1 \\ H_\alpha q_2 \\ H_\alpha q_3 \\ -\frac{1}{2}H'_\alpha \zeta h_\alpha \tilde{\mathbf{e}}_2^\alpha \cdot \vec{\mathrm{q}} \\ \frac{1}{2}H'_\alpha \zeta h_\alpha \tilde{\mathbf{e}}_1^\alpha \cdot \vec{\mathrm{q}} \end{bmatrix} \text{no sum on } \alpha \tag{4.101}$$

In the case where the edge forces (q_1, q_2, q_3) or moments (m_1, m_2) are specified per unit edge length along the edge $\xi = -1$, the nodal forces are computed as follows:

$$\underset{5\times1}{\mathbf{s}_\alpha^*} = \int_{-1}^{+1} \begin{bmatrix} H_\alpha q_1 \\ H_\alpha q_2 \\ H_\alpha q_3 \\ H'_\alpha m_1 \\ H'_\alpha m_2 \end{bmatrix} \left\| \frac{\partial \mathbf{x}}{\partial \eta} \right\| \mathrm{d}\eta \tag{4.102}$$

Other edge surface loading $(\xi = +1, \eta = \pm 1)$ can be treated similarly. Summing the individual contributions of body force, surface force and edge force, the element external force vector is given by

$$\vec{\mathbf{f}} = \vec{\mathbf{b}} + \vec{\mathrm{s}} + \vec{\mathrm{s}}^* \tag{4.103}$$

4.7.9 Fibre numerical integration

Integration along the fibre direction can be evaluated by numerical integration; and if the integrand is a smooth function of ζ, the Gaussian quadrature is most efficient. Taking the mid-surface as the reference surface, the membrane and the transverse shear effects can be approximated by a one-point Gaussian rule. However, at least two points are required to represent the bending behaviour. For shells consisting of one homogeneous layer, an efficient alternative is to perform fibre integration analytically [15, 73].

4.7.10 Stress resultants

Bending moments, membrane forces and transverse shear resultants can be computed at any laminar point (ξ, η) through integration over the ζ direction.

1. Moments

$$m_{IJ}(\xi, \eta) = \int_{-1}^{+1} \hat{\sigma}_{IJ} z \frac{\partial z}{\partial \zeta} \mathrm{d}\zeta \qquad I, J = 1, 2 \tag{4.104}$$

$$z = \tfrac{1}{2}\zeta H_\alpha h_\alpha \quad \Rightarrow \quad \frac{\partial z}{\partial \zeta} = \tfrac{1}{2} H_\alpha h_\alpha = \tfrac{1}{2} h \qquad \left(\tfrac{1}{2} \text{ thickness at } (\xi, \eta)\right)$$

2. Membrane forces

$$n_{IJ}(\xi, \eta) = \tfrac{1}{2} h \int_{-1}^{+1} \hat{\sigma}_{IJ} \, \mathrm{d}\zeta \qquad I, J = 1, 2 \tag{4.105}$$

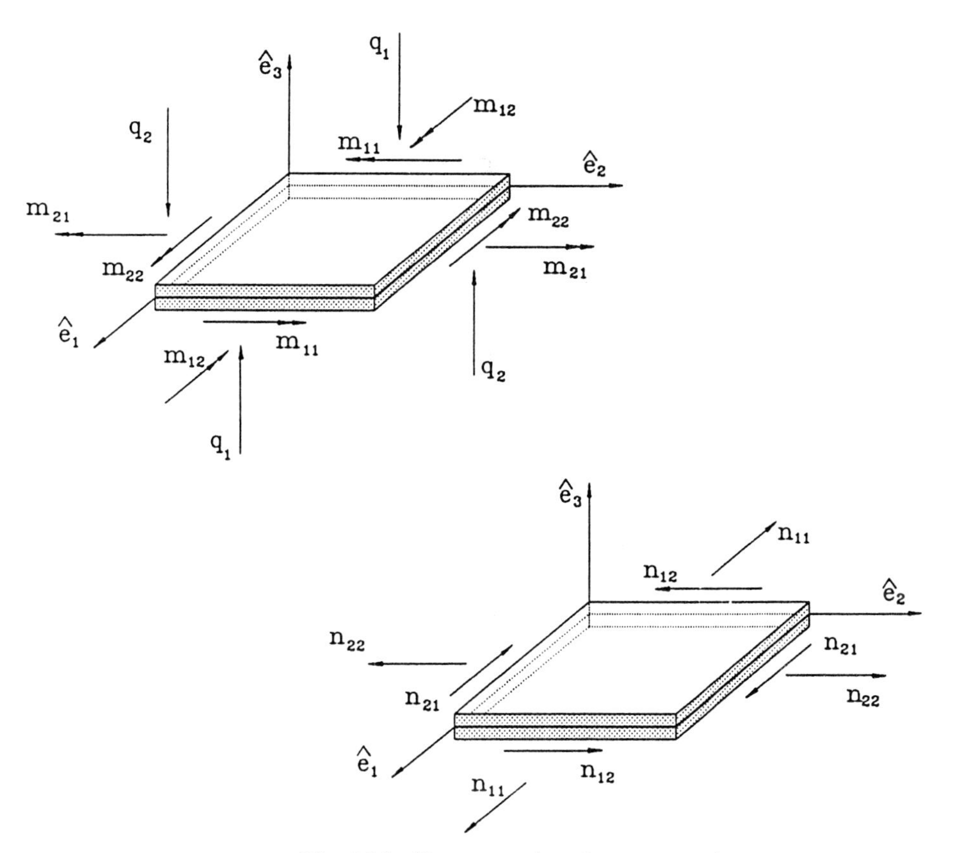

Fig. 4.16 Sign conventions for stress resultants

3. Transverse shear forces

$$q_I(\xi, \eta) = \tfrac{1}{2}h \int_{-1}^{+1} \hat{\sigma}_{I3}\, d\zeta \qquad I = 1, 2 \tag{4.106}$$

The integrations of eqs 4.104–4.106 are performed using two-point Gaussian quadrature over the fibre. The sign conventions for the stress resultants expressed in laminar coordinates are shown in Fig. 4.16.

4.7.11 Curved shell elements

The interpolation functions $H_\alpha(\xi, \eta)$ and quadrature rules for some typical Lagrangian elements are shown in Fig. 4.17.

In each case, the reduced integration rule is one order lower than the normal Gaussian rule. If the normal integration rule and reduced rule are combined, as described in Section 4.2.10, the element is called a selective integration element. Normal integration tends to cause elements to *lock* in the thin shell limits [81, 82]. This phenomenon is especially pronounced for low-order elements but lessens as the order of interpolation is increased. On the other hand, selective and uniform reduced

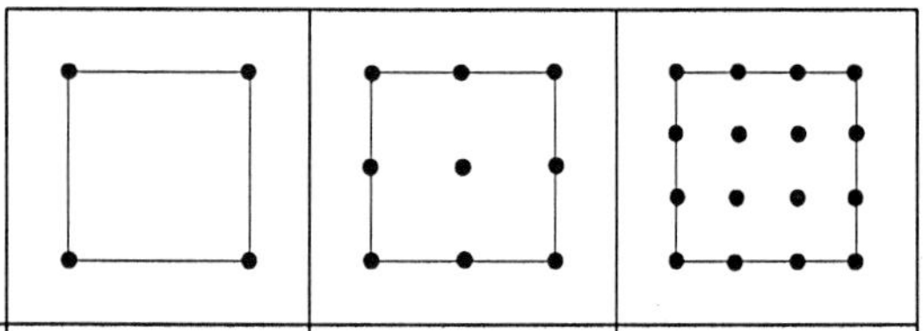

Lamina interpolation functions	Bilinear	Biquadratic	Bicubic
Normal Gaussian rule	2×2	3×3	4×4
Reduced Gaussian rule	1×1	2×2	3×3

Fig. 4.17 Lagrange shell elements

integration elements behave well in the thin shell applications but may sometimes engender rank deficiency (spurious zero-energy mechanism) [72, 79]. The problem is less acute for the selective integration element than for the uniformly reduced integration element. In some situations, the mechanisms are precluded from forming globally by the boundary conditions, but nevertheless, they represent a potentially dangerous deficiency in general applications.

4.8 A SHELL AS AN ASSEMBLY OF FLAT ELEMENTS

The geometry of curved shell structures may be represented approximately by means of facet elements [83, 84]. For instance, consider the shell geometry which is discretized into flat triangular elements. A local Cartesian coordinate system is set up on the plane of each triangle. A plate bending element stiffness and a membrane element stiffness (i.e. plane stress two-dimensional element stiffness) are generated and combined in this coordinate system. There is no coupling between in-plane and out-of-plane effects in the local coordinate system as the plate bending and membrane effects are referred to different degrees of freedom. The element stiffness matrix and force vector are then transformed from local coordinates to global coordinates before assembly.

Let us consider the formulation of a N-node facet element. Suppose that a Cartesian coordinate system is defined by a basis $\{\mathbf{a}_1, \mathbf{a}_2, \mathbf{a}_3\}$, in which $\mathbf{a}_3$ is normal to the element and $\mathbf{a}_1$ and $\mathbf{a}_2$ are in the plane of the element. The global-local transformation is given by

$$\underset{3\times3}{\mathbf{R}} = [R_{ij}] \qquad i, j = 1, 3$$

where $R_{ij} = \mathbf{e}_i \cdot \mathbf{a}_j$, and $\{\mathbf{e}_1, \mathbf{e}_2, \mathbf{e}_3\}$ is the global Cartesian basis.

The membrane stiffness matrix $\mathbf{K}_m(2N \times 2N)$ and the bending stiffness matrix $\mathbf{K}_b(3N \times 3N)$ are placed in an element stiffness matrix of dimension $6N \times 6N$, such that

$$\underset{6N\times6N}{\mathbf{K}} = \begin{bmatrix} \mathbf{K}_m & \mathbf{0} & \mathbf{0} \\ \mathbf{0} & \mathbf{K}_b & \mathbf{0} \\ \mathbf{0} & \mathbf{0} & \mathbf{0} \end{bmatrix} \tag{4.107}$$

In the case where adjacent elements are in the same plane, rank deficiency can occur by virtue of zero stiffness associated with rotation about the normal. Ill-conditioning may also occur if the planes defined by adjacent elements almost coincide.

For these reasons, a fictitious $N \times N$ stiffness $\mathbf{K}_f$ is often added to stabilize the assemblage [90]

$$\underset{6N\times6N}{\mathbf{K}} = \begin{bmatrix} \mathbf{K}_m & \mathbf{0} & \mathbf{0} \\ \mathbf{0} & \mathbf{K}_b & \mathbf{0} \\ \mathbf{0} & \mathbf{0} & \mathbf{K}_f \end{bmatrix} \tag{4.108}$$

The fictitious stiffness resists rotations about the normal to the plane of the element. These are referred to as the '*drilling degrees of freedom*' on account of its action about the normal of the surface. The rows and columns of $\mathbf{K}$ are now reordered so that the three displacement degrees of freedom at each node precede the three rotational degrees of freedom. The element stiffness matrix $\overline{\mathbf{K}}$ referring to global displacement parameters can be obtained by performing the following transformation.

$$\underset{6N\times6N}{\overline{\mathbf{K}}} = \mathbf{R} \cdot \mathbf{K}^* \cdot \mathbf{R}^T \tag{4.109}$$

where $\mathbf{K}^*$ is the re-ordered version of $\mathbf{K}$, and

$$\underset{6N\times6N}{\mathbf{R}} = \begin{bmatrix} \mathbf{R} & \mathbf{0} & \cdot & \cdot & \cdot & \mathbf{0} \\ & \mathbf{R} & \cdot & & & \cdot \\ & & & & & \cdot \\ \text{sym.} & & & & \cdot & \mathbf{0} \\ & & & & \cdot & \mathbf{R} \end{bmatrix} \tag{4.110}$$

Another alternative to the use of fictitious stiffness such that the rotational parameters arise naturally and have a physical significance, is to employ a membrane element that incorporates drilling degrees of freedom. This has been a topic of much recent study by a number of researchers, Allman [85], Bergan and Felippa [86], Cook [87], etc. In a paper by Bergan and Felippa [86], an accurate membrane element with drilling degrees of freedom was developed. Taylor and Simo [88] have also developed a triangular element of this kind which can be linked to the DKT bending element for shell analysis.

4.9 DEGENERATED SHELL ELEMENT L9S

The general expressions for the evaluation of the element stiffness matrix of a degenerated shell element are given in Section 4.7.7. A 9-node degenerated shell element L9S is developed by direct substitution of the same Lagrangian interpolation functions as for the plate bending element L9P described in Section 4.5 into eqs 4.82–4.86. The computer listing for such a shell element L9S can be found in Section 8.6. In the present implementation, two integration schemes can be used in the evaluation of the element stiffness matrix, namely the 2×2 reduced Gaussian rule and the 3×3 normal Gaussian rule.

The performance of the L9S shell element calculated by the 2×2 reduced Gaussian rule is studied using the example of a pinched cylinder. A thin circular cylinder of length L, radius R and thickness h, with ends restrained by rigid diaphragms was subjected to equal and opposite point loads P as shown in Fig. 4.18. The numerical constants used in the finite element analysis are $L = 600$ mm, $R = 300$ mm, $h = 3$ mm, $P = 10$ kN, Young's modulus $E = 300$ kN/mm^2 and Poisson's ratio $\nu = 0.3$.

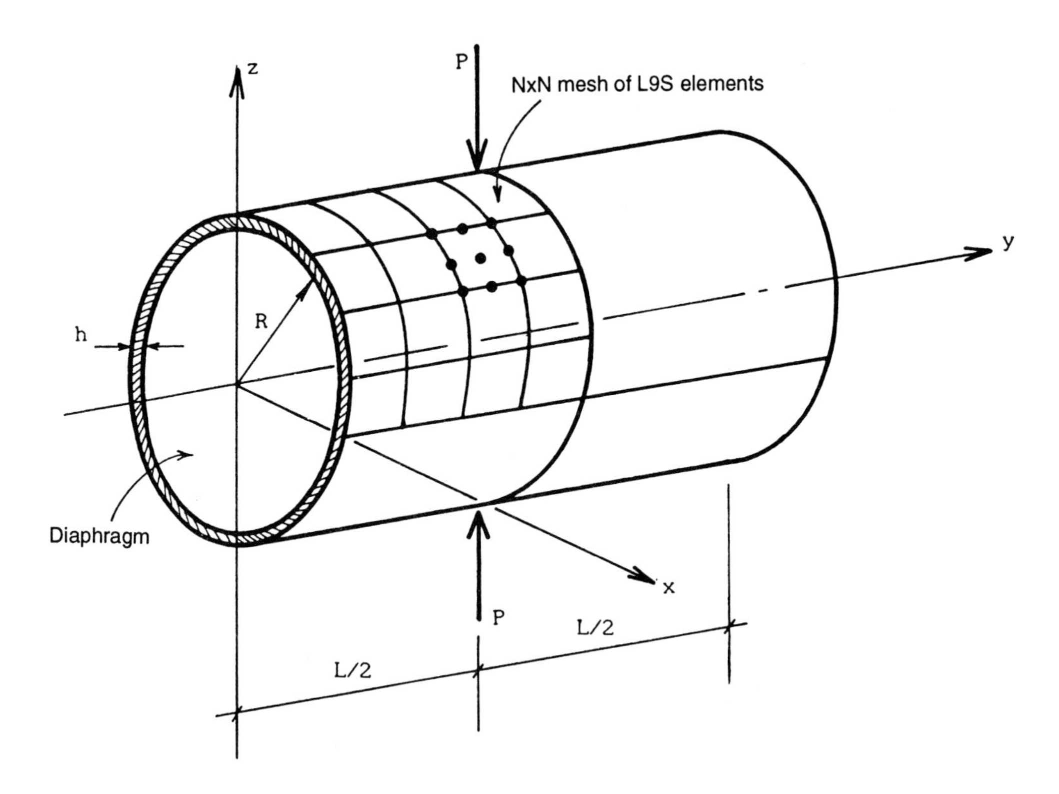

Fig. 4.18 The pinching of a cylindrical shell

Table 4.4 Deflections for different mesh refinements

Mesh	1×1	2×2	3×3	4×4	6×6	8×8	10×10
δ(mm)	0.137 504	1.389 68	1.664 93	1.760 49	1.804 86	1.815 33	1.822 33

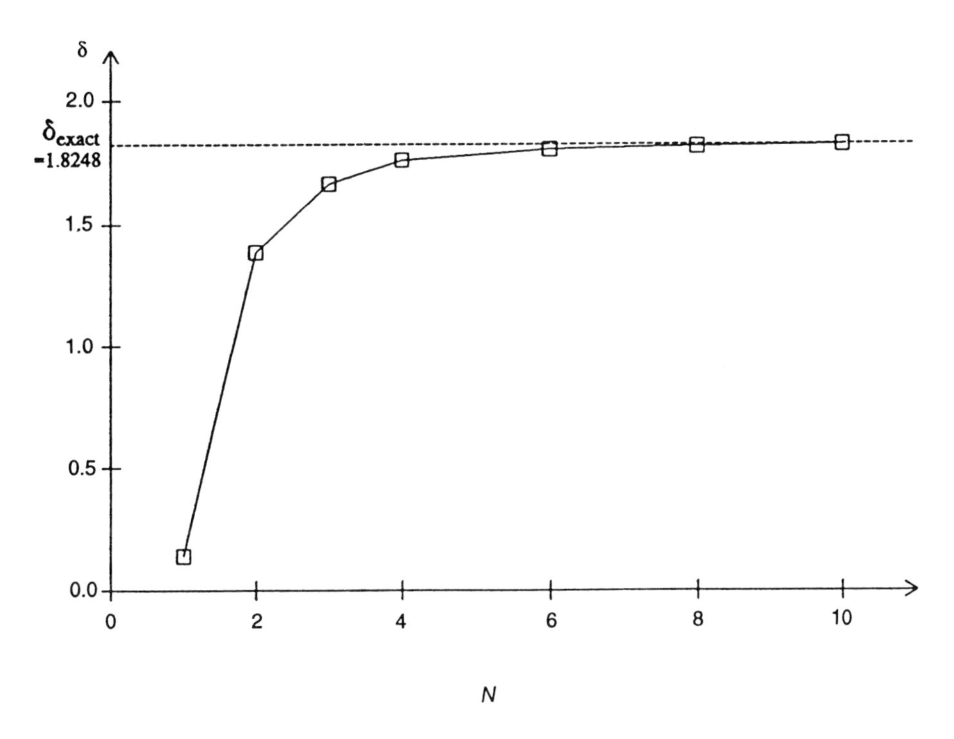

Fig. 4.19 Graph of deflection against number of elements per side N

In the study of convergence, one octant of the cylinder was divided progressively into $N \times N$ L9S shell elements, where N is the number of elements per side. The results are given in Table 4.4, and a graphical plot of the deflection against the number of subdivisions N is shown in Fig. 4.19. From the graph, it is seen that the deflection δ under the point load converges rapidly to the solution of 1.8248 mm quoted in reference [84].

4.10 References to Chapter 4

1. G. H. Ashwell & R. H. Gallagher, (eds) (1976) *Finite Elements for Thin Shells and Curved Members*, Wiley, New York.
2. J. L. Batoz, K. J. Bathe & L. W. Ho (1980) A study of three-node triangular plate bending elements, *Int. J. Num. Methods Engng*, **15**, 1771–812.
3. I. Holand (1969) Stiffness matrices for plate bending elements in *Finite Element Methods in Stress Analysis*, ed. by I. Holland and K. Bell, Tapir, Trondheim, pp 159–78.
4. M. M. Hrabok & T. M. Hrudey (1984) A review and catalogue of plate bending finite elements, *Computers & Structures*, **19**, 479–98.
5. H. Parisch (1979) A critical survey of the 9-node degenerated shell element with special emphasis on thin shell application and reduced integration, *Comp. Methods Appl. Mech. Eng.*, **20**, 323–50.

6. G. P. Bazeley, Y. K. Cheung, B. M. Irons & O. C. Zienkiewicz (1965) Triangular elements in plate bending – conforming and nonconforming solutions, *Proc. Conf. on Matrix Methods in Structural Mechanics*, Wright-Patterson Air Force Base, Dayton, OH, USA.
7. R. W. Clough & C. A. Felippa (1965) A refined quadrilateral element for analysis of plate bending, *Proc. 2nd Conf. on Matrix Methods in Structural Mechanics*, Wright-Patterson Air Force Base, Dayton, OH, USA.
8. R. W. Clough & J. L. Tocher (1965) Finite element stiffness matrices for analysis of plate bending, *Proc. Conf. on Matrix Methods in Structural Mechanics*, Wright-Patterson Air Force Base, Dayton, OH, USA.
9. B. F. de Veubeke (1968) A conforming finite element for plate bending, *Int. J. Solids Struct.*, **4**, 95–108.
10. A. Razzaque (1973) Program for triangular bending elements with derivative smoothing, *Int. J. Num. Methods Engng*, **6**, 333–43.
11. T. H. H. Pian (1964) Derivation of element stiffness matrices by assumed stress distributions, *AIAA Journal*, **2**, 1333–6.
12. T. H. H. Pian & P. Tong (1969) Basis of finite element method for solid continua, *Int. J. Num. Methods Engng*, **1**, 3–29.
13. S. Ahmad, B. M. Irons & O. C. Zienkiewicz (1968) Curved thick shell and membrane elements with particular reference to axi-symmetric problems, *Proc. 2nd Conf. on Matrix Methods in Structural Mechanics*, Wright-Patterson Air Force Base, Dayton, OH, USA.
14. S. Ahmad (1969) *Curved finite elements in the analysis of solid shell and plate structures*, PhD Thesis, University of Wales, Swansea.
15. O. C. Zienkiewicz, J. Too & R. L. Taylor (1971) Reduced integration technique in general analysis of plates and shells, *Int. J. Num. Methods Engng*, **3**, 275–90.
16. S. F. Pawsey & R. W. Clough (1971) Improved numerical integration of thick slab finite elements, *Int. J. Num. Methods Engng*, **3**, 575–86.
17. O. C. Zienkiewicz & R. L. Taylor (1991) *The Finite Element Method*, 4th edn, Vol. II, McGraw-Hill, UK.
18. T. Belytschko, W. K. Liu & J. S. J. Ong (1984) A consistent control of spurious singular modes in the 9-node Lagrange element for the Laplace and Mindlin plate equations, *Comp. Methods Appl. Mech. Engng*, **44**, 269–95.
19. M. A. Crisfield (1984) A quadratic Mindlin element using shear constraints, *Computers & Structures*, **18**(5), 833–52.
20. A. Tessler & T. J. R. Hughes (1983) An improved treatment of transverse shear in the Mindlin-type four-node quadrilateral element, *Comp. Methods Appl. Mech. Engng*, **39**, 311–35.
21. A. Tessler & T. J. R. Hughes (1985) A three-node Mindlin plate element with improved transverse shear, *Comp. Methods Appl. Mech. Engng*, **50**, 71–101.
22. E. Reissner (1945) The effect of transverse shear deformation on the bending of elastic plates, *J. Appl. Mech.*, **67**, 69–77, June.
23. R. D. Mindlin (1951) Influence of rotatory inertia and shear on flexural motions of isotropic, elastic plate, *J. Appl. Mech.*, **18**, 31–38.
24. B. O. Haggblad & K. J. Bathe (1990) Specification of boundary conditions for Reissner/Mindlin plate bending finite elements, *Int. J. Num. Methods Engng*, **30**, 981–1011.

25. M. Rossow (1977) Efficient C^0 finite element solution of simply supported plates of polygonal shape, *J. App. Mech.*, **44**, 347–49.
26. L. R. Scott (1976) A survey of displacement methods for the plate bending problem, *US–Germany Symposium on Formulations and Computational Algorithms in Finite Element Analysis*, Massachusetts Institute of Technology, Cambridge, Massachusetts, USA, 9–13 August.
27. G. Sander (1971) Application of the Dual Analysis Principle in High Speed Computing of Elastic Structures, *Proceedings of IUTAM Symposium*, Liège, Belgium, 167–207.
28. E. D. Pugh, E. Hinton & O. C. Zienkiewicz (1978) A study of quadrilateral plate bending elements with reduced integration, *Int. J. Num. Methods Engng*, **12**(7), 1059–78.
29. T. J. R. Hughes, M. Cohen & M. Haroun (1978) Reduced and selective integration techniques in the finite element analysis of plates, *Nuclear Engineering and Design*, **46**, 203–22.
30. T. Belytschko, C. S. Tsay & W. K. Liu (1981) A stabilization matrix for the bilinear Mindlin plate element, *Comp. Methods Appl. Mech. Engng*, **29**, 313–27.
31. E. Hinton & H. C. Huang (1986) A family of quadrilateral Mindlin plate elements with substitute shear strain fields, *Computers & Structures*, **23**, 409–31.
32. T. J. R. Hughes & M. Cohen (1978) The 'Heterosis' finite element for plate bending, *Computers and Structures*, **9**, 445–50.
33. T. Belytschko & C. S. Tsay (1983) A stabilization procedure for the quadrilateral plate element with one-point integration, *Int. J. Num. Methods Engng*, **19**, 405–19.
34. D. S. Malkus & T. J. R. Hughes (1978) Mixed finite element methods – reduced and selective integration techniques: a unification of concepts, *Comp. Methods Appl. Mech. Engng*, **15**, 63–81.
35. H. C. Huang & E. Hinton (1984) A nine-node Lagrangian plate element with enhanced shear interpolation, *Engineering Computations*, **1**, 369–79.
36. R. H. MacNeal (1982) Derivation of element stiffness matrices by assumed strain distributions, *Nuclear Engineering and Design*, **70**, 3–12.
37. C. Militello & D. H. Cascales (1987) Covariant shear strains interpolation in a nine-node degenerated plate element, *Computers & Structures*, **26**, No. 5, 781–85.
38. T. J. R. Hughes & T. E. Tezduyar (1981) Finite elements based upon Mindlin plate theory with particular reference to the four-node bilinear isoparametric element, *J. Appl. Mech.*, **48**, 587–96.
39. T. J. R. Hughes & R. L. Taylor (1982) The linear triangular bending element, in *Proceedings 4th MAFELP Conference*, ed. by J. R. Whiteman, Academic Press, London, pp. 127–142.
40. H. K. Stolarski & M. Y. M. Chiang (1988) Thin plate element with relaxed Kirchhoff Constraints, *Int. J. Num. Methods Engng*, **26**, 913–33.
41. J. L. Batoz & P. Lardeur (1989) A discrete shear triangular nine d.o.f. element for the analysis of thick to very thin plates, *Int. J. Num. Methods Engng*, **28**, 533–60.
42. J. L. Batoz & M. B. Tahar (1982) Evaluation of a new quadrilateral thin plate bending element, *Int. J. Num. Methods Engng*, **18**, 1655–77.
43. C. Jeyachandrabose, J. Kirkhope & L. Meekisho (1987) An improved discrete Kirchhoff quadrilateral thin-plate bending element, *Int. J. Num. Methods Engng*, **24**, 635–54.

44. J. L. Batoz (1982) An explicit formulation for an efficient triangular plate-bending element, *Int. J. Num. Methods Engng*, **18**, 1077–89.
45. C. Jeyachandrabose & J. Kirkhope (1986) Construction of new efficient three-node triangular thin plate bending elements, *Computers & Structures*, **23**, No. 5, 587–603.
46. B. M. Irons (1976) The Semiloof Shell Element in *Finite Elements for Thin Shells and Curved Members*, ed. by D. G. Ashwell and R. H. Gallagher, John Wiley, London, Chapter 11, pp. 197–222.
47. L. P. R. Lyons (1977) *A general finite element system with special reference to the analysis of cellular structures*, PhD Thesis, Imperial College, London.
48. M. A. Crisfield (1983) A four-noded plate bending element using shear constraints; a modified version of Lyons' element, *Comp. Methods Appl. Mech. Engng*, **38**, 93–120.
49. K. J. Bathe & E. N. Dvorkin (1985) A four-node plate bending element based on Mindlin/Reissner plate theory and a mixed interpolation, *Int. J. Num. Methods Engng*, **21**, 367–83.
50. S. W. Lee & T. H. H. Pian (1978) Improvement of plate and shell finite element by mixed formulation, *AIAA J.*, **16**, 29–34.
51. S. W. Lee & S. C. Wong (1982) Mixed formulation finite elements for Mindlin theory plate bending, *Int. J. Num. Methods Engng*, **18**, 1297–1311.
52. R. L. Spilker & N. I. Munir (1980) The hybrid-stress model for thin plates, *Int. J. Num. Methods Engng*, **15**, 1239–60.
53. T. H. H. Pian (1965) Element stiffness matrices for boundary compatibility and for prescribed boundary stresses, *Proc. First Conf. Matrix Methods in Struct. Mech.*, Wright-Patterson Air Force Base, Ohio, USA, AFFDL-TR-66-**80**, 457–78.
54. T. H. H. Pian & K. Sumihara (1984) Hybrid semiloof elements for plates and shells based upon a modified Hy–Washizu principle, *Computers & Structures*, **19**(1–2), 165–73.
55. T. H. H. Pian & D. P. Chen (1982) Alternative ways for formulation of hybrid stress elements, *Int. J. Num. Methods Engng*, **18**, 1679–84.
56. S. W. Lee & J. C. Zhang (1985) A six-node finite element for plate bending, *Int. J. Num. Methods Engng*, **21**, 131–43.
57. R. L. Spilker (1982) Invariant 8-node hybrid stress elements for thin and moderately thick plates, *Int. J. Num. Methods Engng*, **18**, 1153–78.
58. S. W. Lee, C. C. Dai & C. H. Teom (1985) A triangular finite element for thin plates and shells, *Int. J. Num. Methods Engng*, **21**, 1813–31.
59. R. L. Spikler & N. I. Munir (1980) A serendipity cubic-displacement hybrid-stress element for thin and moderately thick plates, *Int. J. Num. Methods Engng*, **15**, 1261–78.
60. T. H. H. Pian (1985) Finite elements based on consistently assumed stresses and displacements, *J. Finite Elements in Analysis and Design*, **1**, 131–40.
61. S. Ahmad, B. M. Irons & O. C. Zienkiewicz (1970) Analysis of thick and thin shell structures by curved finite elements, *Int. J. Num. Methods Engng*, **2**, 419–51.
62. K. C. Park & G. M. Stanley (1986) A curved C^0 shell element based on assumed natural-coordinate strains, *J. Appl. Mech.*, **53**, 278–90.
63. G. M. Stanley (1985) *Continuum-based shell elements*, PhD Thesis, Division of Applied Mechanics, Stanford University, August.

64. T. Belytschko, W. K. Liu, J. S-J. Ong, & D. Lam (1985) Implementation and Application of a 9-node Lagrange Shell Element with Spurious Mode Control, *Computers & Structures*, **20**, 121–8.
65. K. J. Bathe & E. N. Dvorkin (1986) A formulation of general shell elements – the use of mixed interpolation of tensorial components, *Int. J. Num. Methods Engng*, **22**, 607–722.
66. J. Jang & P. M. Pinsky (1988) Convergence of curved shell elements based on assumed covariant strain interpolations, *Int. J. Num. Methods Engng*, **26**, 329–47.
67. O. C. Zienkiewicz & Y. K. Cheung (1965) Finite element method of analysis for arch dam shells and comparison with finite difference procedures, *Proc. Symp. on Theory of Arch Dams*, Southampton University, 1964, Pergamon Press.
68. R. W. Clough & T. L. Tocher (1965) Analysis of thin arch dams by finite element method in *Proc. Symp. on Theory of Arch Dams*, Southampton University, 1964, Pergamon Press.
69. G. Cantin & R. W. Clough (1968) A curved, cylindrical-shell, finite element, *AIAA J.*, **6**, No. 6, 1057–62.
70. Y. K. Cheung & Chen Wanji (1990) Generalized hybrid degenerated elements for plate and shell, *Computers & Structures*, **36**, No. 2, 279–90.
71. G. M. Stanley, K. C. Park & T. J. R. Hughes (1986) Continuum-based resultant shell elements in *Finite Element Methods for Plate and Shell Structures 1: Element Technology*, ed. by T. J. R. Hughes & E. Hinton, Pineridge Press, Swansea, UK.
72. K. C. Park, G. M. Stanley & D. L. Flaggs (1985) A uniformly reduced four-noded C^0 shell element with consistent rank corrections, *Computers & Structures*, **20**, 129–39.
73. R. V. Milford & W. C. Schnobrich (1986) Degenerated isoparametric finite elements using explicit integration, *Int. J. Num. Methods Engng*, **23**, 133–54.
74. G. R. Heppler & J. S. Hansen (1986) A Mindlin element for thick and deep shells, *Comp. Methods Appl. Mech. Engng*, **54**, 21–47.
75. T. Belytschko, B. L. Wong & H. Stolarski (1989) Assumed strain stabilization procedure for the 9-node Lagrangian shell element, *Int. J. Num. Methods Engng*, **28**, 385–414.
76. E. C. Huang & E. Hinton (1986) Elastic-plastic and geometrically nonlinear analysis of plates and shells using a new nine-node element in *Finite Elements for Nonlinear Problems*, ed by P. Bergan *et al.*, Springer–Verlag, Berlin, pp. 283–97.
77. T. J. R. Hughes & E. Carnoy (1983) Nonlinear finite element shell formulation accounting for large membrane strains, *Comp. Methods Appl. Mech. Engng*, **39**, 69–82.
78. J. Oliver & E. Onate (1984) A total Lagrangian formulation for the geometrically nonlinear analysis of structures using finite elements, Part I, Two-dimensional problems: shell and plate structures, *Int. J. Num. Methods Engng*, **20**, 2253–81.
79. A. Kamoulakos (1988) Understanding and improving the reduced integration of Mindlin shell elements, *Int. J. Num. Methods Engng*, **26**, 2009–29.
80. H. Stolarski & T. Belytschko (1983) Shear and Membrane Locking in Curved C^0 Elements, *Comp. Methods Appl. Mech. Engng*, **41**, 279–96.
81. T. J. R. Hughes (1980) Generalization of selective integration procedures to anisotropic and nonlinear media, *Int. J. Num. Methods Engng*, **15**, No. 9, 1413–18.

82. A. Tessler (1985) A priori identification of shear locking and stiffening in triangular Mindlin elements, *Comp. Methods Appl. Mech. Engng*, **53**, 183–200.
83. N. Carpenter, H. Stolarski & T. Belytschko (1986) Improvement in 3-node triangular shell elements, *Int. J. Num. Methods Engng*, **23**, 1643–67.
84. E. N. Dvorkin & K. J. Bathe (1984) A continuum mechanics based four-node shell element for general nonlinear analysis, *Engineering Computations*, **1**, 77–88.
85. D. J. Allman (1988) Evaluation of the constant strain triangle with drilling rotations, *Int. J. Num. Methods Engng*, **26**, 2645–55.
86. P. G. Bergan & C. A. Felippa (1985) A triangular membrane element with rotational degrees of freedom, *Comp. Methods Appl. Mech. Engng*, **50**, 25–69.
87. R. D. Cook (1987) A plane hybrid element with rotational d.o.f. and adjustable stiffness, *Int. J. Num. Methods Engng*, **24**, 1499–508.
88. R. L. Taylor & J. C. Simo (1985) Bending and membrane elements for analysis of thick and thin shells, *Proceedings of the NUMETA '85 Conference*, ed. by J. Middleton & G. N. Pande, Swansea, 7–11 January, A. A. Balkema, Rotterdam, 587–91.
89. I. Babuska & T. Scapolla (1989) Benchmark computation and performance evaluation for a rhombic plate bending problem, *Int. J. Num. Methods Engng*, **28**, 155–79.
90. O. C. Zienkiewicz, J. Parekh & I. P. King (1968) Arch dams analysed by a linear finite element shell program, *Proc. Symp. Arch Dams*, Inst. Civ. Engrs, London.

5 Substructures, Symmetry and Periodicity

5.1 INTRODUCTION

It is always desirable to reduce the number of unknowns in a finite element analysis. A large structure is usually partitioned into parts of manageable size to facilitate independent design, testing and analysis. These parts are called substructures [1] whose configurations are defined by (i) slave coordinates which are confined to a substructure and (ii) master coordinates which are interconnecting coordinates among substructures. The slave coordinates can be eliminated once and for all and the stiffness matrix of a substructure is given in terms of the master coordinates only.

A (sub)structure is said to be symmetrical if it is unchanged under symmetry operations of translation, rotation, reflection and inversion and their combinations. The order of the stiffness matrix can be reduced in accordance with the degree of symmetry. The same can be done in a continuous manner by Fourier transformation if the (sub)structure possesses the property of continuously rotational symmetry or periodic boundary conditions.

5.2 SUBSTRUCTURING

When a large structural system is partitioned into two substructures 1 and 2, the stiffness equations can be written as

$$\begin{bmatrix} \mathbf{K}_{11} & \mathbf{K}_{12} \\ \mathbf{K}_{21} & \mathbf{K}_{22} \end{bmatrix} \begin{Bmatrix} \mathbf{q}_1 \\ \mathbf{q}_2 \end{Bmatrix} = \begin{Bmatrix} \mathbf{Q}_1 \\ \mathbf{Q}_2 \end{Bmatrix} \quad \text{and} \quad \begin{bmatrix} \mathbf{K}_{33} & \mathbf{K}_{34} \\ \mathbf{K}_{43} & \mathbf{K}_{44} \end{bmatrix} \begin{Bmatrix} \mathbf{q}_3 \\ \mathbf{q}_4 \end{Bmatrix} = \begin{Bmatrix} \mathbf{Q}_3 \\ \mathbf{Q}_4 \end{Bmatrix} \tag{5.1}$$

If $\{\mathbf{q}_2\}$ and $\{\mathbf{q}_4\}$ are identical and are taken as the master coordinate $\{\mathbf{q}_m\} = \{\mathbf{q}_2\} = \{\mathbf{q}_4\}$, then the stiffness equation of the system is given by

$$\begin{bmatrix} \mathbf{K}_{11} & 0 & \mathbf{K}_{12} \\ 0 & \mathbf{K}_{33} & \mathbf{K}_{34} \\ \mathbf{K}_{21} & \mathbf{K}_{43} & \mathbf{K}_{22} + \mathbf{K}_{44} \end{bmatrix} \begin{Bmatrix} \mathbf{q}_1 \\ \mathbf{q}_3 \\ \mathbf{q}_m \end{Bmatrix} = \begin{Bmatrix} \mathbf{Q}_1 \\ \mathbf{Q}_3 \\ \mathbf{Q}_2 + \mathbf{Q}_4 \end{Bmatrix} \tag{5.2}$$

Eliminating $\mathbf{q}_1$ and $\mathbf{q}_3$, we have

$$[\mathbf{K}_m]\{\mathbf{q}_m\} = \{\mathbf{Q}_m\} \tag{5.3}$$

where
$$[\mathbf{K}_m] = [\mathbf{K}_{m1} + \mathbf{K}_{m2}] \qquad \{\mathbf{Q}_m\} = \{\mathbf{Q}_{m1} + \mathbf{Q}_{m2}\}$$
$$\{\mathbf{K}_{m1}\} = [\mathbf{K}_{22} - \mathbf{K}_{21}\mathbf{K}_{11}^{-1}\mathbf{K}_{12}] \qquad [\mathbf{K}_{m2}] = [\mathbf{K}_{44} - \mathbf{K}_{43}\mathbf{K}_{33}^{-1}\mathbf{K}_{34}]$$

$$\{\mathbf{Q}_{m1}\} = \{\mathbf{Q}_2 - \mathbf{K}_{21}\mathbf{K}_{11}^{-1}\mathbf{Q}_1\} \qquad [\mathbf{Q}_{m2}] = \{\mathbf{Q}_4 - \mathbf{K}_{43}\mathbf{K}_{33}^{-1}\mathbf{Q}_3\}$$

Obviously, $[\mathbf{K}_{m1}]$ and $\{\mathbf{Q}_{m1}\}$ are unique to substructure 1 and have no connection with substructure 2, while the reverse is true for $[\mathbf{K}_{m2}]$ and $\{\mathbf{Q}_{m2}\}$. Therefore, the usual procedure for substructuring is as follows.

(i) Partition the stiffness matrix of substructure i according to the slave and master coordinates,

$$\begin{bmatrix} \mathbf{K}_{ss} & \mathbf{K}_{sm} \\ \mathbf{K}_{ms} & \mathbf{K}_{mm} \end{bmatrix} \begin{Bmatrix} \mathbf{q}_{si} \\ \mathbf{q}_{mi} \end{Bmatrix} = \begin{Bmatrix} \mathbf{Q}_{si} \\ \mathbf{Q}_{mi} \end{Bmatrix} \tag{5.4}$$

(ii) Eliminate the slave $\{\mathbf{q}_{si}\}$ according to

$$[\mathbf{K}_{ss}]\{\mathbf{q}_{si}\} = \{\mathbf{Q}_{si} - \mathbf{K}_{sm}\mathbf{q}_{mi}\} \tag{5.5}$$

to give

$$[\mathbf{K}_m^{(i)}]\{\mathbf{q}_{mi}\} = \{\mathbf{Q}_m^{(i)}\} \tag{5.6}$$

where

$$[\mathbf{K}_m^{(i)}] = [\mathbf{K}_{mm} - \mathbf{K}_{ms}\mathbf{K}_{ss}^{-1}\mathbf{K}_{sm}]$$

and

$$\{\mathbf{Q}_m^{(i)}\} = \{\mathbf{Q}_{mi} - \mathbf{K}_{ms}\mathbf{K}_{ss}^{-1}\mathbf{Q}_{si}\}$$

(iii) All substructures can be transformed to the common master coordinate $\{\mathbf{q}_m\}$ from the individual coordinate $\{\mathbf{q}_{mi}\}$ by means of the transformation matrix $[\mathbf{T}_i]$ and then assembled to form the overall stiffness equations

$$\{\mathbf{q}_{mi}\} = [\mathbf{T}_i]\{\mathbf{q}_m\} \tag{5.7}$$

$$\sum_i [\mathbf{T}_i]^T[\mathbf{K}_m^{(i)}][\mathbf{T}_i]\{\mathbf{q}_m\} = \sum_i [\mathbf{T}_i]^T\{\mathbf{Q}_m^{(i)}\}$$

or

$$[\mathbf{K}_m]\{\mathbf{q}_m\} = \{\mathbf{Q}_m\} \tag{5.8}$$

(iv) Solve for the system master coordinate $\{\mathbf{q}_m\}$ from eq. 5.8 for applied system force $\{\mathbf{Q}_m\}$.
(v) Find the substructure master coordinate $\{\mathbf{q}_{mi}\}$ from eq. 5.7.
(vi) Find the substructure slave coordinate $\{\mathbf{q}_{si}\}$ from eq. 5.5.
(vii) Find the internal reactions $\{\mathbf{Q}_{mi}\}$ from eq. 5.6.

It should be noted that the condensation process in eq. 5.6 does not require additional programming effort if a subroutine for the solution of linear equations is available. In fact any method which can reduce $\mathbf{K}_{ms}$ to zero in the augmented matrix will give both $[\mathbf{K}_m^{(i)}]$ and $[\mathbf{Q}_m^{(i)}]$ automatically.

$$\begin{bmatrix} \mathbf{K}_{ss} & \mathbf{K}_{sm} & \mathbf{Q}_{si} \\ \mathbf{K}_{ms} & \mathbf{K}_{mm} & \mathbf{Q}_{mi} \end{bmatrix} \quad \text{becomes} \quad \begin{bmatrix} \mathbf{K}_{ss} & \mathbf{K}_{sm} & \mathbf{Q}_{si} \\ 0 & \mathbf{K}_m^{(i)} & \mathbf{Q}_m^{(i)} \end{bmatrix}$$

Consider the mass-and-spring system with two substructures 1 and 2 (shown in Fig. 5.1) as an example. Coordinate 3 at station D is the master coordinate. Substructure 2 has no slave coordinate and substructure 1 has two slave coordinates which

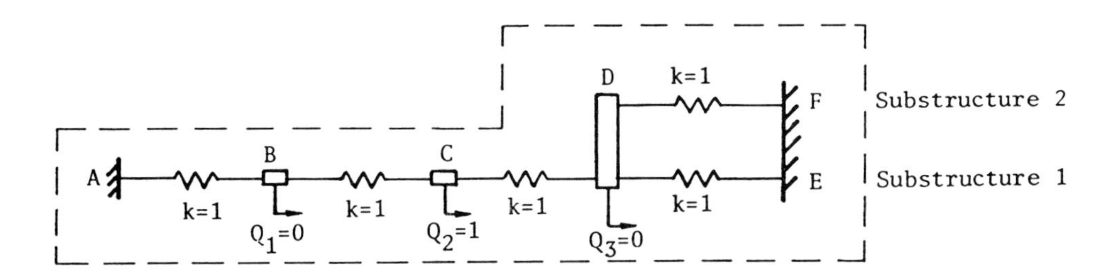

Fig. 5.1 (a) A two-substructure system

Fig. 5.1 (b) Gauss elimination for substructure 1

will be eliminated. Before elimination, the equilibrium equation for substructure 1 is

$$\begin{bmatrix} K_{11}^{(1)} & K_{12}^{(1)} & K_{13}^{(1)} \\ K_{21}^{(1)} & K_{22}^{(1)} & K_{23}^{(1)} \\ K_{31}^{(1)} & K_{32}^{(1)} & K_{33}^{(1)} \end{bmatrix} \begin{Bmatrix} q_1 \\ q_2 \\ q_3 \end{Bmatrix} = \begin{Bmatrix} Q^{(1)} \\ Q^{(2)} \\ Q^{(3)} \end{Bmatrix} \quad \text{or} \quad \begin{bmatrix} 2 & -1 & 0 \\ -1 & 2 & -1 \\ 0 & -1 & 2 \end{bmatrix} \begin{Bmatrix} q_1 \\ q_2 \\ q_3 \end{Bmatrix} = \begin{Bmatrix} 0 \\ 1 \\ 0 \end{Bmatrix} \tag{5.9}$$

Equation 5.9 will be solved by the Gauss elimination method and the unknown coordinates found by backsubstitution.

The forward elimination procedure is now carried out as follows:

(i) Eliminate $K_{21}^{(1)}$ by adding half of the first row to the second row in eq. 5.9

$$\begin{bmatrix} K_{11}^{(2)} & K_{12}^{(2)} & K_{13}^{(2)} \\ 0 & K_{22}^{(2)} & K_{23}^{(2)} \\ 0 & K_{32}^{(2)} & K_{33}^{(2)} \end{bmatrix} \begin{Bmatrix} q_1 \\ q_2 \\ q_3 \end{Bmatrix} = \begin{Bmatrix} Q_1^{(1)} \\ Q_2^{(2)} \\ Q_3^{(3)} \end{Bmatrix} \quad \text{or} \quad \begin{bmatrix} 2 & -1 & 0 \\ 0 & \frac{3}{2} & -1 \\ 0 & -1 & 2 \end{bmatrix} \begin{Bmatrix} q_1 \\ q_2 \\ q_3 \end{Bmatrix} = \begin{Bmatrix} 0 \\ 1 \\ 0 \end{Bmatrix} \tag{5.10}$$

Note that since $K_{31}^{(1)}$ is zero, there is no need to operate on the third row. All the

coefficients below the diagonal $K_{11}^{(2)}$ are now equal to zero.
(ii) Eliminate $K_{32}^{(2)}$ by adding 2/3 of the second row to the third row in eq. 5.10

$$\begin{bmatrix} K_{11}^{(3)} & K_{12}^{(3)} & K_{13}^{(3)} \\ 0 & K_{22}^{(3)} & K_{23}^{(3)} \\ 0 & 0 & K_{33}^{(3)} \end{bmatrix} \begin{Bmatrix} q_1 \\ q_2 \\ q_3 \end{Bmatrix} = \begin{Bmatrix} Q_1^{(1)} \\ Q_2^{(2)} \\ Q_3^{(3)} \end{Bmatrix} \quad \text{or} \quad \begin{bmatrix} 2 & -1 & 0 \\ 0 & \frac{3}{2} & -1 \\ 0 & 0 & \frac{4}{3} \end{bmatrix} \begin{Bmatrix} q_1 \\ q_2 \\ q_3 \end{Bmatrix} = \begin{Bmatrix} 0 \\ 1 \\ \frac{2}{3} \end{Bmatrix} \tag{5.11}$$

Note that the load vector has also been altered after this step. The forward elimination process is now complete and the stiffness matrix has been triangulated. To obtain the solution, a back substitution can be carried out to obtain sequentially $q_3 = \frac{1}{2}$, $q_2 = 1$ and $q_1 = \frac{1}{2}$.

The back substitution process is now complete and the displacement vector is

$$\{q\} = \begin{bmatrix} \frac{1}{2} & 1 & \frac{1}{2} \end{bmatrix}^{\mathrm{T}}$$

In the forward elimination process, only the coefficients to the right and below the dashed lines are altered after each step. It is of interest to note that portions of eqs 5.9–5.11 correspond to the systems (i) to (iii) of Fig. 5.1(b) respectively. In eq. 5.11, the two springs between A and C have been replaced by one spring of $k = \frac{1}{2}$, and in eq. 5.11 the three springs between A and D have been similarly replaced by a single equivalent spring of $k = \frac{1}{3}$ while the load $Q_2^{(2)}$ has been changed to $Q_3^{(3)}$. Thus, finally, substructure 1 with 3 DOF has been reduced to an equivalent system with just one unknown, which is the master coordinate.

When combined with substructure 2, whose governing equation is

$$k\,q_3 = Q \tag{5.12}$$

we have

$$(K_{33}^{(3)} + k)q_3 = Q_3^{(3)} \quad \text{or} \quad \frac{7}{3}q_3 = \frac{2}{3}$$

Therefore, $q_3 = \frac{2}{7}$.

From eqs 5.10 and 5.9, the slave coordinates of substructure 1 can be found as follows

$$\frac{3}{2}q_2 - q_3 = 1 \quad \text{or} \quad q_2 = \frac{6}{7}$$

and

$$2q_1 - q_2 = 0 \quad \text{or} \quad q_1 = \frac{3}{7}$$

Substructure 2 has no slave coordinates and the primary unknowns have all been found. The restoring forces, which are the secondary unknowns, of the springs (elements) are given in the following table where positive restoring force means tension and negative restoring force means compression.

Spring	Restoring force
AB	$kq_1 = \frac{3}{7}$
BC	$k(q_2 - q_1) = \frac{3}{7}$
CD	$k(q_3 - q_2) = -\frac{4}{7}$
DE	$-kq_3 = -\frac{2}{7}$
DF	$-kq_3 = -\frac{2}{7}$

For computational purposes, the Gauss method is the most efficient. However, for theoretical developments, the transformation method is the most convenient. In fact, eq. 5.6 can be derived from eq. 5.4 by the transformation of coordinates

$$\begin{Bmatrix} \mathbf{q}_{si} \\ \mathbf{q}_{mi} \end{Bmatrix} = \begin{bmatrix} -\mathbf{K}_{ss}^{-1} & \mathbf{K}_{sm} \\ & \mathrm{I} \end{bmatrix} \{\mathbf{q}_{mi}\} + \begin{bmatrix} -\mathbf{K}_{ss}^{-1} & \mathbf{Q}_{si} \\ & \mathbf{0} \end{bmatrix} \tag{5.13}$$

No two substructures need be alike. However, due to standardization, many substructures are identical, in which case only one condensation of the representative substructure is required. Also, the modification of one substructure will not affect the other substructures. A new condensed matrix is required for the modified substructure only. Therefore, the substructure method is very suitable for iterative design processes.

The master coordinates only are of interest in the remaining parts of this chapter. When $\{\mathbf{Q}_{si}\} = \{0\}$, then the required transformation is

$$\begin{Bmatrix} \mathbf{q}_{si} \\ \mathbf{q}_{mi} \end{Bmatrix} = \begin{bmatrix} -\mathbf{K}_{ss}^{-1} & \mathbf{K}_{sm} \\ & \mathrm{I} \end{bmatrix} \{\mathbf{q}_{mi}\} \tag{5.14}$$

As stated above, if a structure can be idealized as a repetition of a single substructure, only one condensation is required. For an iterative design process, only the modified substructures need further processing.

5.3 Symmetry

For the partitioned stiffness matrix of a structure (an element, a substructure, or a system)

$$\begin{bmatrix} \mathbf{K}_{11} & \mathbf{K}_{12} \\ \mathbf{K}_{21} & \mathbf{K}_{22} \end{bmatrix} \begin{Bmatrix} \mathbf{q}_1 \\ \mathbf{q}_2 \end{Bmatrix} = \begin{Bmatrix} \mathbf{Q}_1 \\ \mathbf{Q}_2 \end{Bmatrix} \tag{5.15}$$

If a relation between $\{\mathbf{q}_1\}$ and $\{\mathbf{q}_2\}$ can be found so that

$$\{\mathbf{q}_1\} = [\mathbf{T}]\{\mathbf{q}_2\} \tag{5.16}$$

then eq. 5.9 becomes

$$\begin{bmatrix} \mathbf{T}^{\mathrm{T}}\mathbf{K}_{11}\mathbf{T} & \mathbf{K}_{12}\mathbf{T} \\ \mathbf{T}^{\mathrm{T}}\mathbf{K}_{21} & \mathbf{K}_{22} \end{bmatrix} \begin{Bmatrix} \mathbf{q}_2 \\ \mathbf{q}_2 \end{Bmatrix} = \begin{Bmatrix} \mathbf{T}^{\mathrm{T}}\mathbf{Q}_1 \\ \mathbf{Q}_2 \end{Bmatrix}$$

which collapses to

$$[\mathbf{T}^{\mathrm{T}}\mathbf{K}_{11}\mathbf{T} + \mathbf{K}_{12}\mathbf{T} + \mathbf{T}^{\mathrm{T}}\mathbf{K}_{21} + \mathbf{K}_{22}]\{\mathbf{q}_2\} = \{\mathbf{T}^{\mathrm{T}}\mathbf{Q}_1 + \mathbf{Q}_2\} \tag{5.17}$$

The number of unknowns in eq. 5.17 is half that of eq. 5.15.

Some basic symmetry operations are shown in Fig. 5.2. The relation (5.16) for each of these operations is given below where q_x, q_y, q_z are the Cartesian components of displacement at a point and q_θ, q_r, q_z are the polar components of displacement at a point.

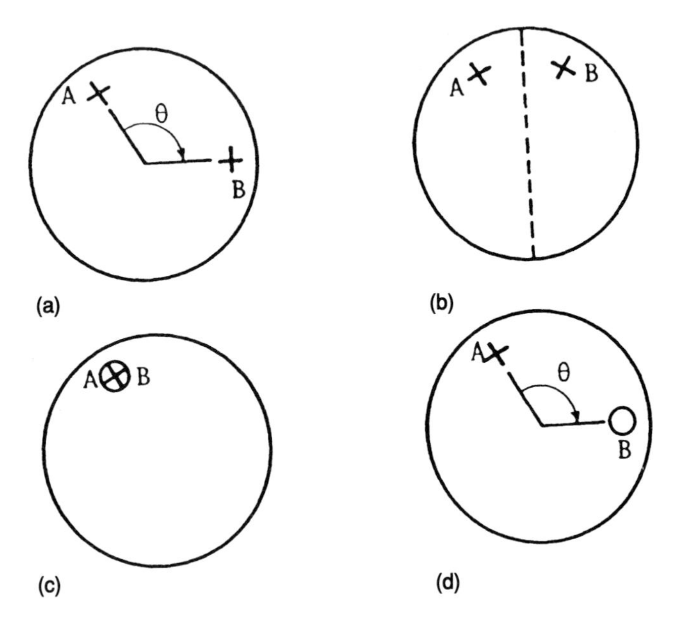

Fig. 5.2 The basic symmetry operations. × denotes a point above the horizontal plane. O denotes a point below the plane. ---- denotes a vertical mirror plane. B is the new position of A after the symmetry operation. (a) Rotation $C_n \theta = \frac{2\pi}{n}$; (b) Reflection about a vertical plane σ_v; (c) Reflection about a horizontal plane σ_h; (d) Improper rotation $S_n \theta = \frac{2\pi}{n}$.

(a) Rotation is the symmetry operation that leaves the magnitudes of the polar components of displacement unchanged.

$$\begin{Bmatrix} q_\theta \\ q_r \\ q_z \end{Bmatrix} = \begin{bmatrix} \cos\theta & \sin\theta & 0 \\ -\sin\theta & \cos\theta & 0 \\ 0 & 0 & 1 \end{bmatrix} \begin{Bmatrix} q_\theta \\ q_r \\ q_z \end{Bmatrix}_{\mathrm{A}}$$

Therefore

$$[\mathbf{T}] = \begin{bmatrix} \cos\theta & \sin\theta & 0 \\ -\sin\theta & \cos\theta & 0 \\ 0 & 0 & 1 \end{bmatrix}$$

(b) Reflection about a vertical plane is the symmetry operation that makes the magnitudes of the Cartesian components unchanged with respect to a vertical mirror.

$$\{\mathbf{q}_A\} = \begin{Bmatrix} q_x \\ q_y \\ q_z \end{Bmatrix} \qquad \{\mathbf{q}_B\} = \begin{Bmatrix} -q_x \\ q_y \\ q_z \end{Bmatrix} \qquad \therefore [\mathbf{T}] = \begin{bmatrix} -1 & & \\ & 1 & \\ & & 1 \end{bmatrix}$$

(c) Reflection about a horizontal plane makes the magnitudes of the Cartesian components of displacement unchanged with respect to a horizontal mirror.

$$\{\mathbf{q}_A\} = \begin{Bmatrix} q_x \\ q_y \\ q_z \end{Bmatrix} \qquad \{\mathbf{q}_B\} = \begin{Bmatrix} q_x \\ q_y \\ -q_z \end{Bmatrix} \qquad \therefore [\mathbf{T}] = \begin{bmatrix} 1 & & \\ & 1 & \\ & & -1 \end{bmatrix}$$

(d) Improper rotation is a compound operation consisting of a rotation followed by a reflection about a horizontal plane.

$$\{\mathbf{q}_B\} = \begin{bmatrix} \cos\theta & \sin\theta & 0 \\ -\sin\theta & \cos\theta & 0 \\ 0 & 0 & 1 \end{bmatrix} \{\mathbf{q}_A\} \qquad \therefore [\mathbf{T}] = \begin{bmatrix} \cos\theta & \sin\theta & 0 \\ -\sin\theta & \cos\theta & 0 \\ 0 & 0 & -1 \end{bmatrix}$$

The symmetry of the structural coordinates will be exploited so that relations similar to eq. 5.16 can be found to reduce the number of unknowns. Let us consider the structure of the stiffness matrix of a substructure having a plane of symmetry. If a substructure has a plane of symmetry P [2], then some of the coordinates, **s**, of $\mathbf{q}_b$ are symmetrical with respect to P and others, **r**, are anti-symmetrical (see Fig. 5.3).

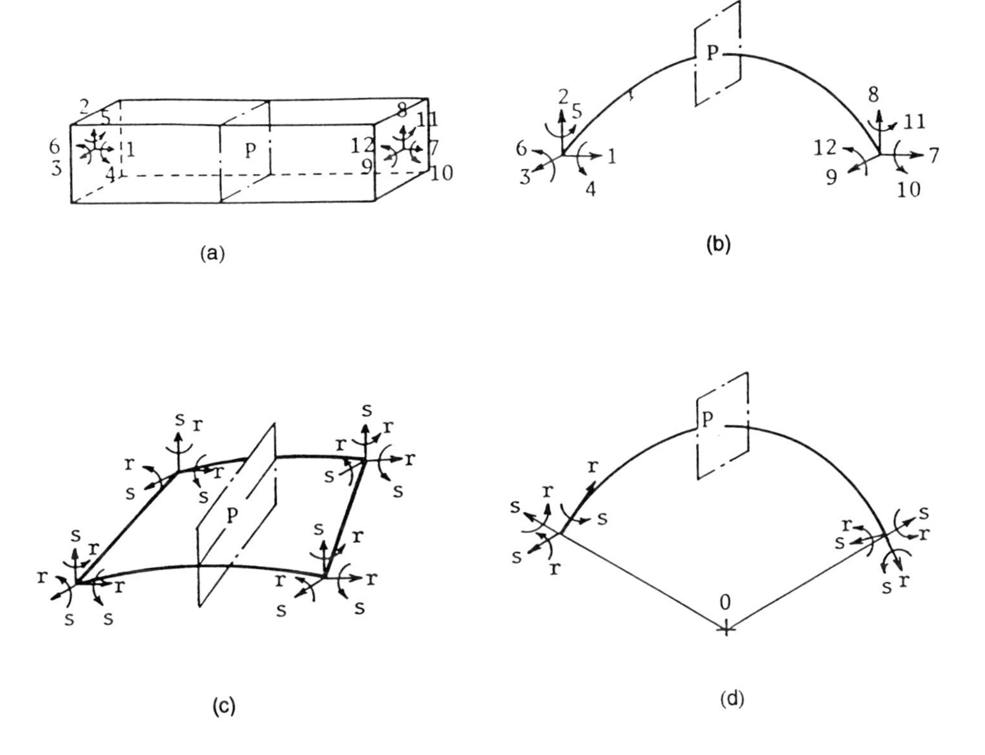

Fig. 5.3 The symmetrical and antisymmetrical coordinates. (a) Beam type substructure; (b) curved substructure; (c) shell type substructure; (d) cyclic symmetric substructure.

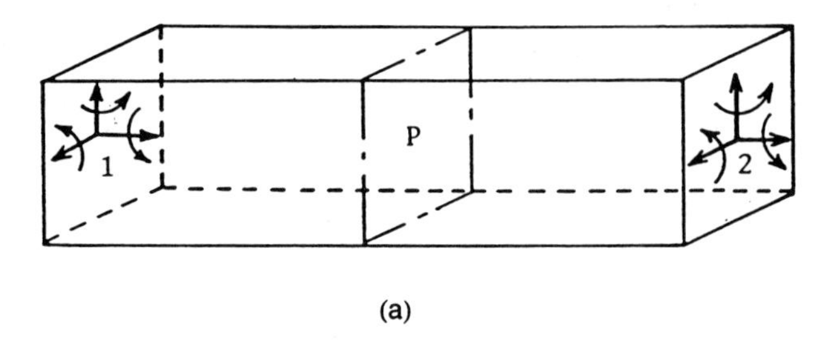

(a)

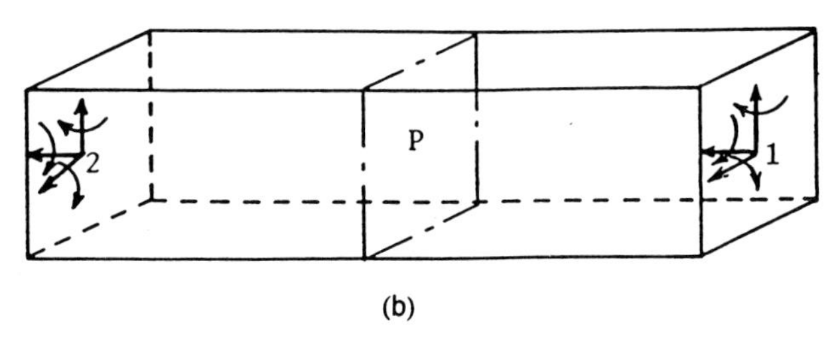

(b)

Fig. 5.4 Mirror image (b) of a substructure (a)

Consider the substructure of Fig. 5.4(a), where the generalized displacements are $\mathbf{s}_1$, $\mathbf{r}_1$, $\mathbf{s}_2$ and $\mathbf{r}_2$.

Here, the subscripts 1 and 2 denote the left- and right-hand ends of P respectively. Figure 5.4(b) is the mirror image of Fig. 5.4(a), with the mirror in parallel with P. If the stiffness equations for Fig. 5.4(a) are

$$\begin{pmatrix} \mathbf{K}_{11} & \mathbf{K}_{12} & \mathbf{K}_{13} & \mathbf{K}_{14} \\ & \mathbf{K}_{22} & \mathbf{K}_{23} & \mathbf{K}_{24} \\ & & \mathbf{K}_{33} & \mathbf{K}_{34} \\ & \text{sym.} & & \mathbf{K}_{44} \end{pmatrix} \begin{Bmatrix} \mathbf{s}_1 \\ \mathbf{s}_2 \\ \mathbf{r}_1 \\ \mathbf{r}_2 \end{Bmatrix} = \begin{Bmatrix} \mathbf{S}_1 \\ \mathbf{S}_2 \\ \mathbf{R}_1 \\ \mathbf{R}_2 \end{Bmatrix} \tag{5.18}$$

then the stiffness equations for Fig. 5.4(b) will be

$$\begin{pmatrix} \mathbf{K}_{11} & \mathbf{K}_{12} & \mathbf{K}_{13} & \mathbf{K}_{14} \\ & \mathbf{K}_{22} & \mathbf{K}_{23} & \mathbf{K}_{24} \\ & & \mathbf{K}_{33} & \mathbf{K}_{34}^{\mathrm{T}} \\ & \text{sym.} & & \mathbf{K}_{44} \end{pmatrix} \begin{Bmatrix} \mathbf{s}_2 \\ \mathbf{s}_1 \\ -\mathbf{r}_2 \\ -\mathbf{r}_1 \end{Bmatrix} = \begin{Bmatrix} \mathbf{S}_2 \\ \mathbf{S}_1 \\ -\mathbf{R}_2 \\ \mathbf{R}_1 \end{Bmatrix} \tag{5.19}$$

Equation 5.19 can be rearranged as

$$\begin{pmatrix} \mathbf{K}_{22} & \mathbf{K}_{12}^{\mathrm{T}} & -\mathbf{K}_{24} & -\mathbf{K}_{23} \\ & \mathbf{K}_{11} & -\mathbf{K}_{14} & -\mathbf{K}_{13} \\ & & \mathbf{K}_{44} & \mathbf{K}_{34}^{\mathrm{T}} \\ & \text{sym.} & & \mathbf{K}_{33} \end{pmatrix} \begin{Bmatrix} \mathbf{s}_1 \\ \mathbf{s}_2 \\ \mathbf{r}_1 \\ \mathbf{r}_2 \end{Bmatrix} = \begin{Bmatrix} \mathbf{S}_1 \\ \mathbf{S}_2 \\ \mathbf{R}_1 \\ \mathbf{R}_2 \end{Bmatrix} \tag{5.20}$$

Since substructures (a) and (b) are identical, then,

$$\mathbf{K}_{12} = \mathbf{K}_{12}^{\mathrm{T}} \quad \mathbf{K}_{34} = \mathbf{K}_{34}^{\mathrm{T}} \quad \mathbf{K}_{11} = \mathbf{K}_{22} \quad \mathbf{K}_{33} = \mathbf{K}_{44} \quad \mathbf{K}_{13} = -\mathbf{K}_{24} \quad \mathbf{K}_{23} = -\mathbf{K}_{14}, \text{ etc.} \tag{5.21}$$

Therefore,

$$\begin{pmatrix} \mathbf{K}_{11} & \mathbf{K}_{12} & \mathbf{K}_{13} & -\mathbf{K}_{23} \\ & \mathbf{K}_{11} & \mathbf{K}_{23} & \mathbf{K}_{13} \\ & & \mathbf{K}_{33} & \mathbf{K}_{34} \\ & \text{sym.} & & \mathbf{K}_{33} \end{pmatrix} \begin{Bmatrix} \mathbf{s}_1 \\ \mathbf{s}_2 \\ \mathbf{r}_1 \\ \mathbf{r}_2 \end{Bmatrix} = \begin{Bmatrix} \mathbf{S}_1 \\ \mathbf{S}_2 \\ \mathbf{R}_1 \\ \mathbf{R}_2 \end{Bmatrix} \tag{5.22}$$

where $\mathbf{K}_{11} = \mathbf{K}_{11}^{\mathrm{T}}$, $\mathbf{K}_{33} = \mathbf{K}_{33}^{\mathrm{T}}$, $\mathbf{K}_{12} = \mathbf{K}_{12}^{\mathrm{T}}$, $\mathbf{K}_{34} = \mathbf{K}_{34}^{\mathrm{T}}$. In conclusion, for a substructure having a plane of symmetry, only six of the ten submatrices are distinct and four of them are symmetrical. The reader is advised to check the stiffness matrices of a plane and a space beam member, which will indeed satisfy conditions (5.22).

Symmetry is not only useful in the formulation of elements (substructures), but also in the reduction of unknowns. Consider the beam in Fig. 5.5 as an example, where the coordinates q_i are decomposed into the symmetric parts s_i and anti-symmetric parts r_i.

$$\begin{pmatrix} k_{11} & k_{12} & k_{13} & k_{14} \\ & k_{22} & k_{23} & k_{24} \\ & & k_{33} & k_{34} \\ & \text{sym.} & & k_{44} \end{pmatrix} \begin{Bmatrix} q_1 \\ q_2 \\ q_3 \\ q_4 \end{Bmatrix} = \begin{Bmatrix} Q_1 \\ Q_2 \\ Q_3 \\ Q_4 \end{Bmatrix} \tag{5.23}$$

$$\begin{pmatrix} k_{11} & k_{12} & k_{13} & -k_{14} \\ & k_{22} & k_{23} & -k_{24} \\ & & k_{33} & -k_{34} \\ & \text{sym.} & & k_{44} \end{pmatrix} \begin{Bmatrix} s_1 \\ s_2 \\ s_3 \\ s_4 \end{Bmatrix} = \begin{Bmatrix} S_1 \\ S_2 \\ S_3 \\ S_4 \end{Bmatrix} \tag{5.24}$$

$$\begin{pmatrix} k_{11} & k_{12} & -k_{13} & k_{14} \\ & k_{22} & -k_{23} & k_{24} \\ & & k_{33} & -k_{34} \\ & \text{sym.} & & k_{44} \end{pmatrix} \begin{Bmatrix} r_1 \\ r_2 \\ r_3 \\ r_4 \end{Bmatrix} = \begin{Bmatrix} R_1 \\ R_2 \\ R_3 \\ R_4 \end{Bmatrix} \tag{5.25}$$

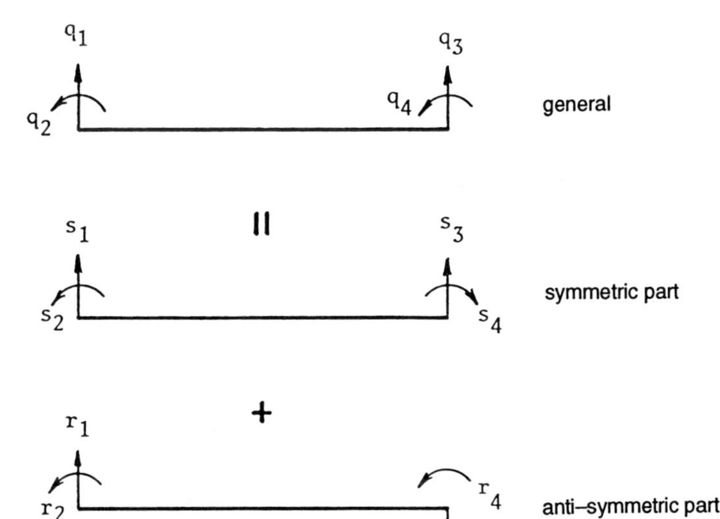

Fig. 5.5 Decomposition of a beam into symmetric parts

Equations 5.24 and 5.25 are derived from eq. 5.23 when certain coordinates reverse their directions. Since by symmetry $s_1 = s_3$, $s_2 = s_4$, $S_1 = S_3$, $S_2 = S_4$, therefore,

$$\begin{bmatrix} k_{11} + k_{33} + 2k_{13} & k_{12} + k_{23} - k_{14} - k_{34} \\ \text{sym.} & k_{22} + k_{44} - 2k_{24} \end{bmatrix} \begin{Bmatrix} s_1 \\ s_2 \end{Bmatrix} = 2 \begin{Bmatrix} S_1 \\ S_2 \end{Bmatrix} \tag{5.26}$$

Similarly,

$$\begin{bmatrix} k_{11} + k_{33} + 2k_{13} & k_{12} - k_{23} - k_{34} + k_{14} \\ \text{sym.} & k_{22} + k_{44} + 2k_{24} \end{bmatrix} \begin{Bmatrix} r_1 \\ r_2 \end{Bmatrix} = 2 \begin{Bmatrix} R_1 \\ R_2 \end{Bmatrix} \tag{5.27}$$

And, finally,

$$\begin{Bmatrix} q_1 \\ q_2 \\ q_3 \\ q_4 \end{Bmatrix} = \begin{Bmatrix} s_1 \\ s_2 \\ s_1 \\ -s_2 \end{Bmatrix} + \begin{Bmatrix} r_1 \\ r_2 \\ -r_1 \\ r_2 \end{Bmatrix} \tag{5.28}$$

where, the force components R_i and S_i are obtained from

$$\begin{Bmatrix} Q_1 \\ Q_2 \\ Q_3 \\ Q_4 \end{Bmatrix} = \begin{Bmatrix} S_1 \\ S_2 \\ S_1 \\ -S_2 \end{Bmatrix} + \begin{Bmatrix} R_1 \\ R_2 \\ -R_1 \\ R_2 \end{Bmatrix} \tag{5.29}$$

Therefore, the solution of one set of 4×4 equations (eq. 5.23) can be obtained by solving two sets of 2×2 equations (eqs 5.26 and 5.27). Since the solution of linear equations requires arithmetic operations proportional to n^3, where n is the order, it is always more advantageous to solve m systems of n equations than to solve one system of $m \times n$ equations. If the system is subject to either symmetric or anti-symmetric load only, then only eq. 5.26 or eq. 5.27 is needed.

Equations 5.23–5.29 can be written in matrix form, as follows,

$$[\mathbf{K}]\{\mathbf{q}\} = \{\mathbf{Q}\} \quad \text{or} \quad \begin{bmatrix} \mathbf{K}_{11} & \mathbf{K}_{12} \\ \mathbf{K}_{21} & \mathbf{K}_{22} \end{bmatrix} \begin{Bmatrix} \mathbf{q}_1 \\ \mathbf{q}_2 \end{Bmatrix} = \begin{Bmatrix} \mathbf{Q}_1 \\ \mathbf{Q}_2 \end{Bmatrix} \tag{5.30}$$

$$\begin{bmatrix} \mathbf{K}_{11} & \mathbf{K}_{12}\mathbf{J} \\ \mathbf{J}\mathbf{K}_{21} & \mathbf{J}\mathbf{K}_{22}\mathbf{J} \end{bmatrix} \begin{Bmatrix} \mathbf{s}_1 \\ \mathbf{s}_2 \end{Bmatrix} = \begin{Bmatrix} \mathbf{S}_1 \\ \mathbf{S}_2 \end{Bmatrix} \qquad [\mathbf{J}] = \begin{bmatrix} 1 & 0 \\ 0 & -1 \end{bmatrix} \tag{5.31}$$

$$\begin{bmatrix} \mathbf{K}_{11} & -\mathbf{K}_{12}\mathbf{J} \\ -\mathbf{J}\mathbf{K}_{21} & \mathbf{J}\mathbf{K}_{22}\mathbf{J} \end{bmatrix} \begin{Bmatrix} \mathbf{r}_1 \\ \mathbf{r}_2 \end{Bmatrix} = \begin{Bmatrix} \mathbf{R}_1 \\ \mathbf{R}_2 \end{Bmatrix} \tag{5.32}$$

$$[\mathbf{K}_{11} + \mathbf{K}_{12}\mathbf{J} + \mathbf{J}\mathbf{K}_{21} + \mathbf{J}\mathbf{K}_{22}\mathbf{J}]\{\mathbf{s}_1\} = \{\mathbf{S}_1\} \tag{5.33}$$

$$[\mathbf{K}_{11} - \mathbf{K}_{12}\mathbf{J} - \mathbf{J}\mathbf{K}_{21} + \mathbf{J}\mathbf{K}_{22}\mathbf{J}]\{\mathbf{r}_1\} = \{\mathbf{R}_1\} \tag{5.34}$$

$$\{\mathbf{q}_1\} = \{\mathbf{s}_1\} + \{\mathbf{r}_1\} \qquad \{\mathbf{q}_2\} = [\mathbf{J}]\{\mathbf{s}_1\} - [\mathbf{J}]\{\mathbf{r}_1\} \tag{5.35}$$

and

$$\{\mathbf{Q}_1\} = \{\mathbf{S}_1\} + \{\mathbf{R}_1\} \qquad \{\mathbf{Q}_2\} = [\mathbf{J}]\{\mathbf{S}_1\} - [\mathbf{J}]\{\mathbf{R}_1\} \tag{5.36}$$

Comparing eqs 5.33 and 5.34 with eq. 5.30, gives

$$\begin{Bmatrix} \mathbf{q}_1 \\ \mathbf{q}_2 \end{Bmatrix} = \begin{bmatrix} \mathbf{I} \\ \mathbf{J} \end{bmatrix} \{\mathbf{s}_1\} = [\mathbf{T}_s]\{\mathbf{s}_1\} \tag{5.37}$$

and

$$\begin{Bmatrix} \mathbf{q}_1 \\ \mathbf{q}_2 \end{Bmatrix} = \begin{bmatrix} \mathbf{I} \\ -\mathbf{J} \end{bmatrix} \{\mathbf{r}_1\} = [\mathbf{T}_r]\{\mathbf{r}_1\} \tag{5.38}$$

And the final results given in eqs 5.35 and 5.36 are therefore

$$\{\mathbf{q}\} = [\mathbf{T}_s]\{\mathbf{s}_1\} + [\mathbf{T}_r]\{\mathbf{r}_1\} \tag{5.39}$$

$$\{\mathbf{Q}\} = [\mathbf{T}_s]\{\mathbf{S}_1\} + [\mathbf{T}_r]\{\mathbf{r}_1\} \tag{5.40}$$

These equations are valid for all structures having one plane of symmetry.

Consider the rectangular space frame shown in Fig. 5.6 as an example of structures having multiple planes of symmetry. The origin 0 is at the centroid and the

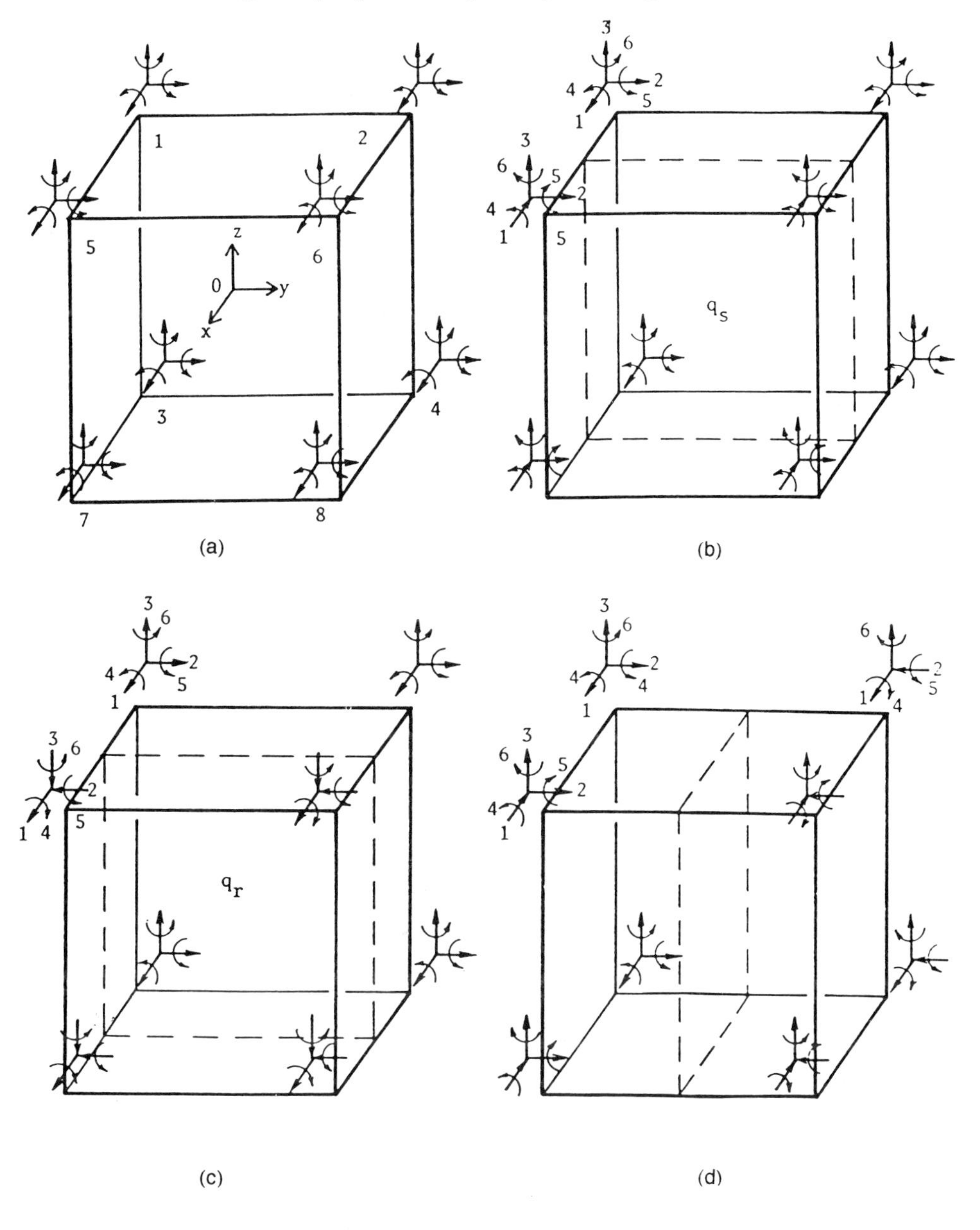

Fig. 5.6 (a) $\{\mathbf{q}\}$; (b) $\{\mathbf{q}_s\}$; (c) $\{\mathbf{q}_r\}$; (d) $\{\mathbf{q}_{ss}\}$; (e) $\{\mathbf{q}_{sr}\}$; (f) $\{\mathbf{q}_{rs}\}$; (g) $\{\mathbf{q}_{rr}\}$; (h) $\{\mathbf{q}_{sss}\}$

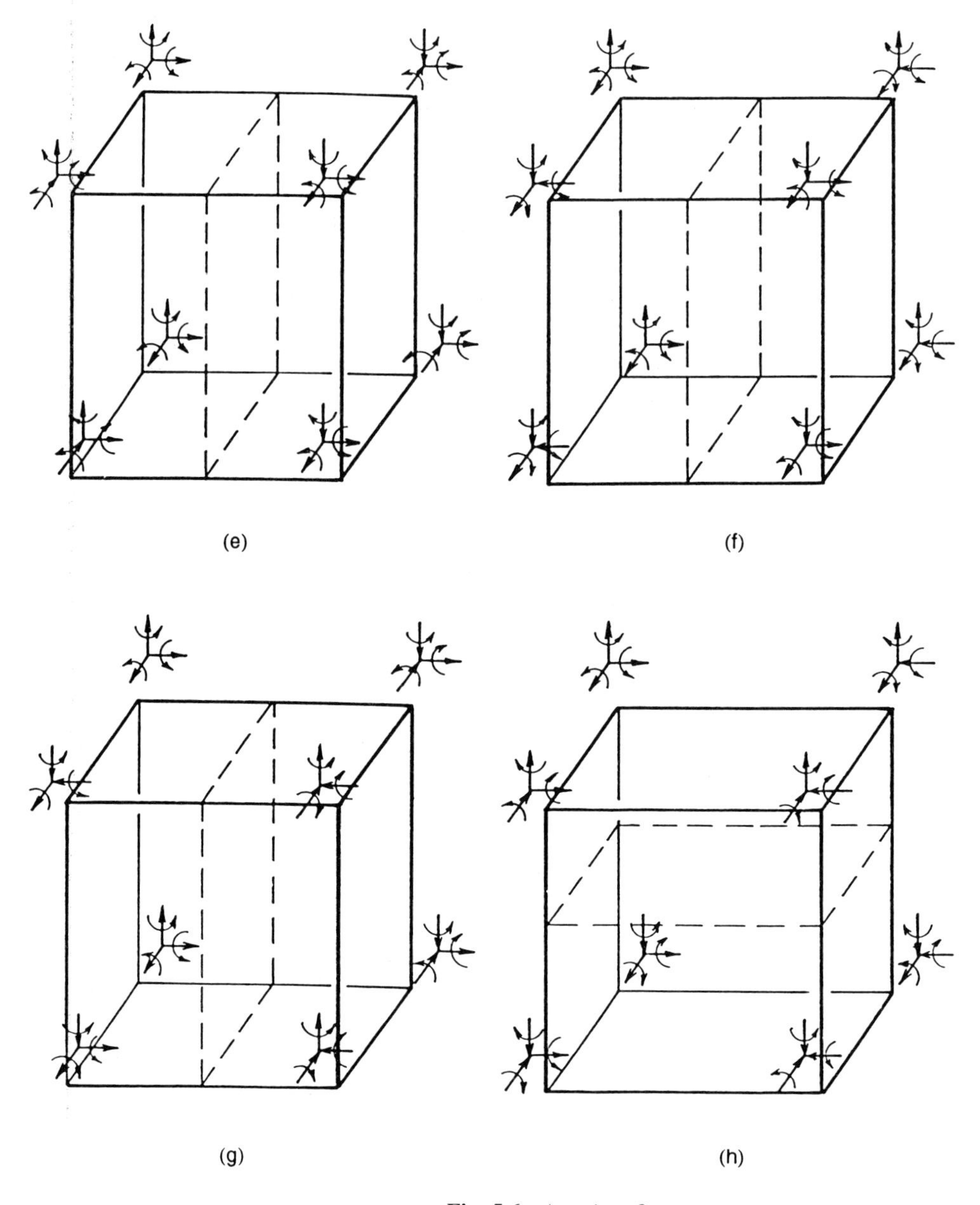

Fig. 5.6 (*continued*)

coordinates x, y, z are along the orthogonal axes. Let the numerical subscripts of $\mathbf{q}$ denote the node numbers.

When the yz plane of symmetry is considered,

$$\{\mathbf{q}\} = \text{col}\{\mathbf{q}_1\mathbf{q}_2 \ldots \mathbf{q}_8\} = \begin{bmatrix} \mathbf{I} \\ \mathbf{J}^{(1)} \end{bmatrix} \{\mathbf{q}_s\} + \begin{bmatrix} \mathbf{I} \\ -\mathbf{J}^{(1)} \end{bmatrix} \{\mathbf{q}_r\} \tag{5.41}$$

where the bracketed superscripts denote the level of symmetry and $\{\mathbf{q}_s\} = \text{col}\{\mathbf{q}_{s1}\mathbf{q}_{s2}\mathbf{q}_{s3}\mathbf{q}_{s4}\}$, $\{\mathbf{q}_r\} = \text{col}\{\mathbf{q}_{r1}\mathbf{q}_{r2}\mathbf{q}_{r3}\mathbf{q}_{r4}\}$, $[\mathbf{J}^{(1)}] = \text{diag}[\mathbf{J}_1, \mathbf{J}_1, \mathbf{J}_1, \mathbf{J}_1]$ where

$[\mathbf{J}_1] = \text{diag}[-1, 1, 1, 1, -1, -1]$, because, as typically shown in Fig. 5.6(b), $q_1^5 = -q_1', q_2^5 = -q_2', q_3^5 = -q_3', q_4^5 = q_4', q_5^5 = -q_5', q_6^5 = -q_6'$, in which the superscript denotes the node number and the subscript the coordinate number. If $[\mathbf{K}]$ is the 48×48 stiffness matrix for the $\mathbf{q}$ system, then the stiffness equations governing $\{\mathbf{q}_s\}$ and $\{\mathbf{q}_r\}$ are given by

$$[\mathbf{K}_s]\{\mathbf{q}_s\} = \{\mathbf{Q}_s\} \quad \text{and} \quad [\mathbf{K}_r]\{\mathbf{q}_r\} = \{\mathbf{Q}_r\} \tag{5.42}$$

where

$$\begin{aligned} &[\mathbf{K}_s] = [\mathbf{T}_s]^{\mathrm{T}}[\mathbf{K}][\mathbf{T}_s] && [\mathbf{K}_r] = [\mathbf{T}_r]^{\mathrm{T}}[\mathbf{K}][\mathbf{T}_r] \\ &[\mathbf{T}_s]^{\mathrm{T}} = [\mathbf{I}, \mathbf{J}^{(1)}] && [\mathbf{T}_r]^{\mathrm{T}} = [\mathbf{I}, -\mathbf{J}^{(1)}] \\ &[\mathbf{Q}_s] = [\mathbf{T}_s]^{\mathrm{T}}[\mathbf{Q}] && [\mathbf{Q}_r] = [\mathbf{T}_r]^{\mathrm{T}}[\mathbf{Q}] \end{aligned}$$

The transformation should take advantage of the fact that both $\mathbf{I}$ and $\mathbf{J}^{(1)}$ are diagonal and that no explicit matrix products are needed. Equations 5.42 are of the order 24 instead of the original 48.

When the xz plane of symmetry is considered,

$$\begin{aligned} [\mathbf{q}_s] &= [\mathbf{I}, \mathbf{J}^{(2)}]^{\mathrm{T}}\{\mathbf{q}_{ss}\} + [\mathbf{I}, -\mathbf{J}^{(2)}]^{\mathrm{T}}\{\mathbf{q}_{sr}\} \\ &= [\mathbf{T}_{ss}]^{\mathrm{T}}\{\mathbf{q}_{ss}\} + [\mathbf{T}_{sr}]\{\mathbf{q}_{sr}\} \end{aligned} \tag{5.43}$$

$$\begin{aligned} [\mathbf{q}_r] &= [\mathbf{I}, \mathbf{J}^{(2)}]^{\mathrm{T}}\{\mathbf{q}_{rs}\} + [\mathbf{I}, -\mathbf{J}^{(2)}]^{\mathrm{T}}\{\mathbf{q}_{rr}\} \\ &= [\mathbf{T}_{rs}]^{\mathrm{T}}\{\mathbf{q}_{rs}\} + [\mathbf{T}_{rr}]\{\mathbf{q}_{rr}\} \end{aligned} \tag{5.44}$$

where $\{\mathbf{q}_{rs}\} = \text{col}\{\mathbf{q}_{rs1}, \mathbf{q}_{rs2}\}$, etc. are of the order 12 and are associated with the first two nodes only, $[\mathbf{J}^{(2)}] = \text{diag}\,[\mathbf{J}_2, \mathbf{J}_2]$ where $[\mathbf{J}_2] = \text{diag}\,[1, -1, 1, -1, 1, -1]$, because, as typically shown in Fig. 5.6(d), $q_1^2 = q_1^1, q_2^2 = -q_2^1, q_3^2 = q_3^1, q_4^2 = -q_4^1, q_5^2 = q_5^1, q_6^2 = -q_6^1$. Obviously, $[\mathbf{T}_{ss}] = [\mathbf{T}_{rs}]$ and $[\mathbf{T}_{sr}] = [\mathbf{T}_{rr}]$. The stiffness equations governing $\{\mathbf{q}_{rs}\}$ etc. are given by

$$[\mathbf{K}_{rs}]\{\mathbf{q}_{rs}\} = \{\mathbf{Q}_{rs}\} \quad \text{etc.} \tag{5.45}$$

where

$$\begin{aligned} &[\mathbf{K}_{rs}] = [\mathbf{T}_{rs}]^{\mathrm{T}}[\mathbf{K}_r][\mathbf{T}_{rs}] && [\mathbf{K}_{sr}] = [\mathbf{T}_{sr}]^{\mathrm{T}}[\mathbf{K}_s][\mathbf{T}_{sr}] \\ &[\mathbf{K}_{rr}] = [\mathbf{T}_{rr}]^{\mathrm{T}}[\mathbf{K}_r][\mathbf{T}_{rr}] && [\mathbf{K}_{ss}] = [\mathbf{T}_{ss}]^{\mathrm{T}}[\mathbf{K}_s][\mathbf{T}_{ss}] \\ &[\mathbf{Q}_{rs}] = [\mathbf{T}_{rs}]^{\mathrm{T}}\{\mathbf{Q}_r\} && [\mathbf{Q}_{ss}] = [\mathbf{T}_{ss}]^{\mathrm{T}}\{\mathbf{Q}_s\} \quad \text{etc.} \end{aligned}$$

Equations 5.45 are of the order 12.

Finally, when the xy plane of symmetry is considered,

$$\begin{aligned} [\mathbf{q}_{ss}] &= [\mathbf{I}, \mathbf{J}^{(3)}]^{\mathrm{T}}\{\mathbf{q}_{sss}\} + [\mathbf{I}, -\mathbf{J}^{(3)}]^{\mathrm{T}}\{\mathbf{q}_{ssr}\} \\ &= [\mathbf{T}_{sss}]^{\mathrm{T}}\{\mathbf{q}_{sss}\} + [\mathbf{T}_{ssr}]^{\mathrm{T}}\{\mathbf{q}_{ssr}\} \\ [\mathbf{q}_{sr}] &= [\mathbf{T}_{srs}]^{\mathrm{T}}\{\mathbf{q}_{srs}\} + [\mathbf{T}_{srr}]^{\mathrm{T}}\{\mathbf{q}_{srr}\} \quad \text{etc.} \end{aligned} \tag{5.46}$$

Obviously, $[\mathbf{T}_{sss}] = [\mathbf{T}_{srs}] = [\mathbf{T}_{rss}] = [\mathbf{T}_{rrs}]$ and $[\mathbf{T}_{ssr}] = [\mathbf{T}_{srr}] = [\mathbf{T}_{rsr}] = [\mathbf{T}_{rrr}]$; $[\mathbf{J}^{(3)}] = \text{diag}[1, 1, -1, -1, -1, 1]$; $\{\mathbf{q}_{sss}\}$, $\{\mathbf{q}_{ssr}\}$, etc. are associated with node number one only. The stiffness equations governing $\{\mathbf{q}_{sss}\}$ etc. are given by

$$[\mathbf{K}_{sss}][\mathbf{q}_{sss}] = [\mathbf{Q}_{sss}] \quad \text{etc.} \tag{5.47}$$

where $[\mathbf{K}_{sss}] = [\mathbf{T}_{sss}]^{\mathrm{T}}[\mathbf{K}_{ss}][\mathbf{T}_{sss}]$ and $[\mathbf{Q}_{sss}] = [\mathbf{T}_{sss}]^{\mathrm{T}}[\mathbf{Q}_{ss}]$ etc. Equations 5.47 are of the order 6 only. Therefore, the original one system of 48 equations is decomposed into eight systems of 6 equations in the form of eq. 5.47. The original coordinate $\mathbf{q}$ is obtained by combining eqs 5.41, 5.43, 5.44 and 5.46.

5.4 Periodic structures

By utilizing the property of periodicity (defined later), the analysis of periodic structures can be greatly simplified. For rotational periodic structures, Thomas [3] used the discrete Fourier transform to change the governing equations for n substructures to n sets of uncoupled equations. Cheung, Chan and Cai [4,5,6] showed that the discrete Fourier transform is equivalent to the U-transformation technique for continuous systems and produced analytical solutions for periodic structures. In this section, geometric (coordinate) transformation will be used to solve the problem.

5.4.1 Periodic boundary conditions

A periodic structure consists of identical substructures coupled together in a regular manner. Depending on the arrangement, the periodicity may be linear, as in long bridges or multi-storey buildings, or circular, as in domes or axisymmetric shells, or may even be extended to two and three dimensions, as in framed roofs and lattices. The analysis of such structures is greatly simplified by utilizing the periodicity property.

When a substructure has a plane of symmetry, the relations in eq. 5.21 also hold. With these relations, n such identical substructures can be assembled by a formal finite element process to give the following difference equations for the kth interface station:

$$\begin{pmatrix} K_{12} & K_{23} \\ -K_{23}^{\mathrm{T}} & K_{34} \end{pmatrix} \begin{Bmatrix} \mathbf{s}_{k-1} \\ \mathbf{r}_{k-1} \end{Bmatrix} + 2 \begin{pmatrix} K_{11} & 0 \\ 0 & K_{23} \end{pmatrix} \begin{Bmatrix} \mathbf{s}_{k} \\ \mathbf{r}_{k} \end{Bmatrix} + \begin{pmatrix} K_{12} & -K_{23} \\ K_{23}^{\mathrm{T}} & K_{34} \end{pmatrix} \begin{Bmatrix} \mathbf{s}_{k+1} \\ \mathbf{r}_{k+1} \end{Bmatrix} = \begin{Bmatrix} \mathbf{S}_{k} \\ \mathbf{R}_{k} \end{Bmatrix} \qquad k = 1, 2, \ldots, n-1 \tag{5.48}$$

$$\begin{pmatrix} K_{11} & K_{13} \\ -K_{13}^{\mathrm{T}} & K_{33} \end{pmatrix} \begin{Bmatrix} \mathbf{s}_{0} \\ \mathbf{r}_{0} \end{Bmatrix} + \begin{pmatrix} K_{12} & -K_{23} \\ K_{23}^{\mathrm{T}} & K_{34} \end{pmatrix} \begin{Bmatrix} \mathbf{s}_{1} \\ \mathbf{r}_{1} \end{Bmatrix} = \begin{Bmatrix} \mathbf{S}_{0} \\ \mathbf{R}_{0} \end{Bmatrix}$$

$$\begin{pmatrix} K_{12} & K_{23} \\ -K_{23}^{\mathrm{T}} & K_{34} \end{pmatrix} \begin{Bmatrix} \mathbf{s}_{n-1} \\ \mathbf{r}_{n-1} \end{Bmatrix} + \begin{pmatrix} K_{11} & -K_{13} \\ K_{13}^{\mathrm{T}} & K_{33} \end{pmatrix} \begin{Bmatrix} \mathbf{s}_{n} \\ \mathbf{r}_{n} \end{Bmatrix} = \begin{Bmatrix} \mathbf{S}_{n} \\ \mathbf{R}_{n} \end{Bmatrix} \tag{5.49}$$

These difference equations are solved for some periodic boundary conditions in the following section and for more general boundary conditions at station $k = 0$ and $k = n$ later. The system is studied by isolating n consecutive substructures. The periodic boundary conditions permit the assumptions,

$$\mathbf{s}_k = \sum_j^{*} \mathbf{e}_j \sin(jk\alpha) \qquad \mathbf{r}_k = \sum_j^{*} \mathbf{g}_j \cos(jk\alpha) \tag{5.50}$$

where $\alpha = \pi/n$, $\mathbf{e}_j$ and $\mathbf{g}_j$ are vectors to be determined, and the range of summation is to be from 0 to n. The summation operator $\overset{*}{\sum}$ is defined by

$$\sum_k^{*} a_k = \frac{1}{2}(a_0 + a_n) + \sum_{k=1}^{n-1} a_k \tag{5.51}$$

Then it can be proved that

$$\sum_k^* \sin(ik\alpha)\sin(jk\alpha) \begin{matrix} = 0 & \text{when} \quad i \neq j \\ = n/2 & \text{when} \quad i = j \end{matrix} \Bigg\} \tag{5.52}$$

$$\begin{aligned} \sum_k^* \cos(ik\alpha)\cos(jk\alpha) &= 0 && \text{when } i \neq j \\ &= n/2 && \text{when } i = j \neq 0 \text{ or } n \\ &= n && \text{when } i = j = 0 \text{ or } n \end{aligned} \tag{5.53}$$

With the aid of expressions 5.52 and 5.53, eq. 5.50 can transform eq. 5.48 to eq. 5.49

$$\mathbf{Kx} = \mathbf{X} \tag{5.54}$$

where $\mathbf{K} = \text{diag}[\mathbf{K}_j]$, $x = \text{col}\{\mathbf{x}_j\}$, $\mathbf{X} = \text{col}\{\mathbf{X}_j\}$, $j = 0.1. \ldots, n$

$$\mathbf{K}_j = \begin{pmatrix} \mathbf{K}_{11} + \mathbf{K}_{12}\cos(j\alpha) & \mathbf{K}_{23}\sin(j\alpha) \\ \mathbf{K}_{23}^{\mathrm{T}}\sin(j\alpha) & \mathbf{K}_{33} + \mathbf{K}_{34}\cos(j\alpha) \end{pmatrix} \quad \mathbf{x}_j = \begin{Bmatrix} \mathbf{e}_j \\ \mathbf{g}_j \end{Bmatrix} \quad \mathbf{X}_j = \begin{Bmatrix} \mathbf{E}_j \\ \mathbf{G}_j \end{Bmatrix} \tag{5.55}$$

$$\begin{aligned} \mathbf{E}_j &= \frac{1}{n}\sum_k^* \mathbf{S}_k \sin(j\alpha) & \mathbf{G}_j &= \frac{1}{n}\sum_k^* \mathbf{R}_k \cos(j\alpha) \\ \mathbf{G}_o &= \frac{2}{n}\sum_k^* \mathbf{R}_k & \mathbf{G}_n &= \frac{2}{n}\sum_k^* (-1)^k \mathbf{R}_k \end{aligned} \tag{5.56}$$

Since $\mathbf{K}$ has been decomposed into block diagonal form, eq. 5.48 may be written in the uncoupled form

$$\mathbf{K}_j\mathbf{x}_j = \mathbf{X}_j \qquad j = 0, 1, \ldots, n \tag{5.57}$$

If a substructure has $2m$ master coordinates, the original system of order $m(n+1)$ is decomposed into $n+1$ systems of m unknowns. The computation is as follows.

(i) Form the force vectors $\mathbf{R}_j$, $\mathbf{S}_j$, find $\mathbf{E}_j$ and $\mathbf{G}_j$ from eq. 5.56 to obtain $\mathbf{X}_j$.
(ii) Form the stiffness matrix of the jth system according to eq. 5.55.
(iii) Solve $\{\mathbf{x}_j\} = \{\mathbf{e}_j, \mathbf{g}_j\}$ from eq. 5.57.
(iv) Find $\mathbf{s}_k$ and $\mathbf{r}_k$, the original unknowns from eq. 5.50.

For cyclic symmetry, $\mathbf{s}_0 = \mathbf{s}_n$ and $\mathbf{r}_0 = \mathbf{r}_n$, the procedure can similarly apply.

5.4.2 General boundary conditions

A solution algorithm for periodic structures having general boundary conditions is presented next. Due to the generally non-periodic boundary conditions, when the boundary conditions at stations 0 and n are not the same, difference calculus and transformation methods are not applicable. The number of operations for a general solution is directly proportional to the number N of substructures involved, which is quite inefficient when N is large as is the case for space structures. A fast algorithm analogous to the fast Fourier transform method is introduced. The number of operations is proportional to log N rather than N. Numerical examples are given to illustrate the effectiveness. Due to the general nature of boundary conditions, a periodic structure can be condensed into a substructure of higher level which can then be connected at the boundaries to another structure or supports.

The structure under consideration is spatially periodic and linear. The assumption of linearity implies material linearity and small deflections so that structural responses are linear functions of the applied loads. The fast static condensation algorithm is applicable to periodic structures of general boundary conditions subject to arbitrary loading cases.

Figure 5.7 shows a finite element model of a structure of one dimensional periodicity with the numbering shown. Each substructure consists of the left-and right-hand substructures respectively. The right station of the ith substructure is named as the ith station. The beginning and ending stations (boundary stations) of the periodic structure are defined as the left station of the first substructure and the right station of the last substructure respectively. A station which is neither a beginning nor an end station is called an interior station. The stiffness equations of a substructure can be partitioned as

$$\begin{bmatrix} \mathbf{K}_{\ell\ell} & \mathbf{K}_{\ell r} \\ \mathbf{K}_{r\ell} & \mathbf{K}_{rr} \end{bmatrix} \begin{Bmatrix} \boldsymbol{\delta}_{\ell} \\ \boldsymbol{\delta}_{r} \end{Bmatrix} = \begin{Bmatrix} \mathbf{F}_{\ell} \\ \mathbf{F}_{r} \end{Bmatrix} \tag{5.58}$$

where $\{\boldsymbol{\delta}_\ell\}$, $\{\boldsymbol{\delta}_r\}$ are displacements of left and right stations of a substructure respectively, and $\{\mathbf{F}_\ell\}$, $\{\mathbf{F}_r\}$ are the corresponding nodal forces; $[\mathbf{K}_{\ell\ell}]$, $[\mathbf{K}_{\ell r}]$, $[\mathbf{K}_{r\ell}]$ and $[\mathbf{K}_{rr}]$ are the partitioned submatrices.

After assembling all the substructures and applying general boundary conditions at the beginning and end stations, the global stiffness equations can be expressed as

$$\begin{bmatrix} \mathbf{K}'_{\ell\ell} & \mathbf{K}'_{c} & & & & \\ & \mathbf{K}_s & \mathbf{K}_c & & 0 & \\ & & \mathbf{K}_s & \mathbf{K}_c & & \\ & & & \ddots & \ddots & \\ & & & & \mathbf{K}_s & \mathbf{K}''_c \\ & & & & & \mathbf{K}''_{rr} \end{bmatrix} \begin{Bmatrix} \boldsymbol{\delta}_0 \\ \boldsymbol{\delta}_1 \\ \vdots \\ \vdots \\ \boldsymbol{\delta}_N \end{Bmatrix} = \begin{Bmatrix} \mathbf{F}_0 \\ \mathbf{F}_1 \\ \vdots \\ \vdots \\ \mathbf{F}_N \end{Bmatrix} \tag{5.59}$$

where N is the number of substructures; $[\mathbf{K}_s] = [\mathbf{K}_{\ell\ell} + \mathbf{K}_{rr}]$ is the stiffness matrix of the interior stations and $[\mathbf{K}_c] = [\mathbf{K}_{\ell r}]$ is the coupling stiffness matrix between

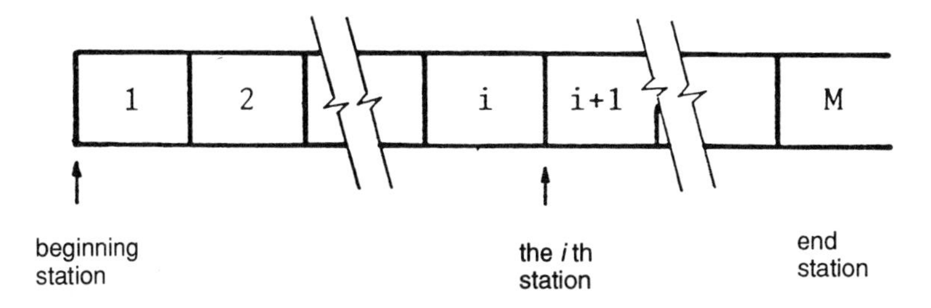

Fig. 5.7 A periodic structure

adjacent stations. $[\mathbf{K}'_{\ell\ell}]$ and $[\mathbf{K}''_{rr}]$ are the stiffness matrices of the beginning and end stations respectively and can be obtained by deleting the corresponding rows and columns of $[\mathbf{K}_{\ell\ell}]$ and $[\mathbf{K}_{rr}]$ respectively, after boundary conditions are applied. $[\mathbf{K}'_c]$ and $[\mathbf{K}''_c]$ are also obtained by deleting the corresponding rows and columns of $[\mathbf{K}_{\ell r}]$, respectively.

Applying a static condensation procedure on the boundary stations, one can modify eq. 5.59 by eliminating $\{\boldsymbol{\delta}_0\}$ and $\{\boldsymbol{\delta}_N\}$ to give

$$\begin{bmatrix} \mathbf{K}_b & \mathbf{K}_c & & & & & \\ & \mathbf{K}_s & \mathbf{K}_c & & 0 & & \\ & & \mathbf{K}_s & \mathbf{K}_c & & & \\ & & & \cdot & \cdot & & \\ & \text{sym.} & & & \cdot & \cdot & \\ & & & & & \mathbf{K}_s & \mathbf{K}_c \\ & & & & & & \mathbf{K}_e \end{bmatrix} \begin{Bmatrix} \mathbf{u}_0 \\ \mathbf{u}_1 \\ \cdot \\ \cdot \\ \cdot \\ \cdot \\ \cdot \\ \cdot \\ \mathbf{u}_n \end{Bmatrix} = \begin{Bmatrix} \mathbf{f}_0 \\ \mathbf{f}_1 \\ \cdot \\ \cdot \\ \cdot \\ \cdot \\ \cdot \\ \cdot \\ \mathbf{f}_n \end{Bmatrix} \tag{5.60}$$

where $n = N - 2$ is the number of substructures remaining and

$$\begin{aligned} [\mathbf{K}_b] &= [\mathbf{K}_s - \mathbf{K}_c^{'\mathrm{T}} \mathbf{K}_{\ell\ell}^{'-1} \mathbf{K}'_c] \\ [\mathbf{K}_e] &= [\mathbf{K}_s - \mathbf{K}''_c \mathbf{K}_{rr}^{''-1} \mathbf{K}_c^{''\mathrm{T}}] \\ \{\mathbf{f}_0\} &= \{\mathbf{F}_1 - \mathbf{K}_c^{'\mathrm{T}} \mathbf{K}_{\ell\ell}^{'-1} \mathbf{F}_0\} \\ \{\mathbf{f}_n\} &= \{\mathbf{F}_{N-1} - \mathbf{K}''_c \mathbf{K}_{rr}^{''-1} \mathbf{F}_N^{''\mathrm{T}}\} \\ \{\mathbf{f}_i\} &= \{\mathbf{F}_{i+1}\} \qquad i = 1, 2, \ldots, n-1 \\ \{\mathbf{u}_i\} &= \{\boldsymbol{\delta}_{i+1}\} \qquad i = 0, 1, \ldots, n \end{aligned} \tag{5.61}$$

Note that each sub-matrix in eq. 5.60 is of the same order. If $\{\mathbf{u}_i\}$, $i = 0.1. \ldots, n$, have been obtained from eq. 5.60, the displacements $\{\boldsymbol{\delta}_i\}$, $i = 0, 1, \ldots, N$, in eq. 5.59 can be obtained as

$$\begin{aligned} \{\boldsymbol{\delta}_0\} &= \{\mathbf{K}_{\ell\ell}^{'-1} \mathbf{F}_0 - \mathbf{K}_{\ell\ell}^{'-1} \mathbf{K}'_c \mathbf{u}_0\} \\ \{\boldsymbol{\delta}_i\} &= \{\mathbf{u}_{i-l}\} \qquad i = 1, 2, \ldots, N-1 \\ \{\boldsymbol{\delta}_N\} &= \{\mathbf{K}_{rr}^{''-1} \mathbf{F}_N - \mathbf{K}_{rr}^{''-1} \mathbf{K}_c^{'\mathrm{T}} \mathbf{u}_n\} \end{aligned} \tag{5.62}$$

Our aim is to develop a Fast Static Condensation Algorithm (FSCA) to solve the stiffness equations of the form given in eq. 5.60. The concept of matrix difference equations will be used throughout and hence eq. 5.60 is expressed alternatively as

$$\mathbf{K}_b \mathbf{u}_0 + \mathbf{K}_c \mathbf{u}_1 = \mathbf{f}_0 \tag{5.63}$$

$$\mathbf{K}_c^{\mathrm{T}} \mathbf{u}_{k-1} + \mathbf{K}_s \mathbf{u}_k + \mathbf{K}_c \mathbf{u}_{k+1} = \mathbf{f}_k \qquad k = 1, 2, \ldots, n-1 \tag{5.64}$$

$$\mathbf{K}_c^{\mathrm{T}} \mathbf{u}_{n-1} + \mathbf{K}_c \mathbf{u}_n = \mathbf{f}_n \tag{5.65}$$

Two cases are of particular interest: (i) when n is even and (ii) when n is odd. These are discussed separately in the following paragraphs.

(a) Even number of substructures
In the case when n is even in eqs 5.63–5.65, and without loss of generality, it can be assumed that $n = 2m$ where m is a positive integer. Equations (5.63–5.65) can therefore be rewritten as

$$\mathbf{K}_b\mathbf{u}_0 + \mathbf{K}_c\mathbf{u}_1 = \mathbf{f}_0 \tag{5.66}$$

$$\mathbf{K}_c^{\mathrm{T}}\mathbf{u}_{2k-1} + \mathbf{K}_s\mathbf{u}_{2k} + \mathbf{K}_c\mathbf{u}_{2k+1} = \mathbf{f}_{2k} \qquad k = 1, 2, \ldots, m-1 \tag{5.67}$$

$$\mathbf{K}_c^{\mathrm{T}}\mathbf{u}_{2m-1} + \mathbf{K}_c\mathbf{u}_{2m} = \mathbf{f}_{2m} \tag{5.68}$$

$$\mathbf{K}_c^{\mathrm{T}}\mathbf{u}_{2k-2} + \mathbf{K}_s\mathbf{u}_{2k-1} + \mathbf{K}_c\mathbf{u}_{2k} = \mathbf{f}_{2k-1} \qquad k = 1, 2, \ldots, m \tag{5.69}$$

From eq. 5.69

$$\mathbf{u}_{2k-1} = \mathbf{K}_s^{-1}(\mathbf{f}_{2k-1} - \mathbf{K}_c^{\mathrm{T}}\mathbf{u}_{2k-2} - \mathbf{K}_c\mathbf{u}_{2k}) \qquad k = 1, 2, \ldots, m \tag{5.70}$$

Thus equations (5.66–5.68) become

$$(\mathbf{K}_b - \mathbf{K}_c\mathbf{K}_s^{-1}\mathbf{K}_c^{\mathrm{T}})\mathbf{u}_0 + (-\mathbf{K}_c\mathbf{K}_s^{-1}\mathbf{K}_c)\mathbf{u}_2 = \mathbf{f}_0 - \mathbf{K}_c\mathbf{K}_s^{-1}\mathbf{f}_1 \tag{5.71}$$

$$(-\mathbf{K}_c^{\mathrm{T}}\mathbf{K}_s^{-1}\mathbf{K}_c^{\mathrm{T}})\mathbf{u}_{2k-2} + (\mathbf{K}_s - \mathbf{K}_c^{\mathrm{T}}\mathbf{K}_s^{-1}\mathbf{K}_c - \mathbf{K}_c\mathbf{K}_s^{-1}\mathbf{K}_c^{\mathrm{T}})\mathbf{u}_{2k}$$
$$+ (-\mathbf{K}_c\mathbf{K}_s^{-1}\mathbf{K}_c)\mathbf{u}_{2k+2} = \mathbf{f}_{2k} - \mathbf{K}_c^{\mathrm{T}}\mathbf{K}_s^{-1}\mathbf{f}_{2k-1} - \mathbf{K}_c\mathbf{K}_s^{-1}\mathbf{f}_{2k+1}$$
$$k = 1, 2, \ldots, m-1 \tag{5.72}$$

$$(-\mathbf{K}_c^{\mathrm{T}}\mathbf{K}_s^{-1}\mathbf{K}_c^{\mathrm{T}})\mathbf{u}_{2m-2} + (\mathbf{K}_c - \mathbf{K}_c^{\mathrm{T}}\mathbf{K}_s^{-1}\mathbf{K}_c)\mathbf{u}_{2m} = \mathbf{f}_{2m} - \mathbf{K}_c^{\mathrm{T}}\mathbf{K}_s^{-1}\mathbf{f}_{2m-1} \tag{5.73}$$

Using

$$\begin{aligned}
\overline{\mathbf{K}}_b &= \mathbf{K}_b - \mathbf{K}_c\mathbf{K}_s^{-1}\mathbf{K}_c^{\mathrm{T}} \\
\overline{\mathbf{K}}_c &= \mathbf{K}_c - \mathbf{K}_c^{\mathrm{T}}\mathbf{K}_s^{-1}\mathbf{K}_c \\
\overline{\mathbf{K}}_s &= \mathbf{K}_s - \mathbf{K}_c^{\mathrm{T}}\mathbf{K}_s^{-1}\mathbf{K}_c - \mathbf{K}_c\mathbf{K}_s^{-1}\mathbf{K}_c^{\mathrm{T}} \\
\overline{\mathbf{K}}_c &= -\mathbf{K}_s^{-1}\mathbf{K}_c \\
\overline{\mathbf{u}}_k &= \mathbf{u}_{2k} \qquad k = 0, 1, 2, \ldots, m \\
\overline{\mathbf{f}}_0 &= \mathbf{f}_0 - \mathbf{K}_c\mathbf{K}_s^{-1}\mathbf{f}_1 \\
\overline{\mathbf{f}}_k &= \mathbf{f}_{2k} - \mathbf{K}_c^{\mathrm{T}}\mathbf{K}_s^{-1}\mathbf{f}_{2k-1} - \mathbf{K}_c\mathbf{K}_s^{-1}\mathbf{f}_{2k+1} \qquad k = 1, 2, \ldots, m-1 \\
\overline{\mathbf{f}}_m &= \mathbf{f}_{2m} - \mathbf{K}_c^{\mathrm{T}}\mathbf{K}_s^{-1}\mathbf{f}_{2m-1} \\
\overline{n} &= m
\end{aligned} \tag{5.74}$$

Equations (5.71–5.73) can be reduced to

$$\overline{\mathbf{K}}_b\overline{\mathbf{u}}_0 + \overline{\mathbf{K}}_c\overline{\mathbf{u}}_l = \overline{\mathbf{f}}_0 \tag{5.75}$$

$$\overline{\mathbf{K}}_c^{\mathrm{T}}\overline{\mathbf{u}}_{k-1} + \overline{\mathbf{K}}_s\overline{\mathbf{u}}_k + \overline{\mathbf{K}}_c\overline{\mathbf{u}}_{k+1} = \overline{\mathbf{f}}_k \qquad k = 1, 2, \ldots, \overline{n}-1 \tag{5.76}$$

$$\overline{\mathbf{K}}_c^{\mathrm{T}}\overline{\mathbf{u}}_{\overline{n}-1} + \overline{\mathbf{K}}_c\overline{\mathbf{u}}_{\overline{n}} = \overline{\mathbf{f}}_{\overline{n}} \tag{5.77}$$

Equations (5.75–5.77) are of the same form as eqs (5.63–5.65) but the number of unknowns is reduced nearly by half. Note that only one inversion of $[\mathbf{K}_s]$ is needed in transformations 5.74, in contrast to the conventional block decomposition which requires about m matrix inversions. Furthermore, the inverse of $[\mathbf{K}_s]$ is not calculated explicitly, and this will be discussed later.

(b) Odd number of substructures

In the case when n is odd in eqs 5.63–5.65, transformations 5.74 cannot be applied directly. However, it can be shown that eqs (5.63–5.65) can be transformed from the odd case to the even case without distorting their forms as follows.

From eq 5.65

$$\mathbf{u}_n = \mathbf{K}_c^{-1}(\mathbf{f}_n - \mathbf{K}_c^{\mathrm{T}}\mathbf{u}_{n-1}) \tag{5.78}$$

Substituting eq 5.78 into eqs 5.63 and 5.64 gives

$$\mathbf{K}_b\mathbf{u}_0 + \mathbf{K}_c\mathbf{u}_1 = \mathbf{f}_0 \tag{5.79}$$

$$\mathbf{K}_c^{\mathrm{T}}\mathbf{u}_{k-1} + \mathbf{K}_s\mathbf{u}_k + \mathbf{K}_c\mathbf{u}_{k+1} = \mathbf{f}_k \qquad k = 1, 2, \ldots, n-2 \tag{5.80}$$

$$\mathbf{K}_c^{\mathrm{T}}\mathbf{u}_{n-2} + (\mathbf{K}_s - \mathbf{K}_c\mathbf{K}_s^{-1}\mathbf{K}_c^{\mathrm{T}})\mathbf{u}_{n-1} = \mathbf{f}_{n-1} - \mathbf{K}_c\mathbf{K}_s^{-1}\mathbf{f}_n \tag{5.81}$$

Using

$$\begin{aligned}
&\overline{\mathbf{K}}_b = \mathbf{K}_b \qquad \overline{\mathbf{K}}_c = \mathbf{K}_s - \mathbf{K}_c\mathbf{K}_s^{-1}\mathbf{K}_c^{\mathrm{T}}\\
&\overline{\mathbf{K}}_s = \mathbf{K}_s \qquad \overline{\mathbf{K}}_c = \mathbf{K}_c\\
&\overline{\mathbf{u}}_k = \mathbf{u}_k \qquad k = 0, 1, 2, \ldots, n-1\\
&\overline{\mathbf{f}}_k = \mathbf{f}_k \qquad k = 0, 1, 2, \ldots, n-2\\
&\overline{\mathbf{f}}_{n-1} = \mathbf{f}_{n-1} - \mathbf{K}_c\mathbf{K}_s^{-1}\mathbf{f}_n\\
&\overline{n} = n-1
\end{aligned} \tag{5.82}$$

Equations (5.79–5.81) can be reduced to

$$\overline{\mathbf{K}}_b\overline{\mathbf{u}}_0 + \overline{\mathbf{K}}_c\overline{\mathbf{u}}_1 = \overline{\mathbf{f}}_0 \tag{5.83}$$

$$\overline{\mathbf{K}}_c^{\mathrm{T}}\overline{\mathbf{u}}_{k-1} + \overline{\mathbf{K}}_s\overline{\mathbf{u}}_k + \overline{\mathbf{K}}_c\overline{\mathbf{u}}_{k+1} = \overline{\mathbf{f}}_k \qquad k = 1, 2, \ldots, \overline{n}-1 \tag{5.84}$$

$$\overline{\mathbf{K}}_c^{\mathrm{T}}\overline{\mathbf{u}}_{\overline{n}-1} + \overline{\mathbf{K}}\overline{\mathbf{u}}_{\overline{n}} = \overline{\mathbf{f}}_{\overline{n}} \tag{5.85}$$

Equations (5.83–5.85) are of the same form as eqs (5.63–5.65) and the odd case is now transformed to the even case as $\overline{n} = n - 1$.

The transformations 5.74 reduce the number of unknowns exponentially (halving the number of equations each time). It is analogous to the power-of-two algorithm in the fast Fourier transform in which the computational cost is proportional to $P \log P$ instead of P^2 in the discrete Fourier transform, where P is the number of discrete points chosen. On the other hand, transformations 5.82 modify the equations without changing their forms so that transformations 5.74 can be applied later. The above procedure is repeated until the number of stations is reduced to two, then the resulting stiffness equations become

$$\begin{bmatrix} \overline{\mathbf{K}}_b & \overline{\mathbf{K}}_c \\ \overline{\mathbf{K}}_c^{\mathrm{T}} & \overline{\mathbf{K}}_c \end{bmatrix} \begin{Bmatrix} \overline{\mathbf{u}}_0 \\ \overline{\mathbf{u}}_1 \end{Bmatrix} = \begin{Bmatrix} \overline{\mathbf{f}}_0 \\ \overline{\mathbf{f}}_1 \end{Bmatrix} \tag{5.86}$$

After $\{\bar{\mathbf{u}}_0\}$ and $\{\bar{\mathbf{u}}_1\}$ in eq. 5.86 have been found, other $\{\bar{\mathbf{u}}_i\}$s can be obtained from eq. 5.70 and eq. 5.78.

The inversion of $[\mathbf{K}_s]$ and $[\mathbf{K}_c]$ in transformations 5.74 and 5.82 is accomplished by the following procedure. The $[\mathbf{K}_s]$ or $[\mathbf{K}_c]$ is first decomposed by making use of its symmetry into $\mathbf{LDL}^{\mathrm{T}}$ and the solution phase of $\mathbf{K}_s^{-1}V$ or $\mathbf{K}_c^{-1}V$ is performed by

$$\mathbf{L}^{-T}\mathbf{D}^{-1}(\mathbf{L}^{-1}\mathbf{V}) \tag{5.87}$$

where $\mathbf{L}$ is a lower triangular matrix, $\mathbf{D}$ is a diagonal matrix and $\mathbf{V}$ is the right-hand vector. It is found that this procedure is very efficient in reducing the round-off errors during matrix inversion.

5.4.3 Solution procedure

Before discussing the solution procedure, the static condensation procedure is summarized here. Consider a system of stiffness equations

$$[\mathbf{K}]\{\mathbf{u}\} = \{\mathbf{f}\} \tag{5.88}$$

which is partitioned into

$$\begin{bmatrix} \mathbf{K}_{\mathrm{MM}} & \mathbf{K}_{\mathrm{MS}} \\ \mathbf{K}_{\mathrm{SM}} & \mathbf{K}_{\mathrm{SS}} \end{bmatrix} \begin{Bmatrix} \mathbf{u}_{\mathrm{M}} \\ \mathbf{u}_{\mathrm{S}} \end{Bmatrix} = \begin{Bmatrix} \mathbf{f}_{\mathrm{M}} \\ \mathbf{f}_{\mathrm{S}} \end{Bmatrix} \tag{5.89}$$

where the subscript M denotes the masters and the subscript S denotes the slaves which are to be eliminated. Applying the static condensation procedure, the stiffness equations are reduced to

$$[\bar{\mathbf{K}}]\{\mathbf{u}_{\mathrm{M}}\} = \{\bar{\mathbf{f}}_{\mathrm{M}}\} \tag{5.90}$$

where

$$[\bar{\mathbf{K}}] = [\mathbf{K}_{\mathrm{MM}} - \mathbf{K}_{\mathrm{MS}}\mathbf{K}_{\mathrm{SS}}^{-1}\mathbf{K}_{\mathrm{SM}}] \tag{5.91}$$

$$\{\bar{\mathbf{f}}_{\mathrm{M}}\} = \{\mathbf{f}_{\mathrm{M}} - \mathbf{K}_{\mathrm{MS}}\mathbf{K}_{\mathrm{SS}}^{-1}\mathbf{f}_{\mathrm{S}}\} \tag{5.92}$$

and

$$\{\mathbf{u}_{\mathrm{S}}\} = \{\mathbf{K}_{\mathrm{SS}}^{-1}\mathbf{f}_{\mathrm{S}} - \mathbf{K}_{\mathrm{SS}}^{-1}\mathbf{K}_{\mathrm{SM}}\mathbf{u}_{\mathrm{M}}\} \tag{5.93}$$

The calculations of $[\bar{\mathbf{K}}]$ by eq. 5.91 and $\{\bar{\mathbf{f}}_{\mathrm{M}}\}$ by eq. 5.92 are called the condensation phase and the equivalent force evaluation phase, respectively. The computation of $\{\mathbf{u}_{\mathrm{M}}\}$ by eq. 5.90 and then $\{\mathbf{u}_{\mathrm{S}}\}$ by eq. 5.93 is called the backward substitution phase.

Similar to the static condensation procedure, the proposed algorithm of FSCA is also divided into condensation, equivalent force evaluation and backward substitution phases. These three phases are discussed separately in the following.

(a) Condensation phase

The condensation phase is concerned with the evaluation of the stiffness matrix in eq. 5.86.

Let N = number of substructures and NSEQ = number of cases performed. IND(I) = 0 if the Ith case is even; and IND(I) = 1 otherwise; $I = 1, 2, \ldots,$NSEQ. The flow diagram is as follows.

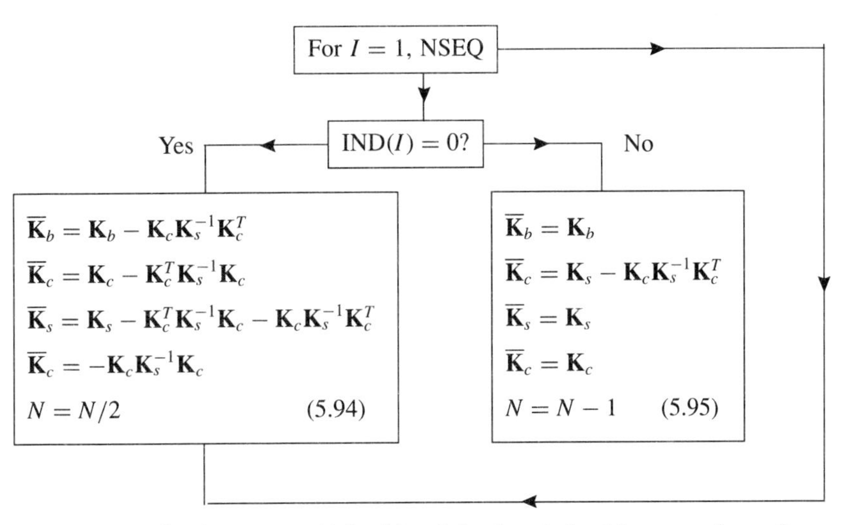

The number of substructures N is either halved or halved less one depending on whether IND(I) is even or odd.

(b) Equivalent force evaluation phase

The equivalent force evaluation phase is to obtain the equivalent force vector in eq. 5.86. Rcset N = number of substructures. The flow diagram is as follows.

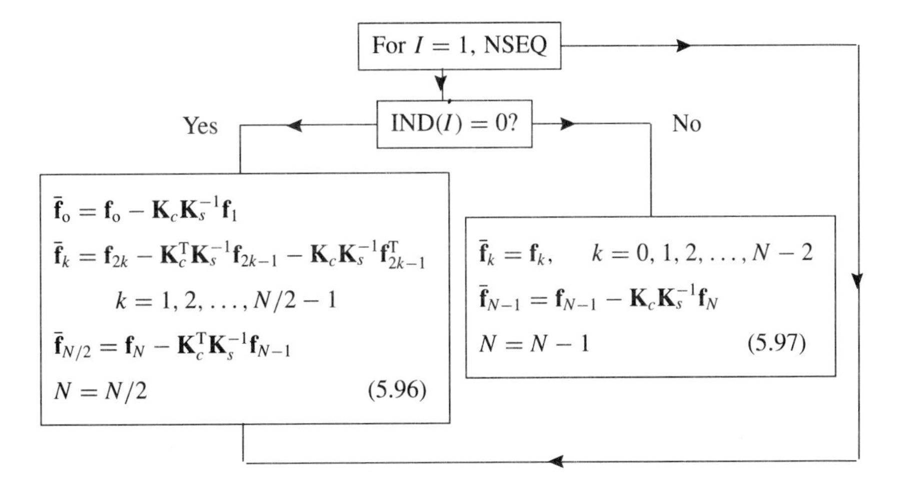

(c) Backward substitution phase

In the backward substitution phase, eq. 5.86 is solved to obtain $\{\mathbf{u}_0\}$ and $\{\mathbf{u}_1\}$ and the other $\{\mathbf{u}_i\}$s are calculated from eqs 5.70 and 5.78. In this phase, the flow diagram is as follows.

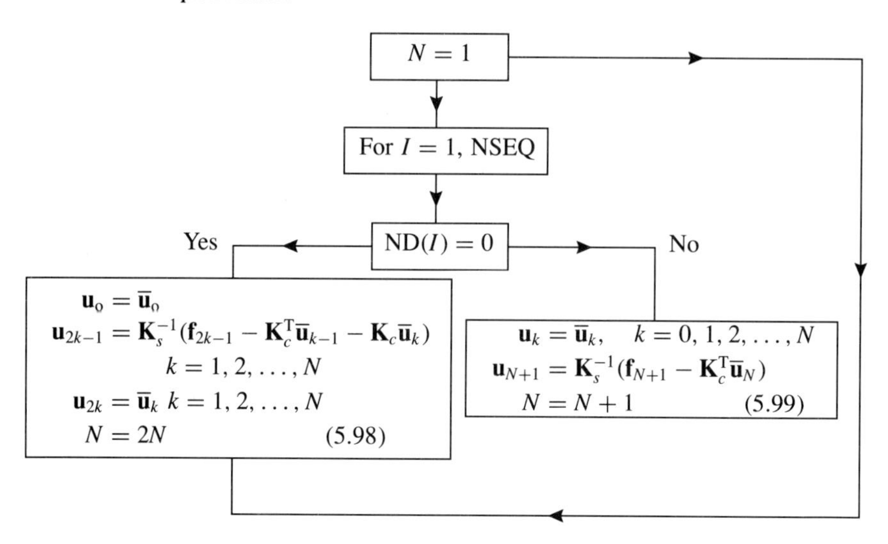

5.4.4 Application to method of substructuring

The concept of substructuring is that all the internal degrees of freedom of a substructure are statically eliminated prior to the element assemblage process. The system consisting of the boundary degrees of freedom associated with the nodes at the boundary only is called a condensed substructure. The slaves and masters are defined as the internal and boundary degrees of freedom, respectively. The substructure with all the slaves already eliminated is then related to other substructures or elements by connecting the masters.

Now, consider the periodic structure shown in Fig. 5.7. The proposed algorithm can condense the periodic structure directly if $N = 2^p$ where p is a positive integer. The stiffness equations corresponding to the periodic structure shown in Fig. 5.7 are

$$\begin{bmatrix} \mathbf{K}_\ell & \mathbf{K}_{\ell r} & & & & 0 & \\ & \mathbf{K}_\ell + \mathbf{K}_{rr} & \mathbf{K}_{\ell r} & & & & \\ & & \mathbf{K}_\ell + \mathbf{K}_{rr} & \mathbf{K}_{\ell r} & & & \\ & & & \cdot & & \cdot & \\ & \text{sym.} & & & & \mathbf{K}_{\ell\ell} + \mathbf{K}_{rr} & \mathbf{K}_{\ell r} \\ & & & & & & \mathbf{K}_{rr} \end{bmatrix} \begin{Bmatrix} \boldsymbol{\delta}_0 \\ \boldsymbol{\delta}_1 \\ \cdot \\ \cdot \\ \cdot \\ \cdot \\ \cdot \\ \cdot \\ \cdot \\ \boldsymbol{\delta}_N \end{Bmatrix} = \begin{Bmatrix} \mathbf{F}_0 \\ \mathbf{F}_1 \\ \cdot \\ \cdot \\ \cdot \\ \cdot \\ \cdot \\ \cdot \\ \cdot \\ \mathbf{F}_N \end{Bmatrix} \tag{5.100}$$

where $[\mathbf{K}_{\ell\ell}]$, $[\mathbf{K}_{\ell r}$ and $[\mathbf{K}_{rr}]$ are sub-matrices from eq. 5.58. These are the partitioned stiffness matrices of a condensed substructure. Since $N = 2^p$, all the slaves of the structure $\boldsymbol{\delta}_1, \boldsymbol{\delta}_2, \ldots, \boldsymbol{\delta}_{N-1}$ which are the masters of the condensed substructures will be eliminated if the transformations 5.74 have been applied p times. Equation 5.100 is thus reduced to eq. 5.86 where

$$[\mathbf{K}_{\text{sub}}] = \begin{bmatrix} \overline{\mathbf{K}}_b & \overline{\mathbf{K}}_c \\ \overline{\mathbf{K}}_c^{\mathrm{T}} & \overline{\mathbf{K}}_c \end{bmatrix} \qquad \{\mathbf{u}_{\mathrm{M}}\} = \begin{Bmatrix} \overline{\mathbf{u}}_0 \\ \overline{\mathbf{u}}_1 \end{Bmatrix} \qquad \{\mathbf{f}_{\mathrm{M}}\} = \begin{Bmatrix} \overline{\mathbf{f}}_0 \\ \overline{\mathbf{f}}_1 \end{Bmatrix} \tag{5.101}$$

are the stiffness matrix, nodal displacements and equivalent force vector of the condensed structure, respectively. The condensed periodic structure can be considered as a substructure at a higher level.

When $N \neq 2^p$ the algorithm cannot be applied directly. Equation 5.100 should be rewritten as

$$\begin{bmatrix} \mathbf{K}_{\ell\ell} & \mathbf{L}^{\mathrm{T}} & 0 \\ \mathbf{L} & \mathbf{K} & \mathbf{R} \\ 0 & \mathbf{R}^{\mathrm{T}} & \mathbf{K}_{rr} \end{bmatrix} \begin{pmatrix} \boldsymbol{\delta}_0 \\ \mathbf{X} \\ \boldsymbol{\delta}_N \end{pmatrix} = \begin{Bmatrix} \mathbf{F}_0 \\ \mathbf{F} \\ \mathbf{F}_N \end{Bmatrix} \tag{5.102}$$

where $[\mathbf{K}]$ has the same form as in eq. 5.60 with $[\mathbf{K}_b] = [\mathbf{K}_c] = [\mathbf{K}_s]$; $\{\mathbf{X}\} = \{\boldsymbol{\delta}_1, \boldsymbol{\delta}_2, \ldots, \boldsymbol{\delta}_{N-1}\}$ and $\{\mathbf{F}\} = \{\mathbf{F}_1, \mathbf{F}_2, \ldots, \mathbf{F}_{N-1}\}$; and $[\mathbf{R}] = \text{col}\,[0, \ldots, 0, \mathbf{K}_{\ell r}]$ and $[\mathbf{L}] = \text{col}\,[\mathbf{K}_{\ell r}^{\mathrm{T}}, 0 \ldots, 0]$. After eliminating $\mathbf{X}$ in eq. 5.102, the stiffness matrix, nodal displacements and equivalent force vector of the condensed structure, respectively, become

$$[\mathbf{K}_{\text{sub}}] = \begin{bmatrix} \mathbf{K}_{\ell\ell} - \mathbf{L}^{\mathrm{T}}\mathbf{K}^{-1}\mathbf{L} & -\mathbf{L}^{\mathrm{T}}\mathbf{K}^{-1}\mathbf{R} \\ -\mathbf{R}^{\mathrm{T}}\mathbf{K}^{-1}\mathbf{L} & \mathbf{K}_{rr} - \mathbf{R}^{\mathrm{T}}\mathbf{K}^{-1}\mathbf{R} \end{bmatrix}, \tag{5.103}$$

$$\{\mathbf{u}_{\mathrm{M}}\} = \begin{Bmatrix} \boldsymbol{\delta}_0 \\ \boldsymbol{\delta}_N \end{Bmatrix} \quad \text{and} \quad \{\mathbf{f}_{\mathrm{M}}\} = \begin{Bmatrix} \mathbf{F}_0 - \mathbf{L}^{\mathrm{T}}\mathbf{K}^{-1}\mathbf{F} \\ \mathbf{F}_N - \mathbf{R}^{\mathrm{T}}\mathbf{K}^{-1}\mathbf{F} \end{Bmatrix}$$

and

$$\{\mathbf{X}\} = \{\mathbf{K}^{-1}\mathbf{F} - \mathbf{K}^{-1}\mathbf{L}\boldsymbol{\delta}_0 - \mathbf{K}^{-1}\mathbf{R}\boldsymbol{\delta}_N\} \tag{5.104}$$

Since $[\mathbf{K}]$ has the same form as in eq. 5.60, the proposed algorithm can be applied. Due to the sparsity of $[\mathbf{R}]$ and $[\mathbf{L}]$, savings in computational effort in the above condensation procedure may be achieved as follows.

The condensation phase on $[\mathbf{K}]$ should be performed only once. The equivalent force evaluation phase on $[\mathbf{K}]^{-1}[\mathbf{L}]$ can be ignored since the resulting force vector in eq. 5.86 can be expressed explicitly as col $[\mathbf{K}_{\ell r}^{\mathrm{T}}, 0]$. On the other hand, the equivalent force evaluation phase on $[\mathbf{K}]^{-1}[\mathbf{R}]$ can be modified so that all the odd numbered blocks of the force vector are deleted without altering the rest in each even case (since all the odd numbered blocks contain zero elements only). Furthermore, great advantage can be taken of the sparsity of $[\mathbf{R}]$ and $[\mathbf{L}]$ in the multiplication with other matrices. Significant savings in computational effort may be possible by taking into account the above considerations in programming. Example 3 in the next section will show the effectiveness of the proposed algorithm as applied to the method of substructuring.

5.5 NUMERICAL EXAMPLES

1 A simply supported/fixed system with 100 identical substructures

Consider the space frame shown in Fig. 5.8, consisting of 100 identical substructures. The structure is simply supported at one end and fixed at the other end and therefore should be treated as a case with general boundary conditions. Each beam element is a steel rod with diameter 0.15 m. The lengths of the beam elements with axes in the X, Y and Z directions are 1.8 m, 2.0 m and 2.5 m, respectively. The inclined members have compatible lengths so that all the members are initially stress free. Young's modulus is taken as 200 kN/mm^2 and the shear modulus is 80 kN/mm^2. A downward point load of 250 kN is acting at the mid-point of the line AA′. The displacements along AA′ are given in Table 5.1.

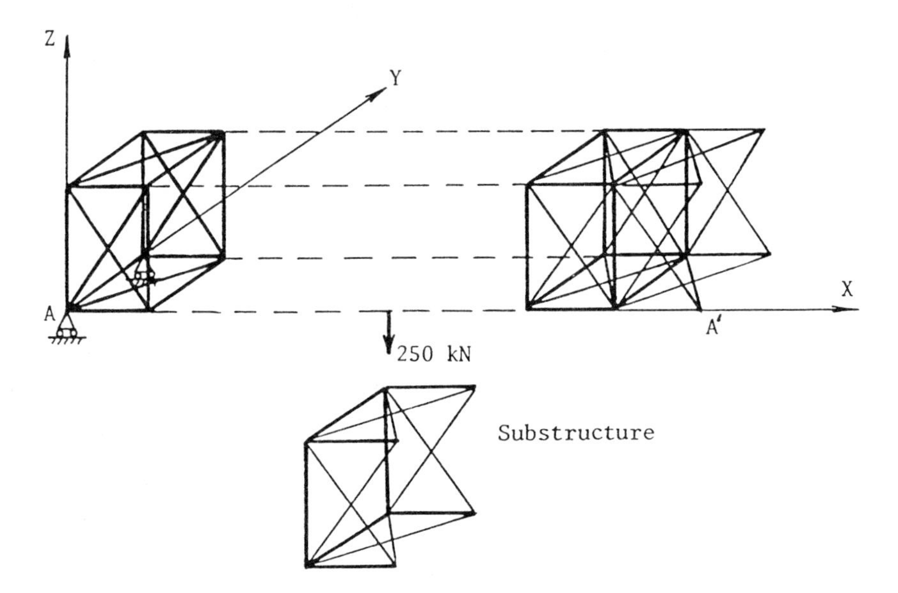

Fig. 5.8 Example 1: A simply supported/fixed system with 100 identical substructures

Table 5.1 Displacements along the line AA′ in Example 1

x	w(m)	$\theta_x(\times 10^{-3}$ rad)	$\theta_x(\times 10^{-3}$ rad)
0	0.000	0.276	9.910
18	−0.174	0.455	9.332
36	−0.331	0.610	7.862
54	−0.452	0.765	5.410
72	−0.520	0.920	1.978
90	−0.517	1.101	−2.453
108	−0.435	0.842	−6.294
126	−0.303	0.610	−7.959
144	−0.161	0.378	−7.465
162	−0.047	0.145	−4.812
180	0.000	0.000	0.000

The structure is analysed using the proposed method as well as the ICE*STRUDL package. It is found that the results are exactly the same. In using the ICE*STRUDL package, the conventional block solver requires about 100 decomposition procedures of a 24 × 24 symmetric matrix while the proposed method requires only eight. The STRUDL was run sometime ago on the SPERRY 1100/61 HI system (mainframe) and the computational time was 855 seconds. The proposed method was run on a smaller Micro-Vax II computer and the computational time was only 28 seconds. The proposed method is about 30 times faster. The in-core array memory and out-of-core storage required in this example are about 12 kilobytes and 100 kilobytes, respectively, for double-precision accuracy. If the conventional solver is used, only the non-zero coefficients in the original equations require about 480 kilobytes of storage space. It is found that both computational time and storage are reduced significantly so that the proposed method is suitable for a microcomputer environment for up to medium size periodic structures. As the number of substructures increases, the advantages of the proposed method are even more obvious.

2 A fixed/fixed system with 50 identical substructures

Consider a space structure consisting of 50 substructures as shown in Fig. 5.9. The structure is fixed at both ends. Young's modulus is taken as 200 kN/mm^2 and the

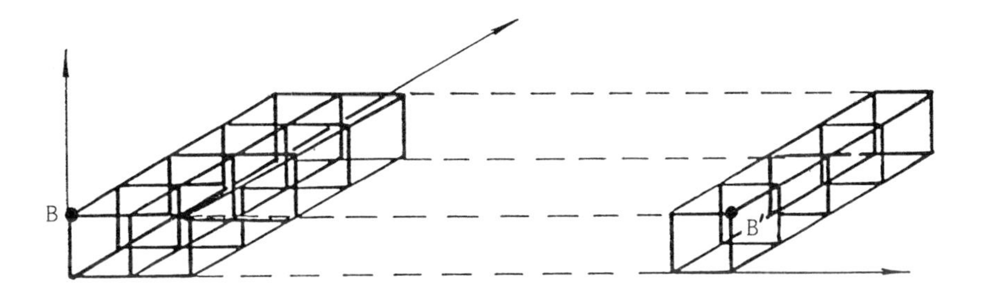

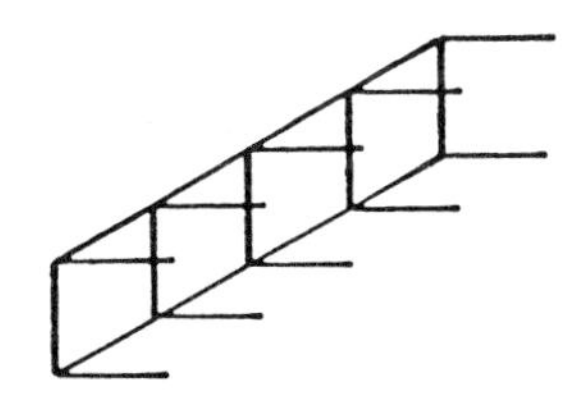

Fig. 5.9 Example 2: A fixed/fixed system with 50 identical substructures

Table 5.2 Displacements along the line BB′ in Example 2

x	w(m)	$\theta_x(\times 10^{-3}$ rad)	$\theta_x(\times 10^{-3}$ rad)
5	−0.055	16.641	7.137
10	−0.102	30.834	5.456
15	−0.136	41.049	3.687
20	−0.157	47.210	1.858
25	−0.164	49.268	0.000

shear modulus is 80 kN/mm^2. All members are of length 1 m and of hollow circular section with 20 mm outside diameter and 3 mm thickness. A downward uniform distributed load of 0.01 kN/m is acting along the line BB′. The displacements along BB′ are given in Table 5.2.

The structure is analysed using the proposed method as well as the ICE*STRUDL package. It is found that the results are again exactly the same. In using the ICE*STRUDL package, the conventional block solver requires about 50 decomposition procedures of a 60×60 symmetric matrix while the proposed method requires only six. The computational time required in STRUDL and the proposed method are 745 seconds and 138 seconds, respectively. The proposed method is about five times faster. The in-core array memory and out-of-core storage required by the proposed method are 73 kilobytes and 360 kilobytes, respectively, for double-precision accuracy. Even a 640 K Ram microcomputer without hard disk can handle a medium size periodic structure (3000 degrees of freedom) such as this without difficulty. However, if the conventional solver is used, the storage required for just the non-zero coefficients in the original equations is about 1450 kilobytes.

3 A periodic structure consisting of 1024 beam elements connected together

Consider a beam element whose stiffness matrix is

$$[\mathbf{K}] = \frac{EI}{12}\begin{bmatrix} 12 & 6\ell & -12 & 6\ell \\ & 4\ell^2 & -6\ell & 2\ell^2 \\ & \text{sym.} & 12 & -6\ell^2 \\ & & & 4\ell^2 \end{bmatrix}$$

where E is Young's modulus, I is the second moment of area and ℓ is the length of the beam. If there are 1024 such beams assembled together, as shown in Fig. 5.10, the resulting stiffness equations will be of the same form as eq. 5.100 with $N = 1024$ and

$$[\mathbf{K}_{\ell\ell}] = \frac{EI}{12}\begin{bmatrix} 12 & 6\ell \\ 6\ell & 4\ell^2 \end{bmatrix} \qquad [\mathbf{K}_{rr}] = \frac{EI}{12}\begin{bmatrix} 12 & -6\ell \\ -6\ell & 4\ell^2 \end{bmatrix}$$

$$[\mathbf{K}_{\ell\ell} + \mathbf{K}_{rr}] = \frac{EI}{12}\begin{bmatrix} 24 & 0 \\ 0 & 8\ell^2 \end{bmatrix} \quad \text{and} \quad [\mathbf{K}_{\ell r}] = \frac{EI}{12}\begin{bmatrix} -12 & 6\ell \\ -6\ell & 2\ell^2 \end{bmatrix}$$

Assume that $EI = 1$, $\ell = 1$ and a unit downward force is acting at node 200.

In this example, node 0 and node 1024 are masters and the rest are slaves. Since $1024 = 2^{10}$, the stiffness matrix and equivalent force vector of the global substructure can be obtained by the proposed method in which transformations 5.74 have been

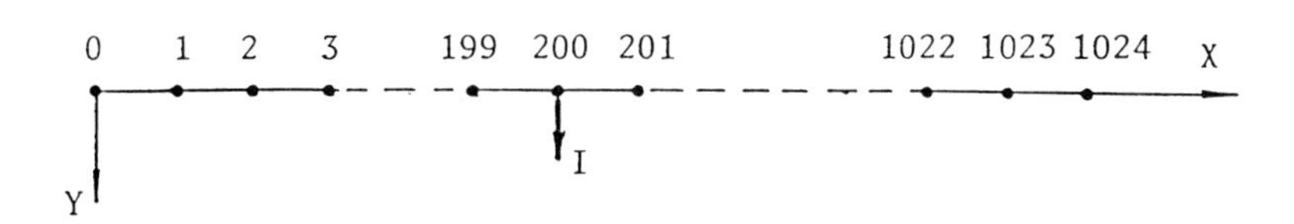

Fig. 5.10 Example 3: A periodic structure consisting of 1024 beam elements connected together

applied ten times only. The stiffness matrix $[\mathbf{K}_{sub}]$ and equivalent nodal force vector $\{\mathbf{f}_M\}$ of the global substructure are listed below

$$[\mathbf{K}_{sub}] = \begin{bmatrix} 0.111\,76 \times 10^{-7} & 0.572\,20 \times 10^{-5} & 0.111\,76 \times 10^{-7} & 0.572\,20 \times 10^{-5} \\ & 0.390\,63 \times 10^{-2} & 0.572\,20 \times 10^{-5} & 0.195\,31 \times 10^{-2} \\ & & 0.111\,76 \times 10^{-7} & 0.572\,20 \times 10^{-5} \\ & & & 0.390\,63 \times 10^{-2} \end{bmatrix}$$

$$\{\mathbf{f}_M\} = \{0.900\,46 \quad 0.129\,50 \times 10^3 \quad 0.995\,40 \times 10^{-1} \quad -0.314\,33 \times 10^2\}^T$$

It is found that the results obtained by the proposed method are exactly equal to a beam of length 1024 units and subject to a unit force at distance 200 units from node 0. If the classical long-hand condensation method were used, the algorithm would fail during the sixteenth element elimination, due to round-off errors.

5.6 References to Chapter 5

1. J. S. Przemienieck (1968) *Theory of Matrix Structural Analysis*, McGraw-Hill, New York.
2. A. Y. T. Leung (1980) Dynamics of periodic structures, *J. Sound & Vibration*, **72**, 451-67.
3. D. L. Thomas (1979) Dynamics of rotationally periodic structures, *Int. J. Num. Methods in Engng*, **14**, 81-102.
4. Y. K. Cheung, H. C. Chan & C. W. Cai (1989) Exact method for static analysis of periodic structures, *ASCE, Engineering Mechanics*, **115**, 415-34.
5. C. W. Cai, Y. K. Cheung & H. C. Chan (1988) Dynamic response of infinite conditions beams subjected to a moving force - an exact method, *J. Sound & Vibration*, **123**, 461-72.
6. C. W. Cai, Y. K. Cheung & H. C. Chan (1990) Uncoupling of dynamic equations for periodic structures, *J. Sound & Vibration*, **139**, 253-63.

6 The Two Level Finite Element Method

6.1 INTRODUCTION

It has been shown in Chapter 5 that if certain relations between unknowns can be found, the computational cost of a finite element analysis can be reduced greatly. Some methods for finding such relations by inspection have been discussed. Very often, these relations can also be found analytically. It is the purpose of this chapter to make good use of these analytical relations. Typical applications include tall building analysis, laminated plate bending, plate subject to concentrated loads and fracture mechanics. A space frame will be used to illustrate the idea.

In order to construct a frame super-element as shown in Fig. 6.1(a), two levels of finite element interpolation are introduced [1,2]. The first (or local) is the interpolation between the two end joints of a beam member. This follows the usual finite element procedure and results in an ordinary 12×12 stiffness matrix (mass matrix or stability matrix) for a beam in space. The second (or global) interpolation is at the frame level. Some joints on the frame are selected first and are called nodes so as to distinguish them from the other joints. Interpolation functions are assumed, so that for given displacements at the nodes, the displacements at other joints are obtained by interpolation. These are called frame interpolation functions and are different from the ordinary finite element interpolation functions in that they are of significance only at the joints of the undeformed frame and are meaningless anywhere else. As an illustration, consider a frame element consisting of two beams and three nodes as shown in Figs 6.1(b) to 6.1(d). If joints 1 and 3 are chosen as nodes, then the frame shape functions are linear. Therefore a unit displacement at joint 3 will make half a unit displacement at joint 2 if the lengths of the beams are identical. If the beam shape functions were not used, then the system would look like Fig. 6.1(b). However, because of the cubic shape functions used for each beam, a unit displacement at joint 3 results in a displacement pattern, as shown in Fig. 6.1(c). Similarly, a unit rotation at joint 3 will produce a deflected shape, as shown in Fig. 6.1(d). With this concept in mind, a frame super-element can be derived as follows.

In accordance with the usual finite element procedure, a fixed number of n reference nodes with displacements $\{\mathbf{a}_i\}$ $(i = 1, \ldots, n)$ are first chosen. Each parameter $\{\mathbf{a}_i\}$ includes three translational and three rotational displacements and corresponds to the displacement vector at the nodes of a space-frame member. The displacements

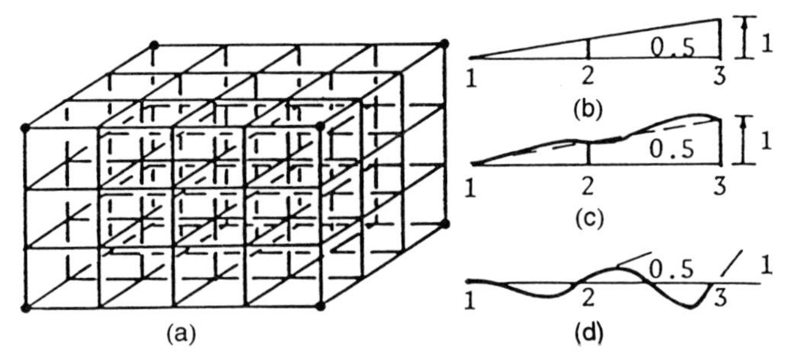

Fig. 6.1 An 8-node frame super-element

{**u**} of all the other joints of the frame super-element are related to $\{\mathbf{a}_i\}$ by

$$\{\mathbf{u}\} = \Sigma[\mathbf{N}_i(x, y, z)]\{\mathbf{a}_i\} = [\mathbf{N}]\{\mathbf{a}\} \tag{6.1}$$

where the shape functions [**N**] are defined as follows:

(i) at joint j with coordinates (x_j, y_j, z_j),

$$[\mathbf{N}_i(x_j, y_j, z_j)] = \delta_{ij}[\mathbf{I}]$$

(ii) $[\mathbf{N}_i]$ is not defined when (x, y, z) are not the coordinates of a joint.

The usual stiffness relation for a frame member is

$$[\mathbf{k}_e]\{\boldsymbol{\delta}_e\} = \{\mathbf{f}_e\} \tag{6.2}$$

where $[\mathbf{k}_e]$, $\{\boldsymbol{\delta}_e\}$ and $\{\mathbf{f}_e\}$ are the stiffness matrix, the displacement vector and the force vector respectively. Now, if a frame member connects joints $\ell(x_\ell, y_\ell, z_\ell)$ and $m(x_m, y_m, z_m)$, it becomes possible to establish the following relationship,

$$\{\boldsymbol{\delta}_e\} = [\mathbf{u}^{\mathrm{T}}(x_\ell, y_\ell, z_\ell) \quad \mathbf{u}^{\mathrm{T}}(x_m, y_m, z_m)]^{\mathrm{T}} \tag{6.3}$$

From eqs 6.1 and 6.3,

$$\{\boldsymbol{\delta}_e\} = \begin{bmatrix} \mathbf{N}(x_\ell, y_\ell, z_\ell) \\ \mathbf{N}(x_m, y_m, z_m) \end{bmatrix} \{\mathbf{a}\} = [\mathbf{T}_e]\{\mathbf{a}\} \tag{6.4}$$

As can be seen, $[\mathbf{T}_e]$ is a transformation matrix connecting the displacement vectors of the frame member to those of the reference nodes of the super-element. It can be a matrix of standard Lagrangian interpolation functions at discrete points as in Fig. 6.1. At this stage, eq. 6.2 can be transformed to

$$[\mathbf{T}_e]^{\mathrm{T}}[\mathbf{k}_e][\mathbf{T}_e]\{\mathbf{a}\} = [\mathbf{T}_e]^{\mathrm{T}}\{\mathbf{f}_e\} \tag{6.5}$$

By assembling the transformed matrices of all the frame members, the stiffness relationship for the super-element is now established as

$$[\mathbf{K}]\{\mathbf{a}\} = \{\mathbf{F}\} \tag{6.6}$$

where $[\mathbf{K}] = \sum_e [\mathbf{T}_e]^{\mathrm{T}}[\mathbf{k}_e][\mathbf{T}_e]$ is the frame stiffness matrix

and $\{\mathbf{F}\} = \sum_e [\mathbf{T}_e]^{\mathrm{T}}\{\mathbf{f}_e\}$ is the force vector.

Thus by solving eq. 6.6 for **a**, the displacements of any joint may be obtained from eq. 6.1. Since the matrices $\mathbf{k}_e$ and $\mathbf{T}_e$ are sparse, eq. 6.5 should be modified when implementing computer programs to avoid full matrix manipulations.

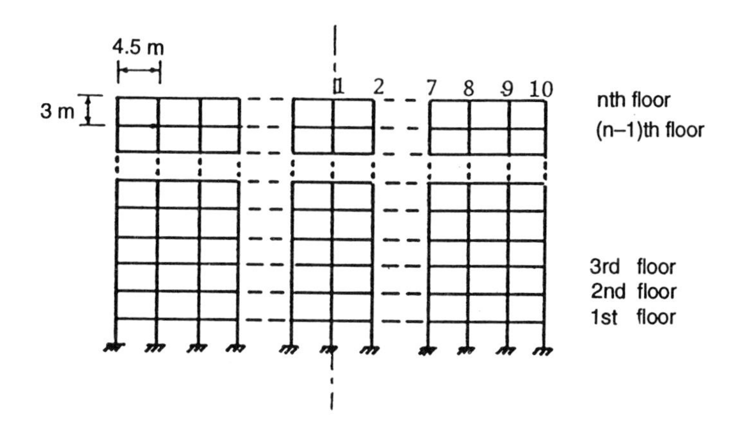

Fig. 6.2 Example 1: $\rho A = EI = 1$

For convenience, very often all the components of displacements are interpolated identically, in which case the shape functions can be written as

$$[\mathbf{N}(x, y, z)] = [\text{row } N_i(x, y, z)\mathbf{I}] \tag{6.7}$$

However, in many cases the variation of the various components of displacement may differ significantly from each other, and it is then not practical to use the form given in eq. 6.7. For example, in the case of multi-storey building frames, the horizontal displacements for all joints at the same floor level can be assumed to be identical by ignoring the axial deformations of the floor beams and slabs. In such cases, it would be computationally advantageous to use different functions for different displacement components.

Example 1

Consider a plane frame structure with 19 columns and 18 beams on each floor as shown in Fig. 6.2. For simplicity it is assumed that $EI = 1$, and that the horizontal displacements of all the joints on the same floor are identical. The frame is loaded in the first case by a concentrated horizontal force of 1 N acting at the top floor and in the second case by uniformly distributed vertical loads of 1 N acting on each floor. By considering each floor as an element and by taking different numbers of reference nodes in each element, the displacements of several frames with different numbers of storeys were computed and they compared very favourably with those given by exact 2-D finite element solutions (Table 6.1). The joint rotations are similarly compared for a 40-storey frame in Table 6.2. Only half of the frame has been used because of skew symmetry.

As can be observed, both the results for joint displacements and rotations converged rapidly. Since joint moments and shear forces are proportional to joint displacements and rotations in a linear analysis, it is expected that the force components will exhibit the same characteristics.

6.2 BUILDING FRAMES [3–5]

When the frame is not orthogonal or the stiffness of the frame members changes suddenly, the second level nodal interpolation by means of standard Lagrangian

Table 6.1 Displacements at every tenth floor (m)

	Total no. of floors	Floor levels	No. of nodes 1	2	3	4	5	Exact
Concentrated load	10	10	2.892	2.900	2.908	2.913	2.915	2.916
	20	10	2.943	2.952	2.960	2.965	2.967	2.968
		20	5.951	5.968	5.986	5.996	5.999	6.001
	30	10	2.943	2.952	2.960	2.965	2.967	2.968
		20	6.008	6.020	6.038	6.048	6.051	6.053
		30	9.010	9.036	9.063	9.079	9.084	9.088
	40	10	2.943	2.952	2.960	2.965	2.967	2.968
		20	6.008	6.020	6.038	6.048	6.051	6.053
		30	9.062	9.088	9.115	9.131	9.136	9.138
		40	12.069	12.104	12.140	12.161	12.168	12.174
Distributed load	10	10	15.04	15.10	15.13	15.16	15.17	15.17
	20	10	44.46	44.60	44.72	44.79	44.83	44.85
		20	60.67	60.89	61.04	61.14	61.19	61.20
	30	10	73.89	74.12	74.32	74.45	74.50	74.52
		20	120.7	121.1	121.4	121.6	121.7	121.7
		30	136.9	137.4	137.7	138.0	138.1	138.1
	40	10	103.3	103.6	103.9	104.1	104.2	104.2
		20	180.7	181.3	181.8	182.1	182.2	182.2
		30	227.5	228.2	228.9	229.3	229.4	229.5
		40	243.7	244.5	245.2	245.6	245.8	245.9

Table 6.2 Computed rotations at joints

Floor levels	No. of nodes	Computed rotations at joints ($\times 10^{-3}$ rad) 1	2	3	4	5	6	7	8	9	10
5–35	1	62.50	62.50	62.50	62.50	62.50	62.50	62.50	62.50	62.50	62.50
	2	59.69	59.92	60.15	60.38	60.61	60.84	61.07	61.30	61.53	61.76
	3	63.01	62.46	61.95	61.50	61.09	60.74	60.43	60.16	59.95	59.78
	4	60.70	61.37	61.89	62.28	62.55	62.70	62.76	62.73	62.63	62.46
	5	62.16	61.65	61.34	61.19	61.17	61.25	61.39	61.57	61.77	61.97
	exact	61.69	61.70	61.70	61.69	61.68	61.69	61.70	61.70	61.71	61.71
40	1	34.55	34.55	34.55	34.55	34.55	34.55	34.55	34.55	34.55	34.55
	2	32.32	32.50	32.68	32.87	33.05	33.23	33.42	33.60	33.78	33.96
	3	35.04	34.56	34.12	33.73	33.38	33.06	32.79	32.56	32.38	32.23
	4	32.92	33.56	34.06	34.44	34.70	34.85	34.91	34.89	34.79	34.63
	5	34.37	33.82	33.48	33.32	33.30	33.38	33.53	33.73	33.94	34.16
	exact	33.94	33.81	33.80	33.84	33.90	33.94	33.96	33.94	33.91	33.86

interpolating functions is not good enough. A method will be introduced to analyze tall buildings of various types: frame, frame-tube, tube-in-tube, bundle-tube, braced frame, macro-frame, etc. The lateral floor displacements are found to be within 1% of the correct solutions in all the cases studied. The final matrix equation is of the order of 21 times the number of storeys. Therefore, the method permits the approximate analysis of a complicated tall building to be performed by a personal computer with acceptable accuracy.

The terms 'local distribution factors', 'global distribution factors' and 'mixing factors' will be discussed specifically next and the computational details of these factors will be given in the subsequent sections.

For a building frame, the stiffness equations of the structure can be written as

$$\begin{bmatrix} \mathrm{K}^{11} & \mathrm{K}^{12} & & & 0 \\ \mathrm{K}^{21} & \mathrm{K}^{22} & \ddots & & \\ & \ddots & \ddots & \ddots & \\ & & \ddots & \mathrm{K}^{n-1,n-1} & \mathrm{K}^{n-1,n} \\ 0 & & & \mathrm{K}^{n,n-1} & \mathrm{K}^{nn} \end{bmatrix} \begin{bmatrix} \boldsymbol{\delta}^1 \\ \boldsymbol{\delta}^2 \\ \cdot \\ \cdot \\ \cdot \\ \cdot \\ \boldsymbol{\delta}^n \end{bmatrix} = \begin{bmatrix} \mathbf{F}^1 \\ \mathbf{F}^2 \\ \cdot \\ \cdot \\ \cdot \\ \cdot \\ \mathbf{F}^n \end{bmatrix} \tag{6.8}$$

where $\{\boldsymbol{\delta}^i\}$ and $\{\mathbf{F}^i\}$ are displacement and force vectors respectively at floor i, $[\mathbf{K}^{ij}]$ is the stiffness matrix associated with floors i and j, and n is the total number of storeys. The dimensions of the vectors $\{\boldsymbol{\delta}^i\}$ may be different due to set back, etc.

The number of unknowns in eq. 6.8 is usually very large. To reduce the number of unknowns without much loss of accuracy, the following assumptions are made.

(i) Rigid in-plane displacement

The six degrees of freedom at a node are separated into two groups: three degrees of freedom associated with out-of-plane displacements and three with in-plane displacements. The in-plane displacements of all the nodes on a particular floor are connected together by the floor slab whose in-plane rigidity is very large compared with the out-of-plane rigidity. Therefore, it is reasonable to assume that the in-plane degrees of freedom are moving horizontally as a rigid plane together with the floor slab. For a floor with m nodes, the number of unknowns of the in-plane displacements is reduced from $3m$ to 3, i.e. u, v, θ of a reference node only.

(ii) Local distribution factors

The remaining $3m$ out-of-plane displacements will not be moving in a completely independent manner but according to certain possible patterns. These patterns are called local distribution factors which are quite independent from floor to floor and can be computed one floor at a time.

(iii) Global distribution factors

There are also certain deformation patterns which cannot be represented by the local distribution factors. An example is the differential elongation of the columns. Therefore, new sets of patterns accounting for global deformation are necessary. These are called the global distribution factors.

(iv) Mixing factors

The actual out-of-plane displacements are obtained by linear combinations of the local and global distribution factors when lateral loads are imposed. The coefficients of the linear combinations are called the mixing factors.

The local distribution factors are unique for a particular floor. If two floors are identical, only one set of distribution factors is required. The global distribution factors are independent of loadings. For different load cases, only the mixing factors are altered. The idea is presented below rigorously.

The nodal displacement vector at floor i, $\{\boldsymbol{\delta}^i\}$, consists of out-of-plane displacements $\{\boldsymbol{\delta}_O^i\}$ and in-plane displacements $\{\boldsymbol{\delta}_I^i\} = [u^i v^i \theta^i]^T$, where u^i, v^i, θ^i are the displacements along the x and y axes and rotation about the z axis, respectively, at a predetermined floor origin. Approximation is introduced if the out-of-plane nodal displacements at floor i having m^i nodes are expressed as

$$\begin{aligned}\{\boldsymbol{\delta}_O^i\} &= [\mathbf{L}^i]\{\boldsymbol{\eta}^i\} + [\mathbf{G}^i]\{\boldsymbol{\xi}^i\} \\ &= [\mathbf{L}^i\ \mathbf{G}^i]\begin{Bmatrix}\boldsymbol{\eta}^i \\ \boldsymbol{\xi}^i\end{Bmatrix}\end{aligned} \qquad (6.9)$$

where $\{\boldsymbol{\delta}_O^i\}$ is of order $3m^i$ consisting of the vertical displacements w_k and the rotations α_k, β_k, about the x and y axes, respectively, at node k, $k = 1, 2, \ldots, m^i$; $[\mathbf{L}^i]$ and $[\mathbf{G}^i]$ are $(3m^i \times 9)$ matrices of local and global distribution factors, respectively; and $\{\boldsymbol{\eta}^i\}$ and $\{\boldsymbol{\xi}^i\}$ are 9-vectors of local and global mixing factors. The dimension of $\{\boldsymbol{\eta}^i\}$ is nine due to three sets of distribution factors of w_k, α_k, β_k corresponding to unit displacements of u^i, v^i, θ^i, respectively, and the dimensions of $\{\boldsymbol{\xi}^i\}$, $[\mathbf{L}^i]$ and $[\mathbf{G}^i]$ are defined in a similar way.

Therefore, the nodal displacement vector $\{\boldsymbol{\delta}^i\}$ can be expressed approximately by

$$\{\boldsymbol{\delta}^i\} = \begin{Bmatrix}\boldsymbol{\delta}_O^i \\ \boldsymbol{\delta}_I^i\end{Bmatrix} = \begin{bmatrix}\mathbf{L}^i & \mathbf{G}^i & \mathbf{0} \\ \mathbf{0} & \mathbf{0} & \mathbf{I}\end{bmatrix}\begin{Bmatrix}\boldsymbol{\eta}^i \\ \boldsymbol{\xi}^i \\ \boldsymbol{\delta}_I^i\end{Bmatrix} = [\mathbf{T}^i]\{\boldsymbol{\zeta}^i\} \qquad (6.10)$$

where $[\mathbf{I}]$ is a 3×3 identity matrix, $\{\boldsymbol{\zeta}^i\}$ is of order $9 + 9 + 3 = 21$ and $[\mathbf{T}^i]$ is a transformation matrix of order $(3m^i + 3) \times 21$. Under the transformation 6.10, eq. 6.8 becomes

$$\begin{bmatrix} \mathrm{K}^{11} & \mathrm{K}^{12} & & & \\ \mathrm{K}^{21} & \mathrm{K}^{22} & \ddots & & 0 \\ & \ddots & \ddots & \ddots & \\ & & \ddots & \mathrm{K}^{n-1,n-1} & \mathrm{K}^{n-1,n} \\ 0 & & & \mathrm{K}^{n,n-1} & \mathrm{K}^{nn} \end{bmatrix} \begin{bmatrix} \zeta^1 \\ \zeta^2 \\ \cdot \\ \cdot \\ \cdot \\ \zeta^n \end{bmatrix} = \begin{bmatrix} \mathbf{f}^1 \\ \mathbf{f}^2 \\ \cdot \\ \cdot \\ \cdot \\ \mathbf{f}^n \end{bmatrix} \qquad (6.11)$$

where $[\mathbf{k}^{ij}] = [\mathbf{T}^i]^T[\mathbf{K}^{ij}][\mathbf{T}^j]$ and $[\mathbf{f}^i] = [\mathbf{T}^i]^T\{\mathbf{F}^i\}$. Equation 6.11 determines the mixing factors $\{\boldsymbol{\eta}^i\}$, $\{\boldsymbol{\xi}^i\}$, and the in-plane floor displacements $\{\boldsymbol{\delta}_I^i\}$. The nodal displacements are obtained from eq. 6.10 afterwards, and the member forces can then be determined. The local distribution factors $[\mathbf{L}^i]$ and the global distribution factors $[\mathbf{G}^i]$ are defined in the following sections.

6.2.1 Local distribution factors

The local distribution factors are obtained by a one-storey model as shown in Fig. 6.3. In the method, only the stiffness associated with floor i is considered

$$[\mathbf{K}^{ii}]\{\boldsymbol{\delta}^i\} = \{\mathbf{F}^i\} \tag{6.12}$$

The stiffness matrix can be rewritten in partitioned form in accordance with the in-plane and out-of-plane displacements as

$$\begin{bmatrix} \mathbf{K}^i_{\mathrm{II}} & \mathbf{K}^i_{\mathrm{IO}} \\ \mathbf{K}^i_{\mathrm{OI}} & \mathbf{K}^i_{\mathrm{OO}} \end{bmatrix} \begin{Bmatrix} \boldsymbol{\delta}^i_{\mathrm{I}} \\ \boldsymbol{\delta}^i_{\mathrm{O}} \end{Bmatrix} = \begin{Bmatrix} \mathbf{F}^i_{\mathrm{I}} \\ \mathbf{F}^i_{\mathrm{O}} \end{Bmatrix}$$

where $\{\mathbf{F}^i_{\mathrm{O}}\}$ includes out-of-plane dead and live loads.

Three sets of displacement patterns for the out-of-plane displacements are needed corresponding to unit in-plane displacements $\{u^i, v^i, \theta^i\}$ one at a time; therefore, collectively,

$$\begin{bmatrix} \mathbf{K}^i_{\mathrm{II}} & \mathbf{K}^i_{\mathrm{IO}} \\ \mathbf{K}^i_{\mathrm{OI}} & \mathbf{K}^i_{\mathrm{OO}} \end{bmatrix} \begin{Bmatrix} \mathbf{I} \\ \mathbf{U}^i_{\mathrm{O}} \end{Bmatrix} = \begin{Bmatrix} \mathbf{R}^i_{\mathrm{I}} \\ \mathbf{R}^i_{\mathrm{O}} \end{Bmatrix} \tag{6.13}$$

where $[\mathbf{I}]$ is a 3×3 identity matrix, $[\mathbf{R}^i_{\mathrm{I}}]$ is an unknown reaction matrix to give unit in-plane displacements; $[\mathbf{R}^i_{\mathrm{O}}] = \text{row } \{\mathbf{F}^i_{\mathrm{O}}\}$; and on solving eq. 6.13, $[\mathbf{U}^i_{\mathrm{O}}]$ consists of the local distribution factors

$$[\mathbf{U}^i_{\mathrm{O}}]^{\mathrm{T}} = \begin{bmatrix} W_{11} & \alpha_{11} & \beta_{11} & W_{12} & \alpha_{12} & \beta_{12} & \ldots & W_{1m^i} & \alpha_{1m^i} & \beta_{1m^i} \\ W_{21} & \alpha_{21} & \beta_{21} & W_{22} & \alpha_{22} & \beta_{22} & \ldots & W_{2m^i} & \alpha_{2m^i} & \beta_{2m^i} \\ W_{31} & \alpha_{31} & \beta_{31} & W_{32} & \alpha_{32} & \beta_{32} & \ldots & W_{3m^i} & \alpha_{3m^i} & \beta_{3m^i} \end{bmatrix}$$

The matrix $[\mathbf{U}^i_{\mathrm{O}}]^{\mathrm{T}}$ can be determined by the second term of eq. 6.13,

$$[\mathbf{U}^i_{\mathrm{O}}] = [\mathbf{K}^i_{\mathrm{OO}}]^{-1}[\mathbf{R}^i_{\mathrm{O}} - \mathbf{K}^i_{\mathrm{OI}}] \tag{6.14}$$

or, if $\{\mathbf{F}^i_{\mathrm{O}}\}$ is zero, and hence $[\mathbf{R}^i_{\mathrm{O}}]$ is zero,

$$[\mathbf{U}^i_{\mathrm{O}}] = -[\mathbf{K}^i_{\mathrm{OO}}]^{-1}[\mathbf{K}^i_{\mathrm{OI}}] \tag{6.15}$$

The local distribution factor matrix $[\mathbf{L}^i]$ in eq. 6.9 is obtained by the rearrangement of $[\mathbf{U}^i_{\mathrm{O}}]$, and it accounts for most of the local effects due to local loads and local

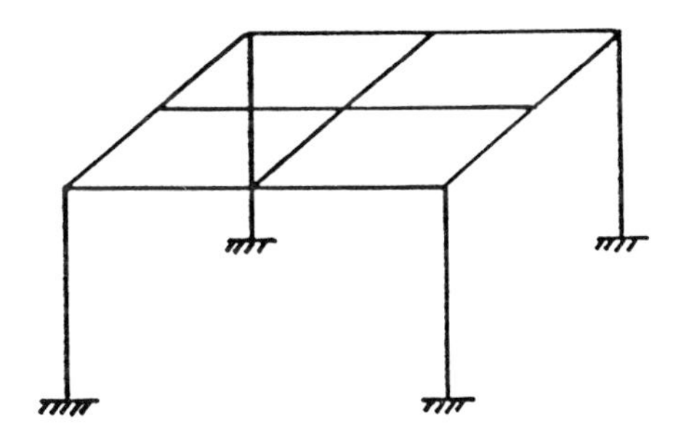

Fig. 6.3 A one-storey model

deformation:

$$[\mathbf{L}^i] = \begin{bmatrix} W_{11} & & & W_{21} & & & W_{31} & & \\ & \alpha_{11} & & & \alpha_{21} & & & \alpha_{31} & \\ W_{12} & & & & & & & & \\ & & \beta_{11} & & & \beta_{21} & & & \beta_{31} \\ & \alpha_{12} & & & & & & & \\ & & & W_{22} & & & W_{32} & & \\ \vdots & & \beta_{12} & \vdots & \alpha_{22} & & \vdots & \alpha_{32} & \\ & \vdots & & & \vdots & \beta_{22} & & \vdots & \beta_{32} \\ W_{1m^i} & & \vdots & W_{2m^i} & & \vdots & W_{3m^i} & & \vdots \\ & \alpha_{1m^i} & & & \alpha_{2m^i} & & & \alpha_{3m^i} & \\ & & \beta_{1m^i} & & & \beta_{2m^i} & & & \beta_{3m^i} \end{bmatrix} \tag{6.16}$$

6.2.2 Global distribution factors

While the local distribution factors are good in representing local deformation, they are incapable of modelling the out-of-plane deformation of floors far away from the ground due mainly to the uneven extension of the columns along the height. One method of compensating for this is to include the three rigid-body out-of-plane movements as global distribution factors. However, as the columns elongate (shorten) unevenly along the height, the rigid-body movements are inadequate to deal with very tall buildings having abrupt changes of stiffness. An alternative method is to adopt the two-level finite element method which predicts the global behaviour of frames.

In essence, the nodal displacements $\{\bar{\boldsymbol{\delta}}\}$ of a few purposely selected master floors (Fig. 6.4) of eq. 6.8 are chosen so that

$$\{\bar{\boldsymbol{\delta}}\} = \text{col}\ \{\bar{\mathbf{w}}, \bar{\boldsymbol{\alpha}}, \bar{\boldsymbol{\beta}}, \bar{\mathbf{u}}, \bar{\mathbf{v}}, \bar{\boldsymbol{\theta}}\} \tag{6.17}$$

where $\{\bar{\mathbf{w}}\} = \text{col}\ \{\bar{w}_j^m\}$ $\quad \{\bar{\boldsymbol{\alpha}}\} = \text{col}\ \{\bar{\alpha}_j^m\}$ $\quad \{\bar{\boldsymbol{\beta}}\} = \text{col}\ \{\bar{\beta}_j^m\}$

$\{\bar{\mathbf{u}}\} = \text{col}\ \{\bar{u}^m\}$ $\quad \{\bar{\mathbf{v}}\} = \text{col}\ \{\bar{v}^m\}$ $\quad \{\bar{\theta}\} = \text{col}\ \{\bar{\theta}^m\}$;

$j = 1, 2, \ldots$, is the number of nodes at floor m; and $m = 1, 2, \ldots$, is the number of master floors. The displacement vector in eq. 6.8 is rearranged so that

$$\{\boldsymbol{\delta}\} = \text{col}\ \{\mathbf{w}, \boldsymbol{\alpha}, \boldsymbol{\beta}, \mathbf{u}, \mathbf{v}, \boldsymbol{\theta}\} \tag{6.18}$$

where $\{\mathbf{w}\} = \text{col}\ \{w_j^i\}$ $\quad \{\boldsymbol{\alpha}\} = \text{col}\ \{\alpha_j^i\}$, $\quad \{\boldsymbol{\beta}\} = \text{col}\ \{\beta_j^i\}$

$\{\mathbf{u}\} = \text{col}\ \{u^i\}$ $\quad \{\mathbf{v}\} = \text{col}\ \{v^i\}$, $\quad \{\boldsymbol{\theta}\} = \text{col}\ \{\theta^i\}$;

$j = 1, 2, \ldots$, is the number of nodes at floor $i (i = 1, 2, \ldots, n)$. A suitable interpolating function $[\mathbf{N}(i)]$ can be choosen so that

$$\begin{aligned} w_j^i &= [\mathbf{N}(i)]\{\bar{\mathbf{w}}_j\} & \alpha_j^i &= [\mathbf{N}(i)]\{\bar{\boldsymbol{\alpha}}_j\} \\ \beta_j^i &= [\mathbf{N}(i)]\{\bar{\boldsymbol{\beta}}_j\} & u^i &= [\mathbf{N}(i)]\{\bar{\mathbf{u}}\} \\ v_j^i &= [\mathbf{N}(i)]\{\bar{\mathbf{v}}\} & \theta^i &= [\mathbf{N}(i)]\{\bar{\boldsymbol{\theta}}\} \end{aligned} \tag{6.19}$$

where $\{\bar{\mathbf{w}}_j\} = \text{col}\ \{\bar{w}_j^m\}$ $\quad \{\bar{\boldsymbol{\alpha}}_j\} = \text{col}\ \{\bar{\alpha}_j^m\}$ $\quad \{\bar{\boldsymbol{\beta}}_j\} = \text{col}\ \{\bar{\beta}_j^m\}$

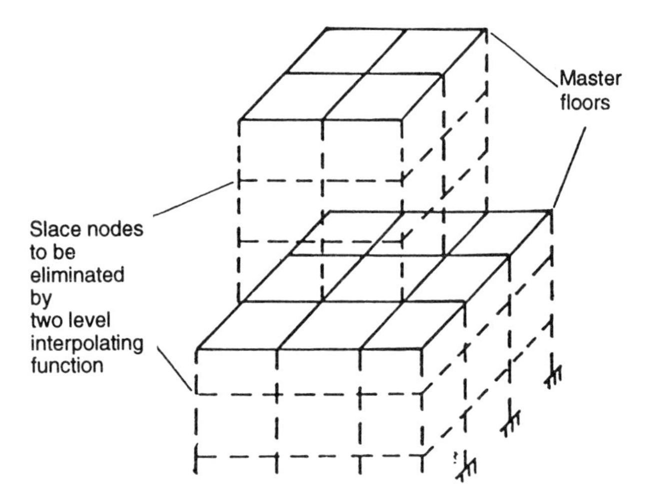

Fig. 6.4

and $m = 1, 2, \ldots,$ is the number of master floors. If the top floor only is taken as the master floor, then $\overline{w}_j$, $\overline{\alpha}_j$, $\overline{\beta}_j\overline{u}$, $\overline{v}$, $\overline{\theta}$ are all scalars and $\mathbf{N}(i) = i/n$, that is, a linear interpolation. It is usual practice to keep the number of master floors to a minimum. If the building stiffness is uniform or changing gradually, the top floor and a middle floor are chosen as masters. For a building having abrupt changes of stiffness at certain floor levels, these floors are taken as masters.

According to transformation 6.19, eq. 6.8 is condensed to a stiffness equation associated with the nodal displacement of the master floors only. Note that the transformation is only required to reduce the number of unknowns, and matrix inversion is not involved at all. Let the top floor be displaced one unit in u, v, θ, one at a time, and from this it is possible to get three sets of solutions similar to the local distribution factors. After the nodal displacements at the master floors are determined, the nodal displacements at other floor levels are obtained by interpolation through eq. 6.19. The nodal displacements obtained in this manner constitute the global distribution factor matrix $[\mathbf{G}^i]$ at floor i. If the three sets of solution are denoted by the first subscripts 4, 5 and 6, respectively, then

$$[\mathbf{G}^i] = \begin{bmatrix} W_{41} & W_{51} & W_{61} \\ \alpha_{41} & \alpha_{51} & \alpha_{61} \\ \beta_{41} & \beta_{51} & \beta_{61} \\ W_{42} & W_{52} & W_{62} \\ \alpha_{42} & \alpha_{52} & \alpha_{62} \\ \beta_{42} & \beta_{52} & \beta_{62} \\ \vdots & \vdots & \vdots \\ W_{4m^i} & W_{5m^i} & W_{6m^i} \\ \alpha_{4m^i} & \alpha_{5m^i} & \alpha_{6m^i} \\ \beta_{4m^i} & \beta_{5m^i} & \beta_{6m^i} \end{bmatrix} \tag{6.20}$$

6.2.3 Mixing factors

After obtaining the local and the global distribution factors, the determination of the mixing factors together with the in-plane floor displacements $\{\zeta\}$ in eq. 6.10 is straightforward. Equation 6.10 is block diagonal having 21 unknowns in each block. The solution is reached comfortably even with a personal computer.

6.2.4 Computational aspects

Since the transformation matrices involved in obtaining the distribution factors and the mixing factors as in eqs 6.10, 6.16, 6.19 and 6.20 are very sparse, a significant saving in computing time and storage is possible if the sparsity is taken into account in the programming. Equation 6.12 is also banded. The solution for the local distribution factors can be obtained without difficulty.

6.2.5 Numerical examples

Four numerical examples are presented. The first three are 20-storey frames of the same floor plan (Fig. 6.5) with 15 columns as numbered. The first one has all members with sections 0.3 m × 0.3 m, the second has strengthened columns of sections 1 m × 1 m at positions 6, 7, 10 and 11 and the third has crossed bracing with sections 0.3 m × 0.3 m along the circumference of the frame between floor 20 and floor 19 (Fig. 6.6). The fourth example is a 23-storey frame with a set back at floor 20. The floor plan from floors 1 to 20 is the same as in the above examples, and that from floors 21 to 23 is shown in Fig. 6.7. All members of the fourth example have sections of 0.3 m × 0.3 m. In all the cases studied, Young's modulus is 210 kN/mm^2 and the shear modulus is 80.8 kN/mm^2. The floor heights are 5 m.

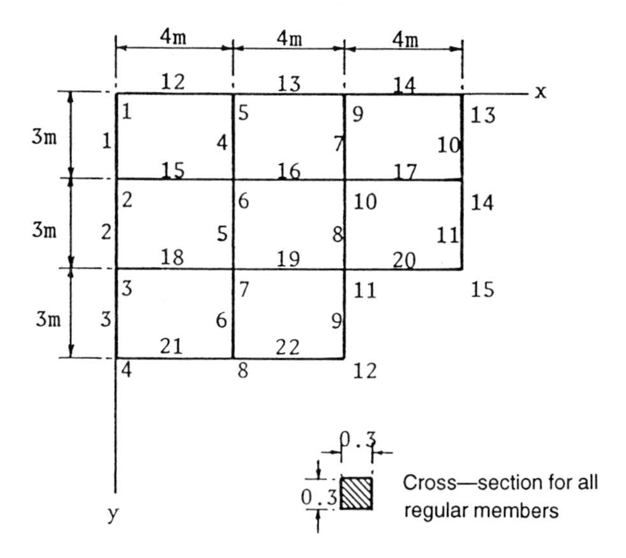

Fig. 6.5 The floor plan for a 20-storey building

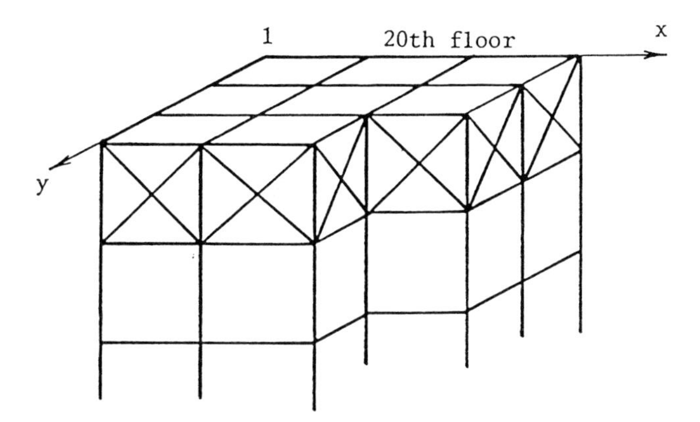

Fig. 6.6 Cross-bracing along the circumference of the frame in Example 3

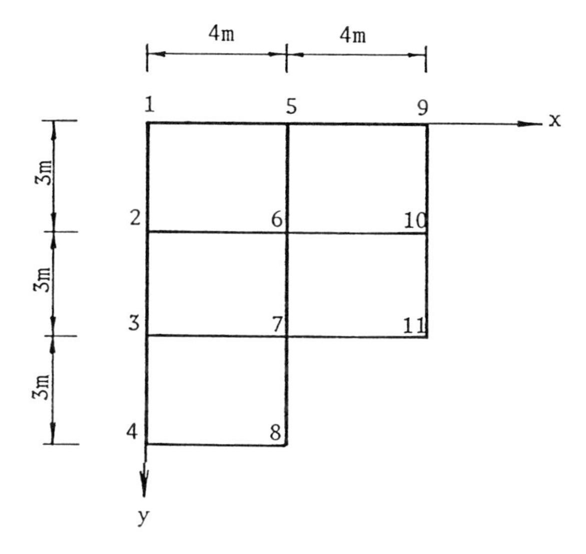

Fig. 6.7 The floor plan of the set back in Example 4

The floor origin is as shown at position 1. The lateral load in the x direction with magnitude 1000 kN is acting at position 1 on each floor level.

The results obtained from the present method are compared with those from full finite element analysis assuming a rigid in-plane floor. The floor displacements and nodal displacements at floor 20 of all examples are listed in Tables 6.3 and 6.4, respectively. The lateral floor displacements are found to be within 1% of the full finite element results in all the cases studied. The final matrix equation is of the order of 21 times the number of storeys.

6.3 PLATE SUBJECT TO CONCENTRATED LOADS [6]

A major difference between the frame problem and the plate problem is that a concentrated load on a frame member does not induce singular stress but a plate

Table 6.3 Floor displacements (units: deflection (m); rotation (rad))

Floor level	*Example 1* Exact u	v	θ	Present u	v	θ	*Example 2* Exact u	v	θ	Present u	v	θ
20	3.997	−1.570	0.307	3.997	−1.570	0.307	2.634	−1.140	0.211	2.632	−1.139	0.211
19	3.938	−1.563	0.304	3.938	−1.562	0.304	2.577	−1.132	0.209	2.575	−1.131	0.209
18	3.863	−1.548	0.300	3.862	−1.548	0.300	2.514	−1.119	0.206	2.512	−1.118	0.206
17	3.771	−1.526	0.294	3.771	−1.525	0.294	2.442	1.101	0.202	2.439	−1.100	0.202
16	3.663	−1.496	0.287	3.663	−1.495	0.287	2.359	−1.077	0.197	2.357	−1.077	0.197
15	3.539	−1.458	0.279	3.539	−1.457	0.279	2.266	−1.048	0.191	2.264	−1.047	0.191
14	3.398	−1.412	0.269	3.398	−1.411	0.269	2.161	−1.013	0.184	2.159	−1.012	0.184
13	3.242	−1.358	0.258	3.242	−1.358	0.258	2.046	−0.971	0.176	2.044	−0.970	0.176
12	3.070	−1.297	0.246	3.069	−1.296	0.245	1.919	−0.924	0.167	1.917	−0.923	0.166
11	2.882	−1.227	0.232	2.881	−1.227	0.231	1.782	−0.871	0.157	1.780	−0.870	0.156
10	2.678	−1.150	0.216	2.678	−1.150	0.216	1.634	−0.812	0.146	1.632	−0.811	0.145
9	2.460	−1.065	0.200	2.460	−1.065	0.199	1.475	−0.747	0.133	1.474	−0.746	0.133
8	2.227	−0.972	0.182	2.227	−0.972	0.181	1.308	−0.676	0.120	1.306	−0.675	0.120
7	1.980	−0.872	0.162	1.980	−0.872	0.162	1.131	−0.599	0.106	1.130	−0.599	0.106
6	1.719	−0.764	0.141	1.719	−0.763	0.141	0.948	−0.516	0.091	0.947	−0.516	0.091
5	1.446	−0.648	0.120	1.445	−0.648	0.120	0.759	−0.428	0.075	0.758	−0.427	0.075
4	1.159	−0.524	0.096	1.159	−0.524	0.096	0.568	−0.334	0.058	0.568	−0.333	0.058
3	0.860	−0.393	0.072	0.860	−0.393	0.072	0.381	−0.235	0.041	0.380	−0.235	0.041
2	0.551	−0.255	0.047	0.551	−0.255	0.047	0.206	−0.136	0.023	0.206	−0.135	0.023
1	0.237	−0.112	0.020	0.237	−0.112	0.020	0.065	−0.046	0.008	0.065	−0.046	0.008

Floor level	*Example 3* Exact u	v	θ	Present u	v	θ	*Example 4* Exact u	v	θ	Present u	v	θ
23	5.531	−2.064	0.413	5.520	−2.061	0.412						
22	5.454	−2.056	0.410	5.445	−2.053	0.409						
21	5.350	−2.038	0.404	5.344	−2.036	0.403						
20	5.227	−2.012	0.396	5.223	−2.011	0.396	3.915	−1.512	0.294	3.911	−1.516	0.294
19	5.100	−1.984	0.389	5.096	−1.982	0.388	3.880	−1.518	0.293	3.876	−1.522	0.294
18	4.956	−1.947	0.380	4.952	−1.946	0.380	3.815	−1.510	0.291	3.812	−1.514	0.291
17	4.796	−1.903	0.370	4.792	−1.902	0.370	3.730	−1.493	0.286	3.727	−1.496	0.287
16	4.620	−1.851	0.358	4.616	−1.850	0.358	3.628	−1.467	0.281	3.625	−1.470	0.281
15	4.428	−1.791	0.345	4.425	−1.790	0.345	3.508	−1.433	0.273	3.506	−1.435	0.273
14	4.220	−1.723	0.331	4.217	−1.722	0.331	3.372	−1.390	0.264	3.370	−1.393	0.264
13	3.997	−1.647	0.315	3.994	−1.647	0.315	3.220	−1.340	0.254	3.218	−1.342	0.254
12	3.759	−1.564	0.298	3.756	−1.563	0.298	3.051	−1.281	0.242	3.049	−1.283	0.242
11	3.506	−1.472	0.279	3.503	−1.471	0.279	2.866	−1.214	0.229	2.865	−1.216	0.229
10	3.238	−1.372	0.259	3.236	−1.372	0.259	2.666	−1.139	0.214	2.664	−1.141	0.214
9	2.956	−1.265	0.238	2.955	−1.264	0.238	2.450	−1.056	0.198	2.449	−1.057	0.198
8	2.661	−1.149	0.215	2.660	−1.149	0.215	2.219	−0.965	0.180	2.218	−0.966	0.180
7	2.353	−1.026	0.191	2.352	−1.025	0.191	1.974	−0.866	0.161	1.973	−0.867	0.161
6	2.032	−0.895	0.166	2.031	−0.894	0.166	1.715	−0.759	0.141	1.714	−0.760	0.141
5	1.699	−0.756	0.140	1.699	−0.756	0.140	1.442	−0.645	0.119	1.442	−0.645	0.119
4	1.356	−0.610	0.112	1.355	−0.609	0.112	1.157	−0.522	0.096	1.156	−0.523	0.096
3	1.001	−0.456	0.084	1.001	−0.456	0.084	0.859	−0.392	0.072	0.859	−0.392	0.072
2	0.638	−0.295	0.054	0.638	−0.295	0.054	0.550	−0.255	0.046	0.550	−0.255	0.046
1	0.273	−0.129	0.023	0.273	−0.129	0.023	0.237	−0.112	0.020	0.237	−0.112	0.020

Table 6.4 Nodal displacements at floor 20 (units: deflection (mm); rotation $\times 10^{-3}$ (rad))

	Example 1						*Example 2*					
	Exact			Present			Exact			Present		
Joint	W (mm)	α	β	W (mm)	α	β	W (mm)	α	β	W (mm)	α	β
1	−33.47	−1.69	−9.36	−33.56	−1.71	−8.62	−26.57	−2.24	−8.66	−27.06	−2.20	−7.91
2	−38.50	−0.96	−9.10	−38.51	−1.10	−8.71	−33.32	−1.27	−8.09	−33.26	−1.34	−8.28
3	−39.36	−0.95	−8.34	−39.34	−1.09	−8.51	−33.88	−0.59	−7.72	−33.66	−0.80	−7.34
4	−44.37	−1.68	−8.08	−44.40	−1.71	−8.65	−36.47	−1.29	−7.27	−36.86	−1.60	−7.51
5	−7.19	0.22	−5.58	−7.17	0.41	−6.47	−4.21	0.17	−4.78	−4.26	0.28	−5.66
6	−7.66	0.12	−6.73	−7.65	0.30	−6.89	−3.66	0.20	−9.61	−3.52	0.20	−9.62
7	−7.60	0.14	−7.25	−7.79	0.29	−6.99	−4.17	0.19	−8.32	−3.93	0.19	−8.33
8	−7.93	0.25	−8.46	−7.69	0.40	−7.22	−7.63	−0.92	−6.73	−7.52	−0.48	−6.36
9	−1.39	3.44	−5.65	−1.24	3.37	−6.52	−1.98	1.74	−4.94	−1.72	1.95	−6.21
10	9.55	3.30	−7.18	9.55	3.18	−7.45	3.01	1.98	−9.62	2.90	1.98	−9.62
11	18.61	2.85	−8.65	18.72	2.78	−8.55	6.70	2.00	−8.35	6.40	2.00	−8.35
12	26.77	2.77	−7.77	26.82	2.88	−8.60	16.59	3.08	−6.42	16.86	3.57	−6.82
13	25.47	5.79	−9.46	25.50	5.64	−8.67	21.79	4.56	−8.89	22.11	4.38	−8.44
14	44.30	6.07	−9.72	44.22	6.15	−9.40	36.46	4.35	−8.80	36.27	3.95	−7.77
15	62.77	5.69	−10.27	62.59	5.56	−10.30	48.39	3.80	−9.93	48.61	4.05	−9.73

(contd.)

Table 6.4 (*continued*)

	Example 3						*Example 4*					
	Exact			Present			Exact			Present		
Joint	W (mm)	α	β	W (mm)	α	β	W (mm)	α	β	W (mm)	α	β
1	−52.45	−4.03	−19.92	−51.62	−3.98	−19.03	−43.59	1.28	−6.63	−43.63	1.31	−6.27
2	−58.21	−2.50	−17.47	−57.94	−2.45	−17.09	−40.18	1.22	−7.09	−39.38	1.28	−7.11
3	−58.57	−2.49	−14.55	−57.76	−2.49	−14.36	−36.98	1.21	−7.03	−36.29	1.19	−6.56
4	−64.29	−4.03	−12.02	−64.12	−4.05	−12.34	−33.52	1.33	−6.81	−33.08	1.35	−6.90
5	−12.89	0.43	−14.52	−10.01	0.18	−13.82	−16.16	1.87	−6.61	−16.47	1.86	−7.02
6	−12.58	0.37	−13.98	−10.25	0.14	−13.83	−9.40	1.63	−6.48	−9.49	1.69	−6.10
7	−11.80	0.30	−12.84	−9.93	0.42	−12.49	−6.63	0.62	−6.94	−6.78	0.76	6.33
8	−12.00	0.35	−12.31	−9.87	0.58	−12.87	−5.57	0.88	−6.95	−5.62	0.84	−7.31
9	2.59	5.92	−14.40	−1.84	5.43	−13.87	10.49	0.68	−6.63	11.04	0.58	−7.09
10	16.64	5.17	−14.36	13.10	5.15	−13.78	12.05	1.46	−6.46	11.64	1.13	−5.90
11	28.31	4.89	−14.27	25.43	5.37	−14.00	19.01	1.61	−6.68	18.24	1.65	−7.10
12	39.38	4.31	−12.65	40.74	5.49	−11.90	22.55	1.10	−6.86	21.69	1.15	−6.36
13	38.69	10.03	−15.71	40.25	9.77	−19.10	38.11	1.31	−6.66	38.57	1.32	−6.53
14	65.45	9.31	−15.94	64.82	8.76	−17.00	42.72	1.37	−7.08	42.77	1.19	−6.91
15	91.74	9.89	−16.52	89.01	9.77	−16.36	47.10	1.23	−6.80	46.83	1.15	−6.81

member does. Direct application of the global nodal interpolating method will not converge to the true solution. Singular global interpolating functions must be assumed in addition to the usual regular interpolating functions.

To fix the idea, a plate under a concentrated load is discretized into conventional finite elements. The normal practice would be to have progressively finer mesh divisions near the load. Analytical solutions [6] are available in the region very near to the point load. Therefore, the nodal displacements of the fine mesh near the point load follow certain patterns. It is reasonable to interpolate the nodal displacements by the known solutions. Since the analytical solutions are valid only at the vicinity of the point load, additional regular interpolating functions are required to take care of the conditions away from the point load. The unknowns are no longer the nodal displacements but are the coefficients (generalized coordinates) of the global interpolating functions of the nodal displacements. The stiffness matrix associated with the generalized coordinates is obtained by matrix transformation. In fact, the transformation is performed at the elemental level so that the order involved is very small. If the regular interpolating functions are chosen such that the generalized coordinates correspond to the nodal displacements of the boundary nodes, an accurate super-element (substructure) which can take good care of the concentrated loads is developed, and it can be connected to the other substructures readily.

Numerical examples for plates of various shapes are compared with the full finite element analyses by the package ICE*STRUDL. STRUDL uses CPT (classical plate theory) plate bending elements whereas the DKT (discrete Kirchhoff theory) elements (Section 4.3.3) [5, 6] are employed in the present study.

6.3.1 Formulation

The usual finite element method suggests the following interpolation for the plate element deflection $w(x, y)$, i.e.

$$w(x, y) = [\mathbf{N}(x, y)]\{\boldsymbol{\delta}_e\} \tag{6.21}$$

where $[\mathbf{N}(x, y)]$ is the shape function matrix and $\{\boldsymbol{\delta}_e\}$ is the nodal displacements (including rotations, etc.) vector. After assembling the elements, the following stiffness equation is obtained

$$[\mathbf{K}]\{\boldsymbol{\delta}\} = \{\mathbf{f}\} \tag{6.22}$$

where $[\mathbf{K}]$ is the global stiffness matrix and $\{\mathbf{f}\}$ is the applied nodal forces vector. An alternative analytical approach is to solve the governing partial differential equation of the plate for the particular solution associated with the loading and for the complementary functions whose coefficients are to be determined by the boundary conditions. The analytical solution is difficult to generalize for plates of arbitrary shape and plate systems, and the conventional finite element method converges slowly for concentrated loads.

Fortunately, the nodal deflections in the vicinity of the concentrated loads correspond closely to the particular solution of the analytical approach. Therefore, many of the unknowns in eq. 6.22 are not independent but come close to the following transformation

$$\{\boldsymbol{\delta}\} = [\mathbf{T}]\{\bar{\boldsymbol{\delta}}\} \tag{6.23}$$

where **[T]** is a transformation matrix consisting of the particular solutions and some regular interpolating functions serving as complementary functions given in the following subsections, and $\{\bar{\boldsymbol{\delta}}\}$ is the coefficients vector associated with the above functions. The order of $\{\bar{\boldsymbol{\delta}}\}$ can be much less than that of $\{\boldsymbol{\delta}\}$. Equation 6.22 is transformed to

$$[\mathbf{T}]^{\mathrm{T}}[\mathbf{K}][\mathbf{T}]\{\bar{\boldsymbol{\delta}}\} = [\mathbf{T}]^{\mathrm{T}}\{\mathbf{f}\} \tag{6.24}$$

which is of a much smaller order than the original.

Equations 6.21 and 6.23 correspond to the local and global interpolations respectively. In this approach, no new element matrices need to be generated. The analytical solution helps to condense the stiffness matrix. In the following sections, the analytical global interpolating functions are discussed first and the transformation process is then studied.

6.3.2 Global interpolating functions

An arbitrary region of a thin plate in bending is shown in Fig. 6.8. Based on the assumptions of classical thin-plate theory, the governing differential equation is

$$\nabla^4 w = \frac{q}{D} \tag{6.25}$$

where q is the distributed load and D is the plate flexural rigidity. Consider a small circular plate subject to a point load P at coordinates (x_j, y_j). The distributed load q in eq. 6.25 is expressed as

$$q = p\,\delta(x - x_j, y - y_j) \tag{6.26}$$

where $\delta(x, y)$ is the Dirac's function. The particular solution is

$$w = \frac{P}{8\pi D} r^2 \ln r \tag{6.27}$$

where $r = \sqrt{[(x - x_j)^2 + (y - y_j)^2]}$

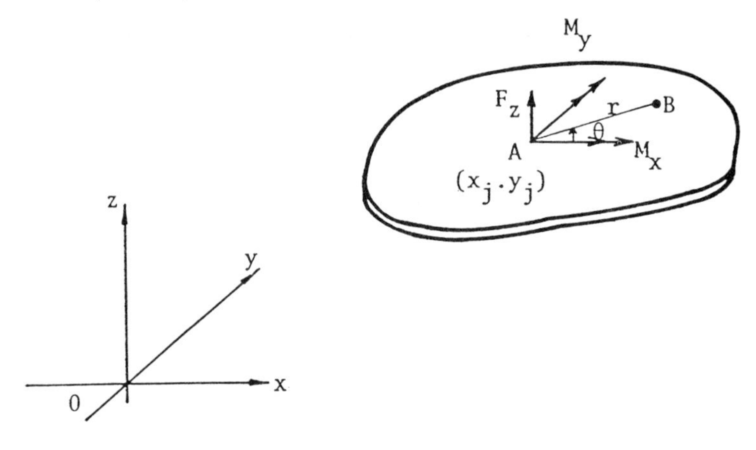

Fig. 6.8 An arbitrary region of a thin plate in bending

Alternatively, when a concentrated moment M_x about the x-axis acts at (x_j, y_j), the particular solution is

$$w = \frac{M_x r(2 \ln r + 1) \sin \theta}{8\pi D} \tag{6.28}$$

where $\theta = \tan^{-1}\left(\frac{y - y_j}{x - x_j}\right)$.

Similarly, if a concentrated moment M_y about the y-axis acts at (x_j, y_j), the particular solution is

$$w = \frac{M_y r(2 \ln r + 1) \cos \theta}{8\pi D} \tag{6.29}$$

Hence, the singular part of the interpolating functions is chosen as follows:

$$g = \begin{cases} r^2 \ln r & \text{for point load in the } z \text{ direction,} \\ r \ln r \sin \theta & \text{for point moment about the } x\text{-axis,} \\ r \ln r \cos \theta & \text{for point moment about the } y\text{-axis.} \end{cases} \tag{6.30}$$

The non-singular part (analogous to complementary functions) of the interpolating function is expressed by the appropriate polynomial in the xy coordinate. In order to fulfil the completeness requirement, a complete polynomial of order m obtained from Pascal's triangle is used.

6.3.3 Transformation

Combining the singular and non-singular parts of the interpolating functions, the expression for deflection becomes

$$w(x, y) = \beta g(x, y; x_j, y_j) + \sum_{i=0}^{m} [(\mathbf{P}_i(x, y)]\{\boldsymbol{\alpha}_i\} \tag{6.31}$$

where $g(x, y; x_j, y_j)$ is the singular function about the point (x_j, y_j); $[\mathbf{P}_i(x, y)] = [x^i, x^{i-1}y, \ldots, y^i]$ is the collection of polynomial terms of order i; β and $\{\boldsymbol{\alpha}\}$ are the generalized coordinates; and m is the order of the complete polynomial.

In matrix form, the deflection for a particular node at (x_i, y_i) is

$$w(x_i, y_i) = [g(x_i, y_i;\ x_j, y_j) \quad 1 \quad \mathbf{P}_1(x_i, y_i) \cdots \mathbf{P}_m(x_i, y_i)] \begin{Bmatrix} \beta \\ \alpha_0 \\ \alpha_1 \\ \vdots \\ \alpha_m \end{Bmatrix} \tag{6.32}$$

The rotations θ_x and θ_y about the x and y axes at the same node can be obtained through differentiating eq. 6.31 with respect to y and x, respectively. Hence,

$$\theta_x = \frac{\partial w}{\partial y} = \left[\frac{\partial g}{\partial y} \; 0 \; \frac{\partial \mathbf{P}_1}{\partial y} \cdots \frac{\partial \mathbf{P}_m}{\partial y}\right] \{\boldsymbol{\alpha}\}$$

$$\theta_y = \frac{-\partial w}{\partial x} = -\left[\frac{\partial g}{\partial x} \; 0 \; \frac{\partial P_1}{\partial x} \cdots \frac{\partial \mathbf{P}_m}{\partial x}\right] \{\boldsymbol{\alpha}\} \tag{6.33}$$

The nodal displacements of node i can be expressed as follows:

$$\{\mathbf{u}_i\} = \begin{Bmatrix} w_i \\ \theta_{x,i} \\ \theta_{y,i} \end{Bmatrix} = [\boldsymbol{\phi}_{-1}(x_i, y_i), \boldsymbol{\phi}_0(x_i, y_i), \boldsymbol{\phi}_1(x_i, y_i), \ldots, \boldsymbol{\phi}_m(x_i, y_i)]\{\boldsymbol{\alpha}\} \tag{6.34}$$

where

$$\{\phi_{-1}\} = \begin{Bmatrix} g \\ \dfrac{\partial g}{\partial y} \\ \dfrac{-\partial g}{\partial x} \end{Bmatrix} \qquad \{\boldsymbol{\phi}_0\} = \begin{Bmatrix} 1 \\ 0 \\ 0 \end{Bmatrix} \qquad \{\boldsymbol{\phi}_k\} = \begin{Bmatrix} \mathbf{P} \\ \dfrac{\partial \mathbf{P}_k}{\partial y} \\ \dfrac{-\partial \mathbf{P}_k}{\partial x} \end{Bmatrix}$$

$$k = 1, 2, \ldots, m \qquad \{\boldsymbol{\alpha}\} = \begin{Bmatrix} \beta \\ \alpha_0 \\ \alpha_1 \\ \vdots \\ \alpha_m \end{Bmatrix}$$

Owing to the singular characteristics of the function g at $\{\mathbf{u}_j\}$, which cannot be transformed by means of $[\mathbf{O}]$ and $[\mathbf{I}]$, the transformation is carried out as follows:

$$\begin{Bmatrix} \mathbf{u}_1 \\ \vdots \\ \mathbf{u}_{j-1} \\ \mathbf{u}_j \\ \mathbf{u}_{j+1} \\ \vdots \\ \mathbf{u}_n \end{Bmatrix} = \begin{bmatrix} \boldsymbol{\phi}_{-1}(x_1, y_1) & \ldots & \boldsymbol{\phi}_m(x_1, y_1) & \mathbf{0} \\ \vdots & & \vdots & \mathbf{0} \\ \boldsymbol{\phi}_{-1}(x_{j-1}, y_{j-1}) & \ldots & \boldsymbol{\phi}_m(x_{j-1}, y_{j-1}) & \\ \mathbf{0} & \mathbf{0} & \mathbf{0} & \mathbf{I} \\ \boldsymbol{\phi}_{-1}(x_{j+1}, y_{j+1}) & \ldots & \boldsymbol{\phi}_m(x_{j+1}, y_{j+1}) & \mathbf{0} \\ \vdots & & \vdots & \\ \boldsymbol{\phi}_{-1}(x_n, y_n) & \ldots & \boldsymbol{\phi}_m(x_n, y_n) & \mathbf{0} \end{bmatrix} \begin{Bmatrix} \boldsymbol{\beta} \\ \boldsymbol{\alpha}_0 \\ \boldsymbol{\alpha}_1 \\ \boldsymbol{\alpha}_2 \\ \vdots \\ \boldsymbol{\alpha}_m \\ \mathbf{u}_j \end{Bmatrix} \tag{6.35}$$

where $\mathbf{I}$ is a 3×3 identity matrix and n is the number of nodes in the plate bending problem. After assembling the stiffness matrices, the resulting finite element equations with respect to the nodal coordinates will be as follows:

$$[\mathbf{K}]\{\boldsymbol{\delta}\} = \{\mathbf{f}\} \tag{6.36}$$

where $\{\mathbf{f}\}$ is the consistent nodal force vector.

According to the conservation of strain energy,

$$\begin{aligned} U &= \tfrac{1}{2}\{\boldsymbol{\delta}\}^{\mathrm{T}}[\mathbf{K}]\{\boldsymbol{\delta}\} \\ &= \tfrac{1}{2}\{\bar{\boldsymbol{\delta}}\}^{\mathrm{T}}[\overline{\mathbf{K}}]\{\bar{\boldsymbol{\delta}}\} \end{aligned} \tag{6.37}$$

where U is the strain energy and $[\overline{\mathbf{K}}] = [\mathbf{T}]^{\mathrm{T}}[\mathbf{K}][\mathbf{T}]$ is the global stiffness matrix with respect to the generalized coordinates.

The original problem with $3n$ degrees of freedom is now reduced to $[(m+1)(m+2)/2] + 4$ for the $m + 1$ polynomials and the singular functions. Since the number of unknowns is reduced, the method is approximate. The resulting two-level finite element equations become

$$[\overline{\mathbf{K}}]\{\bar{\boldsymbol{\delta}}\} = \{\bar{\mathbf{f}}\} \tag{6.38}$$

where $\{\bar{\mathbf{f}}\} = [\mathbf{T}]^{\mathrm{T}}\{\mathbf{f}\}$ is the generalized force vector.

6.3.4 Stress evaluation

The generalized displacements can be obtained by solving eq. 6.38. The nodal displacements are calculated from eq. 6.35. The bending stresses are then obtained according to the following equation:

$$\{\mathbf{M}\} = [\mathbf{D}_b][\mathbf{B}]\{\boldsymbol{\delta}\} \tag{6.39}$$

where $\{\mathbf{M}\}$ are the bending stresses; $[\mathbf{D}_b]$ is the elasticity matrix; and $[\mathbf{B}]$ is the strain-displacement transformation matrix [5].

Around the singular point, the stress can be obtained by direct differentiation of the global interpolating shape functions.

$$M_{xx} = -D\left(\frac{\partial^2 w}{\partial x^2} + \nu\frac{\partial^2 w}{\partial y^2}\right)$$

$$M_{yy} = -D\left(\frac{\partial^2 w}{\partial y^2} + \nu\frac{\partial^2 w}{\partial x^2}\right)$$

$$M_{xy} = -M_{yx} = D(1-\nu^2)\frac{\partial^2 w}{\partial x\,\partial y} \tag{6.40}$$

where $D = \dfrac{Eh^3}{12(1-\nu)^2}$ and $w = \beta g + \sum_{i=0}^{m} \mathbf{P}_i \alpha_i$

6.3.5 Numerical examples

The following examples are used to illustrate the effectiveness of the method.

Example 1

Consider a square clamped plate of length a and subject to a concentrated load at the centre, as shown in Fig. 6.9. One quarter of the plate is used in the analysis and it is divided into 200 triangular elements, as shown in Fig. 6.10.

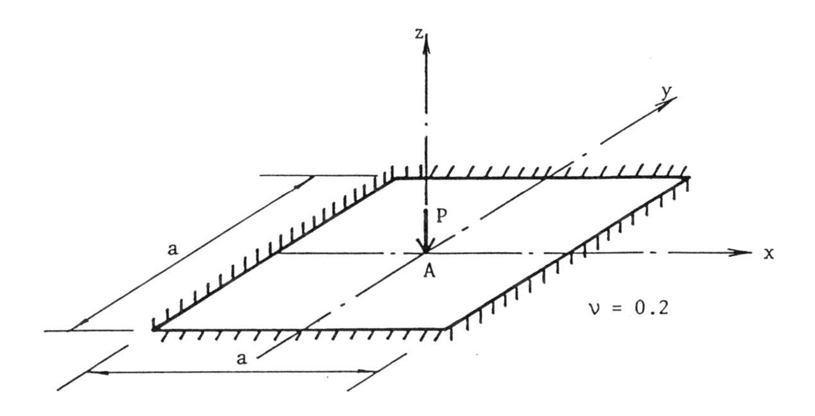

Fig. 6.9 A square clamped plate of length a subject to a concentrated load at the centre

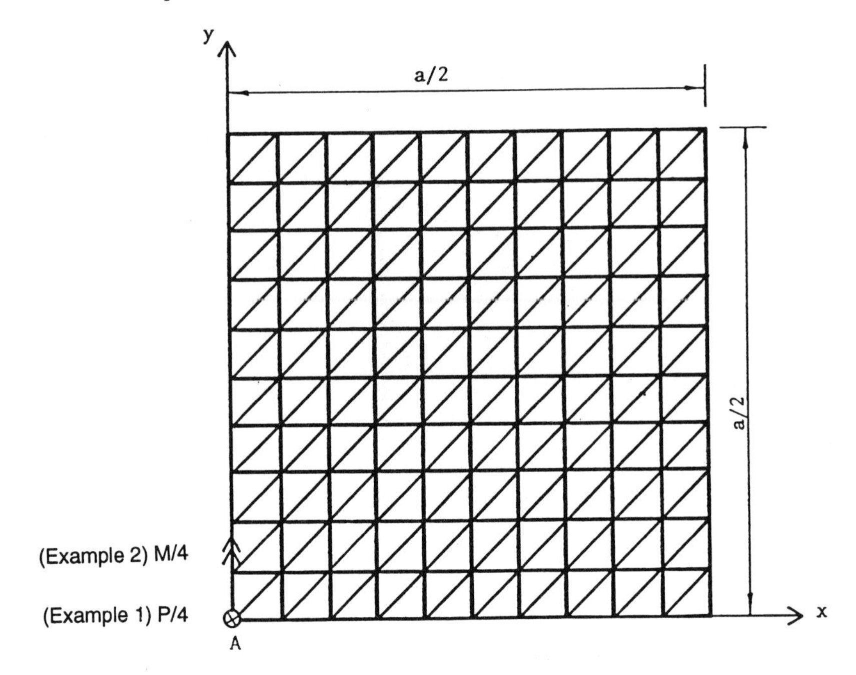

Fig. 6.10 One-quarter of the plate for analysis divided into 200 triangular elements

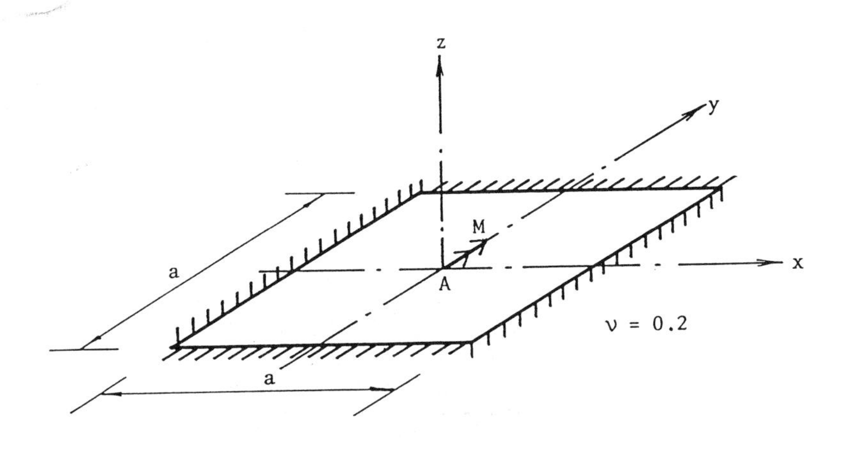

Fig. 6.11 A plate subject to a concentrated moment about the y-axis at the centre

Example 2

Consider the same square plate of Example 1 but subject to a concentrated moment about the y-axis at the centre, as shown in Fig. 6.11. One-quarter of the plate is used in the analysis and it is divided into 200 triangular elements, as shown in Fig. 6.10.

Example 3

Consider a right-angled triangular clamped plate of length a and subject to a concentrated load, as shown in Fig. 6.12. One-half of the plate is used in the analysis and it is divided into 400 triangular elements, as shown in Fig. 6.13.

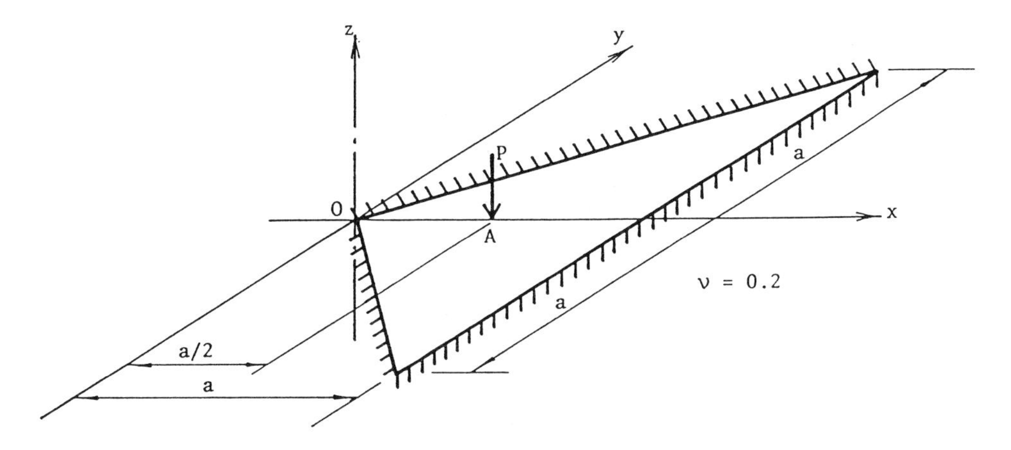

Fig. 6.12 Right-angled triangular clamped plate of length *a* and subject to a concentrated load

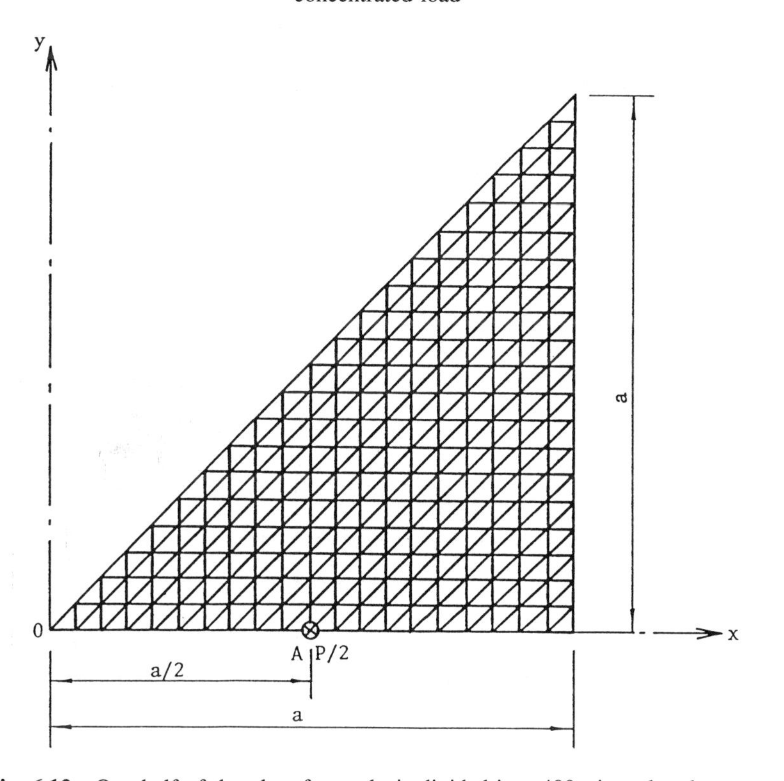

Fig. 6.13 One-half of the plate for analysis divided into 400 triangular elements

The maximum deflections of these examples with different orders of polynomial are shown in Table 6.5. The deflections and stresses of the plate with polynomials up to seventh order are plotted from Figs 6.14–6.19. The results are compared to full finite element analyses obtained from ICE*STUDL's CPT plate bending element with the same finite element mesh. The total numbers of equations required to be solved in these examples are listed in Table 6.6.

Table 6.5 Maximum deflection with different orders of polynomial used

Maximum deflection	*Example 1*	*Example 2*	*Example 3*
order 3	0.524 01	0.879 25	0.143 69
order 5	0.561 47	0.961 49	0.352 43
order 7	0.562 39	0.968 22	0.378 48
STRUDL	0.559 63	0.974 25	0.378 79
Factor	$\frac{Pa^2}{D} \times 10^{-2}$	$\frac{Ma}{D} \times 10^{-2}$	$\frac{Pa^2}{D} \times 10^{-2}$

Table 6.6 Number of equations required

		Two-level finite element method		
Example	*STRUDL*	*order 3*	*order 5*	*order 7*
1	363	14	25	40
2	363	14	25	40
3	693	14	25	40

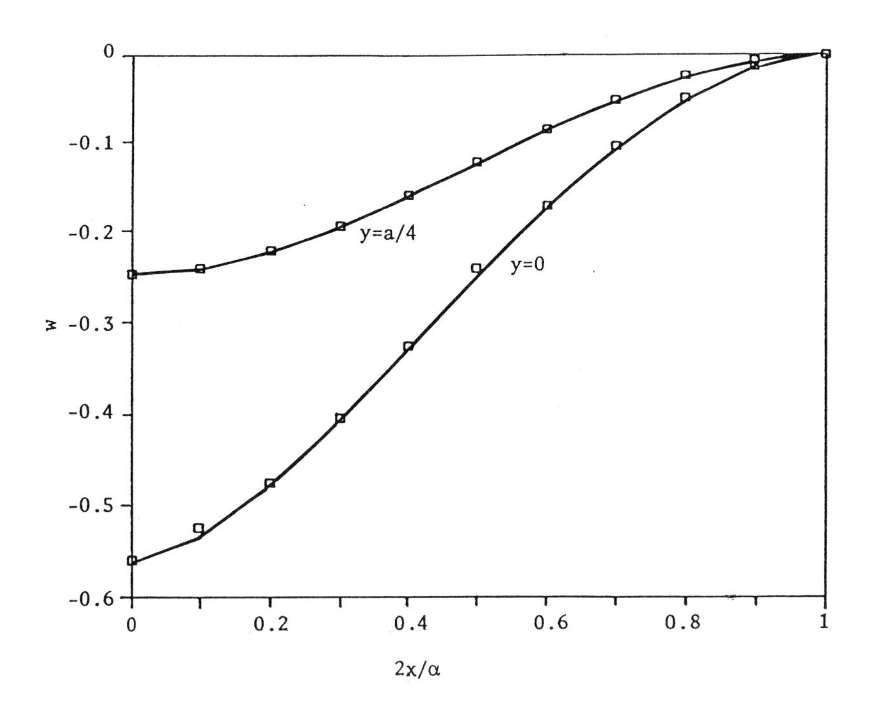

Fig. 6.14 Example 1, deflection. STRUDL; present (order 7).

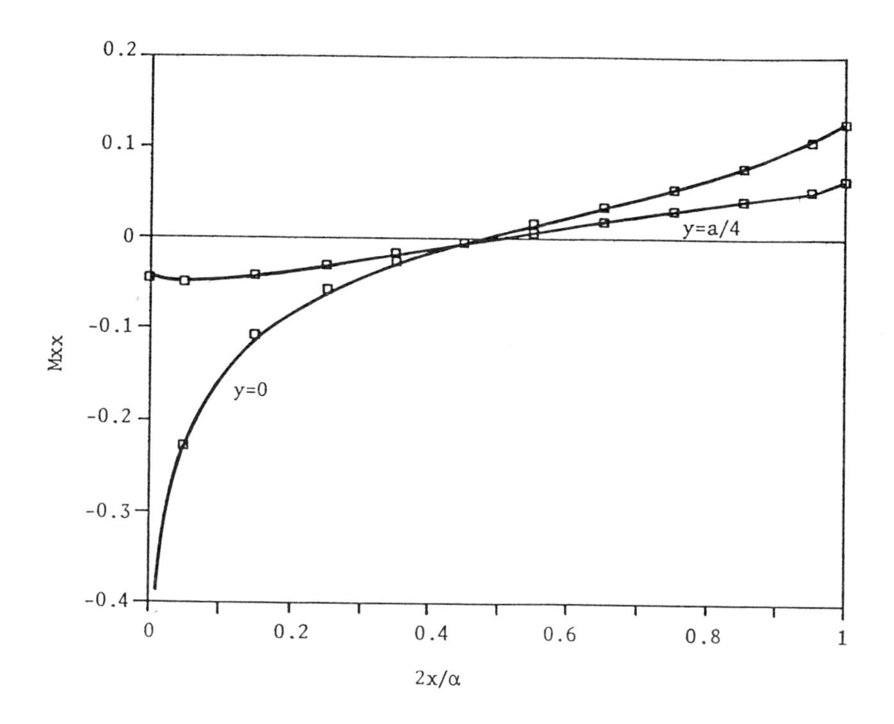

Fig. 6.15 Example 1, stress. STRUDL; present (order 7).

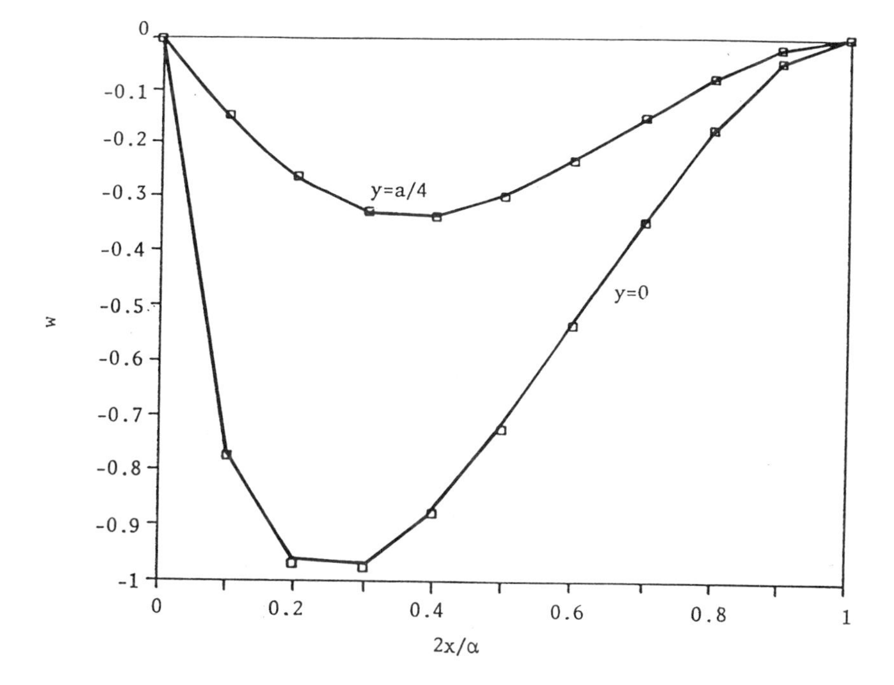

Fig. 6.16 Example 2, deflection. STRUDL; present (order 7).

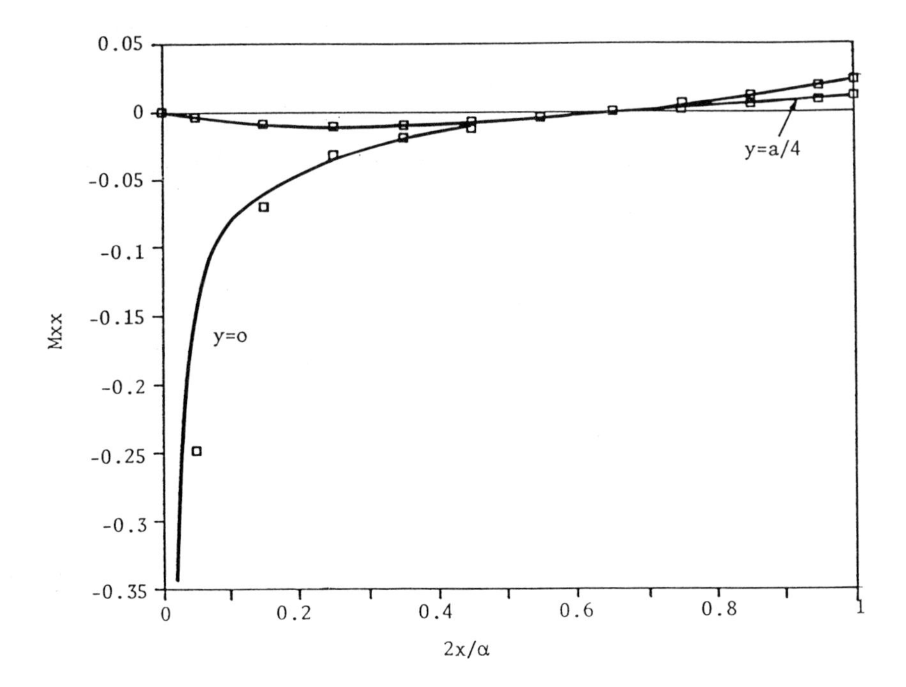

Fig. 6.17 Example 2, stress. STRUDL; present (order 7).

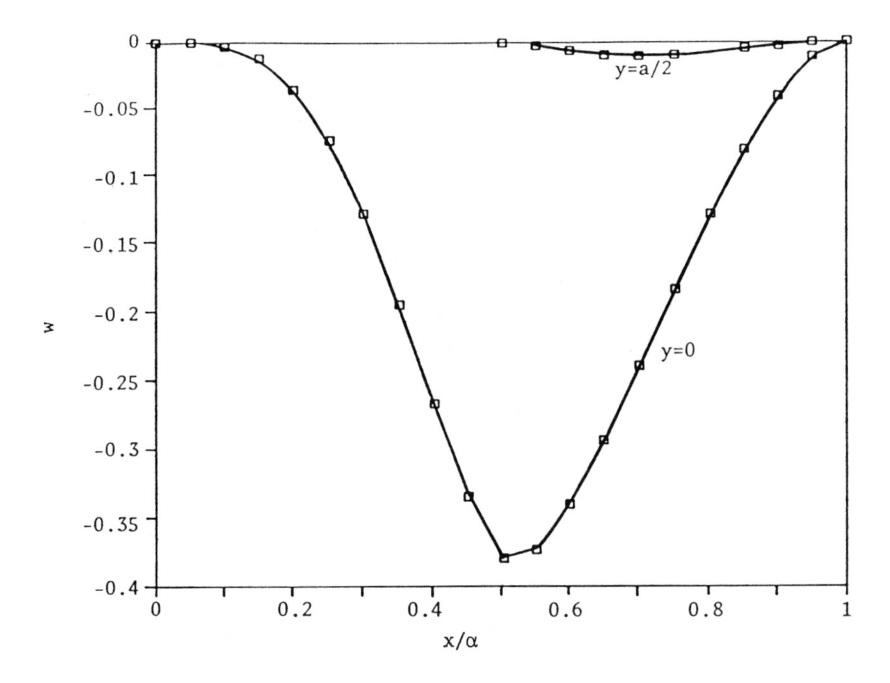

Fig. 6.18 Example 3, deflection. STRUDL; present (order 7).

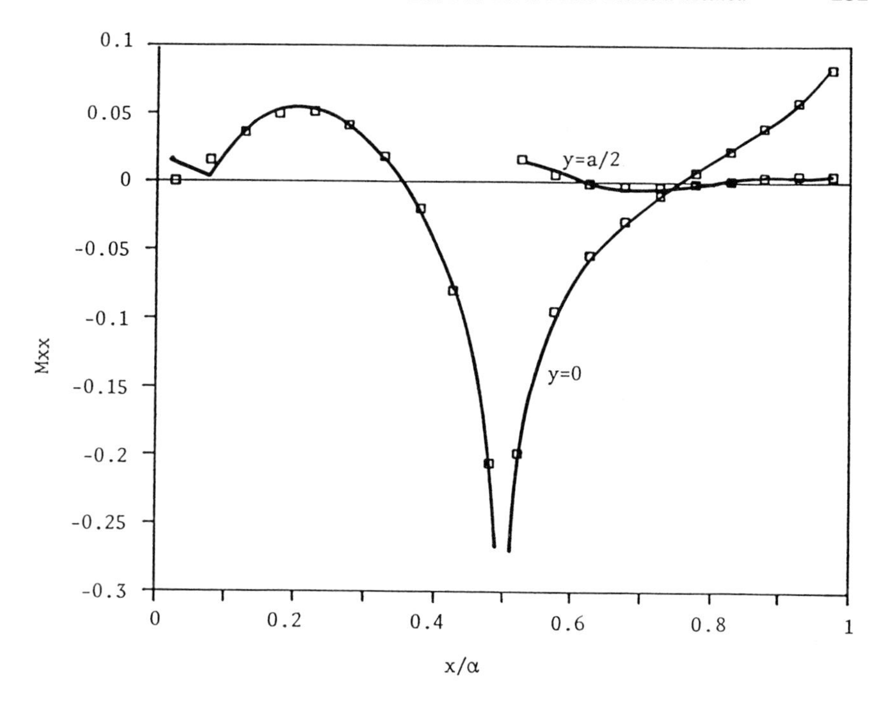

Fig. 6.19 Example 3, stress. STRUDL; present (order 7).

6.4 2D CRACK PROBLEMS [7]

The method is applied to evaluate the stress intensity factors for plates of arbitrary shape using conventional finite elements. The problems of the large number of unknowns and the round-off errors associated with fine meshes are eliminated by means of global interpolating functions. These global interpolating functions are actually the analytical solutions of the displacement pattern near the crack tips. While the analytical solutions do not satisfy the boundary conditions in general, the present method considers the boundary conditions by master nodes. With very few unknowns, accurate results can be predicted and no unconventional finite elements are required. The stress intensity factors can be calculated from the nodal displacements of the super-element in one go. Plates with single- and double-notch edges and a centre crack are taken as examples. The method can be generalized to other crack problems without difficulty.

6.4.1 Formulation

The usual finite element method suggests the following interpolation for the plane displacements $u(x, y)$ and $v(x, y)$

$$\begin{Bmatrix} u(x, y) \\ v(x, y) \end{Bmatrix} = [\mathbf{N}(x, y)]\{\boldsymbol{\delta}_e\} \tag{6.41}$$

where $[\mathbf{N}(x, y)]$ is the shape function matrix and $\{\boldsymbol{\delta}_e\}$ are the nodal displacements in the x and y directions. After assembly of the elements, the following stiffness equation results

$$[\mathbf{K}]\{\boldsymbol{\delta}\} = \{\mathbf{f}\} \tag{6.42}$$

where [**K**] is the global stiffness matrix and {**f**} are the applied nodal forces. The conventional finite element method converges slowly due to the singular characteristic at the crack tip. An alternative approach is the complex function approach of Muskhelishvili, referred to by Leung and Wong [7]. By assuming the complex eigenvalue Goursat functions in series form (which are discussed in the following section), a solution can be obtained in terms of some coefficients which are the generalized coordinates and are to be determined after loading is applied for the given boundary conditions. However, the analytical solution is difficult to generalize for regions of arbitrary shape.

By means of the Goursat solution, all the nodal displacements are expressed in terms of the initial coefficients (generalized coordinates). The generalized coordinates are then related to some suitably chosen master nodes on the boundary to form an accurate super-element. The above-mentioned two-stage transformation equation considered in the previous sections is summarized as follows

$$\{\boldsymbol{\delta}\} = [\mathbf{T}]\{\mathbf{u}_m\} \tag{6.43}$$

where [**T**] is the resultant transformation matrix and $\{\mathbf{u}_m\}$ are the nodal displacements with respect to the master nodes.

The order of $\{\mathbf{u}_m\}$ is much less than that of $\{\boldsymbol{\delta}\}$ because only part of the boundary nodes are chosen as the master nodes. Equation 6.42 is transformed according to the conservation of strain energy to

$$[\mathbf{T}]^{\mathrm{T}}[\mathbf{K}][\mathbf{T}]\{\mathbf{u}_m\} = [\mathbf{T}]^{\mathrm{T}}\{\mathbf{f}\} \tag{6.44}$$

which is of a much smaller order than the original.

Equations 6.41 and 6.43 correspond to the local and global interpolations, respectively. In the present approach, no new element matrices need be generated. The Goursat solution is used in forming matrix [**T**] to condense the stiffness matrix.

In the following sections, the analytical global interpolating functions are discussed first and the transformation process is then studied. Finally, the stress intensity factors are evaluated. The conventional finite element method for the plane strain problem is not discussed, but interested readers can refer to Chapter 2 for details.

6.4.2 Global interpolating functions

Considering a plane crack in an infinite domain, as shown in Fig. 6.20, it can be seen that the crack surface must be free from traction and therefore

$$\sigma_y = \sigma_{xy} = 0 \qquad \text{for } \theta = \pm\pi \tag{6.45}$$

Applying the above boundary conditions to Muskhelishvili's equations, the following general solutions for stresses and displacements can be derived

$$\sigma_x = \sum_{n=1}^{\infty} \frac{n}{2} r^{(n/2-1)} \left\{ a_n \left[\left(2 + \frac{n}{2} + (-1)^n\right) \cos\left(\frac{n}{2} - 1\right)\theta - \left(\frac{n}{2} - 1\right) \cos\left(\frac{n}{2} - 3\right)\theta \right] \right.$$

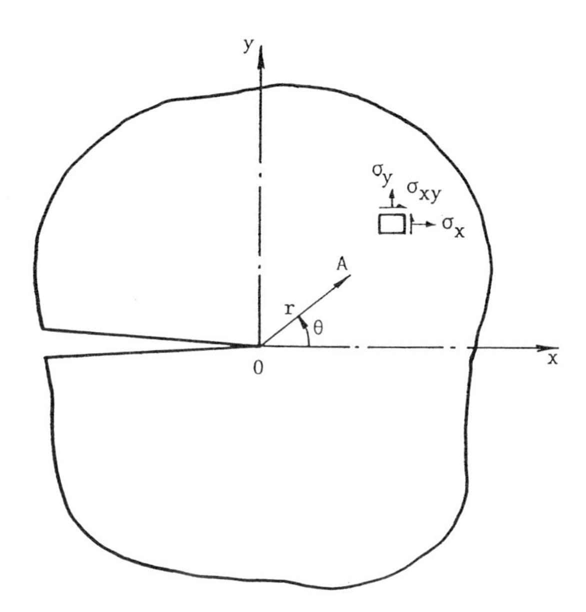

Fig. 6.20 Coordinate axes employed in the stress analysis of a plane crack

$$-b_n\left[\left(2+\frac{n}{2}-(-1)^n\right)\sin\left(\frac{n}{2}-1\right)\theta-\left(\frac{n}{2}-1\right)\sin\left(\frac{n}{2}-3\right)\theta\right]\Big\} \quad (6.46)$$

$$\sigma_y=\sum_{n=1}^{\infty}\frac{n}{2}r^{(n/2-1)}\Big\{a_n\left[\left(2-\frac{n}{2}-(-1)^n\right)\cos\left(\frac{n}{2}-1\right)\theta+\left(\frac{n}{2}-1\right)\cos\left(\frac{n}{2}-3\right)\theta\right]$$

$$-b_n\left[\left(2-\frac{n}{2}+(-1)^n\right)\sin\left(\frac{n}{2}-1\right)\theta+\left(\frac{n}{2}-1\right)\sin\left(\frac{n}{2}-3\right)\theta\right]\Big\} \quad (6.47)$$

$$\sigma_{xy}=\sum_{n=1}^{\infty}\frac{n}{2}r^{(n/2-1)}\Big\{a_n\left[\left(\frac{n}{2}-1\right)\sin\left(\frac{n}{2}-3\right)\theta-\left(\frac{n}{2}+(-1)^n\right)\sin\left(\frac{n}{2}-1\right)\theta\right]$$

$$+b_n\left[\left(\frac{n}{2}-1\right)\cos\left(\frac{n}{2}-3\right)\theta-\left(\frac{n}{2}-(-1)^n\right)\cos\left(\frac{n}{2}-1\right)\theta\right]\Big\} \quad (6.48)$$

$$u=\sum_{n=0}^{\infty}\frac{r^{n/2}}{2\mu}\Big\{a_n\left[\left(x+\frac{n}{2}+(-1)^n\right)\cos\frac{n}{2}\theta-\frac{n}{2}\cos\left(\frac{n}{2}-2\right)\theta\right]$$

$$-b_n\left[\left(x+\frac{n}{2}-(-1)^n\right)\sin\frac{n}{2}\theta-\frac{n}{2}\sin\left(\frac{n}{2}-2\right)\theta\right]\Big\} \quad (6.49)$$

$$v=\sum_{n=0}^{\infty}\frac{r^{n/2}}{2\mu}\Big\{a_n\left[\left(x-\frac{n}{2}-(-1)^n\right)\sin\frac{n}{2}\theta+\frac{n}{2}\sin\left(\frac{n}{2}-2\right)\theta\right]$$

$$+b_n\left[\left(x-\frac{n}{2}+(-1)^n\right)\cos\frac{n}{2}\theta+\frac{n}{2}\cos\left(\frac{n}{2}-2\right)\theta\right]\Big\} \quad (6.50)$$

where a_n and b_n are coefficients to be determined after loading and other boundary conditions are imposed. It should be noted that the first term in eqs 6.46–6.48 contains the $1/\sqrt{r}$ expression which accounts for the singular behaviour at the crack tip. Therefore the relationship between stress intensity factors and coefficients becomes

$$a_1 = \frac{K_{\mathrm{I}}}{\sqrt{(2\pi)}} \quad \text{and} \quad b_1 = \frac{K_{\mathrm{II}}}{\sqrt{(2\pi)}} \tag{6.51}$$

The whole problem is reduced to the determination of a_1 and b_1.

6.4.3 Transformation

Let the total number of master nodes in a super-element be m and that of slave nodes be s. Equations 6.49 and 6.50 can be rewritten into the following matrix form

$$u = [f_1, f_2, \ldots, f_{2m}]\{\mathbf{a}\} \quad \text{and} \quad v = [g_1, g_2, \ldots, g_{2m}]\{\mathbf{a}\} \tag{6.52}$$

where $$\{\mathbf{a}\} = \{a_0, a_1, \ldots, a_{m-1};\ b_0, b_1, \ldots, b_{m-1}\}^{\mathrm{T}} \tag{6.53}$$

are the coefficients defined in eqs 6.49 and 6.50.

After substituting the polar coordinates of the m master nodes into eq. 6.52, the master nodal displacements $\{\mathbf{u}_m\}$ can be obtained as

$$\{\mathbf{u}_m\} = [\mathbf{C}]\{\mathbf{a}\} \tag{6.54}$$

where $[\mathbf{C}]$ is a $2m \times 2m$ square matrix. Similarly, the relationship between slave nodal displacements $\{\mathbf{u}_s\}$ and $\{\mathbf{a}\}$ is

$$\{\mathbf{u}_s\} = [\mathbf{D}]\{\mathbf{a}\} \tag{6.55}$$

where $[\mathbf{D}]$ is a $2s \times 2m$ rectangular matrix. Substituting eq. 6.54 into eq. 6.55 gives

$$\{\mathbf{u}_s\} = [\mathbf{D}][\mathbf{C}]^{-1}\{\mathbf{u}_m\} \tag{6.56}$$

Hence,

$$\{\boldsymbol{\delta}\} = \begin{Bmatrix} \mathbf{u}_s \\ \mathbf{u}_m \end{Bmatrix} = \begin{bmatrix} \mathbf{DC}^{-1} \\ \mathbf{I} \end{bmatrix} \{\mathbf{u}_m\} = [\mathbf{T}]\{\mathbf{u}_m\} \tag{6.57}$$

where $[\mathbf{I}]$ is a $2m \times 2m$ identity matrix and $[\mathbf{T}]$ is the transformation matrix. The conservation of strain energy requires that

$$U = \frac{1}{2}\{\boldsymbol{\delta}\}^{\mathrm{T}}[\mathbf{K}]\{\boldsymbol{\delta}\} = \frac{1}{2}\{\mathbf{u}_m\}^{\mathrm{T}}[\mathbf{K}_m]\{\mathbf{u}_m\} \tag{6.58}$$

where U is the strain energy and

$$[\mathbf{K}]_m = [\mathbf{T}]^{\mathrm{T}}[\mathbf{K}][\mathbf{T}] \tag{6.59}$$

is the condensed stiffness matrix of the super-element. The resulting two-level finite element equation for the super-element becomes

$$[\mathbf{K}_m]\{\mathbf{u}_m\} = \{f_m\} \tag{6.60}$$

where $\{\mathbf{f}_m\} = [\mathbf{T}]^{\mathrm{T}}\{\mathbf{f}\}$ is the equivalent force vector.

6.4.4 Evaluation of stress intensity factor

After connecting the super-element to other finite elements or substructures, the following finite element equations can be established

$$[\overline{\mathbf{K}}]\{\overline{\mathbf{u}}_m\} = \{\overline{\mathbf{f}}\} \tag{6.61}$$

where $[\overline{\mathbf{K}}]$ is the global stiffness matrix and $\{\overline{\mathbf{f}}\}$ is the nodal force vector.

Solving for $\{\overline{\mathbf{u}}_m\}$ and utilising eq. 6.54, the coefficient vector $\{\mathbf{a}\}$ can be obtained by

$$\{\mathbf{a}\} = [\mathbf{C}]^{-1}\{\overline{\mathbf{u}}_m\} \tag{6.62}$$

Therefore, the stress intensity factor can be calculated directly from eqs 6.62 and 6.51

$$K_I = \sqrt{(2\pi)}a_1 \quad \text{and} \quad K_{II} = \sqrt{(2\pi)}b_1 \tag{6.63}$$

6.4.5 Numerical examples

The mesh configuration of the super-element is shown in Fig. 6.21; the circle 'o' represents a master node and the cross × represents a slave node. The total

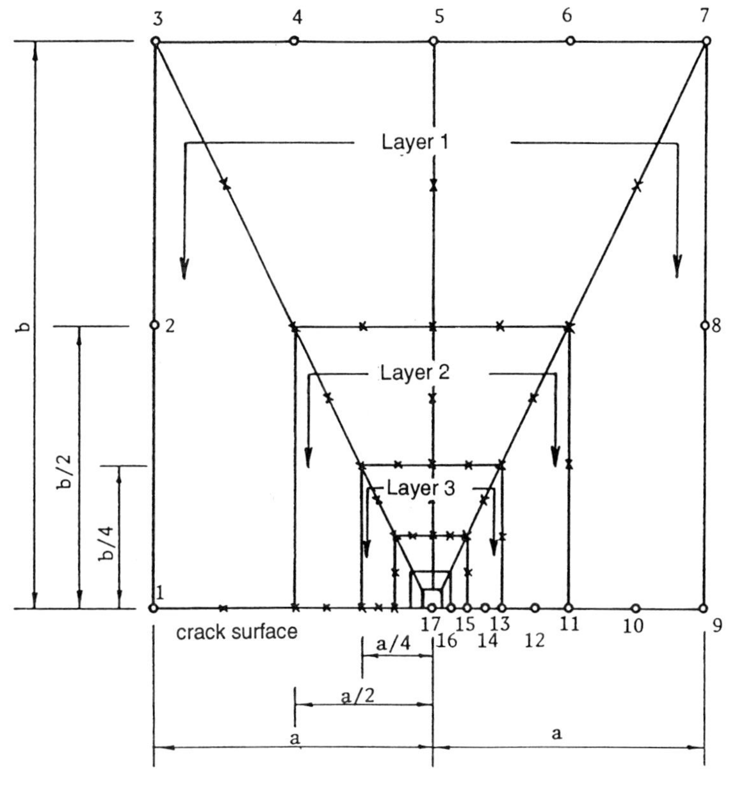

Fig. 6.21 Mesh configuration of the super-element (level = 6)

number of master nodes is 17. The super-element is divided into different layers and each layer contains four 8-node isoparametric plane strain elements. The finite element mesh containing the crack tip is refined by addition of layers, and each new layer is added simply by halving the last layer. The level of refinement is specified by the number of layers in the super-element. The higher the level, the finer the mesh near the crack tip.

Example 1

Consider a strip with a double-edge notch and subject to uniform uniaxial loads, as shown in Fig. 6.22. The connection of the super-element to the other elements is shown in Fig. 6.23. The analytical expression for the mode I stress intensity factor of the double-edge notch problem is expressed as [19]

$$\mathbf{K}_I = \sigma(\pi a)^{1/2} f(a/b) \tag{6.64}$$

where the factor $f(a/b)$ is a function of the crack length a and the plate half-width b. A number of expressions for $f(a/b)$ have been developed in the literature. The one chosen for this study can be found in Reference 19 and is given as follows

$$f\left(\frac{a}{b}\right) = 1.12 + 0.203\left(\frac{a}{b}\right) - 1.197\left(\frac{a}{b}\right)^2 + 1.930\left(\frac{a}{b}\right)^3 \tag{6.65}$$

The equation was derived by Brown based on a least-square fitting to Bowie's results.

The percentage error of the stress intensity factor against the level of division of the super-element (i.e. fineness of mesh) for $a/b = 0.5$ is studied. The results are listed and plotted in Table 6.7 and Fig. 6.24, respectively. The convergence of the

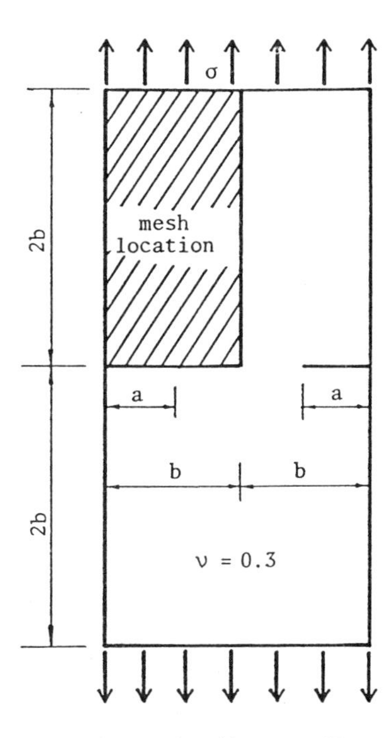

Fig. 6.22 Double-edge notch subject to uniform uniaxial loads

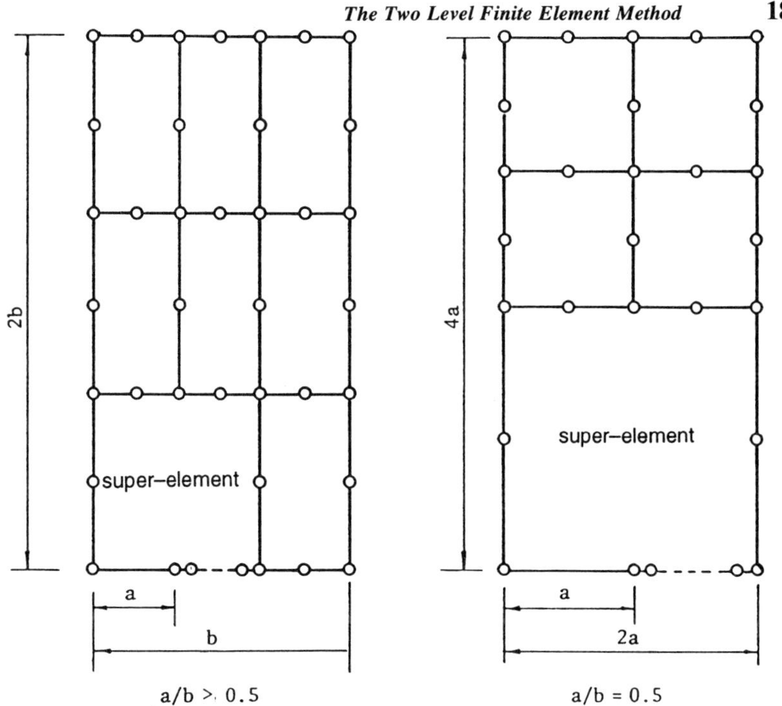

Fig. 6.23 Connection between super-element and other finite elements (mesh orientation) in Example 1 and Example 2

Table 6.7 Double-edge notch

Level	$\frac{\mathbf{K}_I}{\sigma(\pi a)^{1/2}}$	% error	No. of elements	Total DOF
4	1.120 29	3.713	16	142
5	1.139 26	2.083	20	170
6	1.150 55	1.113	24	198
7	1.156 82	0.574	28	226
8	1.159 97	0.303	32	254
9	1.161 62	0.162	36	282
10	1.162 43	0.092	40	310

$a/b = 0.5$, exact $= 1.16350$, no. of masters $= 17$

present method is shown numerically. Although the embedment of the special crack tip elements or quarter-point isoparametric finite elements can certainly improve the accuracy acquired without an excessive number of levels within the super-element, the authors find that the use of conventional finite elements is better in view of their universal availability and convenient implementation.

In the above study it can be seen that the tenth level of division is sufficiently accurate for engineering application. Therefore, the level is chosen as ten in the subsequent analysis. The mode I stress intensity factor $\mathbf{K}_I$, against the a/b ratio is listed in Table 6.8. It has been shown in Table 6.7 that the fourth substructure level gives very accurate results already.

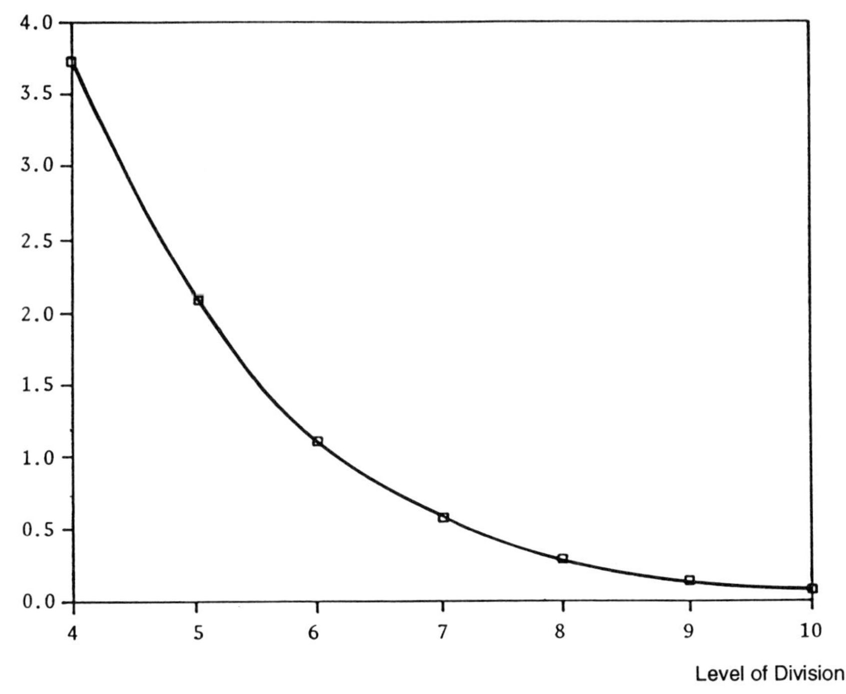

Fig. 6.24 Convergency of the stress intensity factor with respect to the level of division

Table 6.8 Double-edge notch

a/b	$\frac{\mathbf{K}_I}{\sigma(\pi a)^{1/2}}$ Present	Theoretical	% error
0.500 00	1.162 43	1.163 50	0.092
0.416 67	1.131 05	1.136 38	0.469
0.333 33	1.112 59	1.126 15	1.204
0.277 78	1.105 01	1.125 39	1.812
0.238 10	1.100 93	1.126 53	2.272

Example 2

Consider a strip with a centre crack and subject to uniform uniaxial loads, as shown in Fig. 6.25. The connection of the super-element to the other elements is shown in Fig. 6.23. The stress intensity factor for the centre crack problem is the same as that expressed in eq. 6.64. The $f(a/b)$ chosen in Reference 19 is

$$f\left(\frac{a}{b}\right) = \left[\left(\frac{2b}{\pi a}\right) \tan\left(\frac{\pi a}{2b}\right)\right]^{1/2} \tag{6.66}$$

The equation was derived by Irwin and is based on an approximation using a periodic crack solution. Similar to the case of the double-edge notch, the full advantage of symmetry is taken and as a result only one-quarter of the plate was considered in

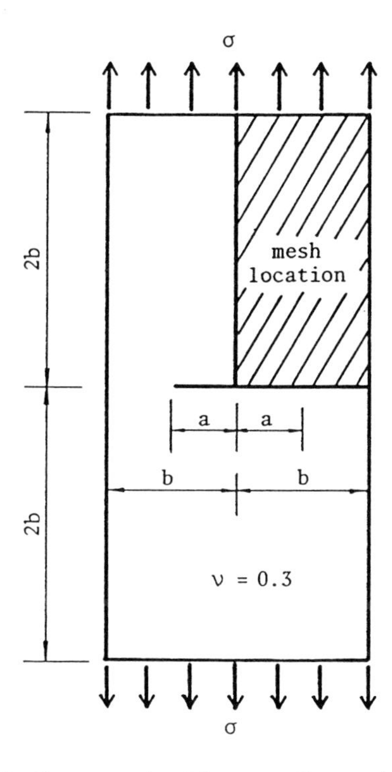

Fig. 6.25 Centre crack subject to uniform uniaxial loads

Table 6.9 Centre crack

a/b	$\frac{\mathbf{K}_I}{\sigma(\pi a)^{1/2}}$ Present	$\frac{\mathbf{K}_I}{\sigma(\pi a)^{1/2}}$ Theoretical	% error
0.500 00	1.158 23	1.128 38	2.645
0.416 67	1.097 94	1.082 77	1.401
0.333 33	1.053 32	1.050 08	0.309
0.277 78	1.030 22	1.033 78	0.344
0.238 10	1.016 10	1.024 40	0.810

the analysis. The mode *I* stress intensity factor $\mathbf{K}_I$ against the a/b ratio is listed in Table 6.9. Good agreement with the theoretical solution is also obtained for $\mathbf{K}_{II}$, as shown in Table 6.10.

6.5 Laminated thick rectangular plates [8]

Next the bending, vibration and buckling of laminated thick plates of rectangular shape will be studied. A thick square plate with arbitrary conditions of support is given in Fig. 6.26. To describe the local behaviour within the cross-section (x–z plane), the usual two-dimensional discretization is employed (Fig. 6.27). The

Table 6.10 Single-edge notch

	$\frac{\sqrt{(\pi a)}}{2P}\mathbf{K}_I$			$\frac{\sqrt{(\pi a)}}{2Q}\mathbf{K}_{II}$		
a/b	Present	Theoretical	% error	Present	Theoretical	% error
0.500 00	5.605 49	5.819 49	3.677	1.474 38	1.559 17	5.438
0.400 00	3.774 84	3.845 73	1.843	1.359 53	1.442 20	5.732
0.333 33	2.944 73	2.982 02	1.250	1.304 11	1.389 86	6.170
0.285 71	2.488 63	2.514 50	1.029	1.275 73	1.361 91	6.328
0.250 00	2.199 15	2.228 63	1.323	1.252 14	1.345 23	6.920

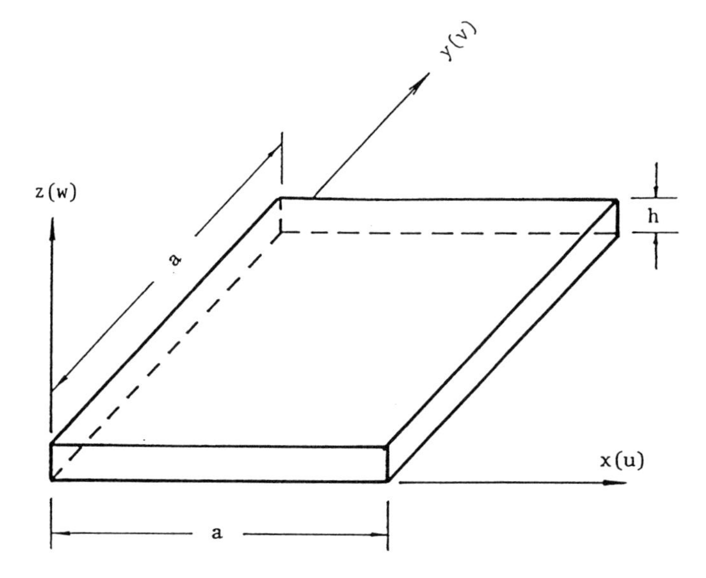

Fig. 6.26 A thick square plate with span length a

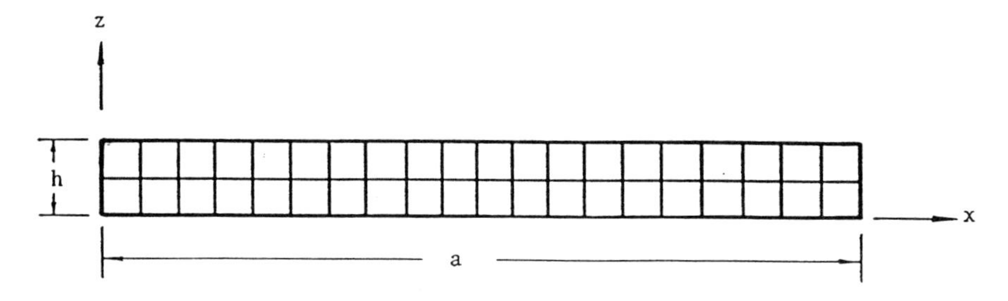

Fig. 6.27 Discretization in the x–z plane

interpolation within each element is defined by the standard isoparametric serendipity shape functions [**N**]. The nodal displacements within the cross-section are then assumed to follow a pattern of variation which is defined by the global functions [$\mathbf{G}(x, z)$]. The local and global behaviour can be incorporated through transformation. The cubic B_3 spline function is chosen to represent the behaviour of the plate along the span, because it possesses the special features of high accuracy and piecewise nature.

The displacements of the rectangular plate can be written as

$$u = \sum_{i=1}^{p}\sum_{j=1}^{q} \alpha_{ij} U_i(x, z) \cdot \Phi_j(y)$$

$$v = \sum_{i=1}^{p}\sum_{j=1}^{q} \beta_{ij} V_i(x, z) \cdot \Phi_j(y)$$

$$w = \sum_{i=1}^{p}\sum_{j=1}^{q} \gamma_{ij} W_i(x, z) \cdot \Phi_j(y)$$

where $U_i(x, z)$, $V_i(x, z)$ and $W_i(x, z)$ comprise the displacements of the nodes within the cross-section computed from the ith member of the global function $[\mathbf{G}(x, z)]$, $\Phi_j(y)$ represents the jth member of the B_3 spline series in the y-direction, and α_{ij}, β_{ij} and γ_{ij} are the unknown displacement parameters. The parameters p and q stand for the total numbers of members in the global function and the spline series respectively. For edges normal to the y-direction, the boundary conditions are taken into account by modifying the local splines [9] at the boundary.

Following the standard Rayleigh–Ritz technique, the final matrix equations can be derived, which give

$$[\mathbf{K}]\{\boldsymbol{\delta}\} = \{\mathbf{f}\} \qquad \text{for static analysis}$$

$$[\mathbf{K}]\{\boldsymbol{\delta}\} = \omega^2[\mathbf{M}]\{\boldsymbol{\delta}\} \qquad \text{for free vibration analysis}$$

$$\text{and} \quad [\mathbf{K}]\{\boldsymbol{\delta}\} = \lambda[\mathbf{G}]\{\boldsymbol{\delta}\} \qquad \text{for linear elastic stability analysis}$$

where $[\mathbf{K}]$, $[\mathbf{M}]$ and $[\mathbf{G}]$ represent the stiffness matrix, the mass matrix and the geometric stiffness matrix, respectively. The load vector and the displacement vector are denoted by $\{\mathbf{f}\}$ and $\{\boldsymbol{\delta}\}$, respectively. During the formation of the matrices $[\mathbf{K}]$, $[\mathbf{M}]$ and $[\mathbf{G}]$, the integration of the spline functions is done in the usual manner while the integration involving the functions $U_i(x, z)$, $V_i(x, z)$, and $W_i(x, z)$ is carried out by a special procedure. This procedure is illustrated with the aid of the following integral equation

$$\int_0^h \int_0^a U_i^{\mathrm{T}} \cdot V_j \, \mathrm{d}x \, \mathrm{d}z = \sum_{r=1}^{t} \{\mathbf{u}_i\}_r^{\mathrm{T}} \left(\int_{A_r} [\mathbf{N}]^{\mathrm{T}}[\mathbf{N}] \, \mathrm{d}A \right) \{\mathbf{v}_j\}_r$$

where t = the total number of elements in the cross-section,
a = width of plate,
h = thickness of plate,
A_r = area of element r.

The vectors $\{\mathbf{u}_i\}_r$ and $\{\mathbf{v}_j\}_r$ contain the nodal displacements of element r computed from the ith and jth members of the global functions $[\mathbf{G}(x, z)]$ respectively. The integral of the shape functions $[\mathbf{N}]$ for the stiffness matrix is evaluated with reduced integration while full integration is used to compute the mass matrix and the geometric stiffness matrix. This equation clearly demonstrates the incorporation of the local-global behaviour with the final stiffness matrix. At this stage, it is worth pointing out that the present method differs from the local-global approach of Leung and Cheung [1] in the sense that the formation of the stiffness matrix of the three-dimensional spline finite prism beforehand is not necessary for carrying out the transformation. Rather, the transformation is first of all carried out to reduce the number of variables, and a much smaller stiffness matrix is then formed. This advantage results in the reduction of the computing time and the computer incore storage requirement. After solving the matrix equation for static analysis, the stresses are calculated at the Gaussian integration points with extrapolation to nodal points via the standard bilinear shape functions. For free vibration and buckling analysis, the eigenpairs are computed by means of the subspace iteration method [10].

6.5.1 Global functions

In the previous literature by Leung and Cheung, different types of global functions have been successfully applied to various classes of problems. These functions can be broadly classified into five major groups, namely, (1) polynomial function [1], (2) analytical solutions [6,7], (3) vibration mode [11], (4) partial solutions [4] and (5) computed shape functions [12,13]. Here, the polynomial functions are employed as the global functions, and they can be expressed as the Kronecker product of two series

$$\mathbf{G}(x, z) = \mathbf{X}(\chi) \cdot \mathbf{Z}(z) = \{\mathbf{X}_i(\chi)\mathbf{Z}_j(z)\}$$

where $\mathbf{X}(\chi) = \{(1-\chi)/2, (1+\chi)/2, (\chi^2-1), (\chi^3-\chi), \ldots, (\chi^m - p), \ldots\}$

with $\chi = 2x/a - 1 \qquad \chi \in [-1, 1]$

$p = 1 \qquad \text{if } m \text{ is even}$

$p = \chi \qquad \text{if } m \text{ is odd}$

and $\mathbf{Z}(z) = \{1, z, z^2, \ldots, z^n, \ldots\} \qquad z \in [0, h]$

By varying the parameters m and n, different orders of polynomials can be used in the x and z directions. Furthermore, it is not necessary to use the same global functions for the three components of displacements, because in some cases their variations may differ significantly from each other. For example, in the case of a moderately thick or thin laminated plate, the transverse displacement component w can be assumed to remain constant through the thickness by ignoring the relatively small quantity of transverse normal strain in the plate.

6.5.2 Numerical examples: static analysis

To establish the accuracy of the present method, we shall consider a series of bidirectional laminated square plates with edges of length a. Since most of the higher-order plate theories are not capable of giving accurate stress predictions for thick laminated plates, emphasis is placed on the analysis of plates with span-to-thickness ratio of 4. Results are compared with the exact elasticity solutions [13–15] and a three-dimensional finite element solution due to Barker *et al.* [16].

In the following examples, the top surface loading is defined as

$$q = \bar{q} \sin\left(\frac{\pi x}{a}\right) \sin\left(\frac{\pi y}{a}\right)$$

where $\overline{q}$ is the maximum intensity of the double sinusoidal load at the centre of the plate. All the edges of the plates are either simply supported or clamped. By symmetry, only a quarter of the plate is analyzed. Each model is divided into four spline sections of equal length along the span. Each ply is composed of orthotropic material with the material principal axis 1 oriented at an angle of 0° or 90° to the global x-axis. The material properties are defined as follows,

$$E_{11} = 25 \times 10^6 \text{ psi} \qquad E_{22} = E_{33} = 1 \times 10^6 \text{ psi}$$
$$G_{12} = G_{13} = 0.5 \times 10^6 \text{ psi} \qquad G_{23} = 0.2 \times 10^6 \text{ psi}$$
$$\nu_{12} = \nu_{23} = \nu_{13} = 0.25$$

where 1 signifies the direction parallel to the fibres, and 2 the transverse direction.

The following normalized quantities are defined for the presentation of results.

$$(\overline{\sigma}_x \overline{\sigma}_y \overline{\tau}_{xy}) = \frac{\sigma_x \sigma_y \tau_{xy}}{\overline{q}s^2}$$

$$(\tau_{yz}\tau_{xz}) = \frac{(\tau_{yz}\tau_{xz})}{\overline{q}s}$$

$$\overline{\sigma}_z = \sigma_z$$

$$\overline{w} = \frac{\pi^4 Qw}{12 s^4 h\overline{q}}$$

$$Q = 4G_{12} + \frac{E_{11} + E_{22}(1 + 2\nu_{22})}{1 - \nu_{12}\nu_{21}}$$

$$s = \frac{a}{h}$$

$$\overline{z} = \frac{z}{h} - \frac{1}{2}$$

Example 1

A three-ply symmetric laminate ($s = 4$) is used to test the convergence of the present method. All layers are of equal thickness with ply angles 0°/90°/0°. Particular attention is focused on the order of the global functions (n) used across the thickness. A fourth-order polynomial ($m = 4$) is used across the width of the plate. The local behaviour is defined by a 6×6 mesh with the thickness of each layer divided into two-element layers. The results are shown in Table 6.11. An approximate solution is obtained when the order n reaches 7 which gives a maximum error of about 5% compared to the exact solution. In this case, the number of unknowns within the cross-section is reduced from 399 nodal variables to 96 generalized parameters, which implies a 76% reduction.

An analysis is also performed on the same plate with reduced thickness ($s = 10$). An aspect ratio of 10 represents a plate of moderate thickness. For this plate, only a quintic polynomial across the thickness is employed as the global functions. Comparisons of the results with the exact solutions are shown in Table 6.12. A good agreement between the two sets of results can be observed. An 82% reduction in the number of unknowns within the cross-section is achieved in this analysis.

Example 2

A five-ply laminate ($s = 2, 4$) with ply angles 0°/90°/0°/90°/0° is analyzed. The total thicknesses of the 0° and 90° layers are the same, whereas layers at the same

Table 6.11 Comparison of results for a three-ply laminate with span-to-thickness ratio of 4

	$\bar{\sigma}_x\left(\frac{a}{2},\frac{a}{2},\pm\frac{1}{2}\right)$	$\bar{\sigma}_y\left(\frac{a}{2},\frac{a}{2},\pm\frac{1}{6}\right)$	$\bar{\sigma}_z\left(\frac{a}{2},\frac{a}{2},\pm\frac{1}{2}\right)$	$\bar{\tau}_{xy}\left(0,0,\pm\frac{1}{2}\right)$	$\bar{\tau}_{xz}\left(0,\frac{a}{2},0\right)$	$\bar{w}\left(\frac{a}{2},\frac{a}{2},0\right)$
$n = 3$	0.750	0.482	1.171	−0.0479	0.210	1.896
	−0.705	−0.504	−0.166	0.0473		
$n = 5$	0.767	0.494	1.010	−0.0490	0.243	1.926
	−0.723	−0.519	−0.005	0.0485		
$n = 7$	0.787	0.502	1.016	−0.0503	0.270	1.961
	−0.743	−0.527	−0.015	0.0498		
exact	0.801	0.534	1.0	−0.0511	0.256	2.02*
	−0.755	−0.556	0.0	0.0505		

*The deflections are compared with the results due to Liou and Sun [20] using a hybrid element approach. In this case, the normalized deflection is defined as $\bar{w} = 100E_{22}w/\bar{q}hs^4$.

Table 6.12 Comparison of results for a three-ply laminate with span-to-thickness ratio of 10

	$\bar{\sigma}_x\left(\frac{a}{2},\frac{a}{2},\pm\frac{1}{2}\right)$	$\bar{\sigma}_y\left(\frac{a}{2},\frac{a}{2},\pm\frac{1}{6}\right)$	$\bar{\sigma}_z\left(\frac{a}{2},\frac{a}{2},\pm\frac{1}{2}\right)$	$\bar{\tau}_{xy}\left(0,0,\pm\frac{1}{2}\right)$	$\bar{\tau}_{xz}\left(0,\frac{a}{2},0\right)$	$\bar{w}\left(\frac{a}{2},\frac{a}{2},0\right)$
$n = 7$	0.586	0.277	0.983	−0.0284	0.340	0.7323
	−0.585	−0.281	−0.022	0.0285		
exact	0.590	0.285	1.0	−0.0289	0.357	0.7546*
	−0.590	−0.288	0.0	0.0289		

*The deflections are compared with the results due to Liou and Sun [20] using a hybrid element approach. In this case, the normalized deflection is defined as $\bar{w} = 100E_{22}w/\bar{q}hs^4$.

Table 6.13 Comparison of results for a five-ply laminate

	$\bar{\sigma}_x\left(\frac{a}{2},\frac{a}{2},\pm\frac{1}{2}\right)$	$\bar{\sigma}_y\left(\frac{a}{2},\frac{a}{2},\pm\frac{1}{3}\right)$	$\bar{\sigma}_z\left(\frac{a}{2},\frac{a}{2},\pm\frac{1}{2}\right)$	$\bar{\tau}_{xy}\left(0,0,\pm\frac{1}{2}\right)$	$\bar{\tau}_{xz}\left(0,\frac{a}{2},0\right)$	$\bar{w}\left(\frac{a}{2},\frac{a}{2},0\right)$
			span-to-thickness ratio = 4			
$n = 7$	0.672	0.612	0.997	−0.0378	0.245	4.070
	−0.638	−0.606	−0.004	0.0369		
exact	0.685	0.633	1.0	−0.0394	0.238	4.291
	−0.651	−0.626	0.0	0.0384		
			span-to-thickness ratio = 2			
$n = 7$	1.286	0.902	1.001	−0.0792	0.226	11.631
	−0.863	−0.776	0.001	0.0595		
exact	1.332	1.001	1.0	−0.0836	0.227	12.278
	−0.903	−0.848	0.0	0.0634		

orientation have equal thicknesses. Under these conditions, the effective laminate stiffnesses in the x and y directions are the same. The cross-section is discretized into a 10×10 mesh with each layer of material divided into two-element layers. Using a global polynomial of orders $m = 4$ and $n = 7$ in this analysis, a reasonable agreement with the exact solutions can be observed (Table 6.13). A total of 1023 nodal variables within the cross-section is contracted down to 96 generalized parameters associated with the global polynomials.

6.5.3 Numerical examples: vibration and buckling analysis

We shall consider another series of cross-ply laminated square plates with available elasticity solutions [17,18] for free vibration and linear elastic stability analysis. All edges of the plates are simply supported. For buckling analysis, the plates are loaded uniaxially. Each ply is composed of orthotropic material with the material principal axis 1 oriented at an angle of 0° or 90° to the global x-axis. The total thickness of the 0° and 90° layers is the same, whereas layers at the same orientation have equal thicknesses. The material properties and the geometry of the plates are defined as follows

$$
\begin{aligned}
&E_{11} = 40\ E_{22} = 40\ E_{33} \\
&G_{12} = G_{13} = 0.6\ E_{22} \qquad G_{23} = 0.5\ E_{22} \\
&\nu_{12} = \nu_{23} = \nu_{13} = 0.25 \\
&\text{span-to-thickness ratio, } s = 5 \quad \text{for vibration analysis} \\
&\qquad\qquad\qquad\qquad s = 10 \quad \text{for buckling analysis}
\end{aligned}
$$

In all cases, only a quarter of the plate is analyzed because of its symmetric properties. Each model is divided into four spline sections of equal length along the span. As we are only interested in the prediction of the global responses in these examples, the global function involves a quartic polynomial ($m = 4$) across the width and a cubic polynomial ($n = 3$) through the depth of the plates.

The following normalized quantities are defined for the presentation of results.

$$\text{Natural frequency,} \qquad \overline{\omega} = 10\omega h\sqrt{\left(\frac{\rho}{E_{22}}\right)}$$

$$\text{Critical buckling load, } \overline{N} = \frac{N_x a^2}{E_{22}h^2}$$

where ρ is the density of the composite material.

Results of comparisons between the present method and the elasticity solutions [17,18] are listed in Table 6.14. Very good agreement between the two sets of results can be seen. The efficiency of the present method is also demonstrated by referring

Table 6.14 Comparison of results for free vibration and buckling analysis of a series of cross-ply laminated square plates

Lamination	Number of layers	Frequency, $\overline{\omega}$		Buckling load, $\overline{N}$	
		Present method	Elasticity solution [17]	Present method	Elasticity solution [18]
Antisymmetric	2	3.4918	3.4250	11.0298	10.8167
	4	4.3289	4.2719	21.7346	21.2796
	6	4.5225	4.5091	23.9024	23.6689
	10	4.6398	4.6498	25.0857	25.3436
Symmetric	3	4.3056	4.3006	23.1172	22.8807
	5	4.5577	4.5374	24.9463	24.5929
	9	4.6602	4.6679	25.5184	25.3436

Table 6.15 Comparison of the number of unknowns within a cross-section between a complete 3-D spline finite prism analysis and the present method

Number of layers	Mesh configuration Number of eight-node elements	NDOF1	NDOF2
2	4×4	195	48
3	8×8	675	48
4	8×8	675	48
5	10×10	1023	48
6	6×6	399	48
9	9×9	840	48
10	10×10	1023	48

NDOF1 = number of degrees of freedom within a cross-section for a complete three-dimensional spline finite prism analysis = number of nodes $\times$ 3
NDOF2 = number of unknowns within each cross-section using the present method with a global polynomial of $m = 4$ and $n = 3$ for all cases

to the data contained in Table 6.15 which compares the amount of reduction in the number of unknowns involved in the analyses.

6.5.4 Conclusions

It has been demonstrated that approximate three-dimensional solutions for a rectangular laminate with arbitrary conditions of support can be obtained by using the global–local approach and the spline function. A compromise between the computational cost and accuracy can be achieved. Regarding computational implementation, the essence of the program relies on a subroutine for carrying out the transformation. Because the assumed displacement functions are constructed on the idea of separation of variables, the integration of the spline function can be done separately. This facilitates the computer programming and accelerates the establishment of the final matrix equation. The saving in computational cost is important when the span-to-thickness ratio of a plate is small and/or when the number of layers involved is large. In these cases, the number of unknowns within the cross-section for a full three-dimensional analysis is extremely large and the global–local approach would be very effective. The method can be applied to other cases such as an angle ply laminate, a thick-walled hollow tube or a cylindrical thick shell subjected to a uniformly distributed load. For a structure with a point loading, this method is still applicable for regions a small distance away from the loaded area. It is also worth noting that other types of functions can be employed as the global functions such as the cubic B_3 spline and the trigonometric series.

6.6 References to Chapter 6

1. A. Y. T. Leung & Y. K. Cheung (1980) A new frame super-element in static and dynamic analysis, *Proc. 7th Australian Conf. in Mech. Struct. & Mat.*, Univ. Western Australia, p. 19–24.

2. A. Y. T. Leung & Y. K. Cheung (1981) Dynamic analysis of frames by a two-level finite element method, *J. Sound & Vibration*, **74**, 1-9.
3. A. Y. T. Leung (1983) Low cost analysis of building frames for lateral loads, *Computers & Structures*, **17**, 475-83.
4. A. Y. T. Leung (1985) Micro-computer analysis of three dimensional tall buildings, *Computers & Structures*, **21**, 639-61.
5. A. Y. T. Leung & S. C. Wong (1988) Local-global distribution factor method for tall building frames, *Computers & Structures*, **29**, 497-502.
6. A. Y. T. Leung & S. C. Wong (1988) Two level finite element method for thin plates subject to concentrated loads, *Microcomputer in Civil Engineering*, **3**, 127-36.
7. A. Y. T. Leung & S. C. Wong (1989) Two level finite element method for plane cracks, *Communication in Applied Numerical Methods*, **5**, 263-74.
8. Y. K. Cheung & J. Kong (1993) Approximate three-dimensional analysis of rectangular thick laminated plates: bending, vibration and buckling, *Computers & Structures*, **47**, No. 2, 193-9.
9. P. C. Shen & J. G. Wang (1987) Static analysis of cylindrical shells by using B spline function, *Computers & Structures*, **25**, 809-16.
10. K. J. Bathe (1982) *Finite Element Procedures in Engineering Analysis*, Prentice-Hall, New Jersey, USA.
11. A. Y. T. Leung & W. K. Wong (1988) Frame mode method for thin-walled structures, *Thin-Walled Structures*, **6**, 81-108.
12. Y. K. Cheung & F. T. K. Au (1992) Finite strip analysis of right box girder bridges using computed shape functions, *Thin-Walled Structures*, **13**, No. 4, 275-98.
13. Y. K. Cheung & J. Kong, Finite strip analysis of flat slab using computed shape function, *Computers & Structures* (to appear).
14. N. J. Pagano (1970) Exact solutions for rectangular bidirectional composite and sandwich plates, *J. Comp. Mat.*, **4**, 20-34.
15. N. J. Pagano & S. J. Hatfield (1972) Elastic behaviour of multilayered bidirectional composites, *AIAA J.*, **10**(7), 931-3.
16. R. M. Barker, F. T. Lin & J. R. Dana (1972) Three-dimensional finite element analysis of laminated composites, *Computers & Structures*, **2**, 1013-29.
17. A. K. Noor (1973) Free vibrations of multilayered composite plate, *AIAA*, **11**, 1038-9.
18. A. K. Noor (1975) Stability of multilayered composite plate, *Fibre Sci. Technol.*, **8**, 81-9.
19. H. Tada, P. C. Paris & G. R. Irwin (1973) *The stress analysis of cracks handbook*, Del Research Corporation, Hellertown, DA, USA.
20. W. J. Liou & C. T. Sun (1987) A three-dimensional hybrid stress isoparametric element for the analysis of laminated composite plates, *Computers & Structures*, **25**, 241-9.

7 Finite Element Mesh Generation

7.1 INTRODUCTION

In finite element analysis, the preparation of input data was at first carried out completely by hand and then later partly manually and partly automatically. The process is painstaking and is prone to human error. Improving the result by carrying out further analyses with refined finite element meshes can only be done to a limited extent because of the manual effort involved.

The advent of geometric modelling technology and the widespread availability of high-speed powerful computers have made possible the automation of analysis within the design cycle. The basic procedure is to develop a geometric model and then to fit in the requirements of the various component entities. This may include boundary conditions, material properties, loading and an indication of allowable error in the analysis. The analysis can be carried out with only minimum user intervention by an adaptive scheme controlling the finite element mesh as a function of the specified numerical error.

Fundamental to this approach is the ability to generate automatically sound computational grids, and this requires the implementation of a robust finite element mesh generator for arbitrary problem domains. A fully automatic mesh generator is defined as one which takes as input a geometric representation and associated mesh control parameters and then automatically produces a valid mesh for finite element analysis. This goal has been realized for two-dimensional analysis, while in three dimensions various approaches to the problem are currently being studied [52–64].

In order to evaluate the relative merits of different mesh generation schemes, it is useful to note the following desirable features of automatic meshing techniques:

1. *Precise modelling of boundaries* – No error beyond the discretization error inherent to the chosen finite element model should be introduced by the meshing process. Boundary nodes should lie precisely on the boundary of the structure. The boundary should be so defined that there is no ambiguity of the region to be discretized. There should be no limitation on the form of the boundary curve that can be accurately modelled.
2. *Good correlation between interior mesh and boundary information* – The curvatures and node spacings on the boundaries of the region should be well represented in the interior of the mesh. This allows the analyst to control the shape of elements in the interior of the region in a predictable fashion. It also permits the analyst to refine the spacing of the mesh where more accurate discretization is required. Unnecessary refinement of the mesh, leading to wasted computations, can also be avoided.

3. *Minimal input effort* – The time and effort required in setting up a finite element model should be minimized by reducing as much as possible the amount of input data. This will also reduce the chances of introducing human error into the analysis. It is best to arrange the input information in a form convenient to the analyst that can be readily communicated to the computer.
4. *Broad range of applicability* – To minimize user learning time, program development time and program size, it is desirable to use a small set of mesh generation techniques that can be applied to a broad range of structural topologies, rather than to use a large set of special purpose mesh generators.
5. *General topology* – The method of meshing should not impose any restriction on the topology of the mesh within a region.
6. *Automatic topology generation* – The mesh generation scheme should be able to create element connectivity without user intervention.
7. *Favourable element shape* – The elements produced by the method should possess shapes that will not produce ill-conditioning in the finite element model.
8. *Optimal numbering pattern* – The numbering of nodes and elements within the structure should be arranged so that favourable conditions are obtained for equation solving in the analysis. Minimum bandwidth, wavefront, and profile are common desirable features.
9. *Computational efficiency* – The method of mesh generation should be simple and inexpensive to minimize cost and to provide good response when applied in an interactive environment.
10. *Graphics support* – Data required for mesh generation should be input by means of a user-friendly interactive procedure in a direct and natural manner. The efforts on the part of the user will be reduced to a minimum and every step of the mesh generation process can be closely monitored.

In the past decade, irregular computational grids have become increasingly popular for a wide variety of numerical modelling applications. This trend demonstrates widespread appreciation of the two computational advantages that irregular grids can provide: they allow points to be situated on the curved boundaries of irregularly shaped domains and to be distributed throughout the interior as desired. As a result, numerical computation on arbitrary domains can be more effectively done using irregular grids. On the other hand, anyone who wants to use an irregular grid must take on the task of grid construction, which may demand as much attention and effort as the computations for which the grid is intended. Buell and Bush [22] have reviewed schemes developed before 1972 for using digital computers to expedite grid constructions. Since that review was written, the repertoire of available grid generators has grown, and a more 'up-to-date' list of references was provided by Thacker [33] in 1980.

Ever since then, more mesh generation strategies have been proposed, in which there are modifications and extensions of old ideas as well as new novel approaches. These techniques include: coordinate transformation – Ghassemi 1982 [43], Crawford *et al.* 1987 [70], Cook 1974 [27]; transfinite mapping – Haber *et al.* 1981 [71]; sweeping function – Wellford and Gorman 1988 [72]; generation in layers – Sadek 1980 [42],

Chorlay *et al.* [63]; division into subregions – Joe and Simpson 1986 [73], Ecer *et al.* 1987 [56]; modified quadtree and octree technique – Yerry and Shephard 1983 [74], 1985 [60]; Delaunay triangulation – Cavendish *et al.* 1985 [17], Lo 1989 [19]; advancing front approach – Lo 1985 [18], 1988 [75]; and others – Lo *et al.* 1982 [76], Imafuku *et al.* 1980 [55], Pissanetzky 1981 [45]. It is likely that other mesh generation schemes exist, which are not known to the authors; and it is quite definite that new mesh generation methods will be developed in the future.

Automatic mesh adaptation techniques are currently receiving much attention as they offer the ability to adjust the mesh according to the results of analysis and thereby improve the solution quality in an optimal manner [6–8,20,61,65,77–84]. The objective of an adaptive remesh program is to control the discretization error by increasing the number of degrees of freedom in regions where the previous finite element model is not adequate. The need for graded meshes can arise from either geometry or physical considerations. In problems where the boundary of the domain has a widely varying curvature, the domain geometry requires a graded mesh. In the case where mesh gradation is needed because of rapid changes of physical quantities such as stresses and temperature, two distinct approaches to mesh generation are possible. In problems where the analyst has some knowledge of the expected solution, a graded mesh with fine node spacing in appropriate parts of the domain can be constructed before the analysis. However, a powerful alternative is to start with a coarse mesh, and then refine the mesh based on the computed results. The latter approach coupled with some suitable error estimators [85–89] would be a fully automatic adaptive mesh generation scheme for obtaining a solution of the required prescribed accuracy.

The adaptive mesh generation techniques existing today can be divided into two main categories. The first category consists of those mesh enrichment strategies in which the number of degrees of freedom is increased in such a way that the original finite element space of a fewer number of degrees of freedom is embedded in the new finite element space of a greater number of degrees of freedom. Such a process of increasing the degrees of freedom is also called an *extension*. There are several possibilities in making extensions. The most common one is the h-version which improves the accuracy of the solution by progressive element subdivision, selectively reducing element sizes at places of greater solution error [78]. The quadtree and octree approaches which spatially decompose the computational domain into rectangular or hexahedral cells are examples of h-version mesh refinement schemes [80,81,83]. The p-version keeps the mesh fixed and accuracy of the solution is achieved by increasing hierarchically the order of the element interpolating functions [88]. The h-p-version is also under investigation, however fully automatic practical implementation of the optimal h-p-version refinement scheme is difficult and has not yet been achieved [8].

The second category consists of those adaptive remeshing schemes in which the entire problem domain is re-discretized based on the existing mesh and in accordance with some node spacing function derived from the error estimates of the solution and the current geometry of the problem domain [20,78,81]. Remeshing is always done in large deformation and metal-forming problems in which the deformed body suffers great changes in geometry and shape, and the elements of the mesh are so severely distorted that solution accuracy using the original mesh can no longer be guaranteed [7]. The remeshing can be repeated whenever necessary, or in parallel with the incremental method, provided that the new node spacing functions derived are based on the computation results and the evolved geometry of the deformed body.

Remesh methods can be considered as more general approaches than the mesh enrichment schemes in the sense that there is no restriction to the topology (element connectivity) of the refined mesh. As a result, they reflect better the required nodal spacing distribution, and are able to cope with the change in shape and geometry as the body deforms. Finally, the highly distorted elements can be removed and regenerated in the remeshing process, and mesh refinement and de-refinement are done in one single process.

Finite element mesh generation on planar domains is discussed in Section 7.2, where the advancing front approach will be explained in detail and illustrated with examples. In Section 7.3, mesh generation over curved surfaces is introduced. The mesh generation method presented is based on the coordinate transformation of elementary developable surfaces. An application of the advancing front technique to construct triangular elements over arbitrary curved surfaces is also described. Three-dimensional mesh generation methods are reviewed in Section 7.4, followed by a discussion on the shortcomings of the Delaunay triangulation as a means of mesh generation in the three-dimensional situation. Once again, the advancing front approach can be shown to be very effective for the triangulation of solid objects. Although the problem of finite element adaptive analysis and remeshing is closely related with the mesh generation problem, a thorough discussion of adaptive mesh generation is beyond the scope of this chapter. Interested readers are advised to consult the cited references at the end of this chapter for further information.

7.2 MESH GENERATION ON PLANAR DOMAINS

7.2.1 Introduction

In the early developments of the automatic mesh generation of finite element meshes for planar regions, attention was focused on the discretization of a planar domain into quadrilateral elements [1]. This is not only due to the fact that the Q8 isoparametric element is a popular element for plane problems, but also because of the use of the isoparametric coordinate transformations in most of the mesh generation algorithms as proposed by Zienkiewicz and Phillips [2]. While the method has been proved to be very useful in partitioning many types of structures, it often requires complex and inconvenient transitions involving triangles in order to change the number of points in one direction. Ghassemi [3] published a technique for mesh generation of quadrilateral zones. Based on a standard rectangular grid, the nodal coordinates are modified step-by-step by changing the angle of each side, arriving at a general quadrilateral shape with four sides of different lengths. Stefanou [4] published a FORTRAN program which can only be used for two-dimensional quadrilateral zones with edges parallel to the coordinate axis. Gordon and Hall [5] proposed the utilization of the *blending function* interpolation to calculate the nodal coordinates for quadrilateral plates and curved surfaces.

Compared to quadrilateral elements, triangular elements are more flexible in filling planar regions with very irregular boundaries and openings. Triangular elements also allow a progressive change of element size without serious distortion, a quality essential for the general use in the automatic mesh adaptation analysis [6–8]. In view of this situation, recent research efforts have concentrated on the generation of triangular meshes for planar domains.

In an early attempt, Fukuda and Suhara [9] suggested an automatic triangulation scheme composed of two distinct phases: (i) interior nodes are inserted randomly within subsquares, the union of which is just large enough to cover the entire domain and (ii) boundary nodes and interior nodes are interconnected to form a triangulation. Based on the work of Fukuda and Suhara, Cavendish [10] made a modification which significantly simplified and clarified many points in the original algorithm. By giving up the concept of generating interior nodes by random trials, Shaw and Pitchen [11] made an improvement over Cavendish's work. This is an important step in the approach towards a more direct generation method. However, the Fukuda–Suhara algorithm and its modified versions are still unnecessarily complicated and time-consuming. In particular, the algorithm requires a great deal of checking to see if a potential triangular element overlaps a previously generated element or the boundary of the domain. In addition, the resulting triangulation is not always the most desirable.

By choosing a better suitability criterion in the construction of triangles, the overlapping check could be completely avoided, and the boundary check limited to the most suitable of the potential elements. The 'max-min' angle criterion is one of the best of such criteria, and the resulting triangulation is the famous Delaunay triangulation [12]. In anticipation of its possible applications in the fields of geophysics, crystallography and statistics, algorithms of different characteristics have been proposed for the construction of Dirichlet tessellation and Delaunay triangulation in an n-dimensional Euclidean space ($n \geq 2$) [13–16]. Recently, further success has been reported in finite element generation based on Delaunay triangulation [17,32]. However, although Delaunay triangulation is a useful tool in generating sound triangular element meshes, its application has primarily been to convex hulls. As for the discretization of non-convex domains, they have to be first decomposed into simpler convex subregions; or in certain particular cases where no element cuts across the domain boundaries, the non-convex domain may be recovered by some special treatment peeling off those elements not belonging to the domain of discretization [17].

One of the most versatile techniques for generating high-quality triangular element meshes over arbitrary irregular planar domains is the *advancing front approach* originally proposed by Lo in 1985 [18,19]. In the mesh generation scheme described by Lo [18], interior nodes are generated prior to triangulation. However, in an adaptive mesh generation procedure based on the same approach presented by Peraire *et al.* [20], interior nodes are generated simultaneously with the construction of triangular elements. It was later found that the two most powerful triangulation techniques, namely the Delaunay triangulation algorithm and the advancing front approach can be integrated together in harmony [19]. The merits of the resulting scheme are its simplicity, efficiency and versatility in generating well-shaped triangular elements.

Since quadrilateral elements can be easily generated from a triangulation as described in Section 7.2.4, generation of quadrilateral elements based on coordinate transformations is only briefly discussed in Section 7.2.2. Triangulation of arbitrary planar domains which becomes increasingly important in finite element adaptive analysis, is discussed in detail in Section 7.2.3. The combined scheme of Delaunay triangulation and advancing front approach is presented in detail in Section 7.2.3(3). Generation of higher order elements will not be explicitly described, as these types of meshes can be easily generated by introducing mid-side nodes to simple triangles and quadrilaterals.

7.2.2 Coordinate transformation

Mesh generation on smooth regular planar domains can be carried out by methods involving coordinate transformation. The given domain is first broken up manually into a number of subregions, over each of which a one-to-one mapping from a standard reference domain can be defined. Three popular finite element mesh generation methods based on these techniques are the Laplacian scheme [21,22], the isoparametric mapping method [2] and the transfinite mapping strategies [5,23,25–27].

(1) Laplacian mesh generation

The Laplacian mesh generation schemes are a family of variations based on the method attributed to Wilson [21,22]. In this method the interior nodes are positioned so that the position vector $\mathbf{P}_i$ of each interior node i satisfies the following equation

$$\mathbf{P}_i = \frac{1}{n}\sum_{k=1}^{n} P_{ik} \tag{7.1}$$

where P_{ik} is a node adjacent to node P_i, and n is the number of such nodes, as shown in Fig. 7.1.

The name of this method is derived from the fact that eq. 7.1 can be interpreted as the Laplacian finite difference operator for the unknowns P_i. Meshes produced by this technique will tend to have fairly uniformly shaped elements. Since these equations represent a set of linear simultaneous equations for all the unknowns P_i, solution can be accomplished through an iterative technique such as Gauss–Seidel or Jacobi iteration [28]. This represents a relatively large computation for mesh generation. Another drawback of the Laplacian method is that the resulting grid does not adequately represent boundary information concerning grid spacing and boundary curvature. A modification of the scheme was developed by Herrmann [29] to alleviate this problem.

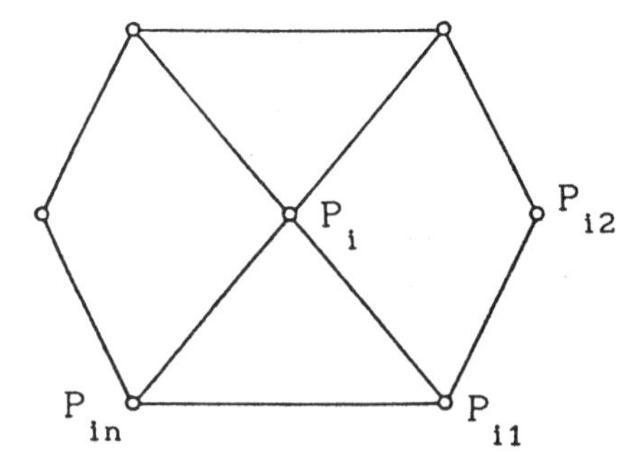

Fig. 7.1 Laplacian scheme

The Laplacian schemes do not possess an inherent method for the generation of interior nodes and element connectivity within a region. This, on the one hand, removes restrictions on mesh topology, but on the other hand, also requires that additional information be supplied by the analyst. In a paper by Denayer [30], an automated technique for generating element connectivity was proposed. The method involves a mapping between an imaginary region defined by an idealized grid composed entirely of regular polygons and the actual region to be meshed. The idealized grid is constructed from boundary curve information, including the number of elements connected to each boundary node.

(2) Isoparametric mapping method

A natural outgrowth of the use of isoparametric mapping to represent curved finite elements is the use of the same mapping to mesh entire regions of a structure. The use of the same isoparametric mappings for grid generation was described by Zienkiewicz and Phillips [2].

In this method, the mapping between the idealized grid and the true grid is first assumed to be of the form

$$\mathbf{P} = \sum_{i=1}^{m} \mathbf{P}_i \phi_i(u, v) \tag{7.2}$$

where $\mathbf{P}_i$ are the known position vectors of the m boundary nodes, ϕ_i are assumed shape functions over the region to be meshed, and (u, v) are the parametric coordinates of the reference domain. Equation 7.2 provides a one-to-one correspondence between the reference domain (usually a unit square or triangle) and the physical domain. The resulting region boundaries are modelled by simple Lagrange polynomials. The curvilinear coordinate system produced by this method provides a natural means of producing element topology automatically. Node points may be located at the intersections of constant coordinate curves in the u and v directions. Connecting the nodes along the curves produces a grid of quadrilateral elements. This grid may be diagonalized to form triangular elements. However, this method imposes a restriction on the topology of the mesh: there must be an equal number of elements on the opposite side of a region. Nevertheless, creation of the mesh connectivity does not require any effort on the part of the analyst. More general topology could be treated at the expense of increased input data by subdividing manually the problem domain into simpler regions as in Fig. 7.2.

(3) Transfinite mapping method

The transfinite mapping techniques are a class of methods for establishing curvilinear coordinate systems in domains bounded by smooth curves. The method was primarily developed by Gordon, Hall and their associates [5,23,25,26] for the approximation of complex surfaces and volumes. The transfinite method describes a surface or volume which will match the desired or true domain exactly on the domain boundary at an infinite number of points. It is this property which gives rise to the term transfinite mapping. This property also contrasts with the isoparametric mappings which match the true surface at only a finite number of points, i.e. the node points used for interpolation.

To describe these mappings it is necessary to introduce the concept of a projector, $\mathcal{P}$. A projector which operates on a region Ω bounded by four curves $f_1(u)$, $f_2(u)$,

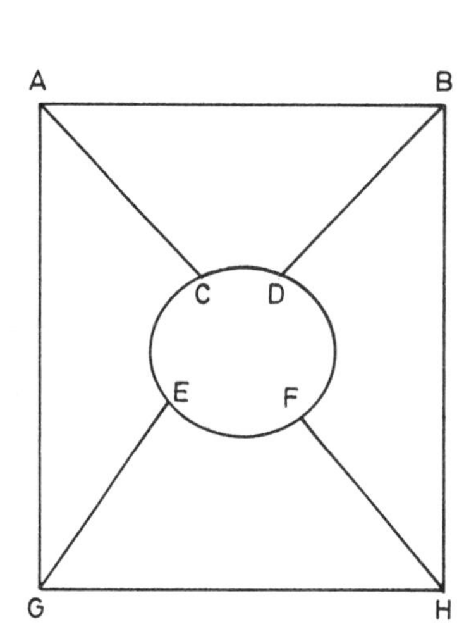

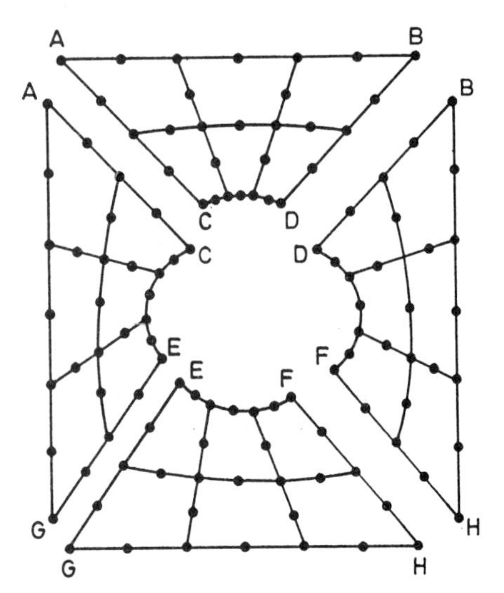

Fig. 7.2 Discretization by the method of isoparametric mapping

$g_1(v)$ and $g_2(v)$ is shown in Fig. 7.3(a). Two linear projectors can be formed, one interpolating in the u-direction and one in the v-direction

$$\mathcal{P}_1(\Omega) \equiv P_1(u, v) = (1 - v)f_1(u) + vf_2(u) \qquad u, v \in [0, 1] \tag{7.3}$$

$$\mathcal{P}_2(\Omega) \equiv P_2(u, v) = (1 - u)g_1(v) + ug_2(v) \qquad u, v \in [0, 1] \tag{7.4}$$

These projectors are shown in Figs 7.3(b) and 7.3(c). The product projector $\mathcal{P}_1\mathcal{P}_2(\Omega) = \mathcal{P}_2\mathcal{P}_1(\Omega)$ is shown in Fig. 7.3(d). This projector matches Ω exactly at the four corners with linear approximation along the four sides. Finally a Boolean sum projector, $\mathcal{P} = \mathcal{P}_1 \oplus \mathcal{P}_2$, can be defined so that Ω is matched exactly on its entire boundary as shown in Fig. 7.3(e).

$$(\mathcal{P}_1 \oplus \mathcal{P}_2)(\Omega) \equiv \mathcal{P}_1(\Omega) + \mathcal{P}_2(\Omega) - \mathcal{P}_1\mathcal{P}_2(\Omega)$$

$$\mathcal{P}(\Omega) = (1 - v)f_1(u) + vf_2(u) + (1 - u)g_1(v) + ug_2(v)$$
$$- (1 - u)(1 - v)P_{00} - (1 - u)vP_{01} - uvP_{11} - u(1 - v)P_{10} \tag{7.5}$$

where P_{00}, P_{01}, P_{11} and P_{10} are the four corner points of domain Ω.

This projector represents a curvilinear coordinate system created by a mapping of the unit square onto Ω. The projector $\mathcal{P}$ may be called the transfinite bilinear Lagrange interpolant of Ω. It is identical to the simplest form of Coons patch [31]. If the boundary curves are defined by Lagrange polynomials, the isoparametric mapping is obtained as a special case of the family of transfinite mappings. Higher order interpolants may be used to force the coordinate curves to pass through specified curves at the interior of Ω (Gordon and Hall [5,23,25,26]).

7.2.3 Automatic mesh generation for arbitrary planar domains

Mesh generation methods which make use of coordinate transformation may be too tedious to be applied to multi-connected regions having very irregular boundaries.

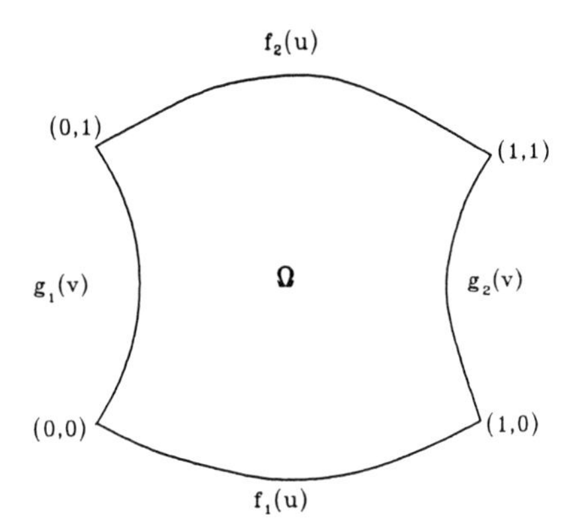

Fig. 7.3(a) Domain bounded by curves $f_1(u)$, $f_2(u)$, $g_1(v)$, and $g_2(v)$

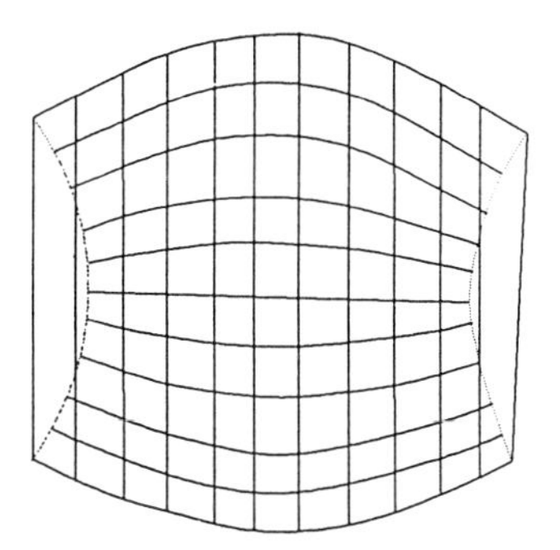

Fig. 7.3(b) Bilinear projector $\mathcal{P}_1$

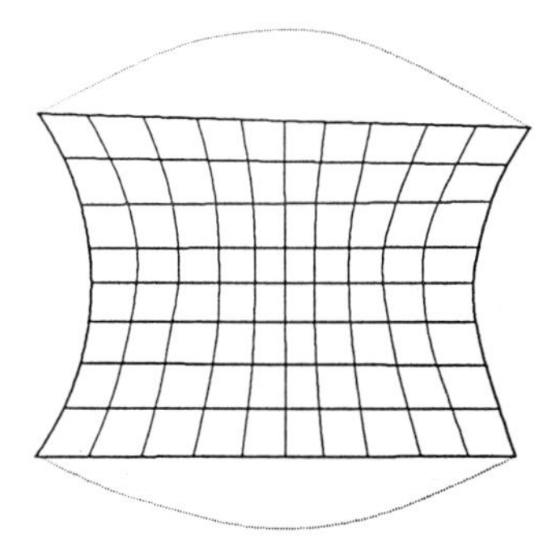

Fig. 7.3(c) Bilinear projector $\mathcal{P}_2$

For difficult complex domains, more general and versatile automatic mesh generation procedures have to be used, which should possess the following characteristics:

(i) No restriction on the shape of the domain boundaries; and the mesh generated is not sensitive to the orientation of the domain.
(ii) Openings within the domain can be handled more or less automatically; manual subdivision into convex polygons is not required.

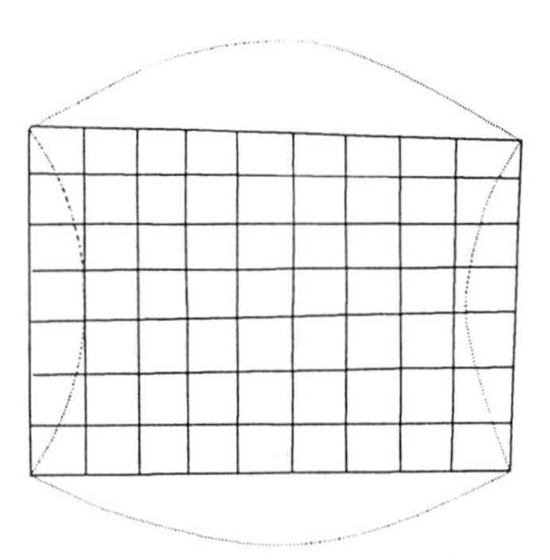

Fig. 7.3(d) Bilinear projector $\mathcal{P}_1\mathcal{P}_2$

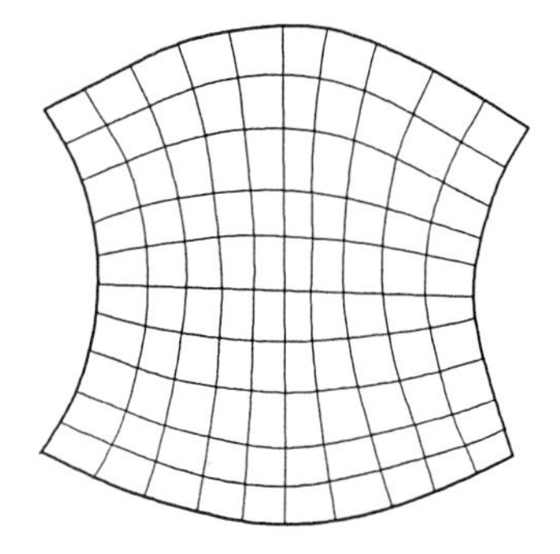

Fig. 7.3(e) Bilinear projector $\mathcal{P}_1 \oplus \mathcal{P}_2$

(iii) Ability to produce a great variety of mesh patterns, i.e. the number of elements around a node is not fixed and the relative position of nodes is not predetermined by mathematical formulae.

The two currently most popular automatic mesh generation schemes are the method of advancing front [18] and the Delaunay triangulation algorithm [17]. However, it should be recognized that these two mesh generation schemes are not incompatible with each other. By generalizing the notion of Delaunay triangulation to non-convex planar domains, an integrated scheme can be formulated which virtually retains all the merits of the two approaches.

(1) Delaunay triangulation

The Delaunay method is based on triangulating a set of points in n-space to produce n-dimensional simplexes for a convex hull. Thus the two major features of Delaunay-based methods are the generation of an initial set of points and the construction of a triangulation which satisfies the Delaunay properties. Although most of the current effort has gone into improving the performance of the triangulation algorithm [15,16,34–36], a reliable method for automatically generating points is not yet available. The fundamental problem here is that the points must be located in such a way as to satisfy the mesh characteristic requirements as well as to produce a valid computational mesh.

The difficulty of generating points on the boundary and in the interior of the geometric model can be dealt with in a number of ways. In two-dimensions, it is possible to create points on the boundary of the object in such a way as to satisfy the mesh density specification. In three-dimensions, the task is somewhat more difficult, as it is not clear how to develop the points on complex surfaces. The triangulation then proceeds based on the boundary points alone, and is then modified according to the quality of the resulting elements and the mesh control attributes. Another possibility is to develop points on the boundary as before, and then fill the interior of the geometrical model with a regular sprinkling of points. The generation of a valid computational grid using the Delaunay-based approach, however, goes beyond the particular methods for point generation and triangulation. A valid mesh must also be topologically compatible and geometrically similar to the physical model. Topological compatibility requires that the mesh topology and the geometric model topology are congruent, while geometric similarity requires that the mesh geometry is an acceptable approximation to the geometry of the model. The two requirements will, in the limit of mesh refinement, produce a mesh which matches the geometric model exactly.

Convex hull of node points

Let $\mathbb{P} = \{P_i, i = 1, N\}$ be a set of N distinct points in the Euclidean plane $\mathbb{R}^2$, and define the set of polygons $\mathbb{V} = \{V_i, i = 1, N\}$ where $V_i = \{X \in \mathbb{R}^2 : \|X - P_i\| < \|X - P_j\| \; \forall j \neq i\}$, and $\| \cdot \|$ denotes Euclidean distance norm. V_i represents a region of the plane $\mathbb{R}^2$ whose points are nearer to node P_i than to any other points. Thus, V_i is an open convex polygon (usually called a Voronoi polygon) whose boundaries are portions of the perpendicular bisectors of the lines joining node P_i to node P_j when V_i and V_j are contiguous. The collection of Voronoi polygons $\mathbb{V}$ is called the Dirichlet tessellation. In general, a vertex of a Voronoi polygon is shared by two other neighbouring polygons, so that connecting the three generating points associated with such adjacent polygons will result in a triangle, say T_k. The set of triangles $\{T_k\}$ is called the Delaunay triangulation. This construction can be shown to be a triangulation of a convex hull for the set of node points $\mathbb{P}$.

An important property of the two-dimensional Delaunay triangulation which makes it suitable for use as a finite element mesh of triangular elements is that its triangles are as close to equilateral as possible for the given set of nodes such that the minimum angle of the triangulation is maximized [12]. Consequently, ill-conditioned and thin triangles are avoided whenever possible. However, a serious drawback of the Delaunay triangulation as a means of finite element mesh generation is that without modification or post-processing, it is only applicable to convex domains [38].

In two dimensions, Watson's algorithm [16] makes use of the simple concept that three given node points form a Delaunay triangle if and only if the circumdisk defined by these nodes contains no other node points in its interior. Figure 7.5 illustrates this property of the Delaunay triangulation for the set of nodes shown in Fig. 7.4. In effect, Watson's approach is to reject, from the set of all possible triangles which might be formed, those with non-empty associated circumdisks. Those triangles not rejected form the Delaunay triangulation for the given system of node points.

The algorithm is initiated by calculating the (x, y) coordinates of three points which form a triangle T_0 that surrounds all the node points to be inserted. The

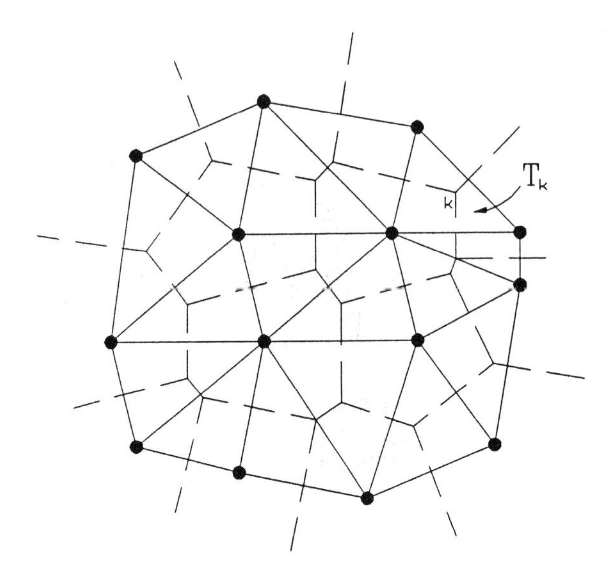

Fig. 7.4 Dirichlet tessellation (dashed lines) and Delaunay triangulation (solid lines) of 13 points

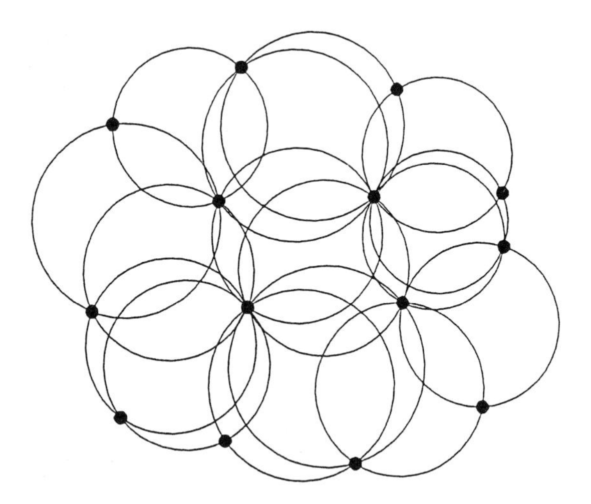

Fig. 7.5 Circumcircles for the points in Fig. 7.4

circumcentre coordinates and circumradius of the circumcircle defined by T_0 are also calculated and recorded. The node points are then introduced one at a time. The algorithm operates by maintaining a list of triplets of nodes which represent completed Delaunay triangles. Associated with each such triangle are the (x, y) coordinates of its circumcentre and its circumradius. For each new node point entered, a search is made on all current triangles to identify those whose circumdisks contain the new point (Fig. 7.6). For each such disk, the associated triangle is flagged to indicate removal. As shown in Fig. 7.7, the union of all such triangles forms what is called an insertion polygon containing the new node point. It can be shown

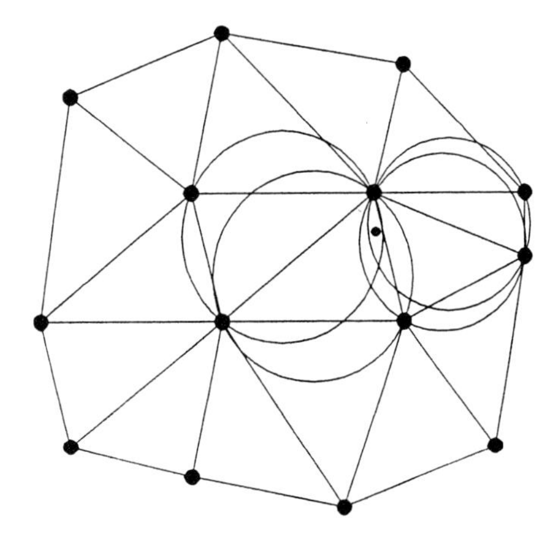

Fig. 7.6 Insertion of a new point

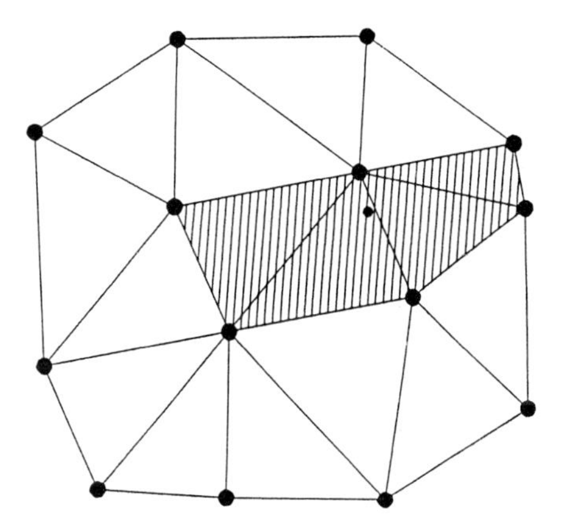

Fig. 7.7 Insertion polygon (shaded region)

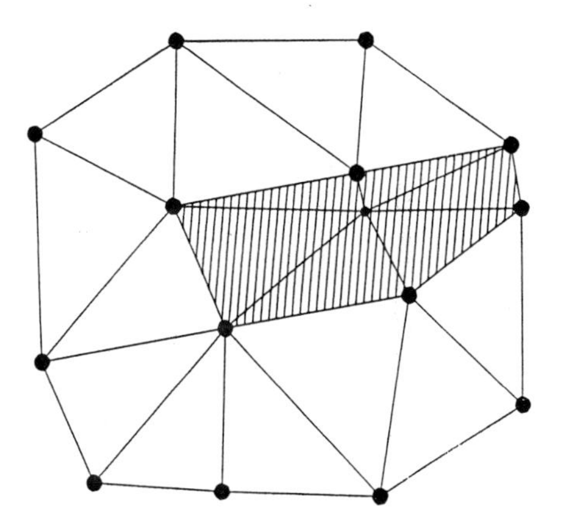

Fig. 7.8 Local triangulation (shaded region)

that no previously inserted node is contained in the interior of the polygon and that each boundary node of the polygon may be connected to the new node by a straight line lying entirely within the polygon. Thus a new triangulation of the region enclosed by the polygon is formed (Fig. 7.8). It can be shown that this local triangulation, when combined with the triangles outside the polygon, forms a new Delaunay triangulation which includes the newly added point. Repeated use of this insertion algorithm permits all node points to be inserted, while ensuring that at each step the triangulation retains its Delaunay properties.

*(2) **Advancing front technique***

The advancing front approach can be considered as one of the most general automatic meshing techniques for the discretization of an arbitrary planar domain into high-quality triangular elements. The procedure was originally described by Lo [18,19], who introduced all the interior nodes prior to the construction of triangles. Peraire *et al.* [20,39], who employed background grids to store information such as node spacing and stretching parameters, differed from Lo [18] by introducing internal nodes and constructing triangular elements simultaneously. A general and efficient remeshing algorithm based on this technique has also been developed for the discretization of planar domains into triangular elements in accordance with a given node spacing function [40]. The contour lines of the node spacing function at suitable calculated levels provide the natural lines of division of the problem domain into subregions where triangular elements of different sizes are generated.

(a) The generation front Γ

The generation front Γ consists of line segments on which triangles are to be created, and it advances each time when a triangle is constructed, as can be seen from Fig. 7.9.

At the start of the generation process, the generation front is exactly equal to the collection of all the boundary segments. While the domain boundary always

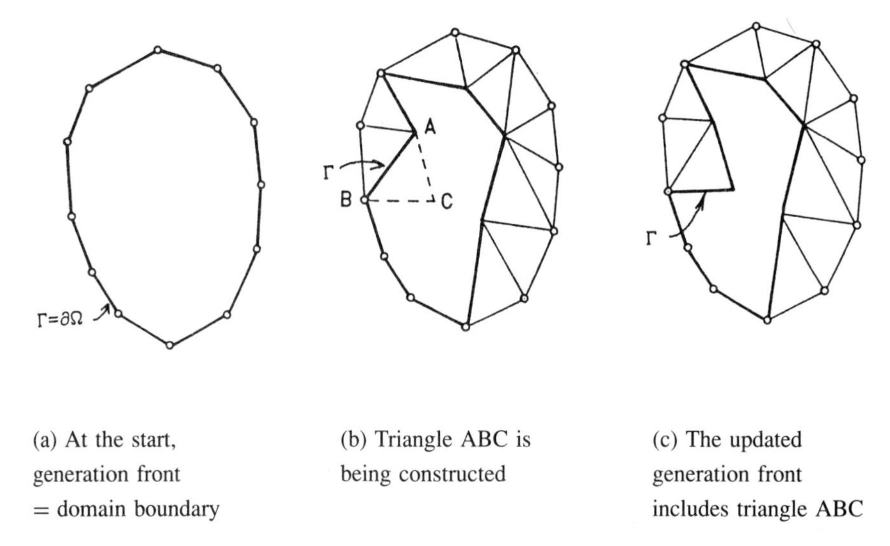

(a) At the start, generation front = domain boundary

(b) Triangle ABC is being constructed

(c) The updated generation front includes triangle ABC

Fig. 7.9 Evolution of the generation front during triangulation

remains the same, the generation front changes continuously throughout the generation process and has to be updated whenever a new element is formed. The updating process consists of removing the faces of the generated element that belong to the front and then adding the new faces with the correct orientation. The orientation of the line segments on the front is essential as it allows the determination of the direction in which the front should move when acceptable elements are generated. The generation process terminates when no more line segments are left in the front.

(b) Generation of interior nodes

Interior nodes within a given domain Ω can be generated by the method proposed in reference [18]. Without going into detail, the following is a brief description of the essential procedures:

(i) Sort out the $y_{\min}$ and $y_{\max}$ of the domain.
(ii) Imaginary horizontal lines at different levels are drawn between $y_{\min}$ and $y_{\max}$ across the domain.
(iii) The spacing between any two imaginary lines is exactly equal to the average element size of the domain.
(iv) Intersection points between each horizontal line and the domain are determined.
(v) Each horizontal line must cut the domain in an even number of points, and the intersection points are arranged in ascending order of magnitude.
(vi) Assuming that there are $2n$ cuts (intersections) between a particular horizontal line and the region concerned, the cuts are considered two by two, beginning with the first and second cuts. Nodes are generated on this horizontal line between the cuts according to the prescribed spacing. This only suggests a series of potential positions where nodes can be generated; however, whether a node is finally generated depends on how close it is from the domain boundary.

Let $\mathbb{S} = \{P_iQ_i, i = 1, N\}$ be the set of line segments representing the region boundary, and X be a potential node point. Let $P_mQ_m \in \mathbb{S}$ be the line segment closest to point X, i.e.

$$\|X - P_m\|^2 + \|X - Q_m\|^2 \leq \|X - P_i\|^2 + \|X - Q_i\|^2 \qquad \forall P_iQ_i \in \mathbb{S}$$

A measure of the quality of triangle ABC, α, can be defined by

$$\alpha(ABC) = 2\sqrt{3}\frac{\hat{k} \cdot \overrightarrow{AB} \times \overrightarrow{AC}}{\|AB\|^2 + \|BC\|^2 + \|CA\|^2} \tag{7.6}$$

in which $\hat{k}$ is a unit vector normal to the planar domain and $2\sqrt{3}$ is a normalizing factor so that equilateral triangles will have a maximum α value equal to 1. Figure 7.10 gives the α values of some typical triangles.

Point X will be accepted if $\alpha(P_mQ_mX) > \alpha_{\min}$, where $\alpha_{\min}$ can be fixed arbitrarily, for instance, $\alpha_{\min}$ can be conveniently set equal to 0.5.

(c) Forming triangular elements

The triangulation process is initiated by selecting the last segment $AB \in \Gamma$. Let Σ be the node points on the generation front Γ and Λ be the set of interior nodes

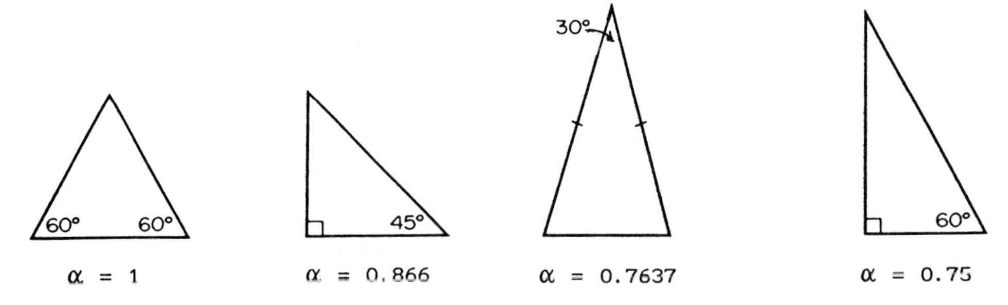

Fig. 7.10 Values of α for some typical triangles

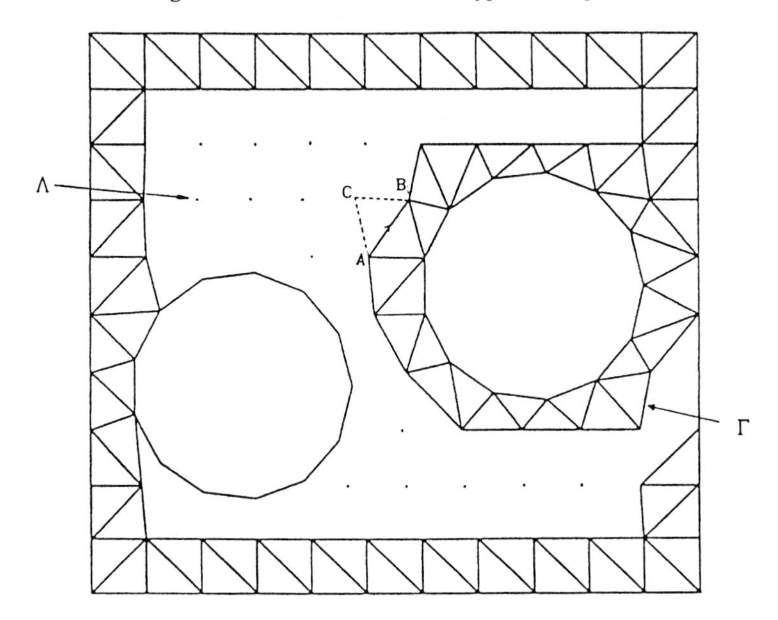

Fig. 7.11 Intermediate stage of triangulation. Node C is selected to form triangle ABC.

remaining inside the generation front (Fig. 7.11). The goal is to determine a node $C \in \Sigma \cup \Lambda$ such that C lies to the left of the directed line segment AB and that triangle ABC is in some sense optimal. The main difference between the advancing front approach and that of Cavendish [10] is in the search of candidate nodes for the construction of triangular elements. In the scheme of Cavendish, every node of the mesh is a candidate node and will be carefully considered. However, in the advancing front approach, candidate nodes are taken from two much restricted sets Σ and Λ, which change continuously during triangulation, and both will be reduced to zero at the end of the triangulation.

The advantages gained by considering nodes on or within the generation front (the set $\Sigma \cup \Lambda$) instead of all the interior and boundary nodes are (i) the set of candidate nodes is much reduced, and (ii) the checking process to see if triangle ABC overlaps any previously generated triangles or the domain boundary can be

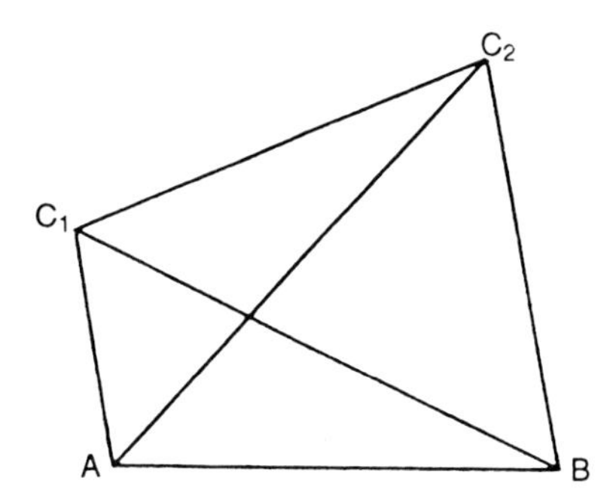

Fig. 7.12 Selection between nodes C_1 and C_2

much more efficiently carried out. It is observed in the majority of cases, owing to the even distribution of the interior nodes, that consideration of the minimum value of the norm $(AB^2 + CB^2)$ is sufficient to determine the point C, ensuring the best triangulation obtainable from the system of interior nodes and boundary nodes.

The choice of node C by the above criterion is simple enough, but it may not be sufficient to guarantee the best triangulations for regions having irregular boundaries. For such cases, it is advisable, as demonstrated in Fig. 7.12, to choose two nodes C_1 and C_2 which are closest to and on the left of the directed line segment AB. The choice between C_1 and C_2 can be made as follows.

Define for node C_1,

$$\alpha_1 = \alpha(ABC_1) \quad \beta_1 = \alpha(C_1BC_2) \quad \gamma_1 = \alpha(AC_1C_2) \quad \text{and} \quad \lambda_1 = \max(\beta_1, \gamma_1)$$

Similarly, define for node C_2,

$$\alpha_2 = \alpha(ABC_2) \quad \beta_2 = \alpha(C_2BC_1) \quad \gamma_2 = \alpha(AC_2C_1) \quad \text{and} \quad \lambda_2 = \max(\beta_2, \gamma_2)$$

Node C_1 will be selected if $\alpha_1\lambda_1 > \alpha_2\lambda_2$ and vice versa. The value α_1 is a better criterion for judging the quality of triangle ABC_1, compared with that suggested by Cavendish [10]. The bigger the α value, the better is the shape of the triangle; a negative value means that such a triangle is not acceptable. The sides of the triangle are not compared in turn, but are done once and for all by a single formula.

The value α_1 represents the quality of triangle ABC_1, while β_1 and γ_1, represent respectively those of triangles C_1BC_2 and AC_1C_2. Hence λ_1 measures the best possible shape of the triangle formed by C_2 and an edge of triangle ABC_1. The values of β_1 and γ_1 will both be negative if node C_2 is contained in triangle ABC_1, and this case will be ruled out automatically since the product $\alpha_1\lambda_1$ will then be negative. Similar considerations are taken into account assuming C_2 is selected; the overall results of the two possibilities are then compared for an ideal selection. Obviously, the method of selection can be easily generalized to consider more than two nodes for a better choice of node C; however, the improvement is not too significant.

*(3) **The combined scheme***

The possibility of integrating the two currently most popular mesh generation strategies, namely the method of advancing front and the Delaunay triangulation algorithm is discussed in Reference 19. The merits of the resulting scheme are its simplicity, efficiency and versatility. With the introduction of 'non-Delaunay' line segments, the concept of using Delaunay triangulation as a means of mesh generation is clarified.

An efficient algorithm is described for the construction of Delaunay triangulation over non-convex planar domains. Interior nodes are first generated within the given planar domain using the method described in Section 7.2.3(2b). These interior nodes and the boundary nodes are then linked up together by the advancing front technique to produce a valid triangulation. In the mesh generation process, the Delaunay property of each triangle is ensured by selecting a node having the smallest associated circumcircle. In contrast to convex domains, intersection between the proposed triangle and the domain boundary has to be checked; this can be done simply by considering only the 'non-Delaunay' segments on the generation front. Through the study of numerous examples of various characteristics, it is found that high-quality triangular element meshes are obtained by this integrated algorithm, and the mesh generation time bears a linear relationship with the number of elements/nodes of the triangulation.

(a) Delaunay triangulation of non-convex planar domains

Let $\partial\Omega$ be the boundary of planar domain Ω, and Λ be the set of interior nodes. A Delaunay triangulation for domain Ω and interior nodes Λ is a collection of triangles $\{T_k\}$ such that

(i) each T_k is formed by three node points belonging to $\partial\Omega \cup \Lambda$;
(ii) each T_k lies completely inside the domain Ω;
(iii) the circumcircle associated with each T_k contains in its interior no other node point which forms a valid triangle with any edge of T_k;
(iv) Ω is totally covered by $\{T_k\}$ and no two triangles of $\{T_k\}$ overlap.

(b) Delaunay and non-Delaunay triangles

(i) Delaunay triangles are those triangles whose associated circumcircles contain no node points on their circumference or in their interior.
(ii) Triangle T_k will be called 'semi-Delaunay' if there are node point(s) lying on the circumference of the circumcircle of T_k.
(iii) Triangle T_k will be called 'non-Delaunay' if there are node point(s) lying inside the circumcircle associated with T_k.

In the sequel, semi-Delaunay triangles will be considered as a special case of non-Delaunay triangles.

(c) Forming triangular elements

The steps taken in the construction of triangular elements are the same as those in the advancing front approach, and the only difference is in the criterion for the selection of the candidate node C for line segment AB on the generation front. Let Γ_1 be the set of non-Delaunay line segments and Γ_2 be the set of Delaunay line segments on the generation front Γ. Since Γ_1 and Γ_2 is a partition of Γ, we always have $\Gamma_1 \cup \Gamma_2 = \Gamma$ and $\Gamma_1 \cap \Gamma_2 = \emptyset$. At the beginning of triangulation, $\Gamma_1 = \Gamma = \partial\Omega$ and $\Gamma_2 = \emptyset$.

The triangulation is initiated by selecting the last segment $AB \in \Gamma_1$. Let Σ be the node points on the generation front Γ, and Λ be the set of interior nodes remaining inside the generation front. The goal is to determine a node $C \in \Sigma \cup \Lambda$, such that

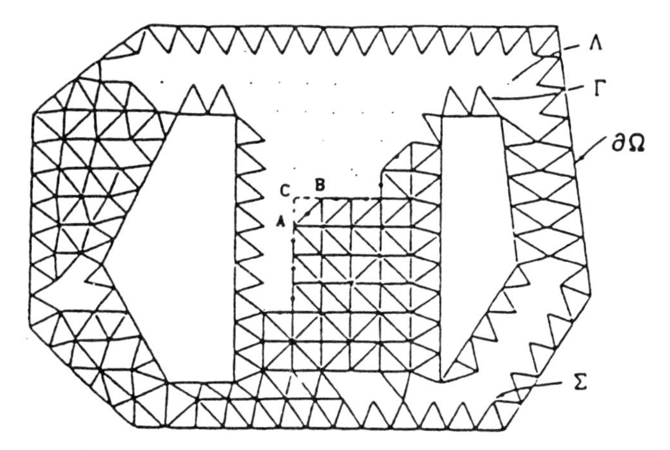

Fig. 7.13 Forming triangle *ABC*. - · - indicates non-Delaunay segment.

triangle ABC lies completely within the domain Ω and its associated circumcircle is the smallest (Fig. 7.13). A node $C_i \in \Sigma \cup \Lambda$ is said to be a candidate node if it satisfies

(i) $\hat{k} \cdot \overrightarrow{C_i A} \times \overrightarrow{C_i B} > 0$
(ii) $C_i A \cap \Gamma_1 \in \{\emptyset, A, \{C_i, A\}, C_i A\}$ and $C_i B \cap \Gamma_1 \in \{\emptyset, B, \{C_i, B\}, C_i B\}$

Let $\mathbb{S} = \{C_i\} \subset \Sigma \cup \Lambda$ be the set of candidate nodes. The node to be selected is a node $C_m \in \mathbb{S}$ such that $C_m \in \mathbf{C}_{ABC}\ \forall C_i \in \mathbb{S}$ where $\mathbf{C}_{ABC_i}$ is the circumcircle of triangle ABC_i, and $C_m \in \mathbf{C}_{ABC_i}$ means that C_m is inside or on the circumference of circumcircle $\mathbf{C}_{ABC}$. Condition (i) ensures that the proposed triangle lies on the left-hand side of the line segment *AB*. As no intersection check is needed for Delaunay segment Γ_2 (property of Delaunay triangulation [12–16]), condition (ii) ensures that the proposed triangle does not cut across the generation front Γ. When there are no more segments left in Γ_1, the construction process continues with segments of Γ_2 until the generation front is reduced to zero, $\Gamma = \emptyset$. Calculations for intersections are kept to a minimum by first exhausting all the segments in Γ_1. Most of the intersection check is done at the beginning of the triangulation process since boundary segments are assumed to be non-Delaunay. From Fig. 7.13, it can also be seen that as more and more Delaunay triangles are constructed, the number of non-Delaunay segments decreases rapidly. Although from time to time semi-Delaunay triangles are encountered, in general, their formation is not so numerous except for the special case of regular rectangular grids.

(d) Updating Γ_1 and Γ_2
The generation front $\Gamma = \Gamma_1 \cup \Gamma_2$ has to be updated whenever a triangular element is formed. Let $\mathbf{C}_{ABC}$ be the circumcircle of triangle ABC, Γ can be updated by:

(i) removing the line segment *AB* from the generation front; and
(ii) if triangle ABC is non-Delaunay, i.e. $P \in \mathbf{C}_{ABC}$ for some $P \in \Sigma \cup \Lambda$, setting $\Gamma_1^* = \Gamma_1 \cup \{AC, CB\}$, otherwise setting $\Gamma_2^* = \Gamma_2 \cup \{AC, CB\}$.

(e) Improving the quality of triangles by iteration
After triangulation, the smoothing process follows. It is done by shifting each interior node to the centre of the surrounding polygon. Let α_i be the α-quality (eq. 7.6) of

the triangles connected to an interior node, and α_i^* be the new α_i values of the same set of triangles taking into account the proposed nodal displacement. To ensure that the geometric mean of the α-values of the triangles is increased for each iteration, the shift is carried out only if the product of α_i^* is greater than the corresponding product of α_i, i.e.

$$\Pi\alpha_i^* > \Pi\alpha_i$$

(f) Examples

Examples of triangulation of planar domains of different boundary characteristics are given in this section. All the examples presented here are planar domains which have been triangulated as a single region without subdivision into simpler regions. These examples were run on computer VAX/VMS 8600 and the resulting meshes were plotted out on a Tektronix T4016 graphics terminal.

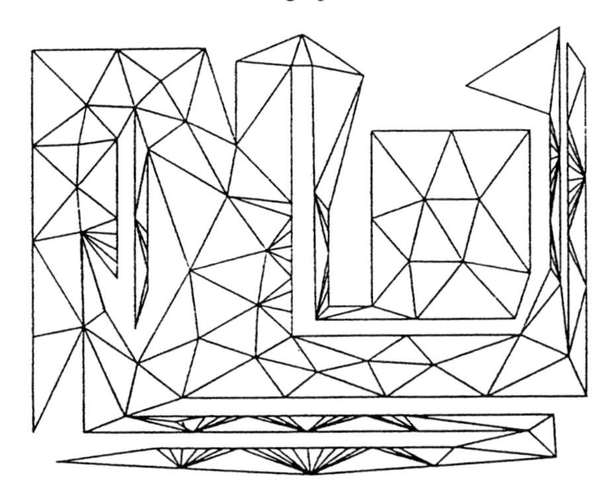

Fig. 7.14 Number of elements (NE) = 176. Number of nodes (NN) = 158. Geometrical mean of element α-qualities, $\overline{\alpha} = 0.4651$

Fig. 7.15 NE = 392, NN = 344, $\overline{\alpha} = 0.7312$

Figure 7.14 shows the triangulation of a domain having very difficult boundary conditions with rapid variation of segment lengths; and there are long strips having nodes closely spaced on one side and sparsely spaced on the other side. Figure 7.15 shows the finite element mesh of a domain having very irregular boundaries. Without serious variation in nodal spacing, the quality of the triangular elements is significantly better than that of the mesh shown in Fig. 7.14. Figure 7.16 shows that excellent-quality triangular elements can be generated for non-convex regions composed of line segments of uniform lengths.

The algorithm works equally well for the triangulation of a domain having many internal openings as shown in Fig. 7.17. Figures 7.18 and 7.19 are two triangular meshes in which the quality of the elements is optimized by two cycles of smoothing. The computer time for the generation of interior nodes, the triangulation process, the smoothing process and the graphic plot-out of the examples is summarized in Table 7.1, and the graph of the total CPU time of mesh generation versus the

Fig. 7.16 NE = 1954, NN = 1284, $\overline{\alpha} = 0.9489$

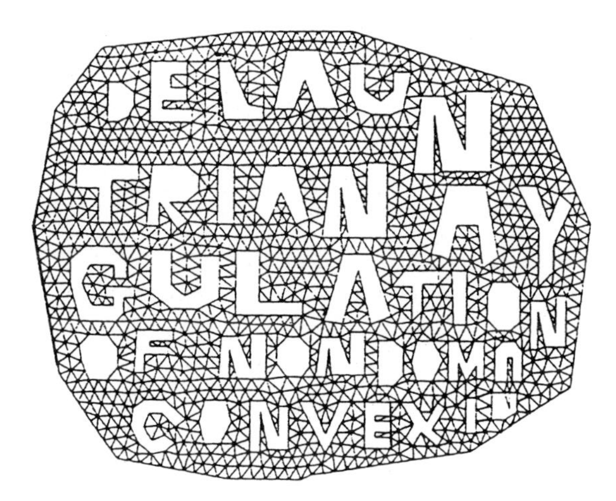

Fig. 7.17 NE = 2197, NN = 1512, $\overline{\alpha} = 0.9227$

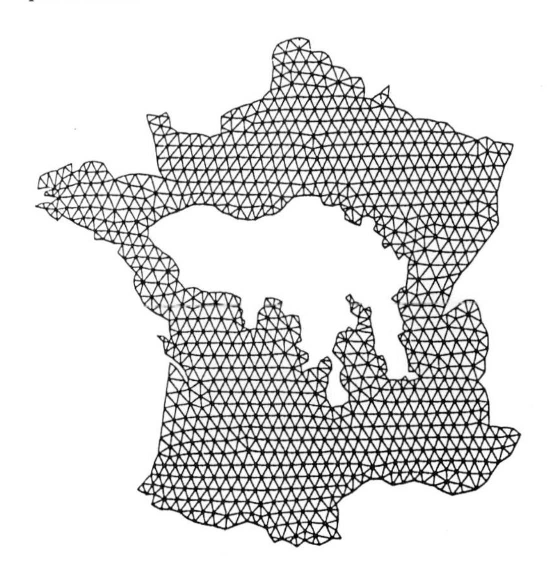

Fig. 7.18 NE = 1811, NN = 1092, $\overline{\alpha} = 0.9653$

Fig. 7.19 NE = 4293, NN = 2354, $\overline{\alpha} = 0.9612$

Table 7.1 Summary of CPU time of triangulation

		CPU time(s) (VAX/VMS 8600)			
Fig.	Generation of interior nodes	Triangulation	Smoothing	Plotting	Total
7.14	0.03	0.21	0.03	0.18	0.45
7.15	0.07	0.26	0.28	0.37	0.98
7.16	0.30	2.55	1.23	1.81	5.88
7.17	0.37	3.80	1.50	2.05	7.72
7.18	0.36	3.15	1.49	2.13	7.13
7.19	0.68	5.89	2.78	3.99	13.33

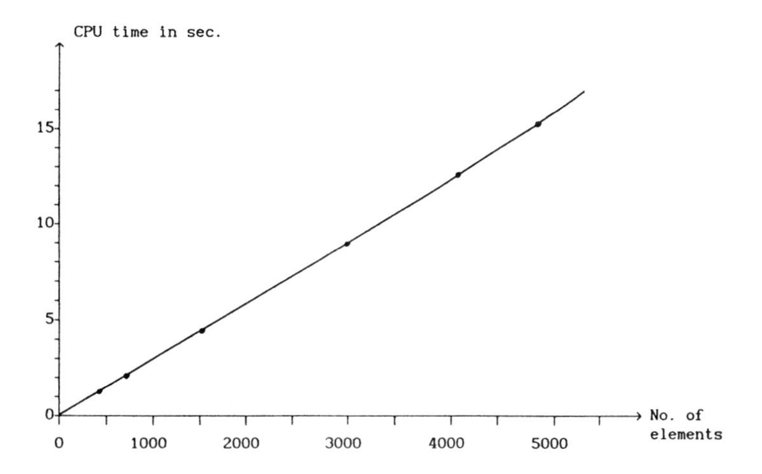

Fig. 7.20 Graph of CPU time versus number of elements

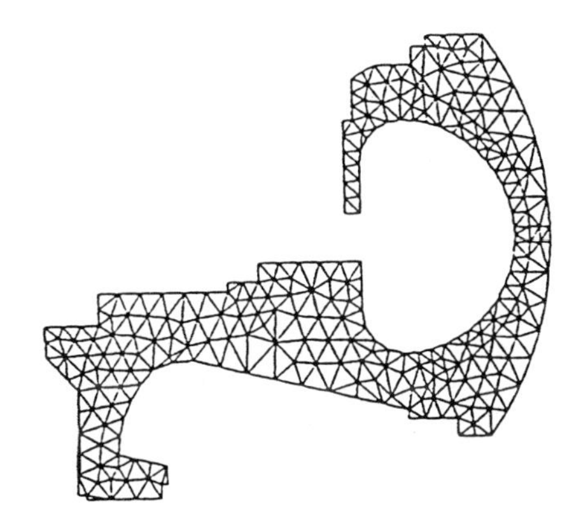

Fig. 7.21 Axisymmetric analysis of a machine component

number of elements is given in Fig. 7.20. Finally, two practical examples are given: Fig. 7.21 shows the finite element mesh for axisymmetric analysis of a machine part, and Fig. 7.22 is the finite element discretization of an underground structure and the neighbouring soil.

7.2.4 Generation of quadrilateral elements

A number of algorithms have been proposed for the generation of triangular element meshes over planar domains [19,41–43]. Nevertheless, the generation of a finite element mesh of quadrilateral elements is still confined to regular *nice* domains which allow the use of such techniques as coordinate transformations, isoparametric mapping, spline functions, drag method, etc. [2–5,44,45]. For the generation of a quadrilateral element mesh over an irregular region, there is still no systematic general approach. In fact, there exist domains which cannot be filled up solely by quadrilaterals, e.g. domains having boundaries composed of an odd number of edges.

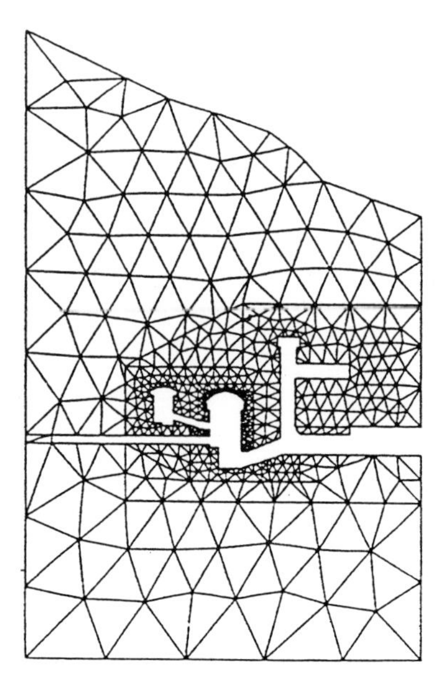

Fig. 7.22 The finite element mesh of an underground structure

As triangular and quadrilateral elements both possess certain advantages over each other, it is very desirable to have a finite element mesh composed of both element types. While there are considerable difficulties in generating simultaneous triangular and quadrilateral elements, it can be stated however, without loss of generality, that the construction of quadrilateral elements from a triangulation will be much easier. In this section, an algorithm based on a clear geometrical concept is described for the construction of quadrilateral elements over a triangular element mesh [24]. The resulting mesh is a homogeneous mixture of triangles and quadrilaterals, whose proportion is optimized by a parameter γ.

(1) The algorithm

The algorithm presented in the following sections works for triangulations on a plane and over curved surfaces. Although the construction procedure is not so obvious, the idea is rather simple; a quadrilateral element will be generated whenever a diagonal between two triangles is removed. The quality of the quadrilaterals produced and hence the quality of the resulting finite element mesh will depend largely upon the way in which the diagonals are removed. In order to determine the exact sequence in which unique-lines (diagonals) are to be removed, there should be some global measure of each unique-line of the mesh on the quality of the quadrilateral that it will generate and how the neighbouring unique-lines would be affected upon its removal. A distortion coefficient β will therefore be attributed to each unique-line which determines the quality of the quadrilateral that will be generated as a result of its removal. A new β^*-value for each unique-line can then be computed by subtracting from the current β-value the β-values of its neighbouring unique-lines. Thus β^* is a combined measure on the value of the diagonal and the influence of its removal on the neighbouring unique-lines. The diagonal which bears a maximum β^*-value will be the first to be removed. After the removal of a diagonal (the generation of a quadrilateral), the mesh as well as the β^*-values of the neighbouring unique-lines has to be updated. Removal of diagonals continues so long as there is a diagonal whose β-value is greater than a certain prescribed γ-value. For a given triangulation, more quadrilateral elements will be generated for smaller γ-values and vice versa; and in the extreme case that $\gamma = 1$, no diagonal will be removed. The generation process stops when no more unique-lines can be removed or the β-values of all unique-lines fall below γ. The resulting mesh will be a hybrid mesh made up of triangular and quadrilateral elements.

(2) The distortion coefficient β of a quadrilateral

From Fig. 7.23, it can be seen that four triangles can be obtained by cutting the quadrilateral $ABCD$ along the diagonals AC and BD. Let $\alpha_1, \alpha_2, \alpha_3$ and α_4 be the α-values, calculated by eq. 7.6, of the triangles arranged in descending order of magnitude, i.e.

$$\{\alpha_1, \alpha_2, \alpha_3, \alpha_4\} = \{\alpha(ABC), \alpha(ACD), \alpha(ABD), \alpha(BCD)\} \qquad \alpha_1 \leq \alpha_2 \leq \alpha_3 \leq \alpha_4$$

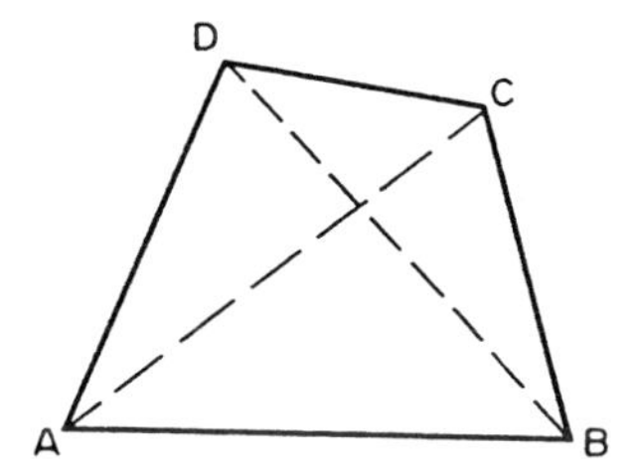

Fig. 7.23 Quadrilateral $ABCD$

A distortion coefficient β for the quadrilateral $ABCD$ can now be defined by

$$\beta = \frac{\alpha_3 \cdot \alpha_4}{\alpha_1 \cdot \alpha_2} \tag{7.7}$$

Obviously, β lies between 0 and 1 for a convex (valid) quadrilateral. For rectangles, β attains a maximum value of 1, whereas for quadrilaterals degenerated to triangles, β approaches 0. The higher the β-value the better is the shape of the quadrilateral compared to the two best triangles of subdivision along one of the diagonals. Hence quadrilaterals can be preferably generated by removing diagonals having large β-values. The β-values of some typical quadrilaterals are shown in Fig. 7.24.

(3) Retrieval of unique-lines from a triangulation

All the edges of the triangles are examined in turn. Suppose k_1 and k_2 are the node numbers of an edge with $k_1 < k_2$. Before registering k_1k_2 as a new line, a check is made on the nodes already connected to k_1. Let $\mathcal{C}_{k_1}$ be the set of nodes already connected to k_1. Then if $k_2 \in \mathcal{C}_{k_1}$, line k_1k_2 is ignored; otherwise, line k_1k_2 is recorded and $\mathcal{C}_{k_1}$ is updated to

$$\mathcal{C}_{k_1} \leftarrow \mathcal{C}_{k_1} \cup \{k_2\}$$

For a given triangular mesh, the average number of nodes connected to a particular node is constant and quite independent of the total number of nodes present in the

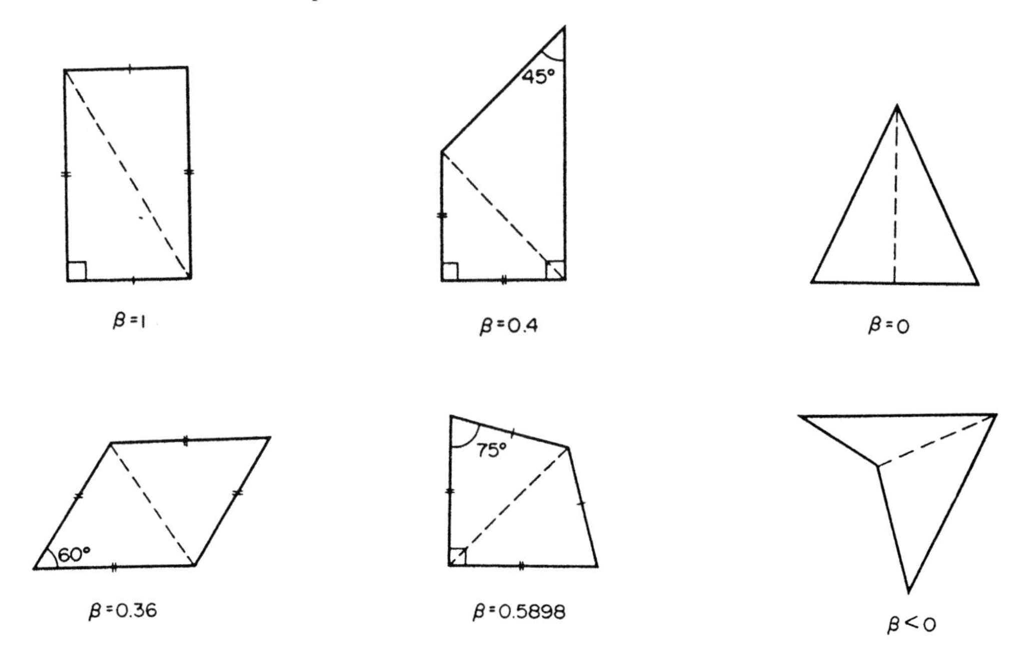

Fig. 7.24 The β-values of some typical quadrilaterals

triangulation. The efficiency of the method will be of order N, where N is the total number of nodes in the mesh. Moreover, the implementation is very simple and no sorting process is involved [37]. With a little additional calculation and care during the unique-line retrieval process, the neighbouring element(s) of each unique-line can be determined, and the three unique-lines of each triangle can be identified. This information is stored and will be useful later in updating the mesh upon the removal of a unique-line.

(4) Removal of diagonals

Before the removal of unique-lines, a check is made on the removability and the β-value of each unique-line. In general, only those lines shared by two triangles can be removed (these removable lines will be called diagonals). Hence, at the beginning of the generation process, all the boundary line segments have to be flagged so as not to be removed. To determine which diagonal is to be removed, the β^*-value of each diagonal is computed, which is equal to the β-value of the diagonal minus the sum of the β-values of the neighbouring diagonals. The diagonal having the greatest β^*-value will be removed, and a quadrilateral element will be generated. If no more unique-lines can be removed or the β-values of the diagonals are less than the prescribed γ-value, the generation process terminates.

(5) Updating the mesh

Whenever a quadrilateral is generated, the mesh has to be updated, and this can be done as follows:

(a) Suppose the line AC in Fig. 7.25 is to be removed. The two triangles ABC and ACD are identified and deleted from the mesh, whereas a quadrilateral element $ABCD$ is added to the mesh.

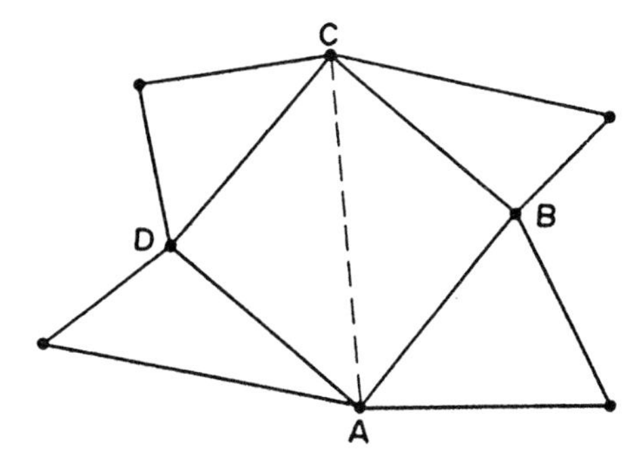

Fig. 7.25 Remove diagonal AC to form quadrilateral $ABCD$

(b) The unique-line associated with the quadrilateral, i.e. the four edges of the quadrilateral AB, BC, CD and DA should be flagged so that they will not be removed as the generation process continues.

(c) The β^*-values of the neighbouring lines are updated; at most eight lines will be affected as shown in Fig. 7.25.

(6) Examples

Figure 7.26 is a triangular element mesh of varying element size. Figures 7.27–7.29 are the resulting hybrid meshes for different γ-values; it can be seen that more quadrilateral elements are generated by choosing a smaller γ-value. Figures 7.30 and 7.31 show that for a regular domain all triangular elements would be converted to quadrilateral elements for any γ-values not equal to 1. Figures 7.32 and 7.33 show that quadrilateral elements are easily recovered at the interior part of the domain. Figures 7.34 and 7.35 show that high-quality finite element meshes can be obtained by the combined use of triangular and quadrilateral elements. The effect of selecting $\gamma = 0.7$ and 0.3 for triangular element meshes over curved surfaces can be very remarkable, as shown in Figs 7.36 and 7.37 (dumb-bell), Figs 7.38 and 7.39 (surfaces of revolution), and Figs 7.40 and 7.41 (conical, cylindrical and spherical surfaces).

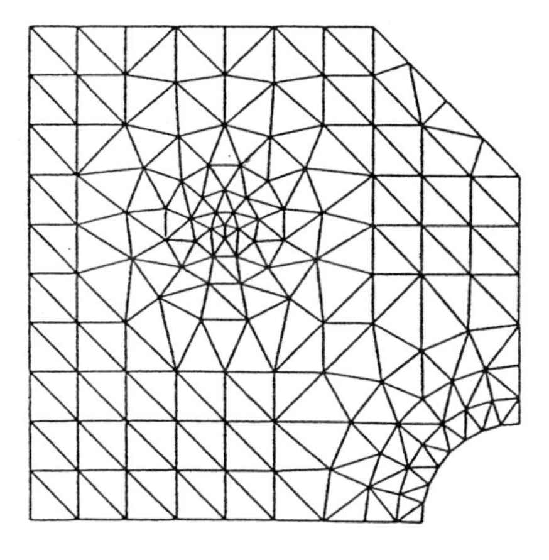

Fig. 7.26 A triangulation of different element sizes

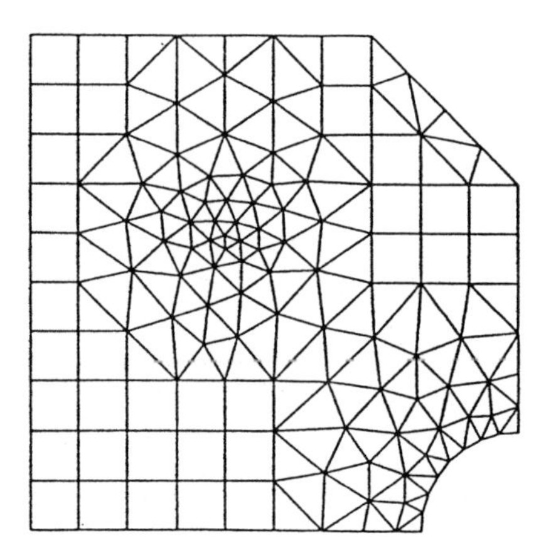

Fig. 7.27 Resulting hybrid mesh for $\gamma = 0.9$

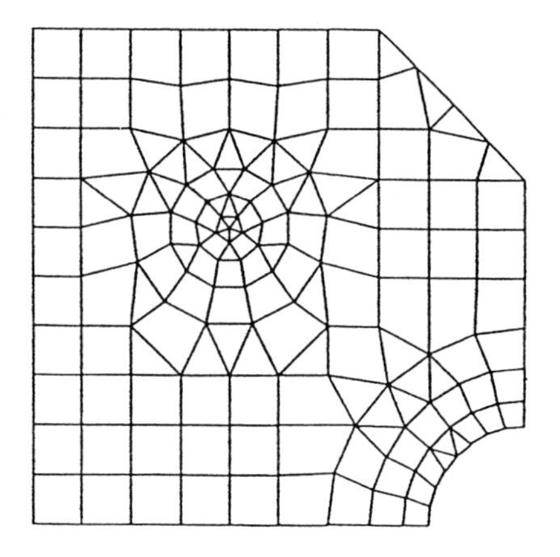

Fig. 7.28 Resulting hybrid mesh for $\gamma = 0.5$

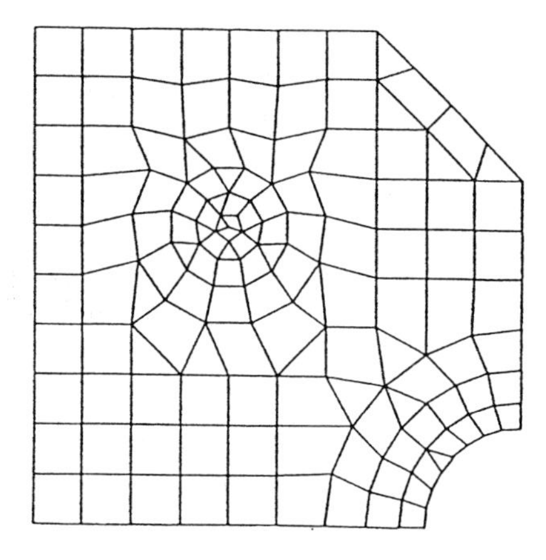

Fig. 7.29 Resulting hybrid mesh for $\gamma = 0.1$

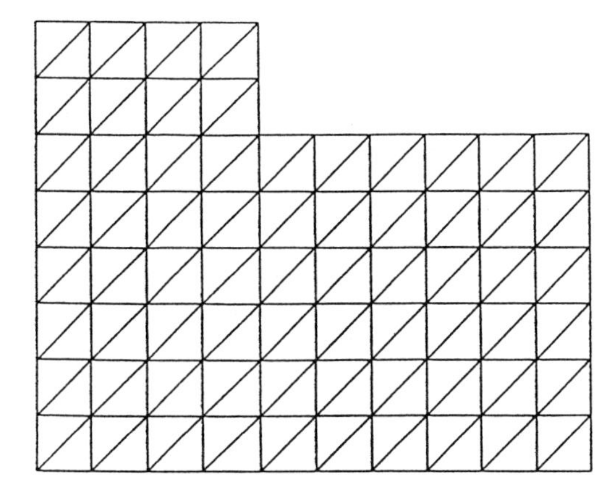

Fig. 7.30 Triangulation of a regular domain

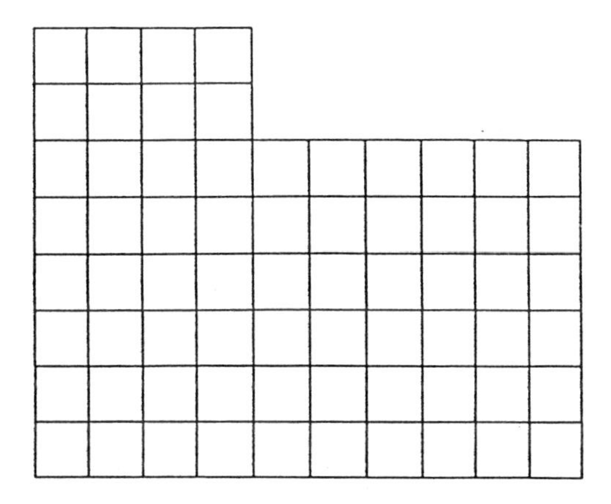

Fig. 7.31 Triangular elements are converted to quadrilateral elements, $\gamma = 0.9$

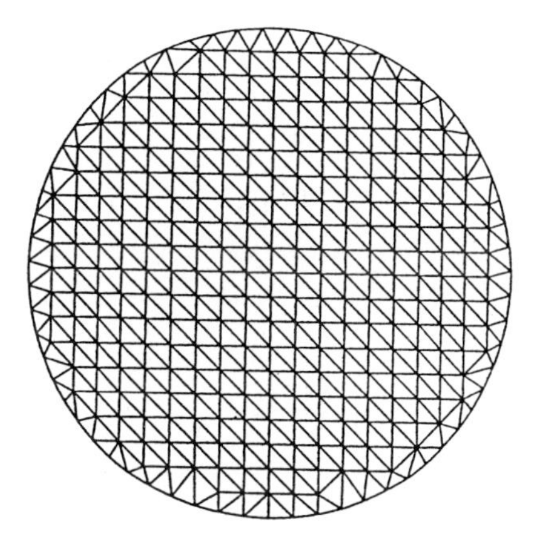

Fig. 7.32 Triangulation of a circular disc

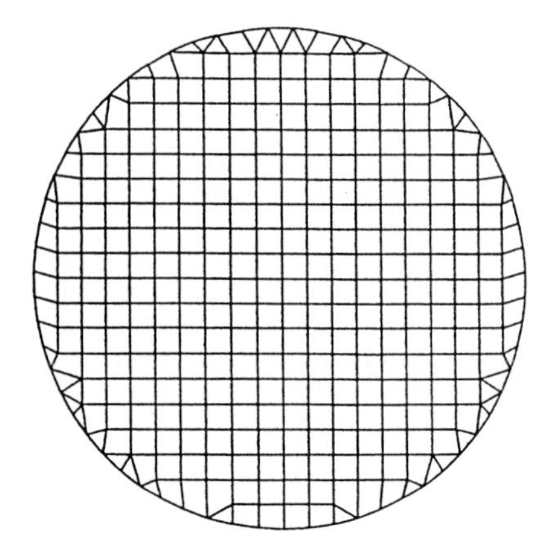

Fig. 7.33 Triangular elements remain only along the boundary of the domain, $\gamma = 0.5$

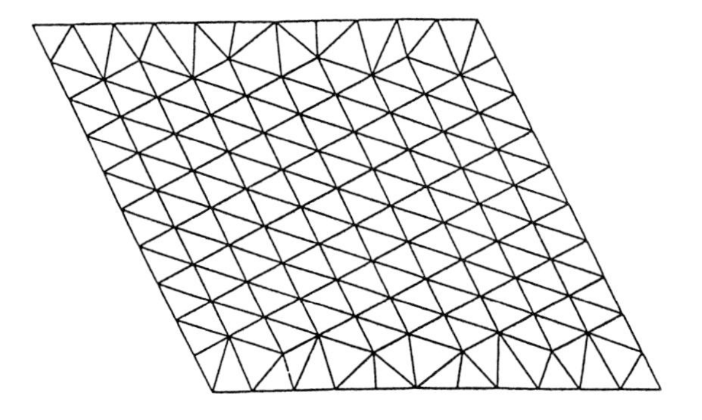

Fig. 7.34 Triangulation of a parallelogram

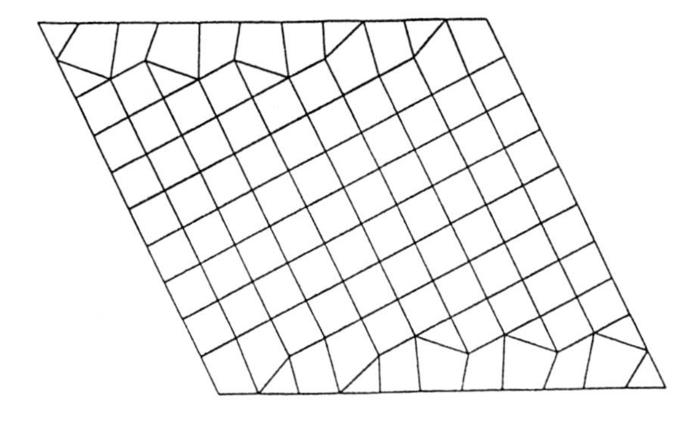

Fig. 7.35 A hybrid mesh of triangular and quadrilateral elements, $\gamma = 0.5$

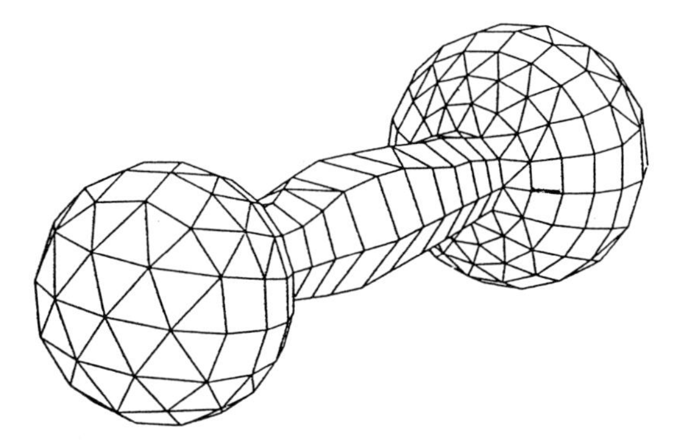

Fig. 7.36 Dumb-bell, $\gamma = 0.7$

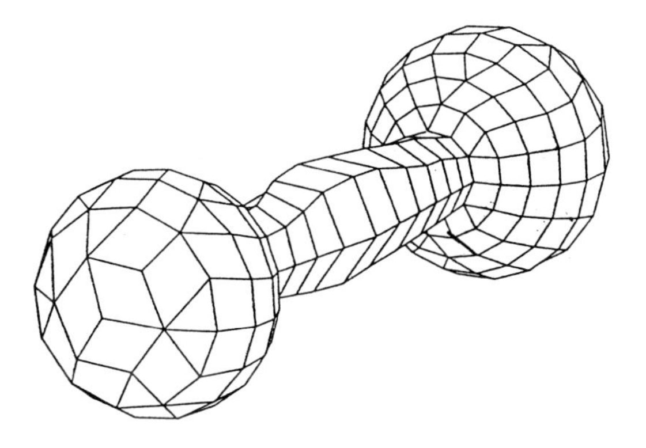

Fig. 7.37 Dumb-bell, $\gamma = 0.3$

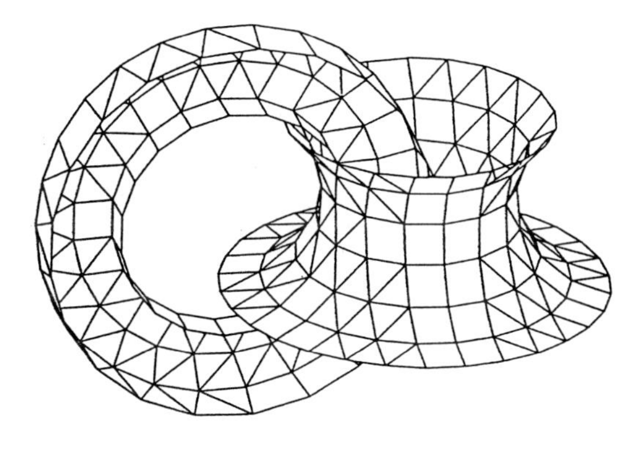

Fig. 7.38 Two surfaces of revolution, $\gamma = 0.7$

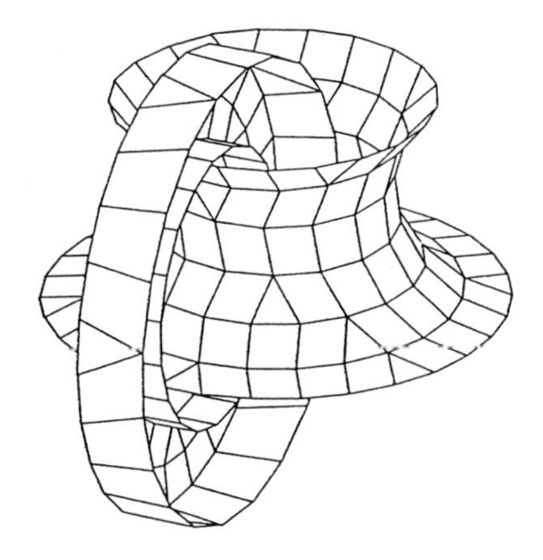

Fig. 7.39 Two surfaces of revolution, $\gamma = 0.3$

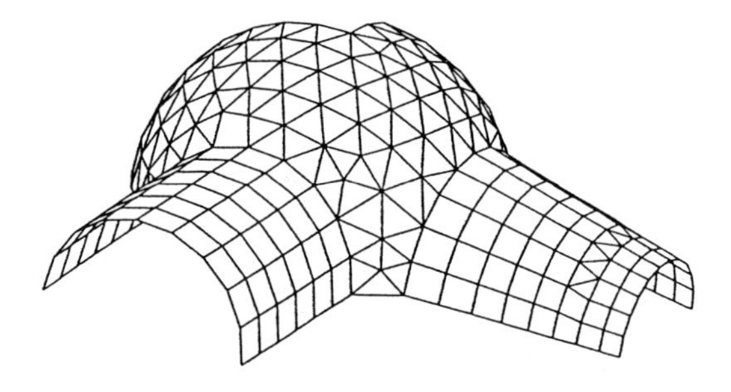

Fig. 7.40 Conical, cylindrical and spherical surfaces, $\gamma = 0.7$

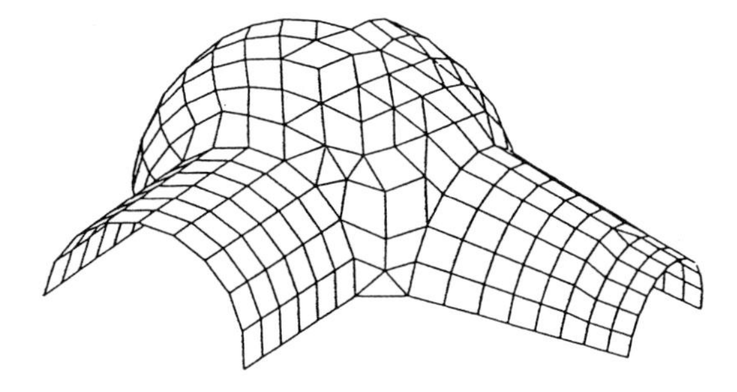

Fig. 7.41 Conical, cylindrical and spherical surfaces, $\gamma = 0.3$

7.3 Mesh generation over curved surfaces

7.3.1 Introduction

Many engineering structures are made up of a spatial assembly of thin flat plates or are shell structures which can be approximated by a large number of flat elements.

Each element is the combination of a bending element and a plane stress element. For the bending elements, an extensive review can be found in References 46 and 47, and it was concluded that DKT (Discrete Kirchhoff Triangle) and HMS (Hybrid Stress Model) elements are the most efficient, cost-effective and reliable elements of their class. These elements are later combined with plane stress elements and extended to shell structure analysis, with considerable success being reported in small- and large-deformation analysis [48,49]. However, if shear deformation is to be taken into account, then the 3-node linear triangular element T_1 discussed in Section 4.3.2 can be used.

Although there is an increasing need of triangulation of shell structures for finite element analyses, algorithms for generating triangular meshes over general curved surfaces are hard to come by [2,5,43,50]. Moreover, all these algorithms use the technique of coordinate transformation in the mesh generation process involving spline function, isoparametric coordinates and transfinite mapping. However, architectural and civil engineering structures are usually made up of simple geometrical surfaces; the cathedral dome, observatory and planetarium are only hemispheres, buildings are rectangular blocks composed of planar facets, a chimney is a cylinder and a cooling tower is a surface of revolution, etc.

For developable surfaces, such as a spatial plate, a cylindrical surface or a conical surface, mesh generation can be done on a plane and put into space by simple transformations. Algorithms of various characteristics have been proposed for mesh generation over planar domains [10,18,41,43,51]; these techniques can be used to triangulate developable surfaces with irregular boundaries and openings. As for spherical surfaces and surfaces of revolution, mesh generation has to be done in space. In the following sections, a method is presented to generate interior nodes over an arbitrary multi-connected spherical surface. Finally, a surface of arbitrary form can also be triangulated. For such a surface, a system of interior nodes is required for defining the geometry of the surface. The node linking scheme proposed in Reference 18 can be extended to interconnect boundary nodes and interior nodes over an arbitrary curved surface to form a triangulation.

7.3.2 Transformation of developable surfaces

(1) Spatial plane

Given a polygon $ABCDE$ in $x-y-z$ space, as shown in Fig. 7.42, let the $u-v$ plane be the plane containing the polygon $ABCDE$, and in particular, the u-axis is made parallel to vector AB. Let $\mathbf{a}$ and $\mathbf{b}$ be unit vectors along the u-axis and the v-axis, respectively. Then

$$\mathbf{a} = \frac{\mathbf{AB}}{\|\mathbf{AB}\|} \qquad \mathbf{b} = \frac{\mathbf{AF}}{\|\mathbf{AF}\|} \tag{7.8}$$

where $\mathbf{AF} = \mathbf{AE} - (\mathbf{AE} \cdot \mathbf{a})\,\mathbf{a}$.

For any point P on the polygon $ABCDE$, its (u, v) coordinates are given by

$$u = \mathbf{AP} \cdot \mathbf{a} \qquad v = \mathbf{AP} \cdot \mathbf{b} \tag{7.9}$$

After having carried out mesh generation on the $u-v$ plane, the inverse transformation is performed on all the generated interior nodes. For a point P having coordinates

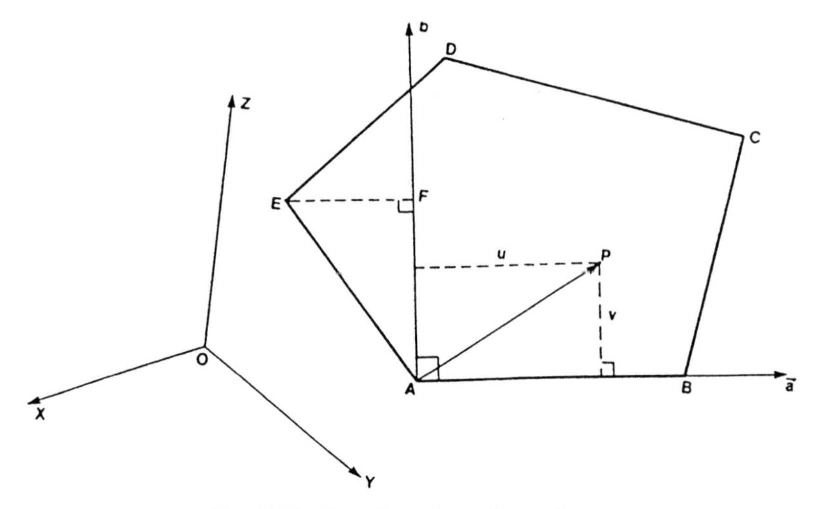

Fig. 7.42 Transformation: plane–plane

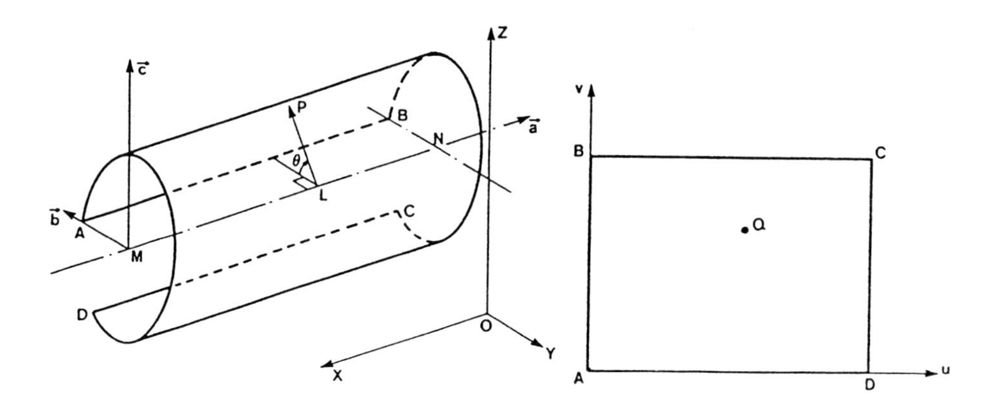

Fig. 7.43 Transformation: plane–cylinder

(u, v), its position vector in $x-y-z$ space is given by

$$\mathbf{OP} = \mathbf{OA} + u\mathbf{a} + v\mathbf{b} \tag{7.10}$$

(2) Cylindrical surface

Figure 7.43 shows a cylindrical surface $ABCD$ with its axis MN and radius $r = \|\mathbf{MA}\|$, where M is the centre of circular arc AD. The unit vector along the axis MN is given by

$$\mathbf{a} = \frac{\mathbf{MN}}{\|\mathbf{MN}\|} \tag{7.11}$$

$$\text{Let } \mathbf{b} = \frac{\mathbf{MA}}{r} \quad \text{and} \quad \mathbf{c} = \mathbf{a} \times \mathbf{b} \tag{7.12}$$

Then $(\mathbf{a}, \mathbf{b}, \mathbf{c})$ forms an orthonormal system having M as origin. Let P be any point on the cylindrical surface and L is a point on MN such that $LP \perp MN$. Then vector

LP is given by

$$\mathbf{LP} = \mathbf{MP} - (\mathbf{MP} \cdot \mathbf{a})\,\mathbf{a} \tag{7.13}$$

The angle between vector **LP** and plane $MNBA$ is given by

$$\theta = a\cos\left(\mathbf{LP} \cdot \frac{\mathbf{b}}{r}\right) \qquad \theta \in [0, 2\pi] \tag{7.14}$$

Hence the (u, v) coordinates of point P are

$$u = r\theta \qquad v = \mathbf{MP} \cdot \mathbf{a} \tag{7.15}$$

As for the inverse transformation of a point Q on the u–v plane, its position vector in x–y–z space is given by

$$\mathbf{MQ} = v\,\mathbf{a} + r\cos\phi\,\mathbf{b} + r\sin\phi\,\mathbf{c} \tag{7.16}$$

where

$$\phi = u/r \tag{7.17}$$

and

$$\mathbf{OQ} = \mathbf{OM} + \mathbf{MQ} \tag{7.18}$$

(3) Conical surface

Similar to the treatment of a cylindrical surface, an orthonormal system (**a**, **b**, **c**) is defined at point M (Fig. 7.44) such that

$$\mathbf{a} = \frac{\mathbf{MN}}{\|\mathbf{MN}\|} \qquad \mathbf{b} = \frac{\mathbf{MA}}{\|\mathbf{MA}\|} \quad \text{and} \quad \mathbf{c} = \mathbf{a} \times \mathbf{b} \tag{7.19}$$

For any point P on the conical surface, we have

$$\mathbf{LP} = \mathbf{MP} - (\mathbf{MP} \cdot \mathbf{a})\mathbf{a} \tag{7.20}$$

$$\theta = a\cos\left(\mathbf{LP} \cdot \frac{\mathbf{b}}{\|\mathbf{LP}\|}\right) \tag{7.21}$$

Hence the (u, v) coordinates of P are

$$u = \|\mathbf{KP}\| \cos\phi \quad \text{and} \quad v = \|\mathbf{KP}\| \sin\phi \tag{7.22}$$

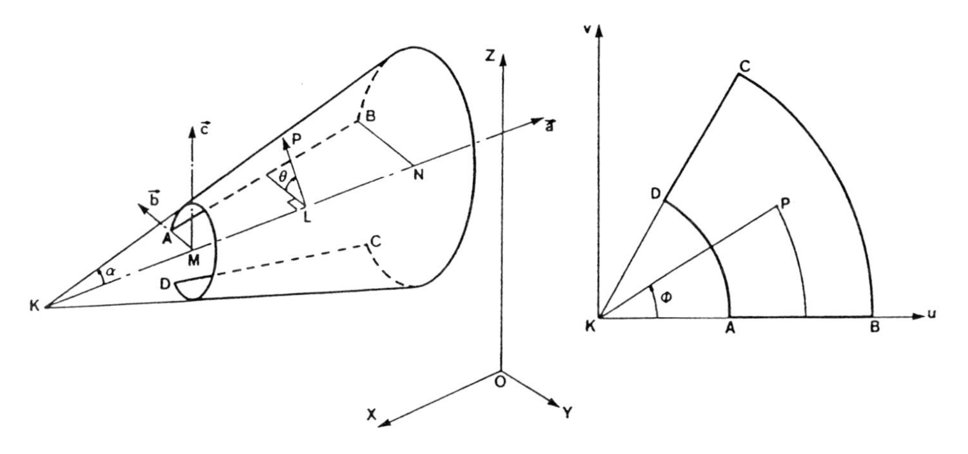

Fig. 7.44 Transformation: plane–cone

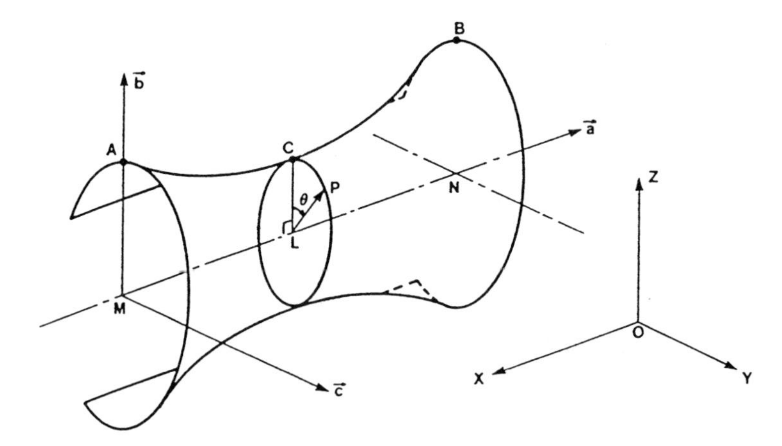

Fig. 7.45 Surface of revolution

where $\|\mathbf{KP}\|\phi = \|\mathbf{LP}\|\theta$. Conversely, if we know the (u, v) coordinates of a point P on the u–v plane, its position vector in space is given by

$$r = (u^2 + v^2)^{1/2}$$

$$\phi = a\cos\left(\frac{u}{r}\right)$$

$$f = r\cos\alpha$$

$$s = r\sin\alpha$$

$$\theta = \frac{r\phi}{s}$$

$$g = s\cos\theta$$

$$h = s\sin\theta$$

$$\mathbf{OP} = \mathbf{OK} + f\mathbf{a} + g\mathbf{b} + h\mathbf{c} \tag{7.23}$$

where α is the semi-vertical angle of the cone.

7.3.3 Surface of revolution

When a given curve AB is made to rotate about an axis MN, a surface of revolution is generated as shown in Fig. 7.45. Again let $(\mathbf{a}, \mathbf{b}, \mathbf{c})$ be an orthonormal system defined at point M such that

$$\mathbf{a} = \frac{\mathbf{MN}}{\|\mathbf{MN}\|} \qquad \mathbf{b} = \frac{\mathbf{MA}}{\|\mathbf{MA}\|} \quad \text{and} \quad \mathbf{c} = \mathbf{a} \times \mathbf{b} \tag{7.24}$$

Let C be any point on curve AB, then

$$\mathbf{ML} = (\mathbf{MC} \cdot \mathbf{a})\,\mathbf{a} \quad \text{and} \quad \mathbf{LC} = \mathbf{MC} - \mathbf{ML} \tag{7.25}$$

The position vector of point P corresponding to angle θ is given by

$$\mathbf{OP} = \mathbf{OM} + \mathbf{ML} + \|\mathbf{LC}\|\cos\theta\,\mathbf{b} + \|\mathbf{LC}\|\sin\theta\,\mathbf{c} \tag{7.26}$$

Nodes generated in layers can be linked together easily to form triangular or quadrilateral elements.

7.3.4 Spherical surface

A spherical surface is defined by its boundary which is a disjoint union of simple closed loops of circular arcs (Fig. 7.46). For simply connected surfaces there is only one closed loop, whereas for multi-connected domains there may be as many internal loops as the number of openings inside the spherical surface.

Generating interior nodes

The method used to generate interior nodes on a spherical surface is a modification of the method described in Section 7.2.3(2b). The following is the description of the generation procedure.

(a) Sort out the ϕ_{min} and ϕ_{max} of the spherical surface, where ϕ is the angle measured from the plane **a**−**b** towards vector **c** (Fig. 7.46).

(b) Imaginary horizontal planes at different levels are drawn across the sphere between ϕ_{min} and ϕ_{max}.

(c) Let d be the average element size and r be the radius of the sphere. Then the number of the horizontal plane is given by

$$N = \text{NINT}\left[\frac{r(\phi_{max} - \phi_{min})}{d}\right] - 1 \tag{7.27}$$

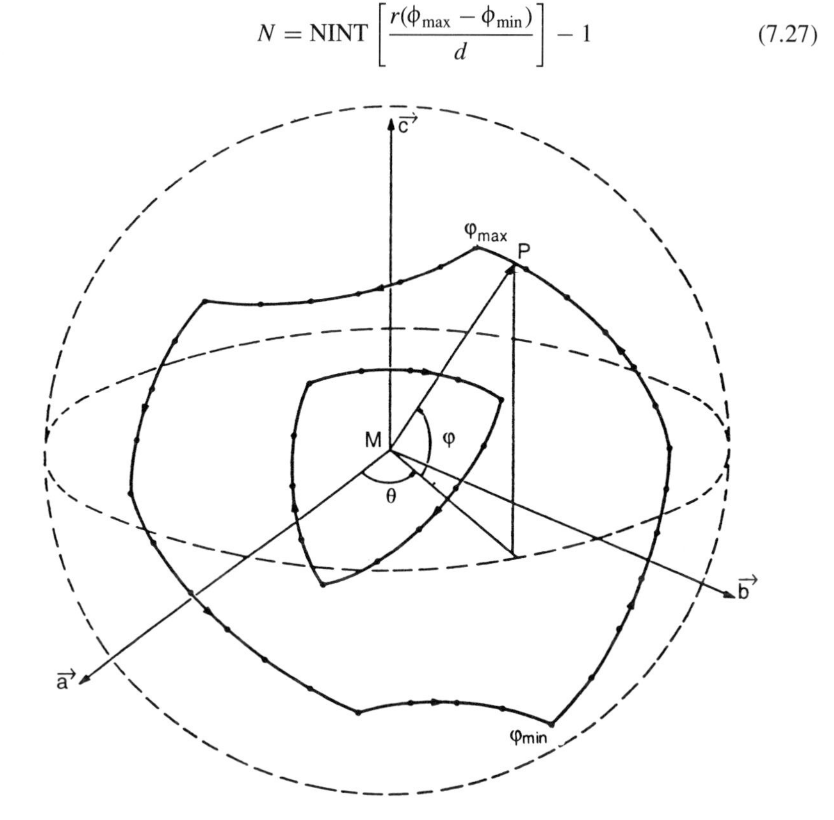

Fig. 7.46 Spherical surface

where NINT [·] means the nearest integer, and the level of the kth horizontal plane z_k can be calculated by

$$z_k = z_0 + r \sin \phi_k \tag{7.28}$$

where $\phi_k = \phi_{\min} + k(\phi_{\max} - \phi_{\min})/(N + 1)$ and z_0 is the z coordinate of the centre of the sphere.

(d) The intersections between the horizontal plane $z = z_k$ and the spherical surface are determined (Fig. 7.47). Let $P_1(\theta_1, \phi_1)$, $P_2(\theta_2, \phi_2)$, be any two adjacent nodes on the boundary of the spherical surface. There is intersection between arc P_1P_2 and plane $z = z_k$ if
(i) $(\phi_1 - \phi_k)(\phi_2 - \phi_1) < 0$
(ii) $(\phi_1 - \phi_k)(\phi_2 - \phi_1) = 0 \quad$ and $\quad \phi_k > \phi_1 \quad$ or $\quad \phi_k > \phi_2$
For any other cases, it is considered that there is no intersection. The point of intersection (θ, ϕ) is given by

$$\theta = \theta_1 + \frac{(\phi_k - \phi_1)(\theta_2 - \theta_1)}{(\phi_2 - \phi_1)} \quad \text{and} \quad \phi = \phi_k \tag{7.29}$$

(e) Each horizontal plane must cut the spherical surface at an even number of points; and the intersection points are arranged in ascending magnitude of θ. If there is no intersection, it means that the intersection of the horizontal plane and the spherical surface is a complete circle.

(f) Assume that there are $2n$ cuts between a particular plane and the spherical surface; the cuts are considered two by two, beginning with the first and the second cuts. Nodes are generated on this horizontal arc according to the prescribed spacing. This only suggests a series of potential positions where nodes can be generated. However, whether a node is finally generated or not depends on how close it is to the domain boundary. This condition can be assured simply by checking that

$$(x - x_i)^2 + (y - y_i)^2 + (z - z_i)^2 < c^2 \qquad \forall i = 1, N$$

where (x, y, z) is a nodal point to be generated, (x_i, y_i, z_i) is a boundary node, N is the number of boundary nodes and c is a constant depending

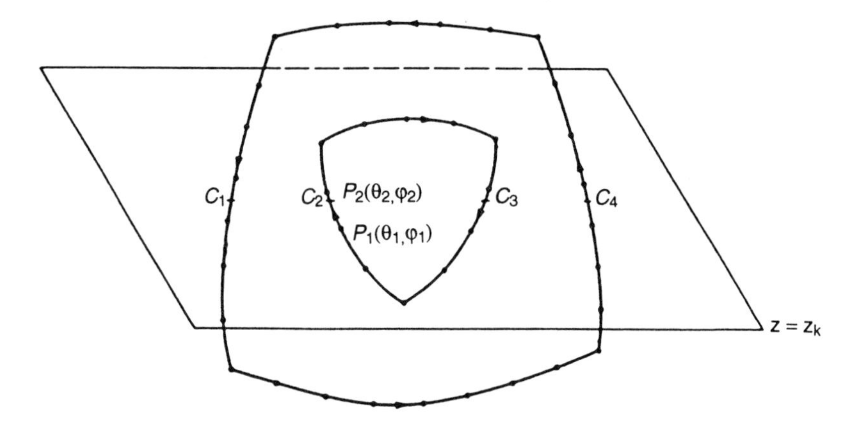

Fig. 7.47 Determination of intersections

on the average element size d of the spherical surface, which can usually be taken as $0.7d$.

(g) After the first two cuts, the process is carried on with the third and fourth cuts in a similar way until it terminates with the $(2n - 1)$th and $2n$th cuts; then proceed with the next horizontal plane.

7.3.5 Arbitrary surfaces

A surface of arbitrary form is defined by its boundary which is a disjoint union of simple closed loops of straight-line segments or circular arcs in space. The interior part of the surface is defined by a system of interior nodes so spaced that there is no abrupt change of curvature. Such an arbitrary surface is shown in Fig. 7.48.

7.3.6 Triangulation of arbitrary curved surfaces

With a slight modification, the advancing front method for planar domains described in Section 7.2.3(2c) can be used to interconnect boundary nodes and interior nodes over a spherical surface or more generally over any arbitrary curved surface. The modification is connected with the definition of $\alpha(ABC)$ given in eq. 7.6, in which the search for a point C in Fig. 7.49 to form a triangular element ABC must now be based on the unit vector $\hat{n}$ normal to the triangular facet ADB, instead of the unit vector $\hat{k}$ along the z-axis. In other words, the product $\hat{k} \cdot \overrightarrow{AB} \times \overrightarrow{AC}$ must be replaced by $\hat{n} \cdot \overrightarrow{AB} \times \overrightarrow{AC}$ in the definition of α.

7.3.7 Examples

In this section, several examples with different characteristics are given. In all cases user-interaction with the program was limited to providing the necessary boundary description and element size of each subregion. Although many of the results appear as hard copies of a plot taken from a graphic display tube, it should be emphasized that all computational work could just as easily have been done without the aid of graphics, which only provide the user with a visual inspection of the various stages of mesh generation. The examples shown here are done on a Norsk Data 570/CX computer, whereas the graphic presentation is done on a Tektronix graphics

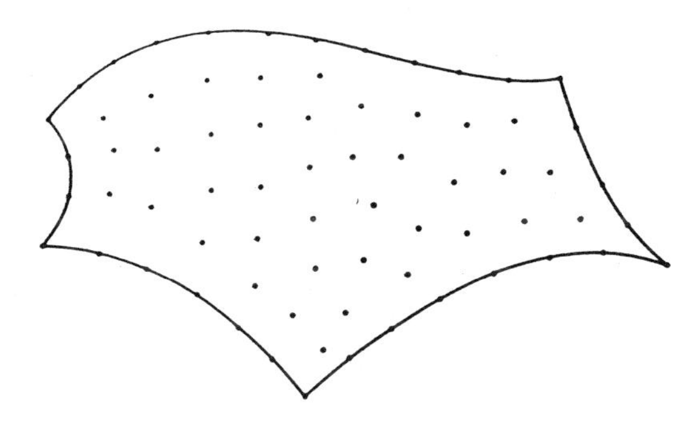

Fig. 7.48 Arbitrary surface

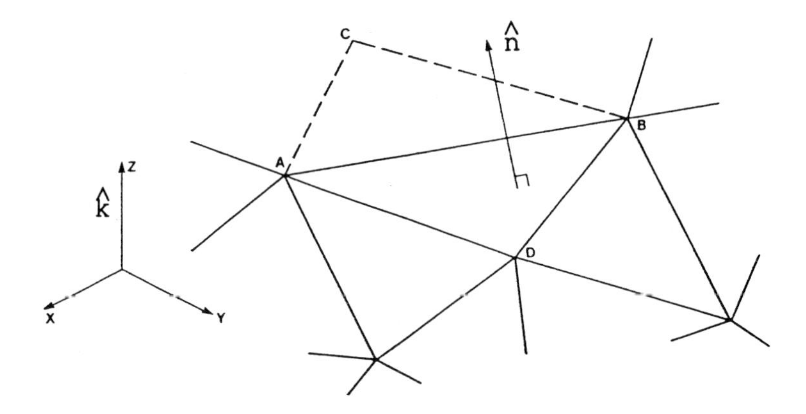

Fig. 7.49 Forming triangular element ABC

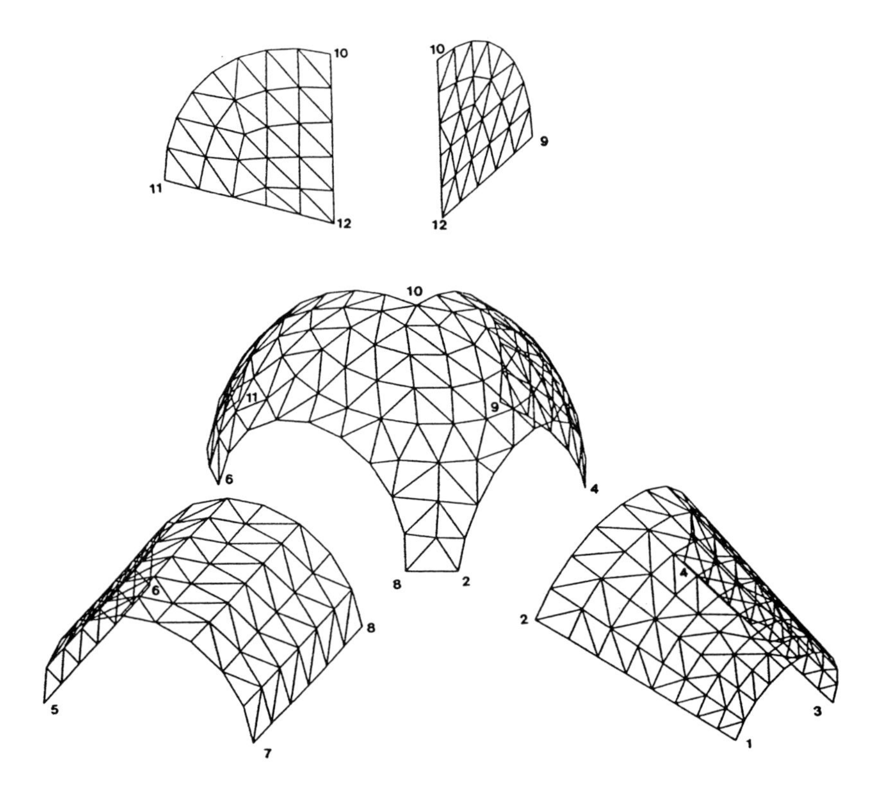

Fig. 7.50 Planes, cylindrical, conical and spherical surfaces are separately meshed

terminal T4014.

Figures 7.50 and 7.51 give an example which demonstrates how planes, cylindrical surfaces, conical surfaces and spherical surfaces are put together to form a single finite element mesh. Figure 7.52 is a dumb-bell which is generated when the contour line on the left is made to rotate about the x-axis. Figure 7.53 shows the different

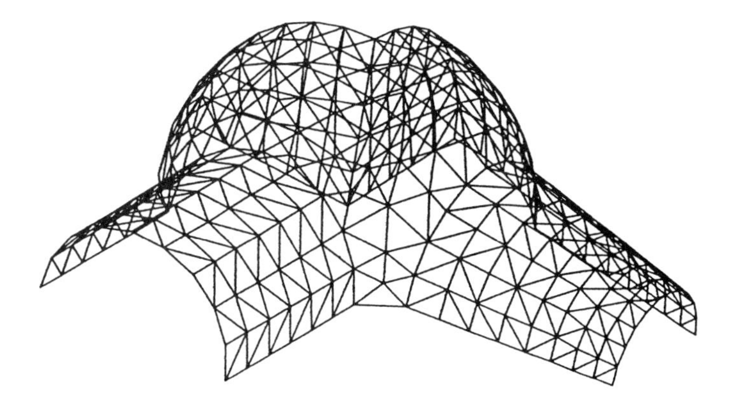

Fig. 7.51 Different surface types are put together to form the complete structure

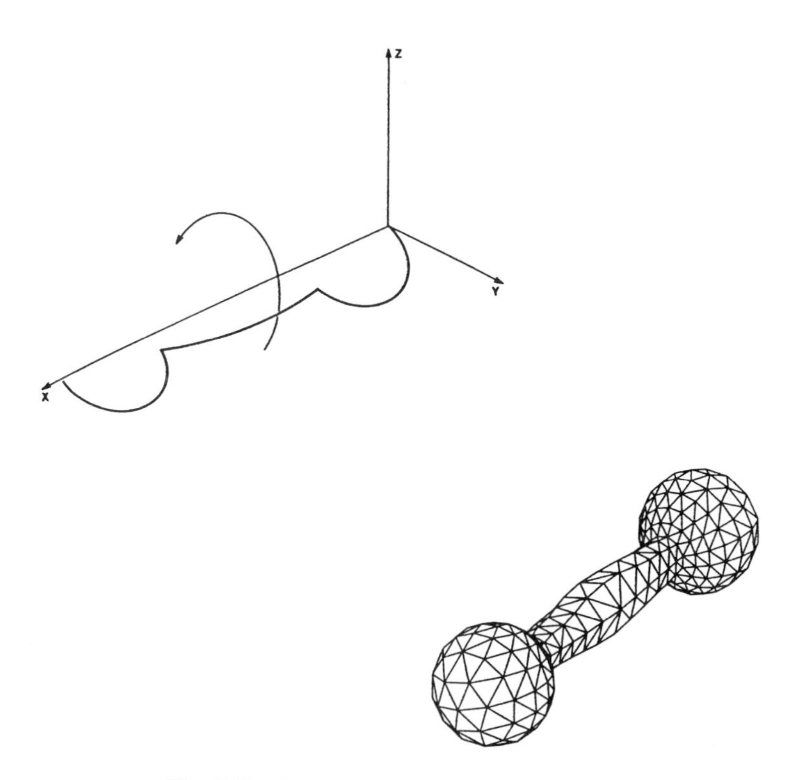

Fig. 7.52 Surface of revolution – a dumb-bell

steps of mesh generation over a conical surface. Figure 7.54 shows part of a pipe junction which is composed of two cylindrical surfaces. Figure 7.55 is a cup with its body made up of two cylindrical surfaces, and the handle comprising two surfaces of revolution. Figure 7.56 is a house which is a combination of 16 spatial planes. Figure 7.57 shows that different element sizes can be used in the discretization of a steel frame structure. Finally, Fig. 7.58 shows more examples of surfaces of revolution, spherical surfaces and an object made up of plane surfaces.

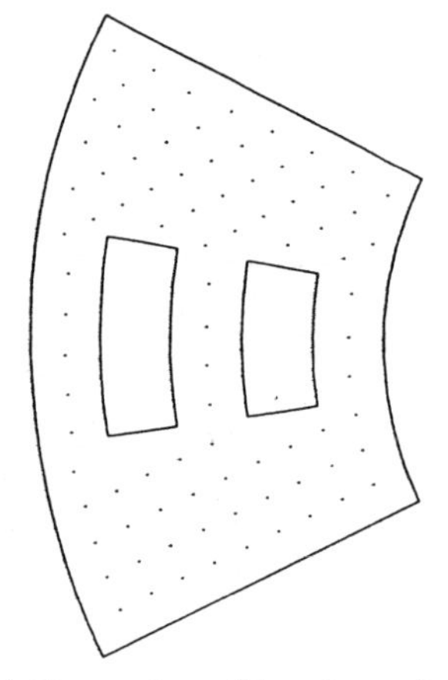

(a) Generation of interior points

(b) Triangulation over a planar domain

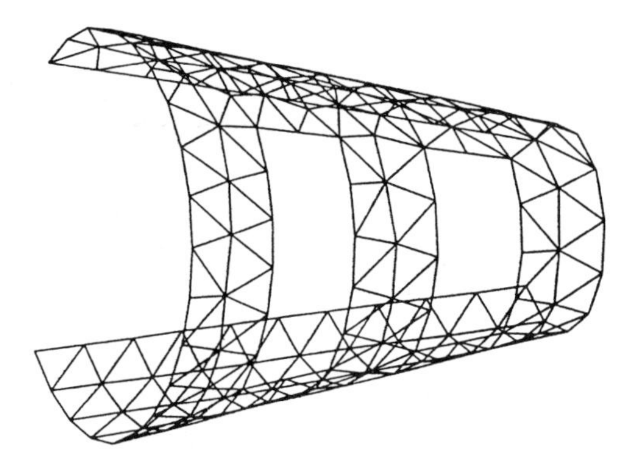

(c) Conical surface after transformation

Fig. 7.53 Procedures of mesh generation over a conical surface

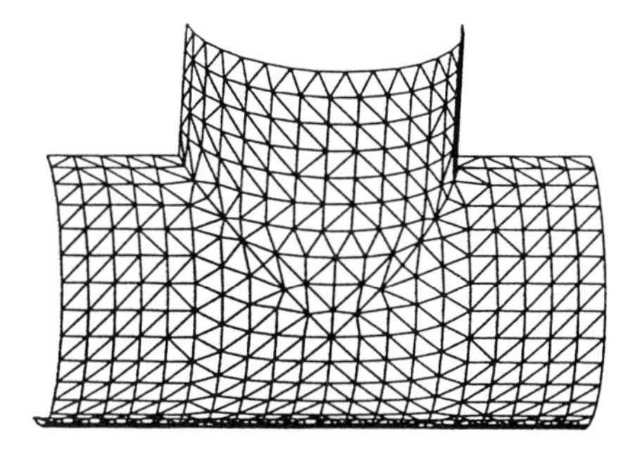

Fig. 7.54 A pipe junction

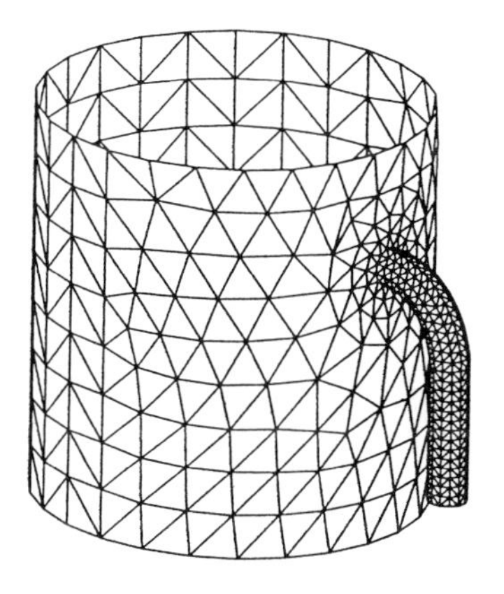

Fig. 7.55 A cup

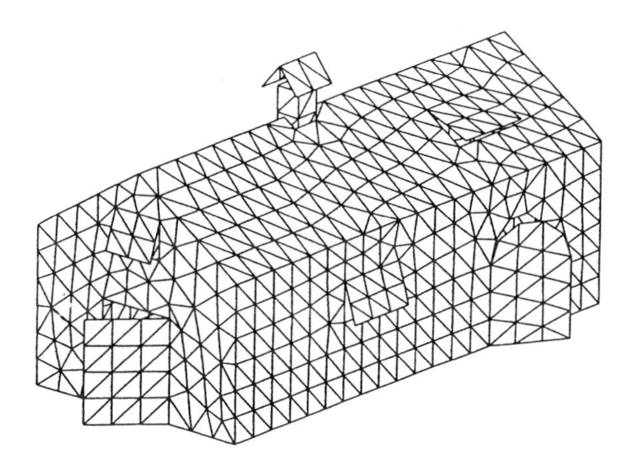

Fig. 7.56 A house

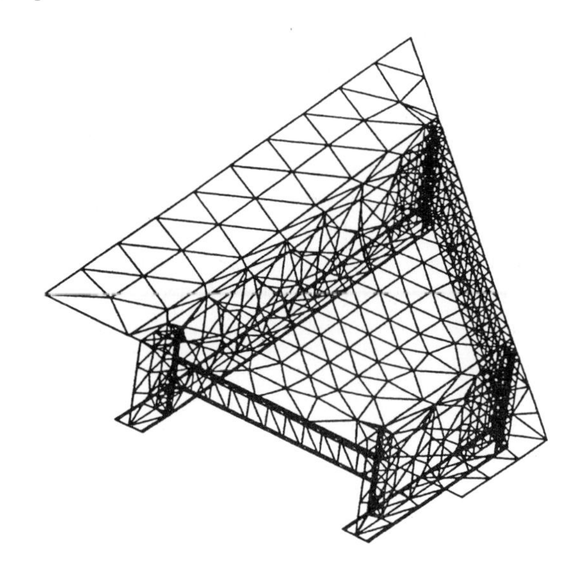

Fig. 7.57 A steel structural frame

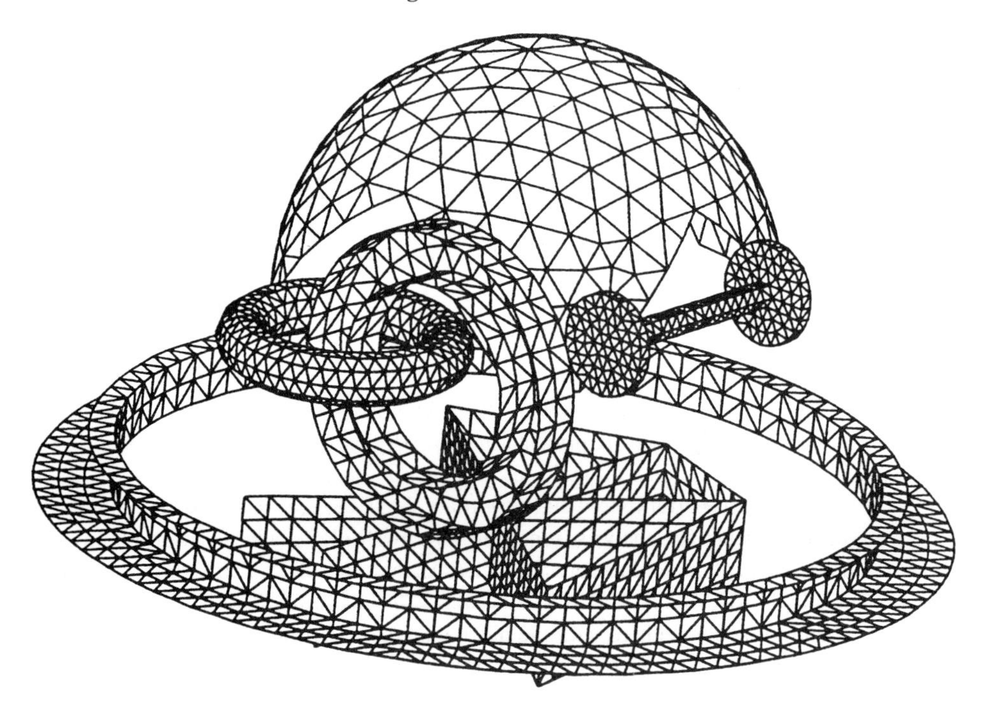

Fig. 7.58 Examples of surfaces of revolution, spherical surfaces and an object made up of plane surfaces

7.3.8 Discussions on mesh generation over surfaces

For finite element mesh generation over curved surfaces, some more work has to be carried out in the future. The most pressing work is to devise an intelligent algorithm for the generation of *well-shaped* triangular element meshes over surfaces piecewise definable by mathematical functions. This represents a large family of surfaces especially useful to the automobile, ship-building and aircraft industries. The main difficulty lies essentially in the generation of interior nodes whose spacing should depend on the surface curvature which may change continuously over the surface. Generation of interior nodes by means of a mapping or projection of any kind may be possible; nevertheless, the change in geometry in applying these mathematical functions has to be carefully taken into account, which in the end will determine the density of the nodes and hence the quality of the mesh.

The advancing front approach may once again prove to be useful in the triangulation of analytical surfaces, and it is still a hot topic in the research of mesh generation. However, it appears that simultaneous generation of interior nodes with the construction of triangular elements is the best strategy in producing a high-quality gradation mesh over a general curved surface.

Realistic physical objects cannot in general be represented by one single surface; they are rather the products of the intersections of several elementary surfaces – spatial plane, surface of revolution, analytical surface, etc. Therefore, it would be extremely valuable to develop an automatic algorithm to generate sound triangular element meshes over surfaces of intersection. The key to success of such an algorithm hinges on the accurate determination of the lines of intersection between surfaces. Once this is done, the boundaries of the elementary surfaces can be modified as two surfaces interact. Individual elementary surfaces with updated boundaries can then be meshed in turn by a standard mesh generation algorithm.

7.4 Mesh generation for volumes (3D)

7.4.1 Introduction

Three-dimensional meshes are traditionally built using hexahedral elements, while tetrahedral elements are reserved for regions in the solid where hexahedra cannot be conveniently fitted. A principal reason why users avoid the tetrahedral element is simply the extreme difficulty of visualizing tetrahedra in a solid mesh. A second objection arises from the well-known fact that linear, constant-strain tetrahedra are poor elements for analysis, requiring fine meshes to produce reasonably accurate finite element approximations. Higher-order versions of the tetrahedron are however available [68,69] and the efficiency of the quadratic tetrahedral element has been found to be competitive with the quadratic hexahedral finite element.

Automatic mesh generation has been successfully realized in 2D situations as discussed in the earlier part of this chapter, while in three dimensions various approaches to the problem are currently being investigated [52–64]. No single scheme has yet been found to be entirely satisfactory, and continuous research efforts are expected in the future. In this chapter, existing methods of automatic mesh generation for 3D solid objects are reviewed. Although the 3D Delaunay triangulation recently aroused much attention, its suitability as a finite element mesh generator is questioned. While in 2D Delaunay triangulation, the 'max–min' angle criterion can be verified over the entire domain [12], no equivalent or similar criterion can be defined for its extension to 3D situations to ensure that tetrahedral elements so generated are well-proportioned and suitable for numerical calculations.

In the following sections, a simple but versatile 3D triangulation scheme based on the advancing front technique for the discretization of arbitrary volumes is presented [67]. To ensure that the tetrahedral elements generated are as equilateral as possible, the ratio of volume of the element to the sum of squares of edges put into a dimensionless form is adopted to judge the quality of a tetrahedral element. The quality of the finite element meshes can thus be assured if the shape of each tetrahedral element is carefully controlled in the mesh construction process. Through the study of numerous examples of various characteristics, it is found that high-quality tetrahedral element meshes can be obtained by this algorithm for a wide range of solid objects.

7.4.2 The existing methods

(1) Super-elements

A 3D solid is divided into a number of subregions by the application of a 20-node isoparametric hexahedral element [2]. These subregions are split up into smaller hexahedra by indicating the number of subdivisions along the local element coordinates. A mesh of tetrahedral elements can be obtained when each hexahedron is finally divided up into five or six tetrahedral elements. The disadvantage of this technique is the large amount of manual work involved in obtaining a sound subdivision of the solid object, and the regularity of the network structure.

(2) Body centre radiation technique

The domain boundary surface consisting of triangular elements is first constructed for a convex 3D solid [52], and then a central point is defined inside the solid. The nodal points on the surface are connected with this interior point to form narrow tetrahedra. The tetrahedra are cut into segments of pentahedra, which are finally divided into tetrahedral elements. By an iteration procedure, the final number of nodal points and their coordinates are then determined. The disadvantage here is that the resulting mesh contains many elongated elements with sharp angles. For general complex objects, a prior subdivision into convex subregions is necessary before this technique can be applied.

(3) Drag method

A generatrix can be defined as a point, line or surface whose motion generates a line, surface or solid [44]. The basis of the drag method is a cut section of 0-dimension (point), 1-dimension (line) and 2-dimension (area) generatrix elements together with a set of displacement (step, increment) control vectors. Starting from its original position, the generatrix set is dragged through a sequence of positions as determined by the succession of displacement control vectors. The first displacement vector is added to the nodes in the original pattern to determine the first transient position, the second displacement vector is added to that position to determine the next position, and so on. Two successive positions of a generatrix element form two opposing faces/edges/ends of the new finite elements, and by connecting the corresponding nodes of the pair, the remaining faces/edges are formed. The displacement vector from one cross-section to another can be expressed by either a simple translation or a rotation. Finite element meshes with relatively simple structure can be generated effectively by this technique. Complex domains, however, must be first broken up into several components, each of which is taken care of in turn by the same drag method. As a result, considerable work will be required in decomposing objects carefully into simpler parts and matching them together again in one single piece later after mesh generation.

(4) Division of the structure into blocks or modules

Finite element mesh on different blocks/modules is independently generated, the blocks/modules are placed in contact in such a way that the desired structure is constructed [53–55]. The merits of this approach are that relatively simple means can be applied for mesh generation in the module level, and that modification of the finite element mesh can be done locally to each individual block. However, the drawbacks associated with this method are that there is a lot of manual work involved in placing various pieces together to form the whole structure, and that compatibility conditions must be checked when two blocks are placed in contact.

(5) Construction of tetrahedral elements around line segments

The procedure of 3D mesh generation consists of forming tetrahedral elements from a number of given points in a defined 3D domain by laying straight lines and plane triangles [57]. Within the 3D objects to be discretized, interior nodes are supplied manually. Tetrahedral elements are constructed around all the line segments between two interior nodes, a boundary node and an interior node, or two boundary nodes. During the generation process, extensive checks must be made to ensure that newly-formed tetrahedra do not pierce or intersect with any existing ones. Mesh generation is completed when tetrahedral elements not belonging to the domain are removed with the aid of user-supplied external auxiliary nodes. This process seems to put too many demands on the user, and requires excessive computation for systems with a large number of nodes.

(6) The modified-octree technique

The octree technique represents a 3D object as a set of cubes of various sizes in much the same manner as the quadtree technique represents a planar object as a set of squares [59,60]. The modifications to the octree that are necessary to obtain an efficient mesh are similar to, but much more extensive than, those used in the modified-quadtree approach. The basic steps in the modified-octree mesh generation process are:

(i) Set up an integer coordinate system that contains the object to be meshed.
(ii) Generate the modified-octree representation of the object accounting for the mesh gradation information specified with the geometrical model.
(iii) Break the modified-octree up into a valid finite element mesh.
(iv) Pull the nodes on the boundary of the modified-octree to the appropriate vertices, edges and faces of the original geometry.
(v) Smooth the locations of the node points to create a better conditioned finite element mesh.

One of the inherent shortcomings of the quadtree or octree decompositions is due to the pre-fixed orientations of the generated zone which do not properly account for preferential directions often dictated by external or internal boundary parts. In order to reduce the number of elements needed to represent curved boundaries, in the modified-octree approach, the concept of 'cut octant' is introduced. To maintain the integer tree storage and to limit the number of cut-octant cases to a manageable number, only the corner and half-points of an octant are used in the cutting process. Nevertheless, cut octant is a tedious and complicated process and has to be done on a case-by-case basis; the number of special cases required in a specific 2D situation is 16, whereas the same situation in three dimensions requires 4096 cases. Other features associated with the modified-octree method are that a one-level transitional rule has to be enforced to ensure a smooth change of element size and transition elements are required in various situations [61]. Since the contact surfaces of neighbouring parts created by the modified-octree technique are in general not compatible, mesh generation for complex objects by means of subdomains is not possible as there is difficulty in matching the boundary of one subdomain with another.

*(7) **Generation by layers***

For the given 3D domain, tetrahedral elements are constructed layer by layer towards the interior of the object until all the volume is filled up [62,63]. The volume to be discretized is bounded by closed surfaces composed of triangular elements. Nodes are generated at a suitable distance normal to each triangular element at the interior part of the volume. Nodes too close to one another, too close to the boundary surface, or outside the volume, are eliminated. Boundary nodes and interior nodes are linked up together to form valid tetrahedral elements which do not intersect any other tetrahedral element already formed. The domain virtually advances towards the centre of the volume by a layer of elements, and as a result, the interior part of the object reduces in size. Using the new layer of elements as the updated domain boundary, the process is repeated until the volume shrinks to zero. This approach can be considered as a particular case of the advancing front technique, in which artificial restrictions have been imposed to limit the movement of the generation front to evolve layer by layer only.

7.4.3 3D Delaunay triangulation

One approach to fully automatic mesh generation in two and three dimensions is to generate and triangulate a set of points within and on the boundary of an object using the property of the Delaunay triangulation [17,58,90,91]. The Delaunay triangulation associated with a set of points in n-dimensional space can be introduced by the following definitions.

Definition 1

Given a set $\mathcal{P}$ of m unique, random points in n-dimensional space, associated with each point $P_i \in \mathcal{P}$ there exists a region V_i such that

$$V_i = \{X : \|X - P_i\| < \|X - P_j\|, \text{ for all } j \neq i\}$$

The collection of regions $\mathbb{V} = \{V_i, i = 1, m\}$ is called the Dirichlet tessellation.

The region V_i can be shown to be the convex intersection of the open half-planes separating the points P_i from P_j, and is the region in n-dimensional space closer to P_i than to any other points. In two dimensions, the V_i are convex polygons, in three dimensions, the V_i are convex polyhedra. Regions which share $(n-1)$-dimensional boundaries are termed neighbouring tiles.

Definition 2

Given a Dirichlet tessellation $\mathbb{V}$ of n-dimensional space, the lines connecting the points P_i to P_j, where V_i and V_j are neighbouring tiles, form the Delaunay triangulation.

In general, the triangulation generates n-dimensional simplexes with the interesting property that a circumscribing hypersphere contains no points other than the $n+1$ points which form the n-dimensional simplex. This property of circumsphere containment is the key to the various algorithms constructing Delaunay triangulation for a given set of points.

In three-dimensions, Watson's algorithm [16] starts with a tetrahedron containing all points to be inserted, and new internal tetrahedra are formed as the points are entered one at a time. At a typical stage of the process, a new point is tested to determine which circumspheres of the existing tetrahedra contain the point. The associated tetrahedra are removed, leaving an insertion polyhedron containing the new point. Edges connecting the new point to all triangular facets of the surface of the insertion polyhedron are created, defining tetrahedra which fill the insertion polyhedron. Combining these with the tetrahedra outside the insertion polyhedron produces a new Delaunay triangulation which contains the newly added point. Triangulation is done when all points are inserted and processed sequentially.

An efficient finite element mesh generation algorithm for the discretization of arbitrary planar domains based on Delaunay triangulation and the advancing front technique has been developed [19]. However, automatic mesh generation schemes for the discretization of general 3D objects using Delaunay triangulation have been less successful. The difficulties involved in 3D Delaunay triangulation as a finite element mesh generator are:

(a) The existence of the so-called degenerate cases

Degenerate cases occurs in practice when a newly inserted node appears to lie on the surface of a circumsphere associated with some existing tetrahedron. The problem becomes apparent whenever the distance from a newly entered nodal point to an existing circumsphere is less than ε, where ε is the expected accumulated computer truncation error. When this happens, there is the danger that an incorrect or inconsistent decision will be made regarding rejection or acceptance of a given 4-tuple. This in turn produces structural inconsistencies in the triangulation, i.e. overlapping tetrahedra or gaps in the mesh.

(b) Creation of 'Sliver'

A serious problem which arises in 3D Delaunay triangulation is the creation of narrow shaped tetrahedra called 'slivers' – badly distorted tetrahedra whose faces are well-proportioned triangles but whose volume can be made arbitrarily small. In a 3D Delaunay triangulation, 'sliver' can account for as much as 10% of the total number of tetrahedra generated [17].

(c) Post-treatments needed for non-convex domains

For general non-convex domains, the aggregate of Delaunay tetrahedra must be post-processed to determine which of the tetrahedra lie inside the solid and are consequently part of the solid finite element mesh. Those that span holes or lie outside the structure must be identified and can be 'peeled' away, but there appears to be no simple way in dealing with the case in which elements cut across domain boundaries.

Nevertheless, the major difficulty of using 3D Delaunay triangulation to yield a valid assembly of tetrahedra well-suited for finite element analysis is that the process does not guarantee the generation of well-proportioned tetrahedral elements. While in 2D Delaunay triangulation, the 'max–min' angle criterion is verified over the entire domain [12], no similar criterion exists for 3D triangulation to ensure that tetrahedral elements so generated are well-proportioned for numerical calculations.

Consider a square based pyramid *ABCDE* as shown in Fig. 7.59. If node *D* is slightly above the plane defined by triangle *ABC*, then by Delaunay triangulation the pyramid will be divided into three tetrahedra *ABCD*, *ABDE* and *BCDE*. However, the volume of tetrahedron *ABCD* can be made as small as possible, rendering the element not suitable for finite element analysis. By inspection, it is easily found that a much better decomposition can be obtained by dividing the pyramid into two tetrahedra *ABCE* and *ACDE*. It is noted that *ABCE* is not a Delaunay tetrahedron since it contains point *D* in its circumscribing sphere. The conclusion that can be drawn from the discretization of this simple pyramid is that 3D Delaunay triangulation in general does not produce the type of tetrahedral element mesh required by a finite element analysis. Unfortunately, this is a fundamental weakness of 3D Delaunay triangulation which occurs throughout the mesh generation process, and there is no remedy unless the whole concept is to be revised.

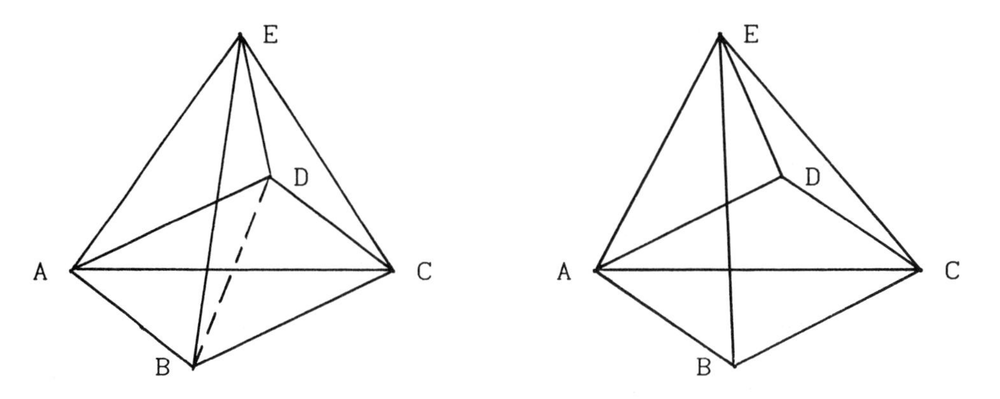

Fig. 7.59 Dividing pyramid *ABCDE* into tetrahedra

7.4.4 3D triangulation based on the advancing front technique

In this section, an efficient 3D triangulation algorithm based on the advancing front technique for the discretization of arbitrary volume is presented. The volume Ω to be discretized is defined by closed surfaces composed of triangular facets. The boundary condition imposed on the volume to be meshed is much more stringent than for many 3D triangulation schemes in which object boundaries are defined loosely by scattered spatial points. Immediately, two advantages can be seen from this boundary definition.

(i) Volumes to be meshed are uniquely defined without ambiguity.
(ii) Whenever necessary, complex objects can be meshed subregion by subregion. The connectivity among individual pieces is guaranteed provided that correct boundary surfaces of subregions are collected for mesh generation in subregion level.

Realizing that 3D Delaunay triangulation does not necessarily produce the type of tetrahedral element mesh required by a finite element analysis, and in order to ensure that tetrahedral elements generated are well-proportioned, the ratio of volume to the sum of squares of edges put into a dimensionless form is used to judge the quality of a tetrahedral element. It can be shown that this ratio, when normalized, attains a maximum value of 1 for regular equilateral tetrahedral elements. The quality of the mesh can thus be assured if the shape of each tetrahedral element is carefully controlled in the mesh construction process.

The procedures of the 3D triangulation by the advancing front approach are very similar to those applied to the triangulation of planar domains. Nevertheless, due to the more complex nature of a 3D problem, many of the steps of the mesh generation process have to be carefully revised, modified or extended to suit new situations.

The algorithm presented here can be used to discretize arbitrary volumes simply-connected or multi-connected. No special treatment is required for internal openings as long as correct orientations are applied when interior boundary surfaces are included as part of the domain boundary. Although the mesh generation program can handle domains composed of any number of subregions, if tetrahedral element mesh of consistent element size is required for a homogeneous domain, no subdivision into simpler subregions is necessary. The mesh generation time mainly depends on the number of elements to be generated and the number of nodes present in the domain. An irregular domain with complex topology and difficult boundary conditions may also require a little more computation time.

The boundary of the body to be discretized is represented by a collection of simple closed surfaces consisting of triangular elements. For simple domains the volume is bounded by one closed surface, whereas for more complicated domains there can be as many internal surfaces as the number of openings inside the domain. A multi-connected body in the form of a torus or rubber tube can also be triangulated as one single region. The triangular facets on the boundary surfaces are oriented in such a way that their normal vectors as defined by the right-hand grip rule are always pointing outwards. The orientation of the surface of triangular facets is routinely done and checked by a data preparation program prior to mesh generation [66]. Node or element numbering on boundary surfaces need not be made in any particular numerical order; this flexibility allows us later to generate tetrahedral element mesh from one subregion to another without bothering to identify the common boundaries between subregions. In contrast to the approach in which interior nodes and elements are created simultaneously [39], the algorithm first generates additional interior nodes according to the average element size of the triangular elements making up the domain boundary surfaces. The program then connects the boundary nodes and interior nodes to form tetrahedral elements in such a way that no elements overlap and the entire region is covered.

(1) Generation of interior nodes

Interior nodes within the given domain Ω are generated layer by layer. Planar cross-sections can be defined when a series of parallel planes cut across the 3D object to be discretized. Interior nodes are generated on these cut sections using the scheme for the generation of interior nodes on an arbitrary planar domain described in Section 7.2.3(2b). The procedures for obtaining parallel cut sections of an arbitrary volume (Fig. 7.60) are as follows:

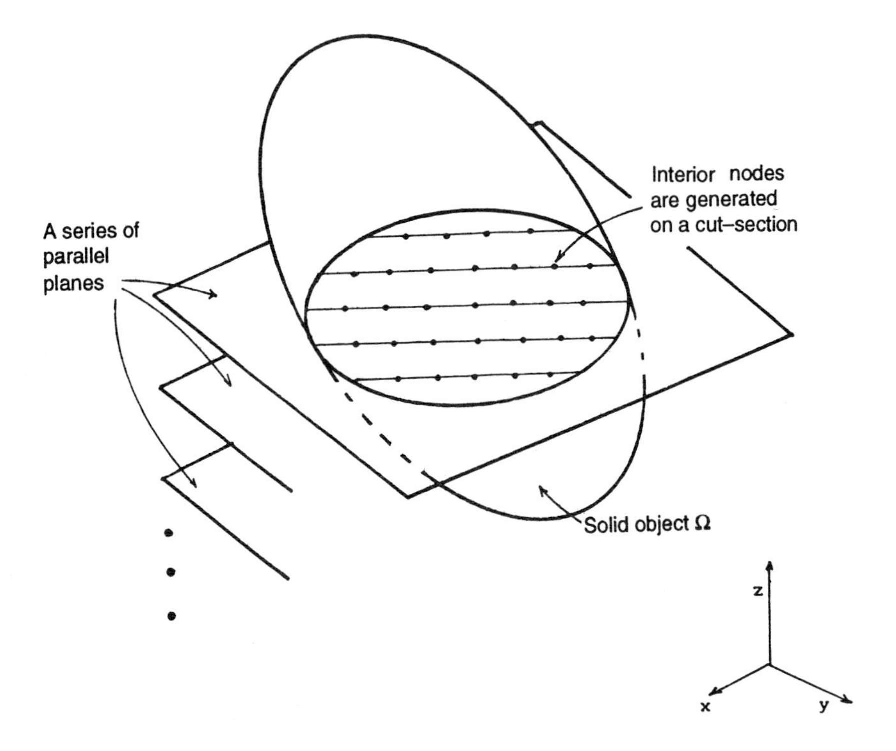

Fig. 7.60 Volume cut across by a series of parallel planes

(a) Sort out the z_{min} and z_{max} of the domain.
(b) Allow imaginary horizontal planes at different levels to cut across the domain.
(c) The spacing between any two imaginary planes is exactly equal to the average element size calculated using the triangular elements on the boundary surfaces.
(d) The intersection between each horizontal plane and the solid object is a plane section bounded by closed loops.
(e) Interior nodes are generated on each cut section by means of the 2D node generation scheme as detailed in Section 7.2.3(2b).

The cut section corresponding to a horizontal plane at level h can be found by considering the intersection between the plane and the triangular facets on domain boundaries. Let $\mathcal{B} = \{\Delta_i, i = 1, N_b\}$ be the set of triangular elements making up the boundary of the solid volume Ω. The contour at a given level h can be obtained by considering the intersection of horizontal plane $z = h$ with all the triangular facets in $\mathcal{B}$ one at a time. Suppose that J_1, J_2, J_3 are the three nodes of triangle Δ_i, and h_1, h_2, h_3 are the heights at nodes J_1, J_2, J_3 respectively. The intersection between the horizontal plane and triangle Δ_i can be determined by considering the intersection of the three edges of the triangle with the horizontal plane as shown in Fig. 7.61.

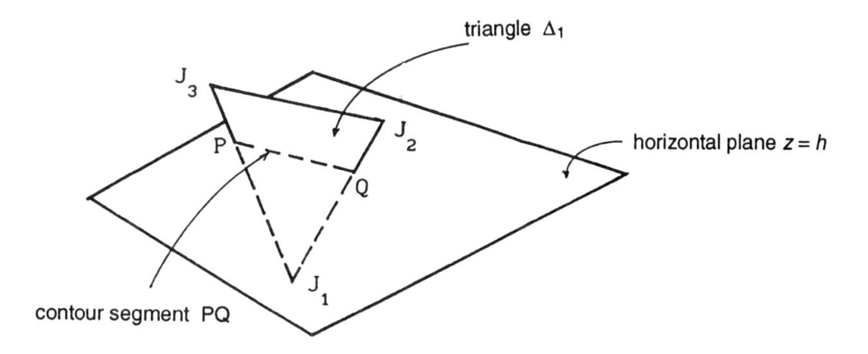

Fig. 7.61 Intersection between triangle Δ_i and horizontal plane $z = h$

Considering edge J_1J_2, there is intersection with the plane if
(i) $(h_1 - h)(h_2 - h) < 0$ or
(ii) $(h_1 - h)(h_2 - h) = 0$ and $(h_1 < h$ or $h_2 < h)$
The point of intersection $P = (u, v)$ is given by

$$u = (1 - t)x_1 + tx_2 \qquad v = (1 - t)y_1 + ty_2$$

where (x_1, y_1) and (x_2, y_2) are the x, y-coordinates of nodes J_1 and J_2 respectively, and parameter t is calculated from

$$t = \frac{h - h_1}{h_2 - h_1}$$

Similar calculations are done to the other two edges J_2J_3 and J_3J_1 for intersections. If there are intersections, the horizontal plane should intersect with exactly two of the three edges of triangle Δ_i. Let P and Q be the two points of intersection, then line segment PQ will be the intersection of the horizontal plane with triangle Δ_i. Having considered all the triangular elements in $\mathcal{B}$ in turn for intersection with the horizontal plane $z = h$ the contour lines at level h are given by the set of all these individual line segments

$$\mathcal{S} = \{P_jQ_j, j = 1, N_s\}.$$

The order of these line segments of intersection, which appears to be quite random, in fact relates directly to the numbering of the triangular elements Δ_i in $\mathcal{B}$. However, since the order of these line segments composing the boundary of the cut section is not important in the generation of internal nodes, retrieval of structural forms in terms of closed loops as defined by the segments in $\mathcal{S}$ by proper renumbering is not required. Points generated on cut sections using the 2D node generation scheme represent only potential positions where nodes can be generated. However, whether a node is finally generated depends also on how close it is from the domain boundary $\mathcal{B}$.

Let X be a potential node point, and $\Delta_i \in \mathcal{B}$ be the triangular facet closest to point X, as shown in Fig. 7.62. A measure of the quality of the tetrahedron formed

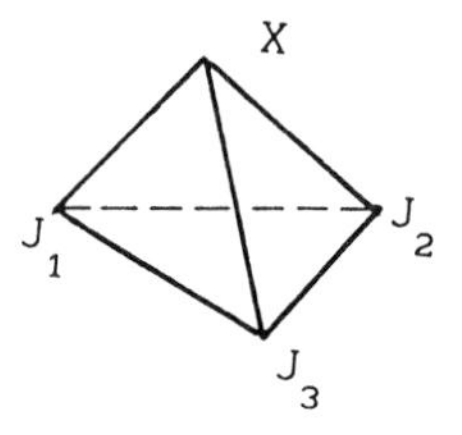

Fig. 7.62

by triangle Δ_i and point X can be defined by

$$\gamma = \frac{72\sqrt{3}(\text{volume of tetrahedron})}{(\text{sum of squares of edges})^{1.5}}$$

$$\gamma(J_1J_2J_3X) = \frac{12\sqrt{3}J_1J_3 \times J_1J_2 \cdot J_1X}{(\|J_1J_3\|^2 + \|J_3J_2\|^2 + \|J_2J_1\|^2 + \|J_1X\|^2 + \|J_2X\|^2 + \|J_3X\|^2)^{1.5}} \tag{7.30}$$

where J_1, J_2, J_3 are the nodes of triangular facet Δ_i and $12\sqrt{3}$ is a normalizing factor so that equilateral tetrahedra will have a maximum value equal to 1.

Point X will be accepted if $\gamma(J_1J_2J_3X) > \gamma_{\min}$, where $\gamma_{\min}$ can be fixed arbitrarily. For instance, $\gamma_{\min}$ can be set equal to 0.5α, in which α is the shape factor of triangle Δ_i (eq. 7.6).

(2) Formation of tetrahedral elements

Assume that a complete nodal system has been generated by the method described in Section 7.4.4(1). By virtue of the orientations of the triangular elements in $\mathcal{B}$, the domain to be triangulated has an interior volume always situated opposite to the normal vectors of the triangular elements. At the beginning of triangulation, the generation front Γ is exactly equal to the collection of domain boundary, $\Gamma = \mathcal{B}$. While the given domain boundary remains the same, the generation front changes continuously throughout the process of triangulation and has to be updated whenever a new tetrahedral element is formed.

In Fig. 7.63, the 3D triangulation process is initiated by selecting the last triangular facet $J_1J_2J_3 \in \Gamma$. Let Σ be the nodal points on the generation front Γ and Λ be the set of interior nodes remaining inside the generation front. The goal is to determine a node $C \in \Sigma \cup \Lambda$ such that the tetrahedron $J_1J_2J_3C$ lies completely within the domain Ω, without cutting across the generation front Γ, and its γ-value is maximized.

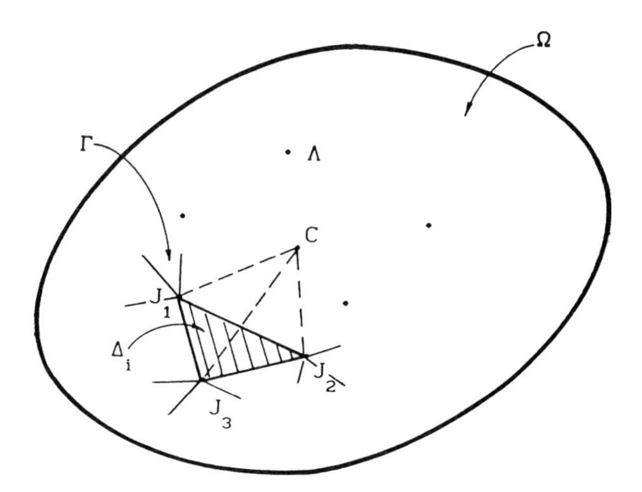

Fig. 7.63 Construction of tetrahedron $J_1J_2J_3C$

(3) Selection of candidate nodes

A node $C \in \Sigma \cup \Lambda$ is said to be a candidate node if it satisfies

(i) $J_1J_3 \times J_1J_2 \cdot J_1C > 0$ and

(ii) $J_iC \cap \Gamma \in \{J_i, \{J_i, C\}, J_iC\} \qquad i = 1, 2, 3$

Condition (i) ensures that the proposed tetrahedron lies in the interior part of the domain Ω, and condition (ii) ensures that the proposed tetrahedron does not cut across the generation front.

Let $\mathbb{C} = \{C_i\} \subset \Sigma \cup \Lambda$ be the set of candidate nodes. The node to be selected is therefore a node $C_m \in \mathbb{C}$ such that the γ-value of tetrahedron $J_1J_2J_3C_m$ is maximized, i.e.

$$\gamma(J_1J_2J_3C_m) \geq \gamma(J_1J_2J_3C_i) \qquad \forall C_i \in \mathbb{C}$$

(4) Shape optimization of tetrahedron elements

The choice of node $C_m \in \mathbb{C}$ by maximizing the γ-value of tetrahedron $J_1J_2J_3C_m$ is simple enough, but it may not be sufficient to guarantee the best triangulation for domains with very irregular boundaries. Under such circumstances, a more sophisticated selection procedure has to be adopted. The idea is that in the selection of node C_i, instead of considering only the γ-quality of tetrahedron $J_1J_2J_3C_i$, the quality of the future tetrahedron elements, generated using triangular facets $C_iJ_2J_3$, $C_iJ_3J_1$, $C_iJ_1J_2$ as the basis (Fig. 7.64), must also be considered.

Let γ_i be the γ-quality of tetrahedron $J_1J_2J_3C_i$, i.e.

$$\gamma_i = \gamma(J_1J_2J_3C_i)$$

Let ξ_i, η_i and ζ_i be the γ-quality of the best tetrahedral element (without cutting into the generation front) that can be constructed using triangular facets $C_iJ_2J_3$, $C_iJ_3J_1$ and $C_iJ_1J_2$ as the basis respectively. Should triangular facet $C_iJ_2J_3$ happen to be on the generation front, $J_3J_2C_i \in \Gamma$, set $\xi_i = 1$; if $J_1J_3C_i \in \Gamma$, set $\eta_i = 1$; and if $J_2J_1C_i \in \Gamma$, set $\zeta_i = 1$. Defining

$$\lambda_i = \gamma_i\xi_i\eta_i\zeta_i$$

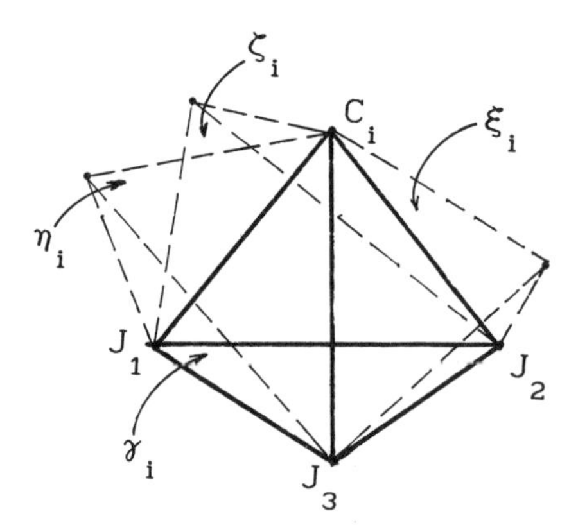

Fig. 7.64 Evaluating tetrahedra associated with node C_i

which measures the overall quality of the tetrahedron formed on triangular facet $J_1J_2J_3$ with node C_i and the future tetrahedra built on the new triangular facets as a consequence that C_i is selected. Hence, in the shape optimization process, with the quality of the tetrahedral elements possibly formed in the subsequent generation taken into account, the node to be selected is the node $C_m \in \mathcal{C}$ such that

$$\lambda_m = \gamma_m \xi_m \eta_m \zeta_m \quad \text{is maximized, i.e.} \quad \lambda_m \geq \lambda_i \qquad \forall\, C_i \in \mathbb{C}$$

(5) Updating the generation front

The generation front Γ has to be updated whenever a tetrahedral element $J_1J_2J_3C_m$ is formed. The four faces of the tetrahedron $J_1J_2J_3C_m$ have to be considered in turn, and Γ can be updated by

(i) removing the triangular facet $J_1J_2J_3$ from the generation front;
(ii) removing the triangular facet $J_3J_2C_m$ if $J_3J_2C_m \in \Gamma$, adding $C_mJ_2J_3$ to Γ, otherwise;
(iii) removing the triangular facet $J_1J_3C_m$, if $J_1J_3C_m \in \Gamma$, adding $C_mJ_3J_1$ to Γ, otherwise;
(iv) removing the triangular facet $J_2J_1C_m$, if $J_2J_1C_m \in \Gamma$, adding $C_mJ_1J_2$ to Γ, otherwise.

(6) Improving the quality of the tetrahedron elements by iteration

After triangulation, the mesh-smoothing process follows. It is done by shifting each interior node P to the centre of the surrounding polyhedron, P^*, which is given by

$$P^* = \frac{1}{n}\sum_{i=1}^{n} P_i$$

where P_i are the nodes connected to P through an edge of a tetrahedral element, and n is the number of points so connected to P.

To accelerate convergence, it is suggested that the shift be made 20% more than the calculated value. Two cycles of smoothing are usually sufficient to obtain a reasonably good distribution of interior nodes. Let γ_k be the γ-qualities of the tetrahedral elements connected to an interior node P, and γ_k^* be the new γ_k-values of the same set of tetrahedral elements associated with node P^*, taking into account the proposed nodal displacement. To ensure that an improvement is made for each iteration, the shift is carried out only if the product γ_k^* is greater than the corresponding product of γ_k, i.e.

$$\Pi\gamma_k^* > \Pi\gamma_k$$

(7) Examples

Examples of 3D triangulation of solid objects of different boundary characteristics are given in this section. All the examples presented are solid objects simply-connected or multi-connected, which have been triangulated as one single region without subdivision into simpler parts. These examples were run on the VAX/VMS 8600 computer, and the resulting meshes were direct copies from a Tektronix T4016 graphics terminal. For the examples given here, the input data consisting of randomly numbered triangular elements defining the boundary surfaces of the volume to be discretized are checked and processed by another independent program described in Reference 66. The computer time required for mesh generation of each example is also quoted for possible comparison.

Figure 7.65 shows the decomposition of a dumb-bell into tetrahedron elements. This is a simply-connected region defined by only a single closed surface. The number of elements in the mesh, $N_e = 521$; the number of node points in the mesh, $N_p = 208$; the geometric mean quality of the tetrahedra, $\bar{\gamma} = 0.7795$; and the CPU

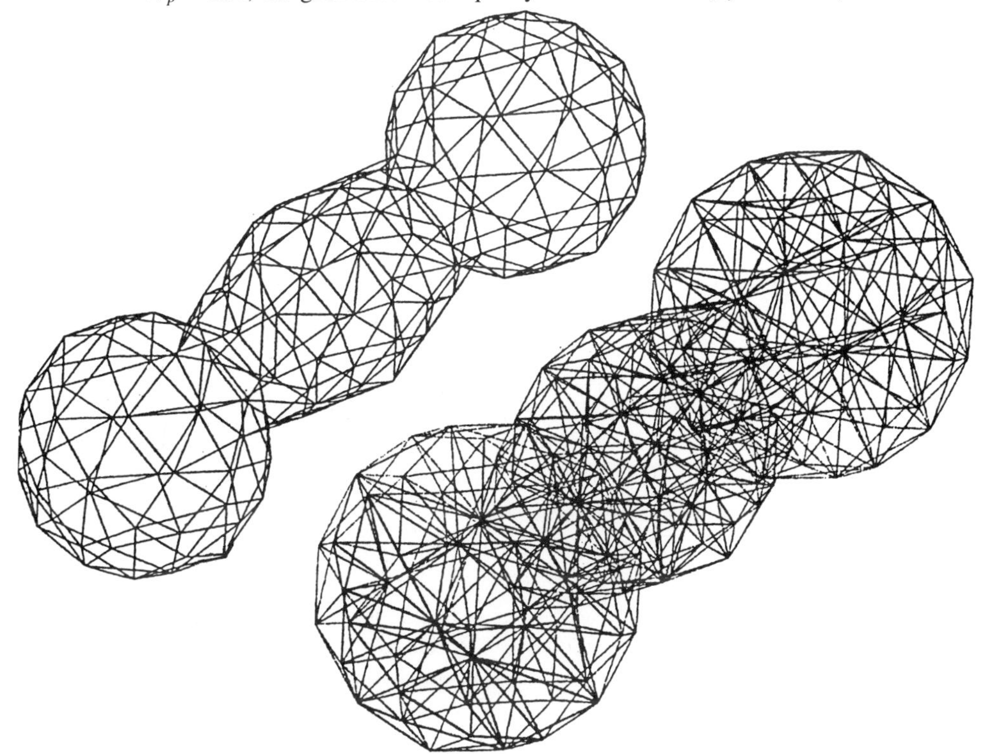

Fig. 7.65 Triangulation of a dumb-bell

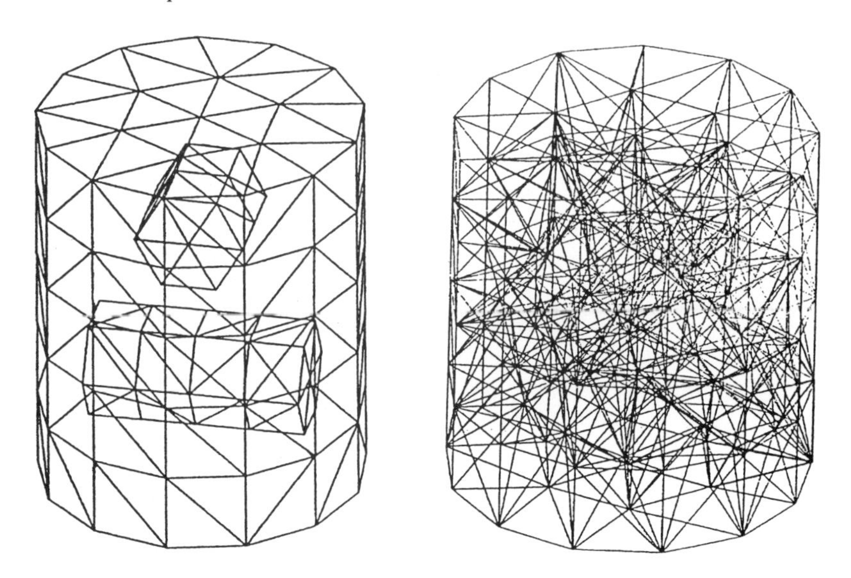

Fig. 7.66 Finite element mesh of a cylinder with two openings

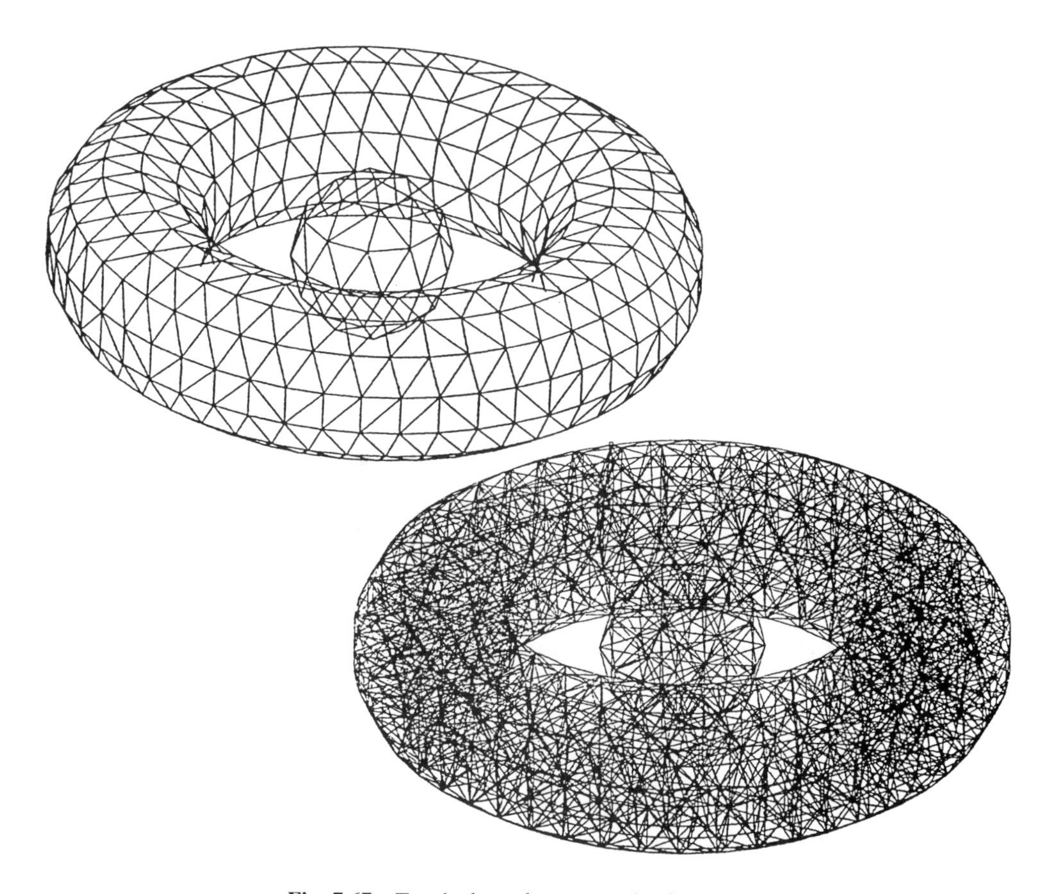

Fig. 7.67 Tetrahedron element mesh of a sphere and a torus

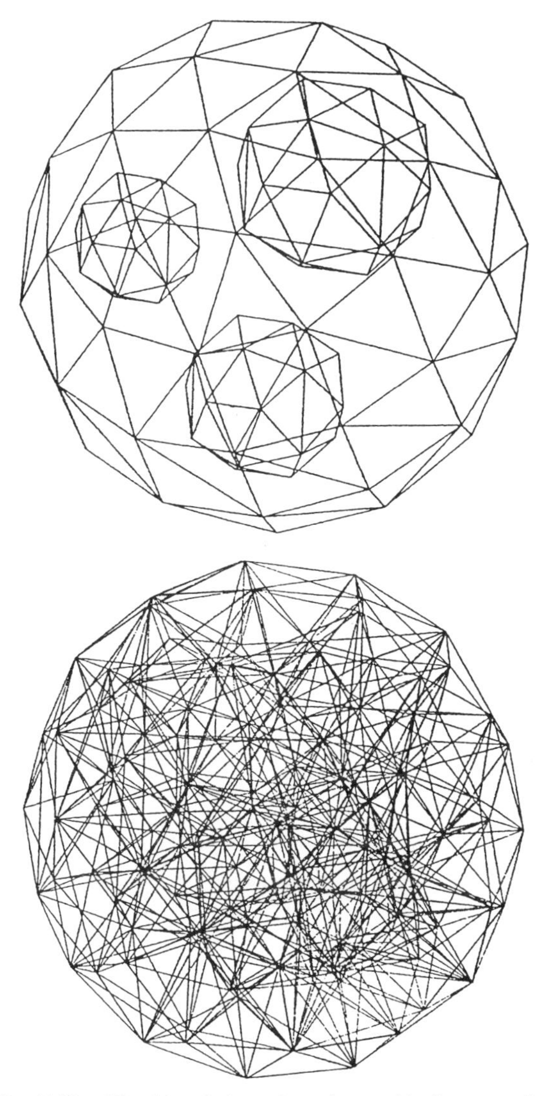

Fig. 7.68 3D triangulation of a sphere with three openings

time required for the generation of interior nodes, 3D triangulation, mesh smoothing and plotting amounted to 66.92 seconds.

In Fig. 7.66, a cylinder with two internal openings is discretized into tetrahedron elements. The presence of interior boundaries did not cause any difficulty in the 3D triangulation process. The characteristics of the resulting mesh are: $N_e = 560, N_p = 177, \overline{\gamma} = 0.6496$, CPU time = 105.38 seconds.

The 3D triangulation of a sphere and a torus is shown in Fig. 7.67. Although the two objects are disjointed, if they are fed into the mesh generator defined by the same set of boundary data, they will be treated as one single volume in the mesh generation process. The characteristics of the mesh are: $N_e = 3184, N_p = 949, \overline{\gamma} = 0.7961$, CPU time = 872.11 seconds.

The triangulation of a sphere with three openings into tetrahedral elements is shown in Fig. 7.68. The diameter of the outer sphere is 10 units whereas the diameters of the internal spherical openings are 2 units, 3 units and 4 units respectively. Again, no particular problem arose in the decomposition of this multi-connected domain. The characteristics of the mesh are: $N_e = 598, N_p = 175, \overline{\gamma} = 0.6902$, CPU time = 113.06 seconds.

The last example, as shown in Fig. 7.69, is a cylinder with three internal openings which are of different sizes and shapes. This is another example of the discretization of a multi-connected domain. The final mesh consists of 1889 tetrahedral elements and 564 nodal points. The geometric mean γ-quality of the mesh is 0.7278 and the total CPU time required for mesh generation and plotting is 371.53 seconds.

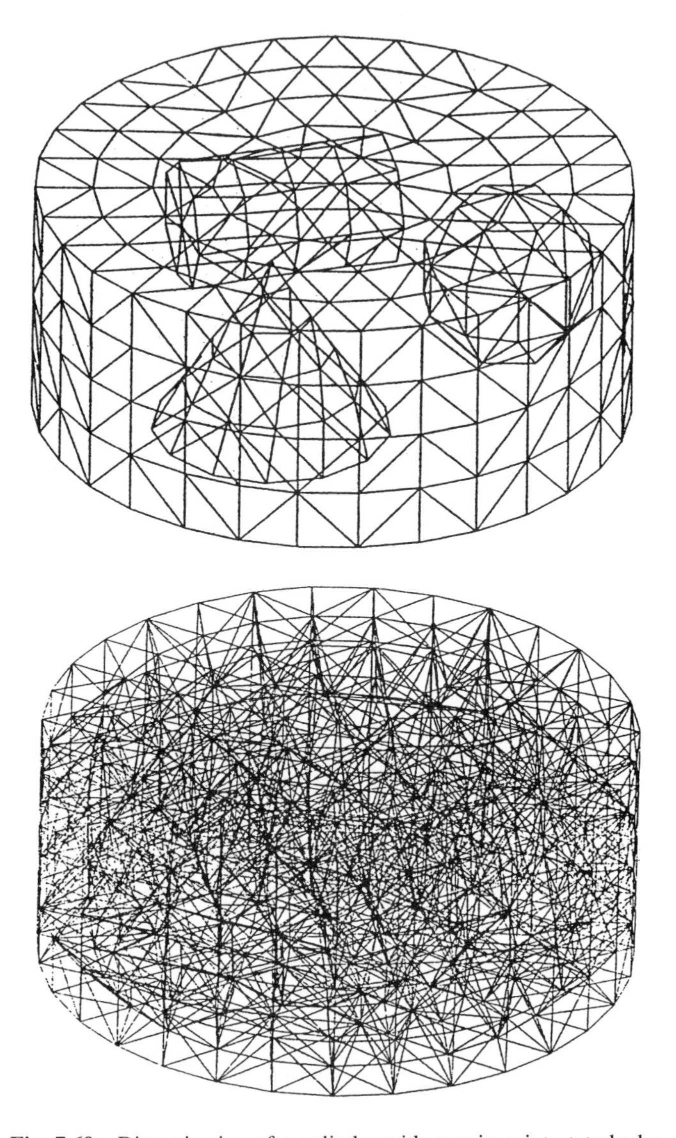

Fig. 7.69 Discretization of a cylinder with openings into tetrahedra

Figure 7.70 shows a transverse section of the finite element mesh shown in Fig. 7.69, revealing cut sections of tetrahedron elements and the three internal openings [64].

Fig. 7.70 Cross-section of the finite element mesh shown in Fig. 7.69

7.5 References to Chapter 7

1. B. Fredriksson & J. Mackerle (1976) *Structural mechanics finite element computer programs: survey and availability*, Rep. LITH-IKP-R-054, Linköping Institute of Technology, Linköping, Sweden.
2. O. C. Zienkiewicz & D. V. Phillips (1971) An automatic mesh generation scheme for plane and curved surfaces by isoparametric coordinates, *Int. J. Num. Meth. Engng*, **3**, 519–28.
3. F. Ghassemi (1978) *Computer aided design techniques in data processing for finite element analysis*, PhD thesis, Imperial College of Science and Technology, University of London.
4. G. D. Stefanou (1980) Automatic triangular mesh generation in flat plates for finite elements, *Computers & Structures*, **11**, 439–64.
5. W. J. Gordon & C. A. Hall (1973) Construction of curvilinear coordinate systems and applications to mesh generation, *Int. J. Num. Meth. Engng*, **7**, 461–77.
6. P. Ladeveze (1987) Optimal mesh for finite element analysis, *Conf. on Automatic Mesh Generation & Adaptation*, Grenoble, France.
7. J.-Ho Cheng (1988) Automatic adaptive remeshing for finite element simulation of forming process, *Int. J. Num. Methods Engng*, **26**, 1–18.

8. O. C. Zienkiewicz, J. Z. Zhu & N. G. Gong (1989) Effective and practical h-p version adaptive analysis procedures for the finite element method, *Int. J. Num. Methods Engng*, **28**, 879–91.
9. J. Fukuda & J. Suhara (1972) Automatic mesh generation for finite element analysis, in *Advances in Computational Methods in Structural Mechanics and Design*, ed. by J. J. Oden *et al.*, UAH Press, Huntsville, Alabama, USA.
10. J. C. Cavendish (1974) Automatic triangulation of arbitrary planar domains for the finite element method, *Int. J. Num. Methods Engng*, **8**, 679–96.
11. R. D. Shaw & R. G. Pitchen (1978) Modification to the Suhara–Fukuda method of network generation, *Int. J. Num. Methods Engng*, **12**, 93–9.
12. R. Sibson (1978) Locally equiangular triangulations, *Comp. J.*, **21**, 243–5.
13. W. Brostow & J. P. Dussault (1978) Construction of Voronoi polyhedra, *J. Comp. Phys.*, **29**, 81–92.
14. J. L. Finney (1979) A procedure for the construction of Voronoi polyhedra, *J. Comp. Phys.*, **32**, 137–43.
15. A. Bowyer (1981) Computing Dirichlet tessellations, *Comp. J.*, **24**, 162–6.
16. D. F. Watson (1981) Computing the n-dimensional Delaunay tessellation with application to Voronoi polytopes, *Comp. J.*, **24**, 167–72.
17. J. C. Cavendish, D. A. Field & W. H. Frey (1985) An approach to automatic three-dimensional finite element mesh generation, *Int. J. Num. Methods Engng*, **21**, 329–47.
18. S. H. Lo (1985) A new mesh generation scheme for arbitrary planar domains, *Int. J. Num. Methods Engng*, **21**, 1403–26.
19. S. H. Lo (1989) Delaunay triangulation of non-convex planar domains, *Int. J. Num. Methods Engng*, **28**, 2695–707.
20. J. Peraire, J. Peiro, K. Morgan & O. C. Zienkiewicz (1987) Finite element mesh generation and adaptation procedures for CFD, *Conf. on Automatic Mesh Generation and Adaptation*, Grenoble, France.
21. E. B. Becker & J. J. Brisbane (1965) *Application of the finite element method to stress analysis of solid propellant rocket grains*, Special Report No. 5–76, Rohm and Haas Co., Huntsville, Alabama, USA, Vol. 1.
22. W. R. Buell & B. A. Bush (1973) Mesh generation – a survey, *J. Eng. Ind., Trans. Amer. Soc. Mech. Engrs*, series B, **95**(1).
23. W. J. Gordon (1971) Blending-function methods of bivariate and multivariate interpolation and approximation, *Siam J. Num. Anal.*, **8**(1), 158–77.
24. S. H. Lo (1989) Generating quadrilateral elements on plane and over curved surfaces, *Computers & Structures*, **31**, No. 3, 421–6.
25. C. A. Hall (1976) Transfinite interpolation and applications to engineering problems, in *Theory of Approximation*, ed. by Law & Sahney, Academic Press, pp. 308–31.
26. E. E. Barnhill, T. Birkhoff & W. J. Gordon (1973) Smooth interpolation in triangles, *J. Approx. Theory*, **8**, 114–28.
27. W. A. Cook (1974) Body oriented (natural) co-ordinates for generating three-dimensional meshes, *Int. J. Num. Meth. Engng*, **8**, 27–43.
28. Carl-Erik Froberg, *Introduction to numerical analysis*, 2nd edn, Addison-Wesley.
29. L. R. Herrmann (1976) Laplacian-isoparamatic grid generation scheme, *J. Engng Mech. Div., ASCE*, **102**, 749–56.

30. A. Denayer (1978) Automatic generation of finite element meshes, *Computers & Structures*, **9**, 359–64.
31. S. A. Coons (1967) *Surfaces for computer-aided design of space forms*, Report MAC-TR-44, MIT, Cambridge, Mass, USA.
32. L. P. Chew (1987) *Guaranteed-quality triangulation for the finite element method*, IEEE Symp. on Foundation of Computer Science.
33. W. C. Thacker (1980) A brief review of techniques for generating irregular computational grids, *Int. J. Num. Meth. Engng*, **15**, 1335–41.
34. R. A. Dwyer (1986) A simple divide-and-conquer algorithm for constructing Delaunay triangulations in O(n loglogn) expected time, ACM 0-89791-194-6/86/0600/0276.
35. P. J. Green & R. Sibson (1978) Computing Dirichlet tessellations in the plane, *Comp. J.*, **21**, 168–73.
36. L. J. Guibas & J. Stolfi (1985) Primitives for the manipulation of general subdivisions and the computation of Voronoi diagrams, *ACM Trans. Graphics*, **4**, 74–123.
37. I. J. Gordon & C. L. Goodzeit (1980) COIFES – An efficient structural graphics program using the hidden-line techniques, *Computers & Structures*, **12**, 699–712.
38. W. J. Schrolder & M. S. Shephard (1988) Geometry-based fully automatic mesh generation and the Delaunay triangulation, *Int. J. Num. Meth. Engng*, **26**, 2503–15.
39. J. Peraire, J. Peiro, L. Formaggia, K. Morgan & O. C. Zienkiewicz (1988) Finite element Euler computations in three dimensions, *Int. J. Num. Meth. Engng*, **26**, 2135–59.
40. S. H. Lo (1991) Automatic mesh generation and adaptation by using contours, *Int. J. Num. Meth. Engng*, **31**, 689–707.
41. A. Bykat (1976) Automatic generation of triangular grids: I. subdivision of general polygon into convex subregions; II. triangulation of convex polygons, *Int. J. Num. Meth. Engng*, **10**, 1329–42.
42. E. A. Sadek (1980) A scheme for the automatic generation of triangular finite elements, *Int. J. Num. Meth. Engng*, **15**, 1813–22.
43. F. Ghassemi (1982) Automatic mesh generation scheme for a two or three dimensional triangular curved surface, *Computers & Structures*, **15**, 613–26.
44. S. Park & C. J. Washam (1979) Drag method as a finite element mesh generation scheme, *Computers & Structures*, **10**, 343–46.
45. S. Pissanetzky (1981) Kubik: an automatic three-dimensional finite element mesh generator, *Int. J. Num. Meth. Engng*, **17**, 255–69.
46. M. M. Hrabok & M. T. Hrudey (1984) A review and catalogue of plate bending finite elements, *Computers & Structures*, **19**, 479–95.
47. J. L. Batoz, K. J. Bathe & L. W. Ho (1980) A study of three-node triangular plate bending elements, *Int. J. Num. Meth. Engng*, **15**, 1771–1812.
48. K. J. Bathe & L. W. Ho (1981) A simple and efficient element for analysis of general shell structure, *Computers & Structures*, **13**, 673–81.
49. N. Carpenter, H. Stolarski & T. Belytschko (1986) Improvements in three-node triangular shell elements, *Int. J. Num. Meth. Engng*, **23**, 1643–67.
50. S. C. Wu & J. F. Abel (1979) Representation and discretization of arbitrary surfaces for finite element shell analysis, *Int. J. Num. Meth. Engng*, **14**, 813–36.
51. L. L. Durocher (1979) A versatile two-dimensional mesh generator with automatic band width reduction, *Computers & Structures*, **10**, 561–75.

52. H. A. Kamel & K. Eisenstein (1971) Automatic mesh generation in two-and three-dimensional interconnected domains, Les congres et colloques de l'universite de Liège, Belgium.
53. H. Ewetz & J. Oppelstrup (1987) *Multiblock mesh generation for 3D CFD applications*, Conference on Automatic Mesh Generation and Adaptation, 1-2 Oct., Grenoble, France.
54. Willy Fritz (1987) *Two-dimensional and three-dimensional block structured grid generation techniques*, Conference on Automatic Mesh Generation and Adaptation, 1-2 Oct., Grenoble, France.
55. I. Imafuku, Y. Kodera, M. Sayawaki & M. Kono (1980) A generalized automatic mesh generation scheme for finite element method, *Int. J. Num. Methods Engng*, **15**, 713-31.
56. A. Ecer, J. T. Spyropoulos & E. Bulbul (1987) *Application of a three-dimensional finite element grid generation scheme for analyzing transonic flow around an aircraft*, Conference on Automatic Mesh Generation and Adaptation, 1-2 Oct., Grenoble, France.
57. V. P. Nguyen (1982) Automatic mesh generation with tetrahedron elements, *Int. J. Num. Methods Engng*, **18**, 273-89.
58. W. J. Schroeder & M. S. Shephard (1988) Geometry-based fully automatic mesh generation and the Delaunay triangulation, *Int. J. Num. Methods Engng*, **26**, 2503-15.
59. M. A. Yerry & M. S. Shephard (1984) Automatic three-dimensional mesh generation by the modified-octree technique, *Int. J. Num. Methods Engng*, **20**, 1965-90.
60. M. A. Yerry & M. S. Shephard (1985) Automatic mesh generation for three-dimensional solids, *Computers & Structures*, **1-3**, 31-9.
61. G. F. Carey, M. Sharma & K. C. Wang (1988) A class of data structures for 2D and 3D adaptive mesh refinement, *Int. J. Num. Methods Engng*, **26**, 2607-22.
62. G. Hubert, *MOSAIC: Strategie et Algorithmes de Maillage*, Report, Compiegne Science Industrie 60200 Compiegne, France.
63. D. Chorlay, G. Hubert & G. Touzot, *GAM 3D: Un mailleur tridimensionnel a partir de la peau, Les systems de Calcul de Structures en CFAO*, ed. by A. Niku-Lari Editions Hermes, 51 rue Rennequin, 75017 Paris, France.
64. S. H. Lo (1990) Visualization of 3D solid finite element mesh by the method of sectioning, *Computers & Structures*, **35**, No. 1, 63-8.
65. W. H. Frey (1987) Selective refinement: A new strategy for automatic node placement in graded triangular meshes, *Int. J. Num. Methods Engng*, **24**, 2183-2200.
66. S. H. Lo (1991) Volume discretization into tetrahedra – I. verification and orientation of boundary surfaces, *Computers & Structures*, **39**, No. 5, 493-500.
67. S. H. Lo (1991) Volume discretization into tetrahedra – II. 3D triangulation by advancing front approach, *Computers & Structures*, **39**, No. 5, 501-11.
68. J. H. Argyris (1965) Tetrahedron elements with linearly varying strain for the matrix displacement method, *J. Royal Aero. Soc.*, 877-80.
69. O. C. Zienkiewicz (1977) *The finite element method*, 3rd edn, McGraw-Hill, London.
70. R. H. Crawford, W. N. Waggenspack & D. C. Anderson (1987) Composite mappings for planar mesh generation, *Int. J. Num. Methods Engng*, **24**, 2241-52.

71. R. Haber, M. S. Shephard, J. F. Abel, R. H. Gallagher & D. P. Greenberg (1981) A general two-dimensional finite element preprocessor utilizing discrete transfinite mappings, *Int. J. Num. Methods Engng*, **17**, 1015–44.
72. L. C. Wellford Jr. & M. R. Gorman (1988) A finite element transitional mesh generation procedure using sweeping functions, *Int. J. Num. Methods Engng*, **26**, 2623–34.
73. B. Joe & R. B. Simpson (1986) Triangular meshes for regions of complicated shape, *Int. J. Num. Methods Engng*, **23**, 751–78.
74. M. A. Yerry & M. S. Shephard (1983) A modified quadtree approach to finite element mesh generation, *I.E.E.E. J.*, 39–46, Jan./Feb.
75. S. H. Lo (1988) Finite element mesh generation over curved surfaces, *Computers & Structures*, **29**, No. 5, 731–42.
76. S. H. Lo, A. Y. T. Leung & Y. K. Cheung (1982) Automatic finite element mesh generation, *Proc. Int. Conf. on Finite Element Method*, Beijing, China, 931–7.
77. H. E. Febres-Cedillo & M. A. Bhatti (1988) A simple strain energy based finite element mesh refinement scheme, *Computers & Structures*, **28**, No. 4, 523–33.
78. O. C. Zienkiewicz, Y. C. Liu & G. C. Huang (1988) Error estimation and adaptivity in flow formulation for forming problems, *Int. J. Num. Methods Engng*, **25**, 23–42.
79. B. Palmerio (1987) *Recent developments in adaptive finite element calculations of compressible flows*, Conference on Automatic Mesh Generation and Adaptation, 1–2 Oct., Grenoble, France.
80. R. A. Ludwig, J. E. Flaherty, F. Guerinoni, P. L. Baehmann & M. S. Shephard (1988) Adaptive solutions of the Euler equations using finite quadtree and octree grids, *Computers & Structures*, **30**, No. 1/2, 327–36.
81. P. L. Baehmann, M. S. Shephard, R. A. Ashley & A. Jay (1988) Automated metal forming modeling utilizing adaptive remeshing and evolving geometry, *Computers & Structures*, **30**, No. 1/2, 319–25.
82. R. Lohner (1988) An adaptive finite element solver for transient problems with moving bodies, *Computers & Structures*, **30**, No. 1/2, 303–17.
83. G. F. Carey, M. Sharma, K. C. Wang & A Pardhanani (1988) Some aspects of adaptive grid computations, *Computers & Structures*, **30**, No. 1/2, 297–302.
84. J. H. Cheng & N. Kikuchi (1985) An analysis of metal forming process using large deformation elasto-plastic formulations, *Comp. Methods Appl. Mech. Engng*, **49**, 71–108.
85. I. Babuska & W. C. Rheinboldt (1978) Error estimates for adaptive finite element computations, *Siam J. Num. Anal.*, **15**, No. 4, 736–54, August.
86. O. C. Zienkiewicz, D. W. Kelly, J. Gago & I. Babuska (1982) *Hierarchical finite element approaches, error estimates and adaptive refinement*, ed. by J. Whiteman, MAFELP IV, Academic Press, New York, 313–46.
87. I. Babuska & W. C. Rheinboldt (1982) A survey of a posteriori error estimators and adaptive approach in the finite element method, *Proc. of China-France Symposium on finite element method*, Gordon and Breach, New York, 1–56.
88. B. A. Szabo (1986) Estimation and control of error based on p convergence, accuracy estimates and adaptive refinement in finite element computations, ed. by I. Babuska *et al.*, 61–78.

89. W. C. Rheinboldt (1985) Error estimates for nonlinear finite element computations, *Computers & Structures*, **20**, No. 1–3, 91–8.

90. J. L. Coulomb (1987) *2D and 3D mesh generation, experiment with the Delaunay's tesselation*, Conference on Automatic Mesh Generation and Adaptation, 1–2 Oct., Grenoble, France.

91. J. Y. Talon (1987) *Algorithmes pour l'amelioration topologique de maillages 2D et 3D*, Conference on Automatic Mesh Generation and Adaptation, 1–2 Oct., Grenoble, France.

8 Implementation

8.1 Introduction

From the theory presented in the previous chapters, we are now in a position to put all the components together and develop a finite element computer program for structural and field problems [8,12–21]. As has already been mentioned in Chapter 1, a finite element analysis can be divided roughly into the following three distinct phases:

1. Data input module and preprocessing.
2. Element evaluation, system assembly and solution.
3. Results output module and postprocessing.

The flow chart of the detailed breakdown of a finite element analysis run is shown in Fig. 8.1. The data input module must supply sufficient information to define the problem without ambiguity, so that a meaningful solution to the problem can be found. The data required for a finite element analysis usually consist of the geometrical definition of the problem domain, the material properties, and the boundary and loading conditions. Due to a substantial difference in the nature of the data types, it is recommended that geometry, material, boundary conditions and loading data are to be input using four separate modules.

Based on the input geometrical data, the structure has to be discretized into finite elements. This is best done by means of an automatic mesh generation program. Automation reduces errors and spares the user a lengthy and tedious chore. Solution accuracy may also be increased because a computer-generated mesh is more regular and is easier to optimize than one prepared manually. The mesh generation algorithms described in Chapter 7 would be useful for users to develop their own routines to generate meshes in both 2D and 3D situations.

The optimization of matrix profile by node renumbering is an important step in the resolution of a large system of simultaneous equations. Node labelling will not only reduce significantly the solution effort in the matrix decomposition, but will also allow the handling of problems of a much bigger size. A discussion on node relabelling schemes is given in Section 8.2, and the complete listing of an efficient computer coding for reduction of matrix profile is also provided.

The solution phase of the finite element analysis consists of five main steps:

1. Evaluation of the element stiffness matrix.
2. Assembly of the system stiffness matrix.
3. Introduction of boundary conditions.
4. Decomposition of the system stiffness matrix.
5. Solving for the system response due to the applied loading.

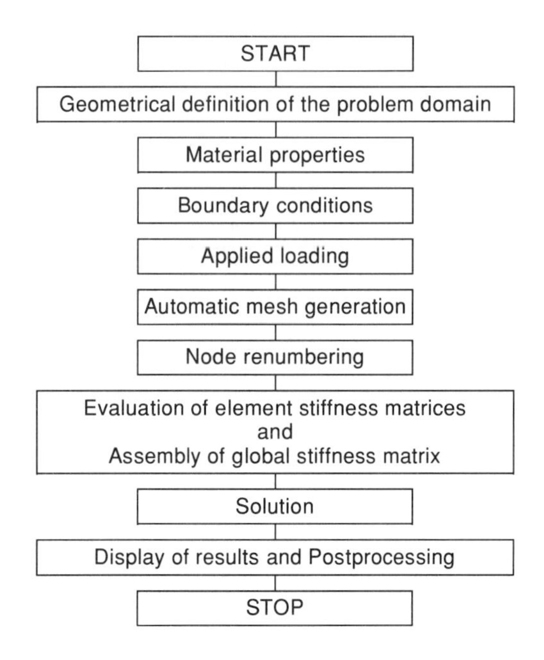

Fig. 8.1 Flow chart of a finite element analysis

The evaluation of the element stiffness matrix for different problem types has been discussed in previous sections. The procedures for the assembly of the system stiffness matrix are given in Section 8.4. The assembly process involving the use of out-of-core disk storage will also be discussed in the same section. The algorithm for the decomposition of the global stiffness matrix and the corresponding computer implementation may have a direct consequence on the efficiency of the finite element program. Accordingly, the problem is properly addressed in Section 8.3. The essential characteristics and the implementation procedures of a profile solver using out-of-core storage are also discussed. Finally, the introduction of boundary conditions and the evaluation of the right-hand side force vector are presented in Section 8.5.

In Section 8.6, a compact finite element program FACILE (Finite-element Analysis Code In Linear Elasticity) is introduced. In order to make things simple, the present version is confined to linear static analysis. It is hoped that through the practical implementation of the theory, readers can have a better understanding of the essential procedures of the finite element analysis. Readers should also realize that the program can be easily extended to dynamic and non-linear calculations by making appropriate modifications.

Owing to the large volume of information involved in the data and the results of a finite element analysis, input and output would best be done through the use of interactive computer colour graphics. An intelligent user-friendly preprocessor and postprocessor would be most desirable, and could greatly enhance the popularity of the finite element package.

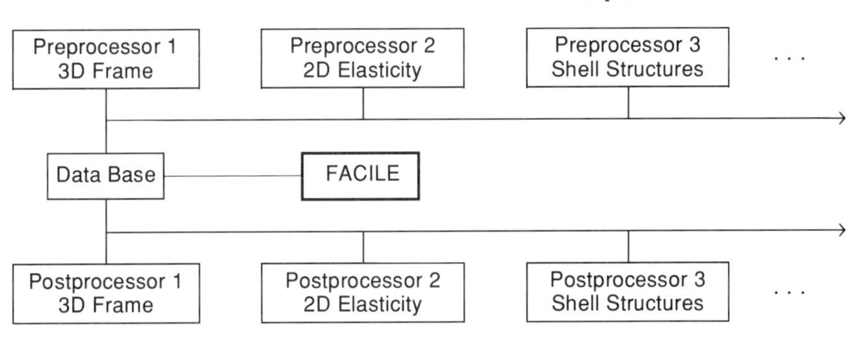

Fig. 8.2 Arrangement of preprocessors and postprocessors with program FACILE

No attempt has been made for a serious discussion on the preprocessing and postprocessing of the finite element analysis, which would, on its own, require a separate book of equal thickness. The only comment that the authors would like to make is that preprocessing and postprocessing should be tailored to a given problem type if they are to be successful and efficient. The geometry and the data required for a shell analysis are quite different from those of a 2D heat conduction problem. As a result, the preprocessing should be quite unique for a given problem type and especially designed to carry out its intended role. The same is also true for postprocessing, as the requirements and the presentation of results in graphical format for various problem types are remarkably different.

Fortunately, for a wide range of physical problems, some part of the analysis can be made common and executed more or less in the same way by a single computer program. This has been achieved in the program FACILE, which can handle all sorts of linear static problems. FACILE can be considered as the core of analysis, which is connected to the preprocessor and postprocessor of each problem type by means of data files in binary form prepared in the format compatible with the program FACILE. Making full use of the independent characteristics of the input and output modules, such a scheme would be a practical approach to build up a complex versatile system for finite element analysis. Figure 8.2 shows the arrangement of such a scheme.

8.2 Optimization of Matrix Profile

8.2.1 Introduction

In the application of the finite element method to the analysis of structural and continuum problems, a large set of linear simultaneous algebraic equations will be ultimately generated, which can be written in matrix form as

$$\mathbf{A}\mathbf{x} = \mathbf{b}$$

where $\mathbf{A}$ is a $n \times n$ matrix, $\mathbf{b}$ is a vector of length n and the unknown vector $\mathbf{x}$ is to be sought. In most applications, the matrix $\mathbf{A}$ is sparse, symmetrical and positive-definite. A method which is widely used in finite element analysis for solving a system of simultaneous equations is the profile solution scheme. In this scheme, the 'skyline' h_j of column j of matrix $\mathbf{A}$ is defined as the row index of the first non-zero entry of column j. Storage and operations of entries for column j are confined to those entries lying below the skyline up to and including the diagonal term.

The bandwidth of matrix **A** may be defined as

$$B = \max_{1 \leq j \leq n} \{b_j\}$$

where $b_j = j + 1 - h_j$, and the quantity

$$P = \sum_{j=1}^{n} b_j$$

is known as the profile of matrix **A**.

The storage requirement of the profile scheme is obviously less than that of the traditional bandwidth solution scheme in which all entries including zeros inside the envelope of the bandwidth are stored and operated on. In addition, the efficiency of the profile solution scheme is higher than that of the bandwidth solution scheme. Since n and B are large numbers, the number of arithmetic operations for the bandwidth solution scheme is roughly of order $\mathcal{O}(nB^2)$ while that for the profile solution scheme is roughly $\mathcal{O}(\sum b_j^2)$ which is always smaller than $\mathcal{O}(nB^2)$ as $b_j \leq B$ for $j = 1, n$. Also in order to optimize the performance of the profile method, P should be minimized. This can be achieved by relabelling the original numbering scheme in such a way that non-zero entries are clustered together towards the matrix diagonal.

It is especially important for meshes which are generated by an automatic mesh generation process [1], for which the original profile is extremely large. For problems in which the number of nodes of the finite element mesh is very large and the connectivity of nodes is high and complicated, it is impractical or just impossible to label the nodes manually. As a result, many automatic node relabelling algorithms have been developed and designed to minimize the matrix profile [2–5]. The best-known algorithms include the one due to Gibbs *et al.* [3], the reverse Cuthill–Mckee algorithm implemented by Everstine [5] and the one proposed by Sloan [6]. The one by Sloan appears to be the most effective algorithm available for profile reduction since it gives not only the best average performance but also yields much less erratic results in its worst performance [6].

8.2.2 Node renumbering program RENUM

The node renumbering program presented in this section is an improved version of Sloan's algorithm, which reduces the profile of a sparse, symmetric matrix. There is no limitation to the type of elements used in the finite element mesh and the number of nodes in an element need not be fixed. A mesh composed of several disjointed pieces can also be relabelled in exactly the same way as though it were a connected domain. Node numbers in the mesh can be quite arbitrary, which serve nothing more than a label to a particular node, and the only requirement is that they should be distinct.

The input and output variables of the program RENUM are as follows:

Input:
1. NN = Total number of nodes in the finite element mesh.
2. NE = Total number of elements in the finite element mesh.
3. ME(*) = Array of element node numbers, length = NME.
4. MP(*) = Index array for ME, length = NE + 1.
5. FN1 = File unit 1 : Error messages.

Output:
1. PV(*) = Permutation vector of node numbers,
   PV(I) = New node number of node I.
2. IPV(*) = Inverse permutation vector of node numbers,
   IPV(J) = Node number of new node J.

NEWME(*) = New element node numbers after renumbering.

4. OLDPRO = Old profile.
5. NEWPRO = New profile.

Working arrays: 1. ADJ(LADJ) = Array of node adjacency information, in which LADJ=NE*NV**2 where NV is the number of nodes in an element.
(Two nodes are adjacent if they are nodes of the same element.)
2. XADJ (MNODE), MNODE = The maximum node number of the mesh.
3. IWORK(*) = Working array of length 3*MNODE + 1.

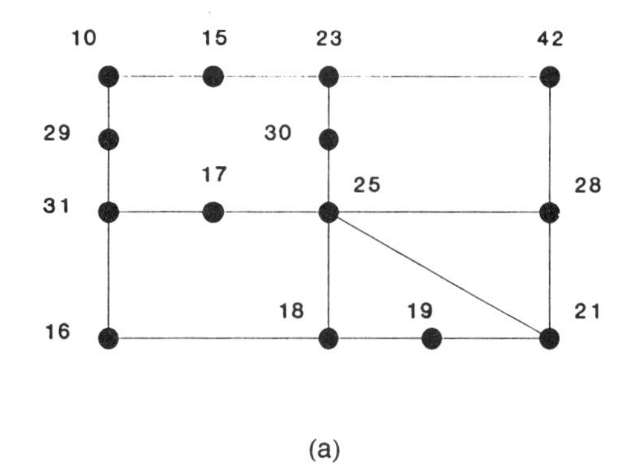

(a)

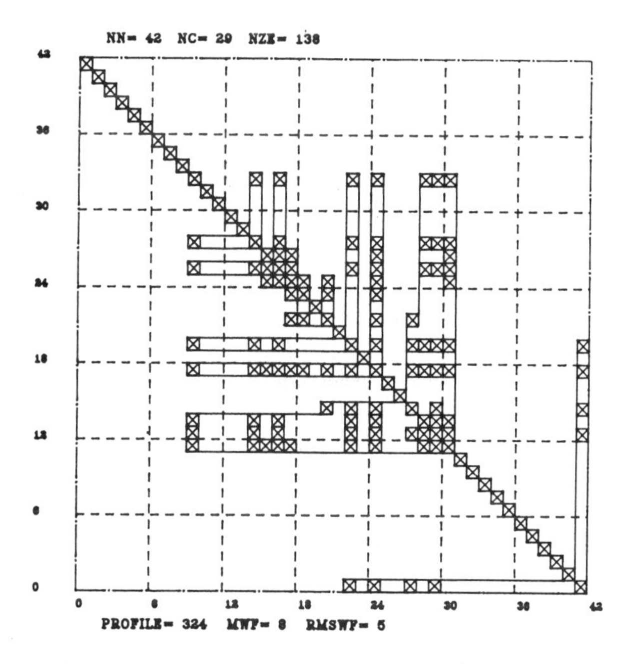

(b)

Fig. 8.3 Finite element mesh and matrix profile before renumbering

8.2.3 Example

An example of node renumbering is given in this section. The example selected is a simple mesh consisting of five elements and fourteen nodes as shown in Fig. 8.3(a), and the corresponding matrix pattern with a profile of 324 is shown in Fig. 8.3(b).

The input data for the relabelling problem are:

NN = 14
NE = 5
MP = [1 9 14 17 22 26]
ME = [10 29 31 17 25 30 23 15 23 30 25 28 42 25 21 23 31 16 18 25
17 18 19 21 25]

The profile of the mesh is reduced to 68 after node renumbering. The mesh of new node numbers is given in Fig. 8.4(a) and the corresponding matrix pattern can

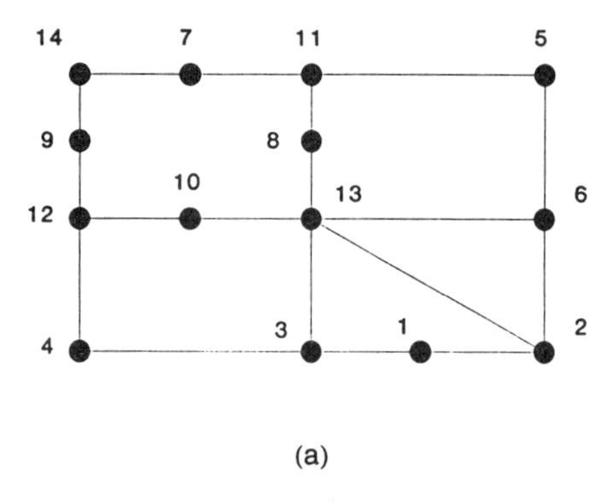

(a)

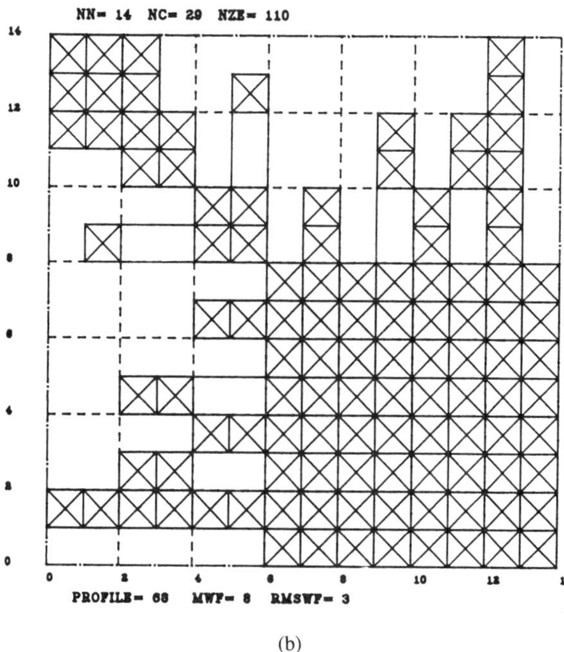

(b)

Fig. 8.4 Finite element mesh and matrix profile after renumbering

be found in Fig. 8.4(b). Since the solution time for the decomposition of a matrix is roughly proportional to the matrix profile raised to the power of 1.5, the effort to decompose the matrix of the relabelled mesh is only one tenth of that of the old mesh.

```
      SUBROUTINE RENUM (NN,NE,ME,NME,MP,FN1,PV,IPV,NEWME,OLDPRO,NEWPRO,
     -                  LADJ,ADJ,XADJ,IWORK)
      INTEGER NN,NE,NME,LADJ,OLDPRO,NEWPRO,MP(NE+1),ME(NME),NEWME(NME),
     -        PV(*),IPV(*),ADJ(LADJ),XADJ(*),IWORK(*),FN1,NOC,IMIN
C
C FUNCTION : MATRIX PROFILE REDUCTION BY RENUMBERING NODES OF A F. E. MESH
C
C INPUT :
C              NN = TOTAL NO. OF NODES   IN THE FINITE ELEMENT MESH
C              NE = TOTAL NO. OF ELEMENT IN THE FINITE ELEMENT MESH
C           ME(*) = ELEMENT NODE NUMBERS, NODES OF ELEMENT I ARE :
C                   ME(J), J=MP(I),....,MP(I+1)-1
C             NME = LENGTH OF ARRAY ME
C           MP(*) = INDEX ARRAY FOR ME, LENGTH = NE+1
C             FN1 = FILE UNIT 1 : ERROR MESSAGE
C
C OUTPUT :
C           PV(*) = PERMUTATION VECTOR OF NODE NUMBERS,
C                   PV(I) = NEW NODE NUMBER OF NODE I
C          IPV(*) = INVERSE PERMUTATION VECTOR OF NODE NUMBERS,
C                   IPV(J) = NODE NUMBER OF NEW NODE NUMBER J
C        NEWME(*) = NEW ELEMENT NODE NUMBERS AFTER RENUMBERING
C          OLDPRO = OLD PROFILE
C          NEWPRO = NEW PROFILE
C
C WORKING ARRAYS:
C                 ADJ(LADJ)  : LADJ = NE*NV**2,
C                              WHERE NV IS THE NUMBER OF NODES IN AN ELEMENT
C               XADJ(MNODE)  : MNODE = THE MAX. NODE NUMBER OF THE MESH
C                  IWORK(*)  = WORKING ARRAY OF LENGTH 3*MNODE +1
C
      MNODE=0
      DO 66 I=1,NME
      IF (ME(I).GT.MNODE) MNODE=ME(I)
   66 CONTINUE
      CALL GRAPH (FN1,MNODE,NE,MP(NE+1)-1,ME,MP,LADJ,ADJ,XADJ)
      CALL LABEL (MNODE,ADJ,XADJ,PV,IWORK,OLDPRO,NEWPRO,NOC)
      DO 11 I=1,MNODE
      IWORK(I)=0
   11 IPV(I)=0
      IMIN=MNODE
      DO 22 I=1,NE
      DO 22 J=MP(I),MP(I+1)-1
      NEWME(J)=PV(ME(J))
      IF (IMIN.GT.NEWME(J)) IMIN=NEWME(J)
      IF (IWORK(ME(J)).EQ.0) IWORK(ME(J))=1
   22 CONTINUE
      IF (IMIN.GT.1) THEN
      IMIN=IMIN-1
      DO 33 I=1,MNODE
      PV(I)=PV(I)-IMIN
      IF (PV(I).LT.0) PV(I)=0
   33 CONTINUE
      DO 44 I=1,NE
      DO 44 J=MP(I),MP(I+1)-1
   44 NEWME(J)=NEWME(J)-IMIN
      ENDIF
      DO 55 I=1,MNODE
      IF (IWORK(I).EQ.1) IPV(PV(I))=I
   55 CONTINUE
      OLDPRO=OLDPRO-MNODE+NN
      NEWPRO=NEWPRO-MNODE+NN
      RETURN
      END
```

```
      SUBROUTINE GRAPH (FN1,N,NE,INPN,NPN,XNPN,IADJ,ADJ,XADJ)
      INTEGER N,NE,NODEJ,NODEK,MSTRT,IADJ,I,J,K,JSTRT,JSTOP,LSTRT,
     -        LSTOP,L,NEN1,MSTOP,M,INPN,XNPN(*),ADJ(*),XADJ(*),NPN(*)
C
C     FUNCTION : FORM ADJACENCY LIST XADJ AND ADJ
C
      DO 11 I=1,IADJ
   11 ADJ(I)=0
      DO 22 I=1,N
   22 XADJ(I)=0
      DO 33 I=1,NE
      JSTRT=XNPN(I)
      JSTOP=XNPN(I+1)-1
      NEN1=JSTOP-JSTRT
      DO 33 J=JSTRT,JSTOP
      NODEJ=NPN(J)
      XADJ(NODEJ)=XADJ(NODEJ)+NEN1
   33 CONTINUE
      L=1
      DO 44 I=1,N
      L=L+XADJ(I)
   44 XADJ(I)=L-XADJ(I)
C
      XADJ(N+1)=L
      DO 55 I=1,NE
      JSTRT=XNPN(I)
      JSTOP=XNPN(I+1)-1
      DO 55 J=JSTRT,JSTOP-1
      NODEJ=NPN(J)
      LSTRT=XADJ(NODEJ)
      LSTOP=XADJ(NODEJ+1)-1
      DO 77 K=J+1,JSTOP
      NODEK=NPN(K)
      DO 88 L=LSTRT,LSTOP
      IF (ADJ(L).EQ.NODEK) GOTO 77
      IF (ADJ(L).EQ.0) GOTO 1
   88 CONTINUE
      WRITE (FN1,10)
      STOP
    1 ADJ(L)=NODEK
      MSTRT=XADJ(NODEK)
      MSTOP=XADJ(NODEK+1)-1
      DO 99 M=MSTRT,MSTOP
      IF (ADJ(M).EQ.0) GOTO 2
   99 CONTINUE
      WRITE (FN1,10)
      STOP
    2 ADJ(M)=NODEJ
   77 CONTINUE
   55 CONTINUE
C
      K=0
      JSTRT=1
      DO 111 I=1,N
      JSTOP=XADJ(I+1)-1
      DO 222 J=JSTRT,JSTOP
      IF (ADJ(J).EQ.0) GOTO 3
      K=K+1
  222 ADJ(K)=ADJ(J)
    3 XADJ(I+1)=K+1
  111 JSTRT=JSTOP+1
   10 FORMAT (//' *** ERROR IN GRAPH ***',
     -        //' CANNOT ASSEMBLE NODE ADJACENCY LIST',
     -        //' CHECK NPN AND XNPN ARRAYS')
      RETURN
      END

      SUBROUTINE LABEL (N,ADJ,XADJ,NNN,IW,OLDPRO,NEWPRO,NOC)
      INTEGER N,I1,I2,I3,I,SNODE,LSTNUM,OLDPRO,NEWPRO,E2,NOC,
     -        XADJ(*),ADJ(*),NNN(*),IW(*)
      E2=XADJ(N+1)-1
      DO 11 I=1,N
```

```
11 NNN(I)=0
   I1=1
   I2=I1+N
   I3=I2+N+1
   LSTNUM=0
   NOC=0
 1 NOC=NOC+1
   CALL DIAMTR (N,E2,ADJ,XADJ,NNN,IW(I1),IW(I2),IW(I3),SNODE,NC)
   CALL NUM (N,NC,SNODE,LSTNUM,E2,ADJ,XADJ,NNN,IW(I1),IW(I2))
   IF (LSTNUM.LT.N) GOTO 1
   CALL PROFIL (N,NNN,E2,ADJ,XADJ,OLDPRO,NEWPRO)
   IF (OLDPRO.LT.NEWPRO) THEN
   DO 22 I=1,N
22 NNN(I)=I
   NEWPRO=OLDPRO
   ENDIF
   RETURN
   END

   SUBROUTINE NUM (N,NC,SNODE,LSTNUM,E2,ADJ,XADJ,S,Q,P)
   INTEGER NC,LSTNUM,JSTRT,JSTOP,ISTOP,NBR,NABOR,I,J,NEXT,ADDRES,
  -        NODE,SNODE,ISTRT,MAXPRT,PRTY,N,W1,W2,E2,Q(NC),XADJ(*),
  -        NQ,ADJ(*),P(N),S(N),ACTIVE,PREACTIVE,INACTIVE
   DATA W1,W2,ACTIVE,PREACTIVE,INACTIVE/1,2,0,-1,-2/
   DO 11 I=1,NC
   NODE=Q(I)
   P(NODE)=W1*S(NODE)-W2*(XADJ(NODE+1)-XADJ(NODE)+1)
11 S(NODE)=INACTIVE
   NQ=1
   Q(NQ)=SNODE
   S(SNODE)=PREACTIVE
 1 ADDRES=1
   MAXPRT=P(Q(1))
   DO 22 I=2,NQ
   PRTY=P(Q(I))
   IF (PRTY.LE.MAXPRT) GOTO 22
   ADDRES=I
   MAXPRT=PRTY
22 CONTINUE
   NEXT=Q(ADDRES)
   Q(ADDRES)=Q(NQ)
   NQ=NQ-1
   ISTRT=XADJ(NEXT)
   ISTOP=XADJ(NEXT+1)-1
   IF (S(NEXT).EQ.PREACTIVE) THEN
   DO 33 I=ISTRT,ISTOP
   NBR=ADJ(I)
   P(NBR)=P(NBR)+W2
   IF (S(NBR).NE.INACTIVE) GOTO 33
   NQ=NQ+1
   Q(NQ)=NBR
   S(NBR)=PREACTIVE
33 CONTINUE
   ENDIF
   LSTNUM=LSTNUM+1
   S(NEXT)=LSTNUM
   DO 44 I=ISTRT,ISTOP
   NBR=ADJ(I)
   IF (S(NBR).NE.PREACTIVE) GOTO 44
   P(NBR)=P(NBR)+W2
   S(NBR)=ACTIVE
   JSTRT=XADJ(NBR)
   JSTOP=XADJ(NBR+1)-1
   DO 55 J=JSTRT,JSTOP
   NABOR=ADJ(J)
   P(NABOR)=P(NABOR)+W2
   IF (S(NABOR).NE.INACTIVE) GOTO 55
   NQ=NQ+1
   Q(NQ)=NABOR
   S(NABOR)=PREACTIVE
55 CONTINUE
```

```
44 CONTINUE
   IF (NQ.GT.0) GOTO 1
   RETURN
   END

   SUBROUTINE DIAMTR (N,E2,ADJ,XADJ,MASK,LS,XLS,HLEVEL,SNODE,NC)
   INTEGER NC,J,SNODE,DEGREE,MINDEG,ISTRT,ISTOP,HSIZE,NODE,JSTRT,
           JSTOP,EWIDTH,I,WIDTH,DEPTH,ENODE,N,SDEPTH,E2,

           XADJ(*),ADJ(*),XLS(N+1),LS(N),MASK(N),HLEVEL(N)
   MINDEG=N
   DO 11 I=1,N
   IF (MASK(I).NE.0) GOTO 11
   DEGREE=XADJ(I+1)-XADJ(I)
   IF (DEGREE.GE.MINDEG) GOTO 11
   SNODE=I
   MINDEG=DEGREE
11 CONTINUE
   CALL ROOTLS (N,SNODE,N+1,E2,ADJ,XADJ,MASK,LS,XLS,SDEPTH,WIDTH)
   NC=XLS(SDEPTH+1)-1
 1 HSIZE=0
   ISTRT=XLS(SDEPTH)
   ISTOP=XLS(SDEPTH+1)-1
   DO 22 I=ISTRT,ISTOP
   NODE=LS(I)
   HSIZE=HSIZE+1
   HLEVEL(HSIZE)=NODE
22 XLS(NODE)=XADJ(NODE+1)-XADJ(NODE)
   IF (HSIZE.GT.1) CALL KISORT (HSIZE,HLEVEL,N,XLS,1)
   ISTOP=HSIZE
   HSIZE=1
   DEGREE=XLS(HLEVEL(1))
   DO 33 I=2,ISTOP
   NODE=HLEVEL(I)
   IF (XLS(NODE).EQ.DEGREE) GOTO 33
   DEGREE=XLS(NODE)
   HSIZE=HSIZE+1
   HLEVEL(HSIZE)=NODE
33 CONTINUE
   EWIDTH=NC+1
   DO 44 I=1,HSIZE
   NODE=HLEVEL(I)
   CALL ROOTLS (N,NODE,EWIDTH,E2,ADJ,XADJ,MASK,LS,XLS,DEPTH,WIDTH)
   IF (WIDTH.GE.EWIDTH) GOTO 44
   IF (DEPTH.GT.SDEPTH) THEN
   SNODE=NODE
   SDEPTH=DEPTH
   GOTO 1
   ENDIF
   ENODE=NODE
   EWIDTH=WIDTH
44 CONTINUE
   IF (NODE.NE.ENODE)
  -   CALL ROOTLS (N,ENODE,NC+1,E2,ADJ,XADJ,MASK,LS,XLS,DEPTH,WIDTH)
   DO 55 I=1,DEPTH
   JSTRT=XLS(I)
   JSTOP=XLS(I+1)-1
   DO 55 J=JSTRT,JSTOP
55 MASK(LS(J))=I-1
   RETURN
   END

   SUBROUTINE ROOTLS (N,ROOT,MAXWID,E2,ADJ,XADJ,MASK,LS,XLS,
  -                  DEPTH,WIDTH)
   INTEGER ROOT,DEPTH,NBR,MAXWID,LSTRT,LSTOP,LWDTH,NODE,NC,WIDTH,N,
  -        JSTRT,JSTOP,I,J,E2,XADJ(*),ADJ(*),MASK(N),XLS(N+1),LS(N)
   MASK(ROOT)=1
   LS(1)=ROOT
   NC=1
   WIDTH=1
   DEPTH=0
   LSTOP=0
   LWDTH=1
```

```
 1 LSTRT=LSTOP+1
   LSTOP=NC
   DEPTH=DEPTH+1
   XLS(DEPTH)=LSTRT
   DO 11 I=LSTRT,LSTOP
   NODE=LS(I)
   JSTRT=XADJ(NODE)
   JSTOP=XADJ(NODE+1)-1

    DO 11 J=JSTRT,JSTOP
    NBR=ADJ(J)
    IF (MASK(NBR).EQ.0) THEN
    NC=NC+1
    LS(NC)=NBR
    MASK(NBR)=1
    ENDIF
11  CONTINUE
    LWDTH=NC-LSTOP
    WIDTH=MAX(LWDTH,WIDTH)
    IF (WIDTH.GE.MAXWID) GOTO 2
    IF (LWDTH.GT.0) GOTO 1
    XLS(DEPTH+1)=LSTOP+1
 2  DO 22 I=1,NC
22  MASK(LS(I))=0
    RETURN
    END

    SUBROUTINE KISORT (NL,LIST,NK,KEY,IDFG)
    INTEGER NL,NK,T,VALUE,LIST(NL),KEY(NK),IDFG
    IF (IDFG.EQ.1) THEN
    DO 11 I=2,NL
    T=LIST(I)
    VALUE=KEY(T)
    DO 22 J=I-1,1,-1
    IF (VALUE.LT.KEY(LIST(J))) GOTO 22
    LIST(J+1)=T
    GOTO 11
22  LIST(J+1)=LIST(J)
    LIST(1)=T
11  CONTINUE
    ELSE
    DO 33 I=2,NL
    T=LIST(I)
    VALUE=KEY(T)
    DO 44 J=I-1,1,-1
    IF (VALUE.GT.KEY(LIST(J))) GOTO 44
    LIST(J+1)=T
    GOTO 33
44  LIST(J+1)=LIST(J)
    LIST(1)=T
33  CONTINUE
    ENDIF
    RETURN
    END

    SUBROUTINE PROFIL (N,NNN,E2,ADJ,XADJ,OLDPRO,NEWPRO)
    INTEGER NEWPRO,I,J,N,JSTRT,JSTOP,OLDPRO,NEWMIN,OLDMIN,E2,
   -         NNN(N),XADJ(N+1),ADJ(E2)
    OLDPRO=N
    NEWPRO=N
    DO 11 I=1,N
    JSTRT=XADJ(I)
    JSTOP=XADJ(I+1)-1
    OLDMIN=ADJ(JSTRT)
    NEWMIN=NNN(ADJ(JSTRT))
    DO 22 J=JSTRT+1,JSTOP
    OLDMIN=MIN(OLDMIN,ADJ(J))
22  NEWMIN=MIN(NEWMIN,NNN(ADJ(J)))
    OLDPRO=OLDPRO+DIM(I,OLDMIN)
11  NEWPRO=NEWPRO+DIM(NNN(I),NEWMIN)
    RETURN
    END
```

8.3 SOLUTION OF A SYSTEM OF LINEAR EQUATIONS

8.3.1 Introduction

The solution of the system $\mathbf{Kx} = \mathbf{f}$ is the most important step in the finite element analysis. The number of unknowns n is directly proportional to the number of nodes and the number of degrees of freedom per node. The accuracy and range of application of the finite element method is limited only by the number of simultaneous linear equations that can be solved economically with today's computers.

Methods of solutions are generally divided into two broad classes:

(a) direct methods which are also called Gaussian elimination methods, and
(b) iterative methods of which the Gauss–Seidel variation is the most popular.

During the early years of development of finite element analysis, iterative techniques were thought to be more effective and easier to program than direct elimination methods. However, further research in algorithmic solutions and computer programming resulted in a near-unanimous choice of elimination methods because of their near-optimum accuracy and speed for general systems.

Gaussian elimination is widely used as a direct solution procedure. The most common variation of the Gaussian elimination techniques divide the operations into two steps.

(a) Triangulation of the matrix coefficients – a systematic series of linear transformations is applied to the system of equations in order to reduce it to a triangular system.
(b) For an upper triangular system, the last equation contains only one unknown and is readily solved. The previous equation contains two unknowns, one of which has just been calculated. By successive substitutions, the remaining unknowns can be found. The process is repeated for each line of the system in sequence from bottom to top.

Gaussian elimination can be reformulated into a two-phase process that does not require a simultaneous modification of $\mathbf{K}$ and $\mathbf{f}$. Such a process, referred to as matrix decomposition, matrix factorization or triangulation is widely used in finite element analysis. In studying the available matrix decomposition algorithms, it is apparent that the amount of stiffness matrix coefficient data which is actively being processed at any time is a very small percentage of the overall coefficient data vector. Indeed, when decomposing coefficients of a given stiffness equation, only those columns having coefficients which intersect the particular equation column are needed [7]. As a result, elimination methods for matrices arising from finite element analyses can be classified as:

(a) methods which apply when all equations can be contained in core storage,
(b) methods which apply if the active coefficients can be contained in core, and
(c) methods which apply when there is insufficient core storage to contain the active coefficients.

In this section, matrix reduction methods belonging to categories (a) and (b) are discussed. The stiffness coefficients are stored in a linear array using the efficient skyline profile storage scheme. In the skyline profile storage method, the matrix coefficients are partitioned in columns and are stored column by column. The coefficients stored in a linear array are addressed indirectly through the use of a pointer vector, which indicates the position of the last element of each matrix column in the one-dimensional storage form.

Without dividing the equations into blocks, the efficiency of the working vector can be much improved by a dynamic storage scheme to hold as many equations as possible at any one time. Compared to the blocked-skyline method, the total number of data transfer from core to disk and vice versa is reduced and the amount of active coefficients being processed in a single data transfer is maximized. The matrix decomposition without using direct-access files is first presented to explain how triangulation is done on a one-dimensional working array using skyline profile storage. Its extension to allow for out-of-core storage is then discussed in detail. Computer codes for the profile solver using out-of-core storage are included in Section 8.3.5.

8.3.2 Matrix decomposition

A square matrix $\mathbf{K}$ can be decomposed into a lower triangular matrix $\mathbf{L}$ and an upper triangular matrix $\overline{\mathbf{U}}$. The matrix $\overline{\mathbf{U}}$ can be further decomposed into a diagonal matrix $\mathbf{D}$ and an upper triangular matrix $\mathbf{U}$ whose diagonal elements are equal to 1. Moreover, if the matrix $\mathbf{K}$ is symmetric, the matrices $\mathbf{L}$ and $\mathbf{U}$ are the transpose of each other.

(1)* $\mathbf{L}\overline{\mathbf{U}}$ *decomposition

By carrying out direct matrix multiplication, it can be easily verified that the matrix $\mathbf{K}$ can be decomposed into a lower triangular matrix $\mathbf{L}$ with diagonal elements equal to unity and an upper triangular matrix $\overline{\mathbf{U}}$ such that $\mathbf{L}\overline{\mathbf{U}} = \mathbf{K}$, i.e.

$$\begin{bmatrix} 1 & & & & & \\ \ell_{21} & 1 & & & & \\ \ell_{31} & \ell_{32} & 1 & & 0 & \\ \vdots & & & \ddots & & \\ \ell_{i1} & \ell_{i2} & \rightarrow & \ell_{i,i-1} & 1 & \\ \vdots & \vdots & & & & \ddots \\ \ell_{n1} & \ell_{n2} & & \cdots & \ell_{n,n-1} & 1 \end{bmatrix} \begin{bmatrix} \bar{u}_{11} & \bar{u}_{12} & \cdots & \bar{u}_{1j} \cdots & \bar{u}_{1n} \\ & \bar{u}_{22} & \cdots & \bar{u}_{2j} \cdots & \bar{u}_{2n} \\ & & \ddots & \downarrow & \vdots \\ & & & \bar{u}_{jj} & \cdot \\ & & & \ddots & \vdots \\ 0 & & & & \bar{u}_{nn} \end{bmatrix}$$

$$= \begin{bmatrix} k_{11} & k_{12} & \cdots & k_{1j} \cdots & k_{1n} \\ k_{21} & k_{22} & \cdots & k_{2j} \cdots & k_{2n} \\ \cdot & & & \cdot & \cdot \\ \cdot & & & k_{ij} & \cdot \\ \vdots & & & \vdots & \vdots \\ & & & \cdot & \cdot \\ k_{n1} & & & & k_{nn} \end{bmatrix}$$

$$\text{where } \bar{u}_{ij} = k_{ij} - \sum_{m=1}^{i-1} \ell_{im}\bar{u}_{mj} \qquad j = i, n; \; i = 1, n \qquad (8.1)$$

$$\ell_{ij} = \left(k_{ij} - \sum_{m=1}^{j-1} \ell_{im}\overline{u}_{mj} \right) / \overline{u}_{jj} \qquad i = j+1, n; \;\; j = 1, n \tag{8.2}$$

(2) **LDU** *decomposition*

Let $\overline{\mathbf{U}} = \mathbf{DU}$, where $\mathbf{D}$ is a diagonal matrix $\lceil d_1, d_2, \ldots, d_n \rfloor$ and $\mathbf{U}$ is an upper triangular matrix whose diagonal elements are equal to 1, then

$$d_i = \overline{u}_{ii} \qquad u_{ij} = \overline{u}_{ij}/d_i \qquad j = i, n; \;\; i = 1, n \tag{8.3}$$

(3) Crout decomposition

By carrying out the matrix decomposition in the last section,

$$\mathbf{K} = \mathbf{LDU} \quad \text{and} \quad \mathbf{K}^{\mathrm{T}} = \mathbf{U}^{\mathrm{T}}\mathbf{D}\mathbf{L}^{\mathrm{T}} \tag{8.4}$$

for a symmetrical matrix $\mathbf{K}$, $\mathbf{K}^{\mathrm{T}} = \mathbf{K}$, and we have

$$\mathbf{U}^{\mathrm{T}}\mathbf{D}\mathbf{L}^{\mathrm{T}} = \mathbf{LDU} \tag{8.5}$$

Since matrix decomposition is unique [23], $\mathbf{U}^{\mathrm{T}} = \mathbf{L}$ and $\mathbf{L}^{\mathrm{T}} = \mathbf{U}$, and a symmetric matrix $\mathbf{K}$ can be expressed as

$$\mathbf{K} = \mathbf{L}\mathbf{D}\mathbf{L}^{\mathrm{T}} = \mathbf{U}^{\mathrm{T}}\mathbf{D}\mathbf{U} \tag{8.6}$$

(4) Cholesky decomposition

If $\mathbf{K}$ is positive-definite, then all the diagonal elements d_i of diagonal matrix $\mathbf{D}$ are greater than zero. Let $\mathbf{C} = \lceil \sqrt{d_1}, \sqrt{d_2}, \ldots, \sqrt{d_n} \rfloor$, then $\mathbf{C}^2 = \mathbf{D}$. It follows that

$$\mathbf{K} = \mathbf{L}\mathbf{D}\mathbf{L}^{\mathrm{T}} = \mathbf{LCCL}^{\mathrm{T}} = (\mathbf{LC})(\mathbf{LC})^{\mathrm{T}} = \mathbf{L}_c\mathbf{L}_c^{\mathrm{T}} \tag{8.7}$$

where the lower triangular Cholesky matrix $\mathbf{L}_c = \mathbf{LC}$.

(5) Reduction column by column

In a finite element analysis, the resulting stiffness matrix $\mathbf{K}$ is always symmetric. Using Crout decomposition, the matrix $\mathbf{K}$ can be decomposed into the product of a lower triangular matrix $\mathbf{L}$, a diagonal matrix $\mathbf{D}$ and an upper triangular matrix $\mathbf{U}$, in which $\mathbf{L}$ and $\mathbf{U}$ are the transpose of each other. As a result, only either $\mathbf{L}$ or $\mathbf{U}$ needs to be considered in the matrix decomposition process. If we concentrate on the upper half of the symmetric matrix $\mathbf{K}$, the matrix decomposition can be conveniently considered as a column by column reduction process. For instance, column j will then consist of all the elements from the top of the matrix down to and including the diagonal term, $k_{1j}, k_{2j}, \ldots, k_{jj}$, as shown in Fig. 8.5.

With $\ell_{im} = u_{mi}$, eq. 8.1, which is used for the reduction of column j, can be written as

$$\overline{u}_{ij} = k_{ij} - \sum_{m=1}^{i-1} u_{mi}\overline{u}_{mj} \qquad i = 1, j-1 \tag{8.8}$$

This equation, which is used for the determination of $\overline{u}_{ij}$, can also be interpreted as the result of equating the dot product of the vectors in columns i and j to coefficient k_{ij} (Fig. 8.5). By putting $i = j$, the equation for the reduction of diagonal element d_j is given by

$$d_j = k_{jj} - \sum_{m=1}^{j-1} u_{mj}\overline{u}_{mj} \tag{8.9}$$

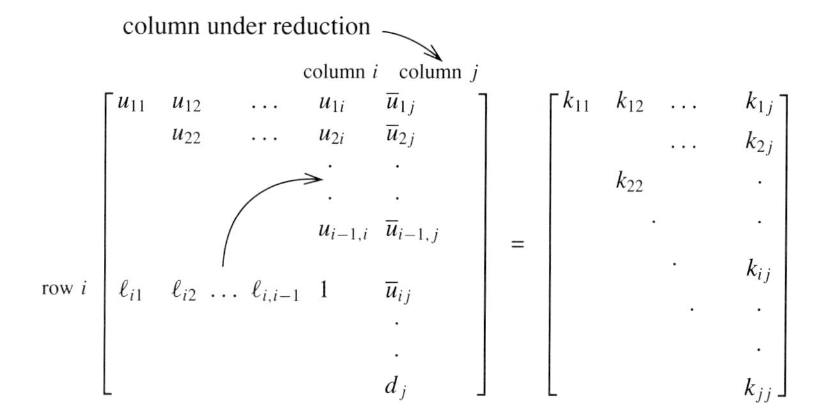

Fig. 8.5 Reduction of upper triangular matrix column by column

The reduction for column j is complete when u_{mj} are obtained from $\overline{u}_{mj}$ by dividing them with the diagonal elements d_m such that

$$u_{mj} = \overline{u}_{mj}/d_m \qquad m = 1, j-1 \tag{8.10}$$

***(6) Solving for* x**

The solution for vector $\mathbf{x}$ in the equation $\mathbf{Kx} = \mathbf{f} \Leftrightarrow (\mathbf{LDU})\mathbf{x} = \mathbf{f}$ is obtained in three steps.

(a) Forward substitution

Let $\mathbf{z} = (\mathbf{DU})\mathbf{x}$, then

$$\mathbf{Lz} = \mathbf{f}$$

for which $\mathbf{z}$ can be solved easily by forward substitution

$$z_i = f_i - \sum_{j=1}^{i-1} \ell_{ij} z_j \qquad i = 1, n$$

$$\begin{bmatrix} 1 & & & & & \\ \ell_{21} & 1 & & & 0 & \\ \vdots & & \cdot & & & \\ \vdots & & & \cdot & & \\ \ell_{i1} & \dots & \ell_{i,i-1} & 1 & & \\ \vdots & & & & \cdot & \\ \vdots & & & & & \cdot \\ \ell_{n1} & & \dots & & \ell_{n,n-1} & 1 \end{bmatrix} \begin{bmatrix} z_1 \\ z_2 \\ \vdots \\ z_i \\ \vdots \\ z_n \end{bmatrix} = \begin{bmatrix} f_1 \\ f_2 \\ \vdots \\ f_i \\ \vdots \\ f_n \end{bmatrix}$$

where $\ell_{ij} = u_{ji}$.

(b) Division by diagonal elements

Writing $\mathbf{y} = \mathbf{Ux}$, then

$$\mathbf{Dy} = \mathbf{z}$$

from which

$$y_i = z_i/d_i \qquad i = 1, n$$

$$\begin{bmatrix} d_1 & & & & & & \\ & d_2 & & & 0 & & \\ & & \cdot & & & & \\ & & & \cdot & & & \\ & & & & d_i & & \\ & & & & & \cdot & \\ & 0 & & & & & \cdot \\ & & & & & & d_n \end{bmatrix} \begin{bmatrix} y_1 \\ y_2 \\ \vdots \\ y_i \\ \vdots \\ y_n \end{bmatrix} = \begin{bmatrix} z_1 \\ z_2 \\ \vdots \\ z_i \\ \vdots \\ z_n \end{bmatrix}$$

(c) Backward substitution
Finally, when vector $\mathbf{y}$ is known, vector $\mathbf{x}$ can be solved from the equation $\mathbf{Ux} = \mathbf{y}$ by backward substitution, such that

$$x_i = y_i - \sum_{j=i+1}^{n} u_{ij} x_j \quad i = n, 1$$

$$\begin{bmatrix} 1 & \cdot & \cdot & \cdot & \cdot & u_{1n} \\ & 1 & & & & \cdot \\ & & \cdot & & & \cdot \\ & & & \cdot & & \cdot \\ & & & 1 & u_{i,i+1} \cdots & u_{in} \\ & 0 & & & \cdot & \vdots \\ & & & & \cdot & \\ & & & & & 1 \end{bmatrix} \begin{bmatrix} x_1 \\ x_2 \\ \vdots \\ x_i \\ \vdots \\ x_n \end{bmatrix} = \begin{bmatrix} y_1 \\ y_2 \\ \vdots \\ y_i \\ \vdots \\ y_n \end{bmatrix}$$

8.3.3 Skyline storage scheme and column reduction

(1) Skyline storage scheme
The most efficient strategy for the storage of non-zero elements of an upper triangular matrix is to store the matrix coefficients column by column of variable lengths marked off by a pointer vector on a linear array. The profile of the skyline is the envelope of the columns of variable heights. Zero terms outside the skyline envelope need not be stored as they will remain zero in the reduction process; however zero terms within the skyline must be stored as these terms are not invariants in the matrix decomposition process. Considering the symmetric 8×8 matrix shown in Fig. 8.6, the number of equations n equals 8 and a linear array $\mathbf{D}$ of dimension 8 is required to store the diagonal elements $k_{11}, k_{22}, \ldots, k_{88}$.

$$\mathbf{D} = [d_1, d_2, \ldots, d_8] = [k_{11}, k_{22}, \ldots, k_{88}]$$

There are 17 off-diagonal elements within the skyline profile, and a linear array $\mathbf{A}$ of dimension 17 is required to hold all these elements as illustrated in Fig. 8.6, with $\mathbf{A} = [a_1, a_2, \ldots, a_{17}] = [k_{12}, k_{23}, \ldots, k_{78}]$.

In addition, an auxiliary pointer $\mathbf{P}$ of dimension $n = 8$ is needed to marked off each column on the linear array $\mathbf{A}$, such that P_i is the number of the off-diagonal element up to column i. Hence, the total memory required to store the symmetric matrix $\mathbf{K}$ equals

$$\begin{aligned} \text{Dim}\ (\mathbf{D}) + \ \text{Dim}\ (\mathbf{A}) + \ \text{Dim}\ (\mathbf{P}) &= n + n + N \\ &= 2n + N \end{aligned}$$

where N is the number of off-diagonal elements under the skyline.

With the aid of the pointer vector $\mathbf{P}$, it would be easy to calculate the corresponding address of coefficients k_{ij} on the linear vector $\mathbf{A}$. Since P_j corresponds to the address of the last element $k_{j-1,j}$ of column j, the address of element k_{ij} is given by

$$k_{ij} = a_{J+i} \tag{8.11}$$

where

$$J = P_j - j + 1 \tag{8.12}$$

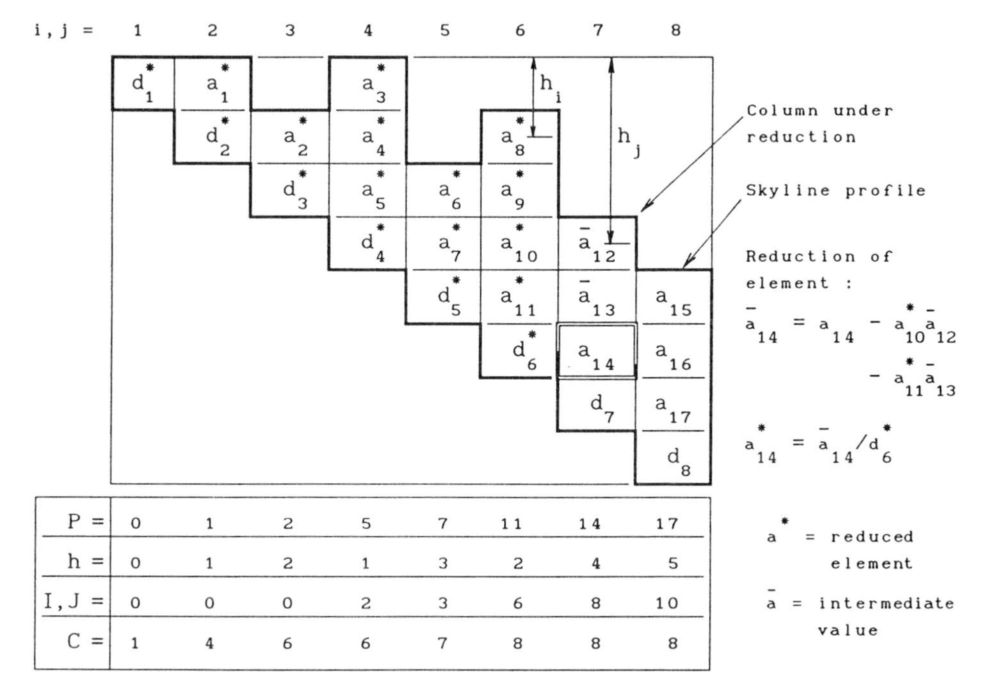

P =	0	1	2	5	7	11	14	17
h =	0	1	2	1	3	2	4	5
I, J =	0	0	0	2	3	6	8	10
C =	1	4	6	6	7	8	8	8

Fig. 8.6 Skyline profile storage scheme – column 7 under reduction

(2) Column reduction

(a) Column reduction within the skyline profile

Since zeros outside the skyline profile are ignored, in the calculation of the sum $\sum_{m=1}^{i-1} u_{mi}\overline{u}_{mj}$, the number of arithmetic operations can be reduced by adjusting the lower summation limit for m. Instead of beginning the summation from $m = 1$, m starts from a value h given by

$$h = \max\,(h_i, h_j) \tag{8.13}$$

where h_i and h_j are the heights measured from the first row to the first non-zero elements in column i and column j respectively, as shown in Fig. 8.6. Heights h_i and h_j can be calculated from pointer vector **P**,

$$h_i = P_{i-1} - P_i + i \qquad h_j = P_{j-1} - P_j + j \tag{8.14}$$

In the reduction of column j, we now have

$$\left.\begin{aligned} &\overline{u}_{ij} = k_{ij} - \sum_{m=h}^{i-1} u_{mi}\overline{u}_{mj} \qquad u_{ij} = \overline{u}_{ij}/d_i \qquad i = h_j, j-1 \\ &d_j = k_{jj} - \sum_{m=h_j}^{j-1} u_{mj}\overline{u}_{mj} \end{aligned}\right\} j = 1, n \tag{8.15}$$

(b) Sequence of column reduction

To optimize storage space and arithmetic operations, the same storage will be used to store the matrix coefficients before and after column reduction. The diagonal and off-diagonal elements are reduced in the following sequence:

$$\left.\begin{aligned}
\overline{u}_{ij} &= u_{ij} - \sum_{m=h}^{i-1} u^*_{mi}\overline{u}_{mj} && i = h_j, j-1 \\
d^*_j &= d_j - \sum_{m=h_j}^{j-1} \overline{u}^2_{mj}/d^*_m && \\
u^*_{ij} &= \overline{u}_{ij}/d^*_i && i = h_j, j-1
\end{aligned}\right\} j = 1, n \qquad (8.16)$$

where d_j, u_{ij} are the initial values of the matrix coefficients, $\overline{u}_{ij}$ are the intermediate values, and d^*_j, u^*_{ij} are the reduced values.

(c) Column reduction using a linear array

The initial values u_{ij}, which are in fact the coefficients of the upper triangular part of stiffness matrix k_{ij} within the skyline profile, are stored in a linear array **A**. Their corresponding address can be calculated by

$$u_{ij} = A_{J+i} \qquad u_{mj} = A_{J+m} \qquad u_{mi} = A_{I+m} \qquad (8.17)$$

where $I = P_i - i + 1$ and $J = P_j - j + 1$.

In terms of the elements of the linear arrays **A** and **D**, the equations for column reduction take the form

$$\left.\begin{aligned}
\overline{A}_{J+i} &= A_{J+i} - \sum_{m=h}^{i-1} A^*_{I+m}\overline{A}_{J+m} && i = h_j, j-1 \\
D^*_j &= D_j - \sum_{m=h_j}^{j-1} \overline{A}^2_{J+m}/D^*_m && \\
A^*_{J+i} &= \overline{A}_{J+i}/D^*_i && i = h_j, j-1
\end{aligned}\right\} j = 1, n \qquad (8.18)$$

8.3.4 Using out-of-core memory

If the work is not being done on a virtual memory operating system, and the size of stiffness matrix **K** is so large that the number of coefficients within the skyline profile exceeds the available central core memory, the linear array **A** must be stored column by column on a mass storage device such as a disk drive. The number of columns (equations) to be read from disk to core depends on the size of the working vector **W** whose minimum length is to be determined in advance prior to the matrix reduction.

***(1) Length of working vector* W**

Vector **W** provides a working place for the reduction of the elements in vector **A**, and as many equations as possible will be read from disk to vector **W** for reduction. Having reduced all the coefficients held in vector **W**, a certain number of equations may leave the working vector **W** to make way for the reduction of other equations. Columns leaving and entering the working area are determined by a control vector **C** which indicates those columns no longer required in the subsequent reduction. The elements of vector **C** are

$$C_i = \max_{j=1,n}\{j; h_j < i\} \qquad (8.19)$$

The significance of C_i is that column i can leave the working area if column C_i has entered the working area and has been reduced. Hence working vector **W** must have at least a length that can accommodate columns from i to C_i for any $i = 1, n$. The number of elements contained in columns i to C_i is given by

$$P_{C_i} - P_{i-1} \tag{8.20}$$

Thus the minimum length for the working vector **W** is

$$\ell_{\min} = \max_{i=1,n}\{P_{C_i} - P_{i-1}\} \tag{8.21}$$

(2) Reading data from the direct-access file

The number of columns and the amount of data to be read from the direct-access file are controlled by the currently available space in the working vector **W**. Let $L \geq \ell_{\min}$ be the length of the working vector **W**, and L_1 be the space occupied by the equations(coefficients) remaining in the working area, then the space available L_2 will be

$$L_2 = L - L_1 \tag{8.22}$$

Supposing that columns j_0 to j_1 remain in the working area, and the j_1th column is the last column reduced, then data from disk up to column j_2 will be read in such that

$$L_2 \geq P_{j_2} - P_{j_1} \tag{8.23}$$

Now on working vector **W**, with the first element of **W** corresponding to the $(P_{j_0-1} + 1)$th element of vector **A**, columns $j_1 + 1$ to j_2 are reduced according to eq. 8.18. This is always possible as columns are not allowed to leave the working area if they are still needed in the subsequent reduction of the forthcoming equations(columns). To determine which columns can leave the working area after the reduction down to column j_2, the C_i-values of the columns are checked. Those columns with C_i-values less than or equal to j_2 can now leave the high-speed core; these reduced coefficients are transferred back to the direct-access file for later forward and backward substitutions in solving for the unknown vector **x**.

$$\text{column leaving } = \{i = j_0, j_2;\ C_i \leq j_2\} \tag{8.24}$$

$$\text{or up to and including column } j = \max\ \{i = j_0, j_2;\ C_i \leq j_2\} \tag{8.25}$$

The cycle can be repeated by putting new values of j_0 and j_1 to

$$j_0^* = j + 1 \quad \text{and} \quad j_1^* = j_2 \tag{8.26}$$

Taking the matrix of Fig. 8.6 as an example, the values of P_i, C_i, P_{c_i}, and $P_{c_i} - P_{i-1}$ for columns 1 to 8 are listed in Table 8.1. From the table, the maximum value in the $P_{c_i} - P_{i-1}$ column is 10; hence the length L of the working array should be at least equal to 10. If L is greater than or equal to 17, the working array is large enough to hold all the off-diagonal coefficients and the matrix can be decomposed in one go. However, if L is between 10 and 17, the data transfer process has to be broken down into a number of stages.

Let us see what will happen if the matrix decomposition is to be done on a working array of the minimum required length of 10. The matrix decomposition can only be completed in four stages. The j_0, j_1, L_1, L_2 and j_2 values and the columns

Table 8.1

Column i	P_i	C_i	P_{c_i}	$P_{c_i} - P_{i-1}$
1	0	1	0	0
2	1	4	5	5
3	2	6	11	10
4	5	6	11	9
5	7	7	14	9
6	11	8	17	10
7	14	8	17	6
8	17	8	17	3

Table 8.2

Cycle	Columns leaving	j_0	j_1	L_1	L_2	j_2	Columns entering	j
1		1	1	0	10	5	1 to 5	2
2	1, 2	3	5	6	4	6	6	4
3	3, 4	5	6	6	4	7	7	5
4	5	6	7	7	3	8	8	8

leaving and entering the working array for each cycle of data transfer are given in Table 8.2.

8.3.5 Program for matrix decomposition and solution of unknowns

The Fortran subroutines for the matrix triangulation, forward and backward substitutions using direct-access files (REDUC2, SOLVE2) are listed in this section, whereas the version (REDUC1, SOLVE1) without using out-of-core storage can be found in Section 8.6 as part of the finite element program FACILE.

(1) Version 1 – Without using out-of-core memory
SUBROUTINE **REDUC1** (N, IP, D, A)
N is the number of equations(columns), vectors D and A contain diagonal and off-diagonal elements of the entire stiffness matrix K. IP is the pointer vector indicating the address of the last element of each column. N and vector IP will not be altered in the reduction process; and the matrix coefficients after reduction are stored in the same memory locations as occupied by the initial values in vectors D and A.
SUBROUTINE **SOLVE1** (N, IP, D, A, B)
B is the right-hand side vector, and the solution will again be put back to vector B.

(2) Version 2 – Using out-of-core memory
SUBROUTINE **REDUC2** (N, IP, IC, D, A, LAW, N1, N2, IREC)
Two direct-access files are used to store vector A before and after matrix reduction. The vector D of diagonal elements is however always maintained in the core memory. Apart from the arguments needed in subroutine REDUC1, subroutine REDUC2 takes the following additional arguments:

LAW = Allowable length of working vector W.

N1, N2 = Channel numbers of direct-access files where A is stored before and after decomposition.

IREC (I=1, N) = Pointer vector indicating the beginning record of each column in direct-access file N1.

IC (I = 1, N) = Vector controlling the reading and writing (entering and leaving) of matrix coefficients from and to direct-access files N1 and N2. Vector IC is calculated in the subroutine REDUC1 itself from vector IP.

```
      SUBROUTINE REDUC2 (NEQ,MP,MQ,D,A,N1,N2,LAW,IREC)
      IMPLICIT DOUBLE PRECISION (A-H,O-Z)
      DIMENSION MP(*),D(*),A(*),MQ(*),IREC(*)
C
C     FUNCTION : TO REDUCE A SYMMETRIC MATRIX TO LOWER AND UPPER
C                TRIANGULAR MATRICES.
C                OUT OF CORE PROFILE SOLVER
C                CROUT DECOMPOSITION A=LDU, WHERE U=TRANSPOSE OF L
C                (SKY-LINE STORAGE SCHEME IS USED)
C
C     INPUT   : NEQ = NUMBER OF EQUATIONS
C               MP(K)  = NUMBER OF OFF-DIAGONAL ELEMENTS UPTO COLUMN K
C               D(K)  = Kth DIAGONAL ELEMENT
C               A(L)  = Lth OFF-DIAGONAL ELEMENT
C               LAW = MAXIMUM LENGTH OF WORKING VECTOR A ALLOWED
C               IREC(K)  = READ COLUMN K FROM THIS RECORD IN D. A. FILES
C               N1,N2 = CHANNEL NUMBERS OF D.A. FILES WHERE A IS STORED
C                       BEFORE AND AFTER DECOMPOSITION
C
C     OUTPUT : D(*),A(*)
C
C     WORKING ARRAY : MQ(K), COLUMN K CAN LEAVE THE HIGH SPEED CORE MEMORY
C                            AS SOON AS COLUMN MQ(K) IS REDUCED
C
C     Calculate vector MQ(NEQ)
C
      DO 22 K=3,NEQ
      K1=K-1
      KH1=MP(K1)-MP(K)+K+1
      DO 22 J=KH1,K
   22 MQ(J)=K
C
C     Ensure length of array A is not greater than the allowable length LAW
C
      LMAX=0
      DO 11 K=2,NEQ
      IF (MQ(K).LE.K) GOTO 11
      L=MP(MQ(K))-MP(K-1)
      IF (L.GT.LMAX) LMAX=L
   11 CONTINUE
      IF (LMAX.GT.LAW) THEN
      WRITE (6,30) LMAX,LAW
   30 FORMAT (///' ERROR...LENGTH OF WORKING VECTOR A REQUIRED =',I7,/
     -            '        GREATER THAN THE MAXIMUM LENGTH ALLOWED =',I7)
      STOP
      END IF
C
      LST=0
      KS=1
      KT=1
      MP(NEQ+1)=100000000
      MQ(NEQ+1)=100000000
      REWIND 1
C
```

```
C     Read columns KQ+1 to KT from direct access file channel N1
C
    2 KQ=KT
      LT=MP(KT)+LAW-LST
    1 KT=KT+1
      IF (MP(KT).LE.LT) GOTO 1
      KT=KT-1
      DO 44 K=KQ+1,KT
      L=MP(K)-MP(K-1)
      IR=IREC(K)
      READ (N1,REC=IR) (A(I),I=LST+1,LST+L)
      LST=LST+L
   44 CONTINUE
C
C     Reducing columes KQ+1 to KT
C
      DO 111 K=KQ+1,KT
      K1=K-1
      LK=MP(K)-K1
      KH=MP(K1)-LK+1
      LK1=LK-LP
      S=D(K)
      DO 222 J=KH+1,K1
      J1=J-1
      LJ=MP(J)-J1
      JH=MP(J1)-LJ+1
      LJ1=LJ-LP
      IF (KH.GT.JH) JH=KH
      T=A(LK1+J)
      DO 333 M=JH,J1
  333 T=T-A(LJ1+M)*A(LK1+M)
  222 A(LK1+J)=T
      DO 444 J=KH,K1
      L=LK1+J
      T=A(L)
      A(L)=T/D(J)
  444 S=S-T*A(L)
  111 D(K)=S
C
      IF (KT.EQ.NEQ) THEN
      J=0
      DO 66 K=KS+1,NEQ
      L=MP(K)-MP(K-1)
      WRITE (N2) (A(I),I=J+1,J+L)
      J=J+L
   66 CONTINUE
      RETURN
      ENDIF
C
C     Write columns KR+1 to KS to direct access file channel N2
C
      KR=KS
    3 KS=KS+1
      IF (MQ(KS).LE.KT) GOTO 3
      KS=KS-1
      LP=MP(KS)
      LRS=LP-MP(KR)
      LST=MP(KT)-LP
      J=0
      DO 55 K=KR+1,KS
      L=MP(K)-MP(K-1)
      WRITE (N2) (A(I),I=J+1,J+L)
      J=J+L
   55 CONTINUE
      DO 33 I=1,LST
   33 A(I)=A(LRS+I)
      GOTO 2
      END
```

SUBROUTINE **SOLVE2** (N, IP, D, A, B, N2, IREC)

Using SOLVE2, reduced coefficients stored in direct-access file N2 can be called upon for the solution of any right-hand side vector B.

```
      SUBROUTINE SOLVE2 (NEQ,MP,D,A,B,N2,IREC)
      DIMENSION MP(*),D(*),A(*),B(*),IREC(*)
C
C     SOLVE FOR LDUx = b, USING OUT-OF-CORE STORAGE IN D.A. FILE CHANNEL N2
C
      REWIND 2
      DO 11 J=2,NEQ
      J1=J-1
      L=MP(J)-MP(J1)
      READ (N2) (A(I),I=1,L)
      JH1=J1-L
      T=B(J)
      DO 22 M=1,L
   22 T=T-A(M)*B(JH1+M)
   11 B(J)=T
C
      DO 33 K=1,NEQ
   33 B(K)=B(K)/D(K)
C
      DO 44 K=NEQ,2,-1
      K1=K-1
      L=MP(K)-MP(K1)
      IR=IREC(K)
      READ (N2,REC=IR) (A(I),I=1,L)
      KH1=K1-L
      T=B(K)
      DO 44 J=1,L
   44 B(KH1+J)=B(KH1+J)-T*A(J)
      RETURN
      END
```

8.4 ASSEMBLY OF SYSTEM STIFFNESS MATRIX

In the finite element procedure, individual element stiffness matrices and applied load vectors are calculated separately and then assembled to form the overall structural stiffness matrix and the right-hand side force vector of the system. Nodal compatibility is the basis for the assembly process. This simple requirement states that all elements adjacent to a particular node must have the same displacement at that node. Here, our notion of displacement is a generalized one which includes translations and rotations for plate bending problems and temperature values for heat conduction problems, etc. When an element stiffness is added arithmetically to the global stiffness matrix, physically it represents the contribution of that element towards the whole structural system.

8.4.1 Procedure of the assembly process

The general assembly procedure for a system containing n degrees of freedom can be summarized in the following steps:

1. Set up an $n \times n$ null matrix and an $n \times 1$ null vector, where n = number of system nodal variables.
2. Starting with one element, transform the element equations from local to global coordinates if these two coordinate systems are not coincident.
3. Using the established correspondence between local and global numbering schemes, change the subscript indices of the coefficients in the square element matrix and the single subscript index of the terms in the element force vector to the global indices.
4. Insert these terms in the locations designated by their indices. Each time a term is placed in a location where another term has already been placed, it is added to whatever value is there.
5. Return to step 2 and repeat this procedure for one element after another until all the elements have been processed. The result will be an $n \times n$ system stiffness matrix $\mathbf{K}$ of influence coefficients and an $n \times 1$ right-hand side force vector $\mathbf{f}$ of resultant nodal actions.

The complete system of equations is then

$$\mathbf{Kx} = \mathbf{f}$$

where $\mathbf{x}$ is the unknown nodal displacement vector.

The above assembly procedure is completely general for all types of problems and elements, as only the element stiffness matrices are used without specific reference to the elements themselves in the formation of global system matrices. Hence, once a computer program for the assembly process has been developed for the solution of a particular class of problems, it is equally applicable to other classes of problems as well.

8.4.2 An example of the assembly process

Consider the following example in which there are five elements interconnected, as shown in Fig. 8.7. The element node numbers of each element are shown in the table below.

Element	Node numbers
1	1 3 4
2	1 4 2
3	2 5
4	3 6 7 4
5	4 7 8 5

For each element, the element stiffness and element force vector can be calculated from the given element geometry, material and loading data. Individual elements and their separate contribution to the system matrix are shown respectively in Figs 8.8(a) and 8.8(b). The step-by-step result of the assembly process of individual element contributions is depicted in Fig. 8.8(c). The force vector is assembled in a similar way except that the result is a column vector instead of a square matrix as shown in Fig. 8.8(d). It is noted that in Fig. 8.7, each cross $\times$ associated with a node in fact represents a sub-matrix whose dimension is equal to the DOF of that node.

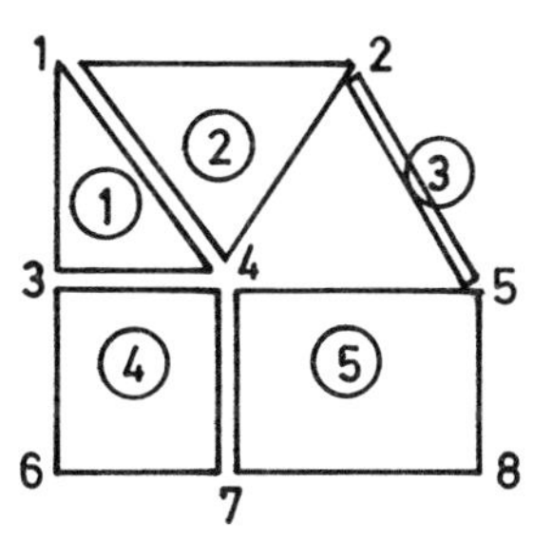

$$
\begin{bmatrix}
\times & \times & \times & \times & & & & \\
\times & \times & & \times & \times & & & \\
\times & & \times & \times & & \times & \times & \\
\times & \times & \times & \times & \times & \times & \times & \times \\
 & \times & & \times & \times & & \times & \times \\
 & & \times & \times & & \times & \times & \\
 & & \times & \times & \times & \times & \times & \times \\
 & & & \times & \times & & \times & \times
\end{bmatrix}
\begin{bmatrix}
\\ \\ \\ \\ \\ \\ \\ \\
\end{bmatrix}
=
\begin{bmatrix}
\times \\ \times \\ \times \\ \times \\ \times \\ \times \\ \times \\ \times
\end{bmatrix}
$$

$$\mathrm{K} \times \mathbf{x} = \mathbf{f}$$

Fig. 8.7 The structure and the corresponding system stiffness matrix

As the system matrix is symmetrical, only the upper half above the diagonal needs to be formed. All the non-zero coefficients are confined within a profile which can be calculated *a priori* from the mesh connectivity information. In finite element programming, only entries in the upper half of the profile need to be stored in a linear array using the 'skyline' storage scheme.

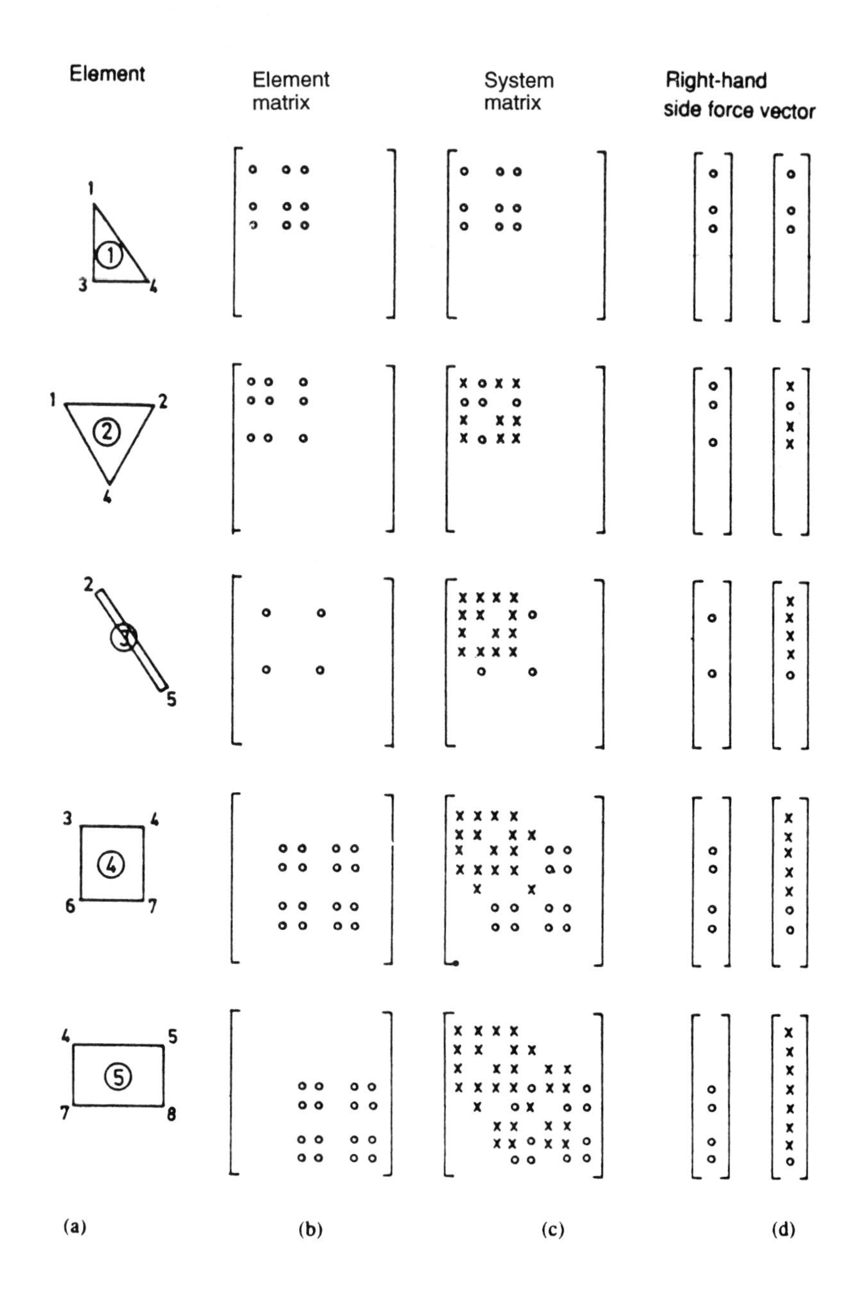

Fig. 8.8 The assembly process of the system matrices

8.4.3 Matrix assembly involving out-of-core storage

If a virtual memory operating system is not being used, and the profile of the system matrix exceeds the available random access memory (RAM) of the computer, out-of-core storage is required to hold the global stiffness matrix. Let L_a be the space available for the assembly of the system matrix, and P be the profile of the system matrix. Then if P is greater than L_a, disk storage is required to hold the global stiffness matrix.

The steps of the assembly process using out-of-core storage are:

1. Calculate the smallest and largest node number of each element.

$$\text{Smallest node number of element } i, S_i = \min_{j=1,N_v} \{N_{ij}\}$$

$$\text{Largest node number of element } i, M_i = \max_{j=1,N_v} \{N_{ij}\}$$

   where N_{ij} is the jth node number of element i, and N_v is the number of nodes in the element.

2. Calculate the length of the working array required.

$$L_r = \max_{i=1,N_e} \{P(k_2) - P(k_1), k_1 = N_f \times (S_i - 1) + 1, k_2 = N_f \times M_i\}$$

   where N_e = number of elements in the mesh,
   N_f = number of degrees of freedom per node, and
   $P(k)$ is the number of off-diagonal entries up to column k (Section 8.3.3).

It is required that the available space L_a has to be greater than or equal to L_r for efficient element assembly using out-of-core storage.

3. Elements are assembled in the sequence as given by the smallest node number S_i; and elements having the same smallest node number can be ordered arbitrarily. Whenever the address of the coefficients of an incoming element goes beyond the available length of the working array, part of the working array has to be transferred to the disk storage to make room for the new entries. Entries of the working array up to but excluding address I_p will be transferred to disk, where I_p is the address on the global stiffness vector of the starting position of the incoming

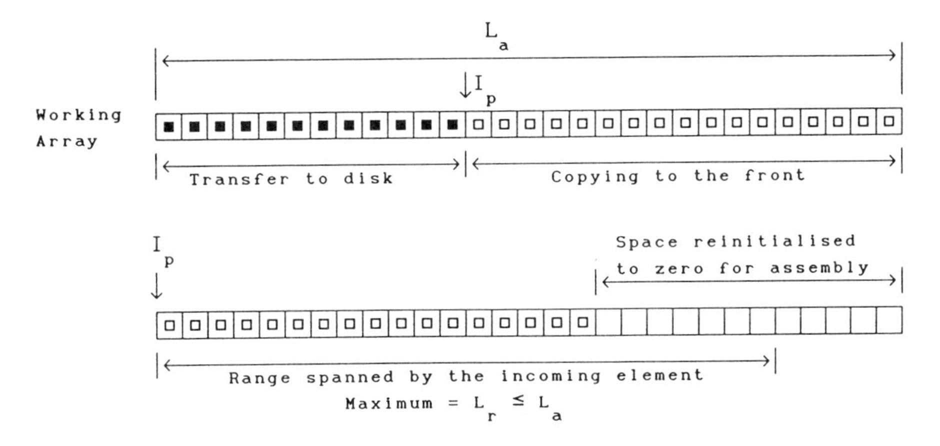

Fig. 8.9 Assembly of system stiffness matrix using out-of-core storage. □ = entries in the process of assembly, ■ = entries fully assembled.

elements (Fig. 8.9). The corresponding address k on the global stiffness vector of element j on the working array can be calculated easily by

$$k = I_p + j - 1$$

8.5 BOUNDARY AND LOADING CONDITIONS

After the assembly of the system stiffness matrix and the determination of the global force vector, we are ready to solve for the response of the structure. However, without the substitution of a minimum number of prescribed displacements to prevent rigid-body movement of the structure, it is impossible to solve the structural problem, because the displacements cannot be uniquely determined in such a situation. This physically obvious fact will be reflected in the matrix $\mathbf{K}$ being singular, i.e. not possessing an inverse. The prescription of appropriate displacement boundary conditions after the assembly stage will permit a unique solution to be obtained. It is also noted that in the solution of the system of equations of the structure, either the force or the displacement has to be specified at each degree of freedom.

8.5.1 Introduction of displacement boundary conditions

As stated previously, the global stiffness matrix $\mathbf{K}$ is singular, as no attention has been paid to the boundary conditions when matrix $\mathbf{K}$ is assembled. Solution to the system $\mathbf{Kx} = \mathbf{f}$ is only possible when proper displacement boundary conditions are imposed to prevent rigid-body motions. Boundary conditions of prescribed displacement value, $x_i = \overline{x}_i$ can be introduced in a number of ways.
(a) Using a large number on the diagonal term
Diagonal term k_{ii} is multiplied by α which is a very large number with respect to all the off-diagonal terms k_{ij}; and the corresponding value of the right-hand side vector is changed to $f_i = \alpha k_{ii}\overline{x}_i$. Consider the modified equation

$$k_{i1}x_1 + k_{i2}x_2 + \cdots + \alpha k_{ii}x_i + \cdots + k_{in}x_n = \alpha k_{ii}\overline{x}_i$$

and dividing the equation throughout by αk_{ii} gives

$$x_i + \frac{k_{i1}x_1 + k_{i2}x_2 + \cdots + k_{in}x_n}{\alpha k_{ii}} = \overline{x}_i$$

However, if the sum $k_{i1}x_1 + k_{i2}x_2 + \cdots + k_{in}x_n$ is small compared to αk_{ii}, then a very good approximation of $x_i \approx \overline{x}_i$ can be obtained.

This method is very simple to code and gives, after solution, all the unknowns with no loss of accuracy provided that a large value for α, say 10^{20}, is used. The system matrix retains its symmetry since only the diagonal term k_{ii} is modified. Another advantage of this method is that the original matrix can be recovered simply by dividing the corresponding diagonal terms where displacements are prescribed by the same number α. This flexibility allows the same structure to be studied under a complete new set of displacement boundary conditions without the formation of the stiffness matrix for a second time.

(b) Modifying rows and columns

In this method, all the terms of the right-hand side force vector are changed to $f'_j = f_j - k_{ji}\overline{x}_i$, except for the term f_i which is set equal to $\overline{x}_i$. The terms on the ith row and the ith column are put to zero, and the term k_{ii} is set equal to 1. This effectively eliminates equation i from the system without changing the size and symmetry of matrix **K**.

$$\begin{bmatrix} k_{11} & \cdots & k_{1,i-1} & 0 & k_{1,i+1} & \cdots & k_{1n} \\ \cdot & & \cdot & \cdot & \cdot & & \cdot \\ \cdot & & \cdot & \cdot & \cdot & & \cdot \\ k_{i-1,1} & \cdots & k_{i-1,i-1} & 0 & k_{i-1,i+1} & \cdots & k_{i-1,n} \\ 0 \cdot & \cdots & 0 & 1 & 0 \cdot & \cdots & 0 \\ k_{i+1,1} & \cdots & k_{i+1,i-1} & 0 & k_{i+1,i+1} & \cdots & k_{i+1,n} \\ \cdot & & \cdot & \cdot & \cdot & & \cdot \\ \cdot & & \cdot & \cdot & \cdot & & \cdot \\ k_{n1} & \cdots & k_{n,i-1} & 0 & k_{n,i+1} & \cdots & k_{nn} \end{bmatrix} \begin{bmatrix} x_1 \\ \cdot \\ \cdot \\ x_{i-1} \\ x_i \\ x_{i+1} \\ \cdot \\ \cdot \\ x_n \end{bmatrix} = \begin{bmatrix} f_1 - k_{1i}\overline{x}_i \\ \cdot \\ \cdot \\ f_{i-1} - k_{i-1,i}\overline{x}_i \\ \overline{x}_i \\ f_{i+1} - k_{i+1,i}\overline{x}_i \\ \cdot \\ \cdot \\ f_n - k_{ni}\overline{x}_i \end{bmatrix}$$

(c) Elimination of equations with specified value $\overline{x}_i$

This effectively requires a complete reshuffling of the matrix coefficients. The load vector is modified as in the previous case, and row i and column i are both completely eliminated, thereby reducing the size of matrix **K**. To avoid costly matrix manipulations, the assembling process can be modified to take into account the specified displacement and prevent the construction of such equations altogether.

8.5.2 Example: Imposing boundary conditions by different methods

Displacement boundary condition $x_1 = \overline{x}_1$ is to be introduced to the following 4×4 system:

$$\begin{bmatrix} k_{11} & 0 & 0 & k_{14} \\ 0 & k_{22} & k_{23} & k_{24} \\ 0 & k_{23} & k_{33} & 0 \\ k_{14} & k_{24} & 0 & k_{44} \end{bmatrix} \begin{bmatrix} x_1 \\ x_2 \\ x_3 \\ x_4 \end{bmatrix} = \begin{bmatrix} f_1 \\ f_2 \\ f_3 \\ f_4 \end{bmatrix}$$

(a) Dominating diagonal term ($\alpha = 10^{20}$)

$$\begin{bmatrix} 10^{20}k_{11} & 0 & 0 & k_{14} \\ 0 & k_{22} & k_{23} & k_{24} \\ 0 & k_{23} & k_{33} & 0 \\ k_{14} & k_{24} & 0 & k_{44} \end{bmatrix} \begin{bmatrix} x_1 \\ x_2 \\ x_3 \\ x_4 \end{bmatrix} = \begin{bmatrix} 10^{20}k_{11}\overline{x}_1 \\ f_2 \\ f_3 \\ f_4 \end{bmatrix}$$

(b) Modifying rows and columns

$$\begin{bmatrix} 1 & 0 & 0 & 0 \\ 0 & k_{22} & k_{23} & k_{24} \\ 0 & k_{23} & k_{33} & 0 \\ 0 & k_{24} & 0 & k_{44} \end{bmatrix} \begin{bmatrix} x_1 \\ x_2 \\ x_3 \\ x_4 \end{bmatrix} = \begin{bmatrix} \overline{x}_1 \\ f_2 \\ f_3 \\ f_4 - k_{14}\overline{x}_1 \end{bmatrix}$$

(c) Elimination of equation 1

$$\begin{bmatrix} k_{22} & k_{23} & k_{24} \\ k_{23} & k_{33} & 0 \\ k_{24} & 0 & k_{44} \end{bmatrix} \begin{bmatrix} x_2 \\ x_3 \\ x_4 \end{bmatrix} = \begin{bmatrix} f_2 \\ f_3 \\ f_4 \quad -k_{14}\overline{x}_1 \end{bmatrix} \qquad x_1 = \overline{x}_1$$

8.5.3 Transformation of variables

Assume that constraints must be imposed on some unknowns $\mathbf{x}'$ which are related to $\mathbf{x}$ through the transformation

$$\mathbf{x} = \mathbf{R}\mathbf{x}' \tag{8.27}$$

where $\mathbf{R}$ is an $n \times n$ square matrix of constants.

$$\mathbf{K}\mathbf{x} = \mathbf{f} \quad \Rightarrow \quad \mathbf{K}(\mathbf{R}\mathbf{x}') = \mathbf{f} \quad \Rightarrow \quad (\mathbf{R}^T\mathbf{K}\mathbf{R})\mathbf{x}' = \mathbf{R}^T\mathbf{f} \quad \Rightarrow \quad \mathbf{K}'\mathbf{x}' = \mathbf{f}' \tag{8.28}$$

where $\quad \mathbf{K}' = \mathbf{R}^T\mathbf{K}\mathbf{R} \quad$ and $\quad \mathbf{f}' = \mathbf{R}^T\mathbf{f}$

Such a transformation may be necessary for the following reasons:

(a) to change the reference system of some variables,
(b) to force a linear constraint between some variables.

8.5.4 Example: Rotation of a reference system at a node

Consider the assemblage of two triangular elements in a 2D finite element mesh, as shown in Fig. 8.10. A change of coordinate system for node 2 is needed to impose a frictionless rolling boundary condition on the inclined plane.

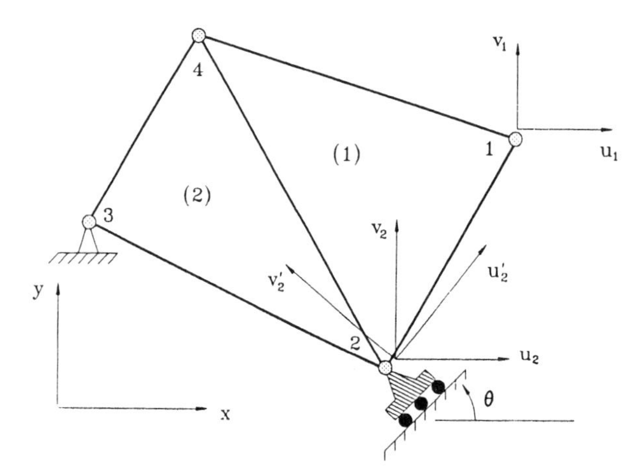

Fig. 8.10 Rolling boundary condition on an inclined plane

The system to be solved is

$$\underset{8\times 8}{\mathbf{K}} \; \underset{8\times 1}{\mathbf{x}} = \underset{8\times 1}{\mathbf{f}}$$

where the vector of unknowns is

$$\mathbf{x} = [u_1, v_1, u_2, v_2, u_3, v_3, u_4, v_4]$$

The boundary condition at node 2 is $v_2' = 0$, where v_2' is the component of displacement perpendicular to the rolling inclined plane.

$$\begin{bmatrix} u_2 \\ v_2 \end{bmatrix} = \begin{bmatrix} c & -s \\ s & c \end{bmatrix} \begin{bmatrix} u_2' \\ v_2' \end{bmatrix}$$

where $c = \cos\theta$ and $s = \sin\theta$.

The transformation matrix $\mathbf{R}$ is given by

$$\underset{8\times 8}{\mathbf{R}} = \begin{bmatrix} 1 & & & & & & & \\ & 1 & & & & & & \\ & & c & -s & & & & \\ & & s & c & & & & \\ & & & & 1 & & & \\ & & & & & 1 & & \\ & & & & & & 1 & \\ & & & & & & & 1 \end{bmatrix}$$

and the new vector of variables is

$$\mathbf{x}' = [u_1, v_1, u_2', v_2', u_3, v_3, u_4, v_4]$$

The matrices $\mathbf{K}'$ and $\mathbf{f}'$ are obtained from eq. 8.28. The new system $\mathbf{K}'\mathbf{x}' = \mathbf{f}'$ can now be solved with conditions $u_3 = v_3 = v_2' = 0$. After solving for $\mathbf{x}', \mathbf{x}$ can be computed using the relation $\mathbf{x} = \mathbf{R}\mathbf{x}'$.

8.5.5 Linear constraints between variables

Transformation 8.28 can also be used to introduce linear constraints between variables:

$$\alpha_i x_i + \alpha_j x_j + \alpha_k x_k + \ldots = x_i' = \beta$$

where $\alpha_i, \alpha_j, \alpha_k$, etc. and β are prescribed constants.

The transformation matrix $\mathbf{R}$ between old variables $[x_i, x_j, x_k, \ldots]$ and new variables $[x_i', x_j, x_k, \ldots]$ is given by

$$\underset{n\times n}{\mathbf{R}} = \begin{bmatrix} 1 & & & & & & & & \\ & 1 & & & & & & & \\ & & \cdot & & & & & & \\ & & & \cdot & & & & & \\ & -\dfrac{\alpha_j}{\alpha_i} & \cdot & \cdot & \dfrac{1}{\alpha_i} & \cdot & \cdot & -\dfrac{\alpha_k}{\alpha_i} & \\ & & & & & & \cdot & & \\ & & & & & & & 1 & \\ & & & & & & & & 1 \end{bmatrix} \leftarrow \text{row } i$$

The matrix transformation $\mathbf{K}' = \mathbf{R}^T\mathbf{K}\mathbf{R}$ can be conveniently carried out in two steps:

Step 1: $\overline{\mathbf{K}} = \mathbf{K}\mathbf{R}$

Column i of $\overline{\mathbf{K}} = \frac{1}{\alpha_i} \times$ column i of $\mathbf{K}$;

column j of $\overline{\mathbf{K}} =$ column j of $\mathbf{K} - \frac{\alpha_j}{\alpha_i} \times$ column i of $\mathbf{K}$;

the other columns of $\overline{\mathbf{K}}$ are identical to the corresponding columns of $\mathbf{K}$.

Step 2: $\mathbf{K}' = \mathbf{R}^T\overline{\mathbf{K}} = \mathbf{R}^T\mathbf{K}\mathbf{R}$

Row i of $\mathbf{K}' = \frac{1}{\alpha_i} \times$ row i of $\overline{\mathbf{K}}$;

row j of $\mathbf{K}' =$ row j of $\overline{\mathbf{K}} - \frac{\alpha_j}{\alpha_i} \times$ row i of $\overline{\mathbf{K}}$;

the other rows of $\mathbf{K}'$ are identical to the corresponding rows of $\overline{\mathbf{K}}$.

Vector **f** is also modified as follows:

$$\mathbf{f}' = \mathbf{R}^T\mathbf{f} \quad \Rightarrow \quad f_i' = \frac{1}{\alpha_i} f_i \quad \text{and} \quad f_j' = f_j - \frac{\alpha_j}{\alpha_i} f_i$$

The other terms of $\mathbf{f}'$ are identical to the corresponding terms in **f**. The new system $\mathbf{K}'\mathbf{x}' = \mathbf{f}'$ can be solved under the condition that $x_i' = \beta$.

8.5.6 Example: Linear constraint between variables

Consider the solution of the following system of linear equations subject to a linear constraint $2x_1 + 3x_2 + 4x_3 = 5$.

$$\mathbf{Kx} = \mathbf{f} \quad \text{or} \quad \begin{bmatrix} 1.25 & 2 & 3 \\ 2 & 3 & 5 \\ 3 & 5 & 7 \end{bmatrix} \begin{bmatrix} x_1 \\ x_2 \\ x_3 \end{bmatrix} = \begin{bmatrix} -21 \\ -35 \\ -49 \end{bmatrix}$$

Let $x_1' = 2x_1 + 3x_2 + 4x_3 = 5$ and $\mathbf{x}' = [x_1', x_2, x_3]$, then

$$\mathbf{R} = \begin{bmatrix} 0.5 & -1.5 & -2 \\ 0 & 1 & 0 \\ 0 & 0 & 1 \end{bmatrix}$$

$$\mathbf{K}' = \mathbf{R}^T\mathbf{KR} = \begin{bmatrix} 0.3125 & 0.0625 & 0.25 \\ 0.0625 & -0.1875 & 0.25 \\ 0.25 & 0.25 & 0 \end{bmatrix} \quad \text{and} \quad \mathbf{f}' = \mathbf{R}^T\mathbf{f} = \begin{bmatrix} -10.5 \\ -3.5 \\ -7 \end{bmatrix}$$

Solving the system $\mathbf{K}'\mathbf{x}' = \mathbf{f}'$ under the condition $x_1' = 5$, gives $x_2 = -33$ and $x_3 = -40$. Finally, using the relation $\mathbf{x} = \mathbf{R}\mathbf{x}'$ yields $x_1 = 132$.

8.5.7 Introduction of specified values to variables

Irrespective of whether matrix transformation to introduce linear constraints between variables is applied or not, fixed values have to be specified on some variables of the system matrix before solution is possible. In the skyline storage scheme described in Section 8.3.3, diagonal terms are stored separately in a vector **D**. As a result, the simplest method for the introduction of boundary conditions of specified displacements would be the scheme presented in Section 8.5.1(a).

8.5.8 Forming the system right-hand side force vector

From Section 3.5.4, the global nodal force vector **f** is composed of four parts, i.e.

$$\mathbf{f} = \mathbf{r} + \mathbf{b} + \mathbf{s} + \mathbf{p}$$

where **r** is the initial stress and strain force vector,
b is the body force vector,
s is the surface force vector, and
p is the concentrated nodal point forces.

Through the use of element interpolating functions, distributed loading at the interior or on parts of the surface of the domain may be conveniently expressed as equivalent element nodal forces which can be easily assembled to form the system right-hand side vector. As for concentrated point forces, they have to be added directly to the corresponding nodal positions of the system force vector.

8.6 FINITE ELEMENT PROGRAM FACILE

8.6.1 Introduction

A self-contained finite element program FACILE written in FORTRAN 77 is described in this section. While the present version is limited to linear static analysis, it can be easily extended to the realms of dynamic and non-linear analyses. The range of problems that can be handled is, however, quite broad and includes 2D and 3D elasticity and field problems. FACILE provides a list of practical and reliable finite elements for the chosen class of problems. Mixed finite element types can also be used in the analysis, provided that the degrees of freedom per node are the same for all elements.

In FACILE, the finite element adopted for plate bending analysis is the 9-node Lagrangian plate element using selective integration. Although it may not be the most efficient element in plate bending, it is however one of the simplest elements and is easy to implement and to use. Of course, other plate elements discussed in Section 4.1 can also be included in FACILE in a similar way following the implementation procedures of the Lagrangian element. For shell structure problems, the 9-node Lagrangian shell element has been chosen, for which either a 4-point or a 9-point integration formula can be used in the evaluation of the element stiffness matrix. Readers are encouraged to develop their own shell elements and incorporate them in FACILE so as to gain understanding and make improvements to the program.

Although the matrix decomposition techniques using out-of-core storage are thoroughly discussed in Section 8.3, an out-of-core profile solver has not been included in FACILE in order to keep the program simple. However, it would be a good exercise to enhance the program with this capability following the description in Section 8.3.4. Multiple loading cases can be dealt with easily by a simple modification, which consists of the reconstruction of the right-hand side load vector and resolution for the corresponding displacements using the decomposed system matrix.

The input data are divided into three groups, and are read from three separate ASCII files in free format. The nodal displacements output to files are also in ASCII form. Readers should understand that the capacity of the program can be much enhanced when it is run in conjunction with a preprocessor and a postprocessor employing extensively interactive computer graphics. FACILE will then become the core of finite element analysis, and files in binary format should be used as a means of communication between programs. The sole purpose of using readable ASCII format for input and output in this version is to help readers to understand the structure of FACILE.

FACILE is a concise program and it is also an excellent education tool, as only essential fundamental features of the finite element analysis are retained. Nevertheless, FACILE can be easily upgraded to a practical powerful analysis package, since it is written in modular form and state-of-the-art numerical techniques have been used in the evaluation of element stiffness matrices, the system matrix decomposition and the arrangement of variable arrays.

8.6.2 Types of problems and finite elements supported by FACILE

FACILE can be applied to analyze a wide range of structural and field problems both in two and three dimensions. As the basic structure of the program is in modular form, FACILE can be easily extended to cover other problem types simply by adding a new finite element in the program for the intended problem. The following lists the types of problems that can be handled and the types of finite elements available in FACILE:

Problem type	*Finite elements*
1. Spatial frame	1. Spatial beam element
2. Plane strain	2. Isoparametric 6-node triangle
3. Plane stress	3. Isoparametric 8-node quadrilateral
4. Axisymmetric	4. 6-node infinite quadrilateral
5. 2D field	5. Isoparametric 10-node tetrahedron
6. 3D elasticity	6. Isoparametric 15-node pentahedron
7. 3D field	7. Isoparametric 20-node hexahedron
8. Plate bending	8. 9-node Lagrangian plate element
9. Shell structure	9. 9-node Lagrangian shell element

In the development of the program FACILE, there is no intention to maintain a long list of finite elements in the program. On the contrary, our philosophy is to keep only a few general efficient finite elements in FACILE to cover as many problem types as possible. Hence only those finite elements which have been proved to be suitable for the selected class of problems in terms of versatility, efficiency and accuracy have been included. Numerical integration will be used exclusively in the evaluation of element stiffness matrices. Each type of finite element is given a standard integration formula, which is usually the optimal integration order for that element [11]. However, users are allowed to modify the integration scheme by a very simple procedure to suit their own purposes.

It is of course entirely possible to analyze a particular problem using different finite elements. For example, a plane stress problem can be analyzed using T6 or Q8 elements. On the other hand, a special element can also be used in several different problem types. For instance, a Q8 element can be used to solve plane strain problems, axisymmetric problems and 2D field problems. Table 8.3 summarizes the relationship between problem types and finite elements of the program FACILE.

8.6.3 Program structure of FACILE

FACILE is written in modular form which allows maximum flexibility for later modification and extension. FACILE is a compact program, which consists of 26 subroutines stored in six files (Fig. 8.11).

***(1) File* FACILE.FOR**

This file contains the program FACILE, whose functions are:

(a) Fix the default parameters.

(b) Open input and output files.

Table 8.3 Types of problems and finite elements in FACILE

Problem	IT	Name	NV	NIP	Program	Finite elements
1. Spatial frame	1	BEAM	2	0	EKBEAM	Spatial beam, 12 DOF
2. Plane strain	2	T6	6	3, 4, 7		2D isoparametric triangle
3. Plane stress 4. Axisymmetric	3	Q8	8	2 × 2 7 3 × 3	EK2D EKAXIS	2D isoparametric quadrilateral
5. 2D field	4	I6	6	2 × 2 7 3 × 3	EK2DF	2D infinite quadrilateral element
6. 3D elasticity	5	T10	10	4, 5		3D isoparametric tetrahedron
7. 3D field	6	P15	15	2 × 3 2 × 4	EK3D	3D isoparametric pentahedron
	7	H20	20	2 × 2 × 2 14	EK3DF	3D isoparametric hexahedron
8. Plate bending	8	L9P	9	2 × 2 3 × 3	EKL9P	Lagrangian plate element
9. Shell structure	9	L9S	9	2 × 2 3 × 3	EKL9S	Lagrangian shell element

NV = number of nodes in the finite element
NIP = number of integration points
IT is the element type

(c) Read in problem definition parameters.
(d) Calculate the problem size and addresses of variable arrays.

***(2) File* EXECUT.FOR**

This file contains the subroutine EXECUT, which is the execution segment carrying out the following operations in sequence:
(a) Read in geometrical data from file FACILE. 1.
(b) Read in element material property data from file FACILE. 2.
(c) Calculate element matrices and assemble system stiffness matrix.
(d) Impose displacement boundary conditions stored in file FACILE. 3.
(e) Decompose system stiffness matrix stored in linear arrays A and D.
(f) Read in loading and construct the right-hand side force vector.
(g) Solve for the global displacement vector.
(h) Output the nodal displacements to file FACILE. 8.

***(3) File* LIB1.FOR**

LIB is the short form for LIBRARY. File LIB1.FOR contains seven subroutines which are directly called by EXECUT.

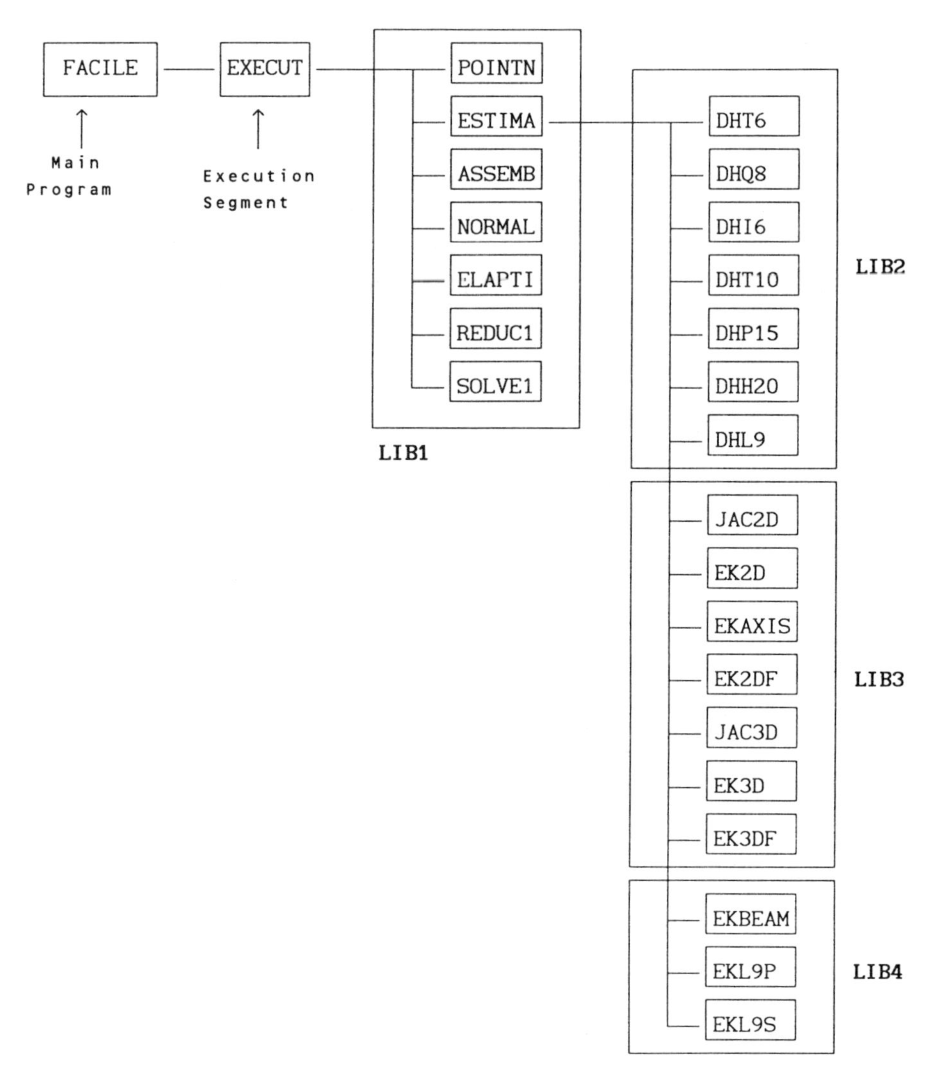

Fig. 8.11 Program structure of FACILE

(a) POINTN – Calculate the pointer vector MP from mesh information stored in array ME.
(b) ESTIMA – Evaluate the element stiffness matrix.
(c) ASSEMB – Assemble element stiffness matrix to system stiffness vectors A and D.
(d) NORMAL – Calculate the base vectors at each node of a Lagrangian shell element.
(e) ELAPTI – Log on the CPU time elapsed.
(f) REDUC1 – Perform Crout decomposition on the system stiffness matrix stored in vectors A and D.
(g) SOLVE1 – Solve for any right-hand side vector after Crout decomposition.

***(4) File* LIB2.FOR**

This file contains subroutines which calculate the values and the derivatives of interpolating functions at integration points for various element types.

(a) DHT6 – 2D 6-node isoparametric triangular element, T6.
(b) DHQ8 – 2D 8-node isoparametric quadrilateral element, Q8.
(c) DHI6 – 2D 6-node infinite quadrilateral element, I6.
(d) DHT10 – 3D 10-node isoparametric tetrahedron element, T10.
(e) DHP15 – 3D 15-node isoparametric pentahedron element, P15.
(f) DHH20 – 3D 20-node isoparametric hexahedron element, H20.
(g) DHL9 – 2D 9-node Lagrangian element, L9.

***(5) File* LIB3.FOR**

This file contains subroutines which evaluate the stiffness matrix of different element types.

(a) JAC2D – Calculate the Jacobian transformation for 2D isoparametric elements.
(b) EK2D – Evaluate the element stiffness matrix for 2D elasticity problems.
(c) EKAXIS – Evaluate the element stiffness matrix for axisymmetric elasticity problems.
(d) EK2DF – Evaluate the element stiffness matrix for 2D field problems.
(e) JAC3D – Calculate the Jacobian transformation for 3D isoparametric elements.
(f) EK3D – Evaluate the element stiffness matrix for 3D elasticity problems.
(g) EK3DF – Evaluate the element stiffness matrix for 3D field problems.

***(6) File* LIB4.FOR**

This file contains subroutines which calculate the element stiffness matrix for special structural problems.

(a) EKBEAM – Evaluate the element stiffness matrix of a 3D beam element.
(b) EKL9P – Evaluate the element stiffness matrix of a 9-node Lagrangian plate element.
(c) EKL9S – Evaluate the element stiffness matrix of a 9-node Lagrangian shell element.

8.6.4 Variables used in FACILE

A single linear array **V** is partitioned to store all the data arrays, variable and working arrays, e.g. nodal coordinates, displacements, system stiffness, etc. Each array indicated in Fig. 8.12 is dynamically dimensioned to the size and precision required for the problem in hand by using a set of address pointers established in the main control program FACILE. Using this scheme, no space is wasted in data storage and a maximum amount of space is reserved to store the global arrays. With the employment of such an automatic dimensioning scheme, it is not possible to know in advance the absolute values for maximum numbers of material sets, nodes or elements allowed. However, before an analysis is performed, the program will check whether sufficient space exists to solve the given problem; and if not, an error message will be printed. The total capacity of the program is controlled by the dimension of the linear array **V** in the main program FACILE.

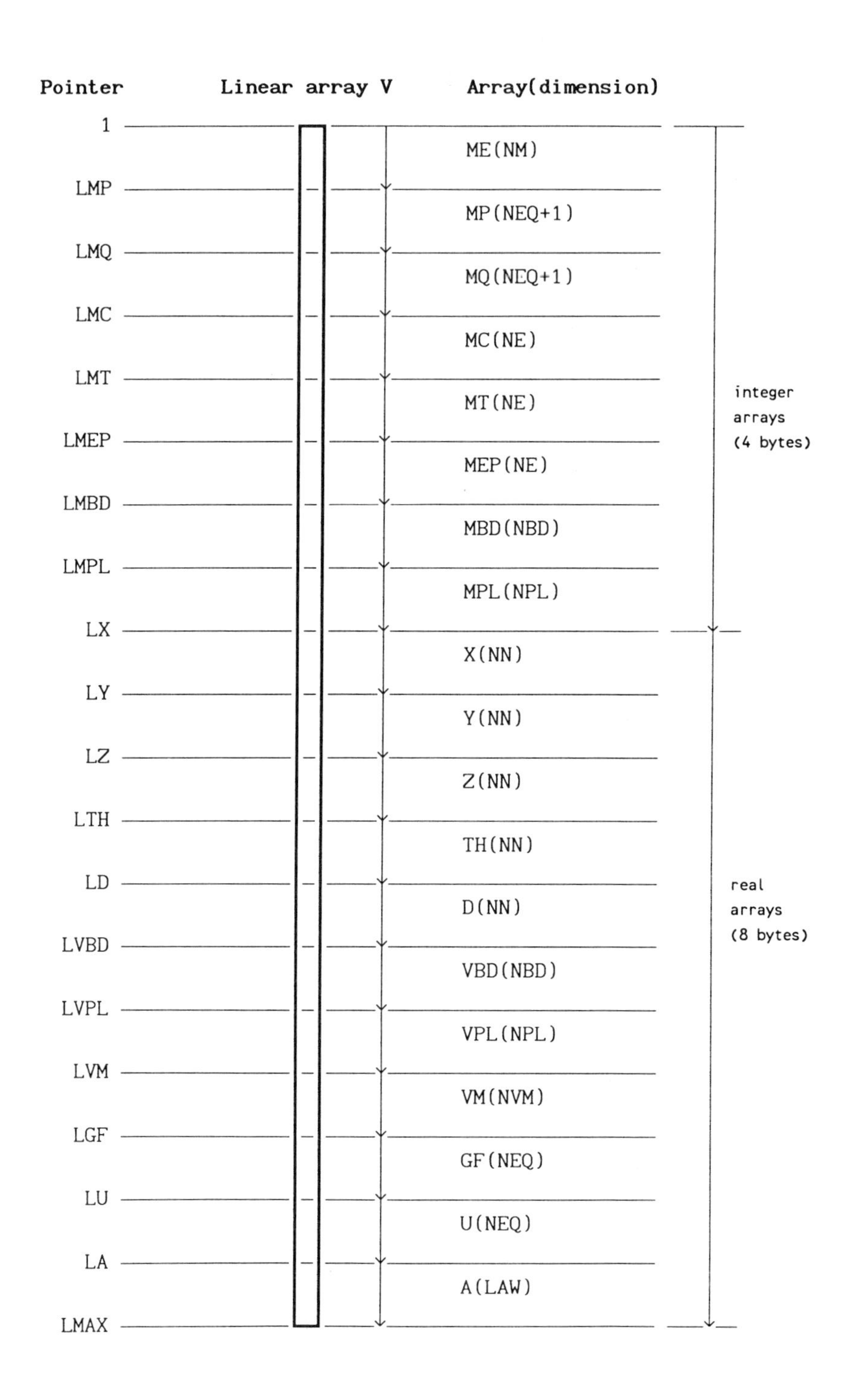

Fig. 8.12 Partition of linear array **V** in FACILE

Important *scalars* used in FACILE:

JP = Problem type identifier.
NN = Number of nodes in the system.
NE = Number of elements in the system.
NM = Length required for array ME.
NC = Number of element types requiring a change in the default number of integration points.
NBD = Number of specified displacements.
NPL = Number of specified forces.
NVM = Length of material constant array VM.
LMAX = Length of global linear array V.
NF = Number of DOF per node.
NEQ = Number of equations (DOF in the whole system).
LAW = Space remaining for global stiffness vector A.
NA = Profile of the global stiffness matrix.
NV = Number of nodes in an element.
NP = Number of integration points used.
NEF = Number of DOF in an element.
IP = Element node number start position on array ME.

Important *arrays* used in FACILE:

ME(*) = Array of element node numbers.
MP(*) = Pointer array of system stiffness vector A, MP(I) = the number of off-diagonal coefficients up to equation I.
MQ(*) = Working array.
MC(*) = Array of element material group numbers.
MT(*) = Array of element types.
MEP(*) = Pointer array, MEP(I) = node number start position of element I on array ME.
MBD (I), VBD(I) = Equation number and displacement value of the Ith displacement boundary condition.
MPL(I), VPL(I) = Equation number and force magnitude of the Ith loading condition.
X(I), Y(I), Z(I) = x, y, z-coordinates of node I.
TH(I) = Plate or shell thickness at node I.
D(*) = Diagonal elements of global stiffness matrix.
A(*) = Array of off-diagonal elements of global stiffness matrix.
VM(*) = Array of material constants.
GF(*) = Global force vector.
U(*) = Global displacement vector.
MV(I) = Number of nodes in an element type I.
MIP(I) = Default number of integration points for element type I.
MF(JP) = Number of DOF per node for problem type JP.
MM(JP) = Number of material parameters for problem type JP.
TOP(JP) = Title of problem type JP.
EK(*,*) = Element stiffness matrix.
GG(*), HH(*) = Values of interpolation functions at integration points.

DH1(*), DH2(*), DH3(*) = Derivatives of interpolation functions with respect to
DG1(*), DG2(*) element coordinates at integration points.
DHX(*), DHY(*), DHG(3,*) = Derivatives of interpolation functions with respect
DGX(*), DGY(*) to global coordinates at integration points.
DET(*), DT4(*) = Determinant of Jacobian at integration points.
WT(*) = Weight of integration at Gaussian points.

8.6.5 Input data for program FACILE

Input data for program FACILE are to be read from three data files FACILE.1, FACILE.2 and FACILE.3. FACILE.1 contains the geometrical data of the problem, FACILE.2 stores the material constants, and FACILE.3 provides the boundary condition information.

Input		Condition	File
JP, NN, NE, NM, NC, NBD, NPL			File **FACILE.1** Geometrical data
N1, [N2, N3]			
MT(I), I=1, NE			
MEP(I), I=1, NE			
ME(I), I=1, NM			
X(I), Y(I)	I=1, NN	if $2 \leq JP \leq 5$	
X(I), Y(I), Z(I),	I=1, NN	if otherwise	
[TH(I), I=1, NN]		if JP = 9	

Input	Condition	File
MC(I), I=1, NE		File **FACILE.2** Material data
VM(I), I=1, NVM		
[IT, NP]	if $NC \neq 0$	

Input	File
NBD, NPL	File **FACILE.3** Boundary condition data
MBD(I), VBD(I), I=1, NBD	
MPL(I), VPL(I), I=1, NPL	

JP = Problem type identifier; NN = No. of nodes; NE = No. of elements.
NM = Required length of ME; NC = No. of changes in integration scheme.
NBD = No. of specified displacements; NPL = No. of specified forces.
MT(I) = Element type; MEP(I) = Node number start position on array ME.
ME(*) = Array of element node numbers.
X(I), Y(I), Z(I) = x, y, z-coordinates of node I.
TH(I) = Thickness at node I; MC(I) = Material identifier of element I.
VM(I) = Vector of material characteristics.
MBD(I), VBD(I) = DOF and specified value of the Ith displacement B. C.
MPL(I), VPL(I) = DOF and specified value of the Ith traction force B. C.

Note: N2 and N3 are only required for spatial frame problems, JP = 1.
For plate bending problems, thickness at node I, TH(I) input as Z(I).

Remarks
(1) Definition of constants for various problem types.
The nodal degrees of freedom and the material constants used in different types of problems are given in Table 8.4:

Table 8.4 Material constants and nodal degrees of freedom

JP Problem	Material constants	Nodal DOF	Remarks
1. Spatial frame	$[A_2, B_2, C_2]$, $[E, G]$, $[A, I_2, I_3, J, F_2, F_3]$,	u, v, w, $\theta_1, \theta_2, \theta_3$	11 parameters divided into 3 groups
2. Plane strain 3. Plane stress	$\lambda_1, \lambda_2, \lambda_3, \lambda_4$	u, v	Orthotropic material
4. Axisymmetric	$\lambda_1, \lambda_2, \lambda_3, \lambda_4, \lambda_5$	u, v	Stratified material
5. 2D Field	k_1, k_2, c	ϕ	c = coefficient of convection
6. 3D Elasticity	λ, μ	u, v, w	Lamé constants
7. 3D Field	k_1, k_2, k_3	ϕ	Coefficients of conductivity in 3 directions
8. Plate bending	E, ν	w, θ_1, θ_2	Isotropic material
9. Shell structure	E, ν	u, v, w θ_1, θ_2	Isotropic material

More general material types other than those pre-determined in the program can be used in the analysis. To include more general material types in a particular problem, the user has to change the number of default material parameters (array MM) for that problem, and to supply accordingly the right number of parameters to define the new material type in FACILE.2. For instance, to define an orthotropic plate element, four material constants will be needed [9], instead of two as in the case of an isotropic plate. Modifications on the element stiffness subroutine EKL9P will also be required.

(2) Arrangement of material constants on array VM

Material constants have to be read in sequentially to fill material array VM from file FACILE.2. The data arrangement of VM for 3D spatial frame problems ($JP = 1$) is quite different from the other problem types.

For $JP = 1$,

NVM

VM =

$[A_2, B_2, C_2] \times N_1$ $\quad [E, G] \times N_2$ $\quad [A, I_2, I_3, J, F_2, F_3] \times N_3$

$$\text{NVM} = 3 \times N_1 + 2 \times N_2 + 6 \times N_3$$

where N_1 = number of orientation vectors,

N_2 = number of material types,

N_3 = number of section property groups.

The first element base vector can be defined by the two end nodes of a beam. A second base vector is needed to fix without ambiguity the orientation of the beam element in space, as shown in Fig. 8.13(b).

In case a zero vector (0,0,0) is entered as the orientation vector, FACILE will automatically calculate the second base vector. For vertical members, the second

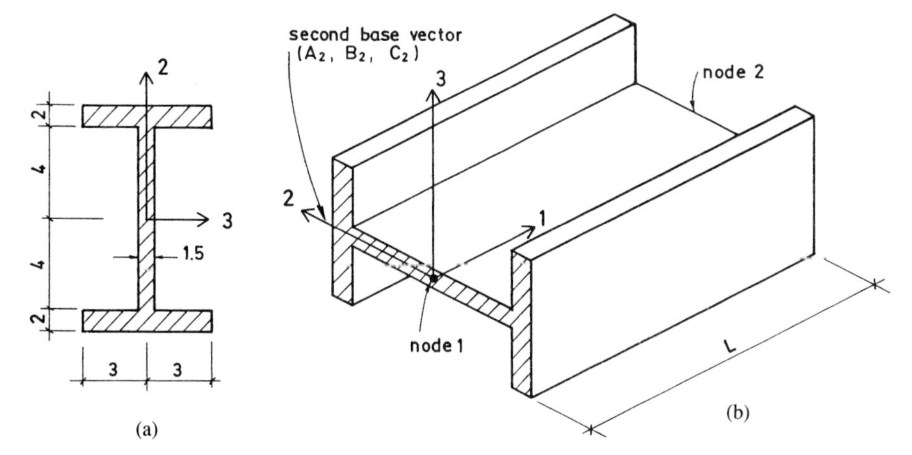

Fig. 8.13 (a) Section of a 3D beam; (b) orientation of a beam element

base vector = (1,0,0), whereas for non-vertical members, the second base vector lies in the vertical plane. The other parameters of a beam element are

E = Young's modulus

ν = Poisson's ratio

G = shear modulus

A = cross-sectional area

I_2 = second moment of area about element axis 2

I_3 = second moment of area about element axis 3

J = equivalent polar moment of area

F_2= shear factor along element axis 2, $F_2 = \dfrac{12EI_3}{GA_2L^2}$

F_3= shear factor along element axis 3, $F_3 = \dfrac{12EI_2}{GA_3L^2}$

where L is the length of the beam element. A_2 and A_3 are the effective cross-sectional areas in shear along axes 2 and 3 respectively. For rectangular sections, $A_2 = A_3 = \frac{5}{6}\times$ cross-sectional area A; and for non-rectangular sections, A_2 and A_3 can be approximated by the area of the web in the directions along the respective axes. When shear effects are to be neglected, F_2 and F_3 are given the value zero.

Take for example the section shown in Fig. 8.13(a),

$$E = 3 \times 10^6 \quad \nu = 0.25 \quad G = 1.2 \times 10^6$$

$$L = 100 \quad A = 2 \times 2 \times 6 + 8 \times 1.5 = 36$$

$$I_2 = 2 \times \frac{2 \times 6^3}{12} + \frac{8 \times 1.5^3}{12} = 74.25 \qquad I_3 = \frac{6 \times 12^3}{12} - 2 \times \frac{2.25 \times 8^3}{12} = 672$$

$$J = \frac{1}{3}(2 \times 6 \times 2^3 + 10 \times 1.5^3) = 43.25$$

$$A_2 \approx 10 \times 1.5 = 15 \qquad F_2 = \frac{12 \times 3 \times 10^6 \times 672}{1.2 \times 10^6 \times 15 \times 100^2} = 0.1344$$

$$A_3 \approx 2 \times 2 \times 6 = 24 \qquad F_3 = \frac{12 \times 3 \times 10^6 \times 74.25}{1.2 \times 10^6 \times 24 \times 100^2} = 0.009\,28$$

The material identifier MC(I) of 3D beam element I is a three-digit number, in which the first digit refers to the orientation vector, the second digit is the material set number, and the last digit is the group number of section properties. As a result, in spatial frame problems, the number of orientation vectors N_1, the number of material sets N_2, and the number of section property groups N_3 can only range from 1 to 9, i.e.

$$1 \leq N_1 \leq 9, \qquad 1 \leq N_2 \leq 9 \quad \text{and} \quad 1 \leq N_3 \leq 9$$

For example, MC(I) = 342 means that beam element I will take the 3rd orientation vector as the second base vector, material set number 4 and section properties of group 2.

For JP $\neq$ 1, number of material constants N_2 depends on the problem type as given in Table 8.4.

VM = [ ]

$$\text{NVM} = [N_2 \text{ material parameters}] \times N_1$$

where N_1 = number of material sets.

(3) Change of numerical integration schemes

If NC $\neq$ 0, there is a change in the number of integration points for certain element types. The user has to supply in file FACILE.2, after the material constants, NC pairs of IT and NP values, where IT is the element type and NP is the revised number of integration points.

(4) Equation number associated with a given nodal DOF

For example, it is necessary to know the equation number corresponding to displacement w at node 16 in a shell structural problem. Since there are five DOF at each node, and w is the third DOF of a node according to Table 8.4,

$$\text{equation number} = (16 - 1) \times 5 + 3 = 78$$

The equation number associated with a force in a certain direction at a node can also be calculated in a similar way.

8.6.6 Sample runs

Ten simple examples, each taken from a particular problem type, are included in this section. These examples will help readers familiarize themselves with the input data format for FACILE and the characteristics of the program when various element types are applied to different problems. Each example starts with a general description, is followed by the mesh of idealization, and finally the corresponding input data. A report on the characteristics and important parameters of the problem along with warning and error messages can be found in the file FACILE.7. The results of the analysis output in file FACILE.8 are the nodal displacements listed node by node in the same order of equation numbers of the system.

The examples are:

1. Analysis of a spatial frame structure.
2. A long cylinder under internal pressure.
3. The bending of a cantilever beam.
4. The bending of a circular plate.
5. Temperature distribution on a circular disc.
6. 3D analysis of a cantilever beam.
7. 3D heat conduction problem.
8. The bending of a rectangular platc.
9. The pinching of a cylindrical shell.
10. A cube under simple tensile force.

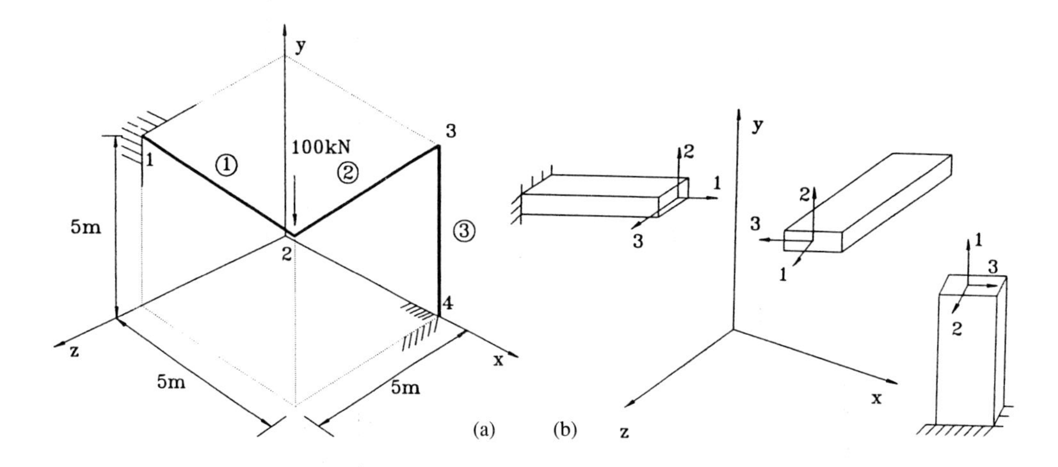

Fig. 8.14 (a) Spatial frame with 3 members. (b) Element local coordinate system.

FACILE.1

```
1 4 3 6 0 12 1
2 1 1
1 1 1
1 3 5
1 2 3 2 4 3
0 5000 5000 5000 5000 5000
5000 5000 0 5000 0 0
```

FACILE.2

```
111 111 211
0 1 0 0 0 1
400 160
10000 2.5E9 7.5E8 1.875E9
0 0
```

FACILE.3

```
12 1
1 0 2 0 3 0 4 0 5 0 6 0 19 0 20 0 21 0 22 0 23 0 24 0
8 -100
```

Displacements	u (mm)	v (mm)	w (mm)	θ_1	θ_2	θ_3
node 2	−0.003 04	−5.996 51	0.876 87	0.001 13	−0.000 24	−0.001 51
node 3	0.957 06	−0.045 36	0.911 29	0.000 75	−0.000 16	−0.000 37

Example 1: Analysis of a spatial frame structure

Figure 8.14(a) shows a 3D frame supporting a point load of 100 kN. In the structure, there are three beam elements and four nodes, in which node 1 and node 4 are encastré. The material and sectional properties of the beam elements are all the same such that $L = 5000$ mm, $E = 400$ kN/mm^2, $G = 160$ kN/mm^2, $A = 10\,000$ mm^2, $I_2 = 2500 \times 10^6$ mm^4, $I_3 = 750 \times 10^6$ mm^4 and $J = 1875 \times 10^6$ mm^4. The shear effect on the structure is neglected. The element local coordinate system in relation to the global coordinate system is given in Fig. 8.14(b).

Example 2: A long cylinder under internal pressure

This is a plane strain problem. From Fig. 8.15, the inner and outer radii of the cylinder are 5 cm and 10 cm respectively. The material constants used are $E =$

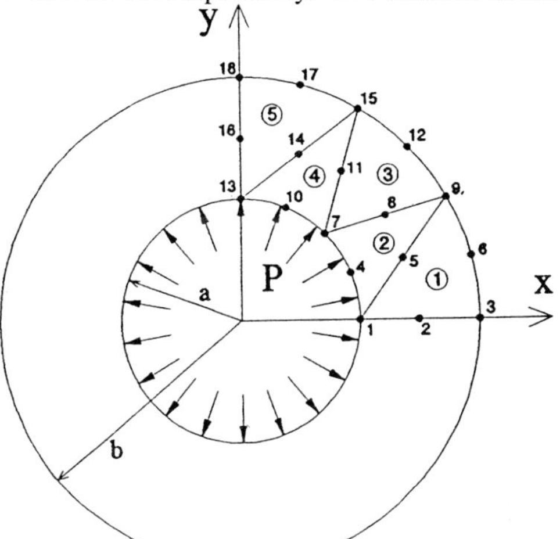

Fig. 8.15 Cross-section of cylinder

```
 3    18     5    30     0     6    10
 1
 2     2     2     2     2     2
 1     7    13    19    25    31
 1     3     9     6     5     2
 1     9     7     8     4     5
 7     9    15    12    11     8
 7    15    13    14    10    11
13    15    18    17    16    14
 5.000000    0.000000
 7.500000    0.000000
10.000000    0.000000
 4.619398    1.913417
 6.830127    2.500000
 9.659258    2.588190
 3.535534    3.535534
 6.097894    4.267767
 8.660254    5.000000
 1.913417    4.619398
 4.267767    6.097894
 7.071068    7.071068
 0.000000    5.000000
 2.500000    6.830127
 5.000000    8.660254
 0.000000    7.500000
 2.588190    9.659258
 0.000000   10.000000
```

FACILE.1

```
 6    10
 2    0.0
 4    0.0
 6    0.0
25    0.0
31    0.0
35    0.0
 1    .686356E+02
 2    .965691E+00
 7    .235702E+03
 8    .976311E+02
13    .970654E+02
14    .970654E+02
19    .976311E+02
20    .235702E+03
25    .965691E+00
26    .686356E+02
```

FACILE.3

```
    1     1     1     1     1     1
 .134615E+04   .134615E+04   .384615E+03   .576923E+03
```

FACILE.2

Node	1	2	3	7	12	13	16	18
u_r (FEM)	0.9244	0.7013	0.5854	0.9387	0.5948	0.9244	0.7013	0.5854
u_r (Ana.)	0.9533	0.7078	0.6067	0.9533	0.6067	0.9533	0.7078	0.6067

1000 kN/cm^2 and $\nu = 0.3$. A pressure of 100 kN/cm^2 is applied on the inner surface of the cylinder. One quarter of a typical cross-section of the cylinder is divided into five T6 triangular elements for analysis.

The analytical solution of radial displacement u_r at a distance r from the centre is given by

$$u_r = \frac{Pa^2}{E(b^2 - a^2)}\left[(1 - 2\nu)(1 + \nu)r + \frac{(1 + \nu)b^2}{r}\right]$$

Example 3: The bending of a cantilever beam

This is a plane stress problem. The dimensions, boundary and loading conditions of the cantilever beam are shown in Fig. 8.16. The finite element mesh for analysis consists of five Q8 elements and 28 nodes. The deflection δ at the tip from elementary beam theory is 4.1667 mm for $E = 1200$ N/mm^2 and $G = 480$ N/mm^2.

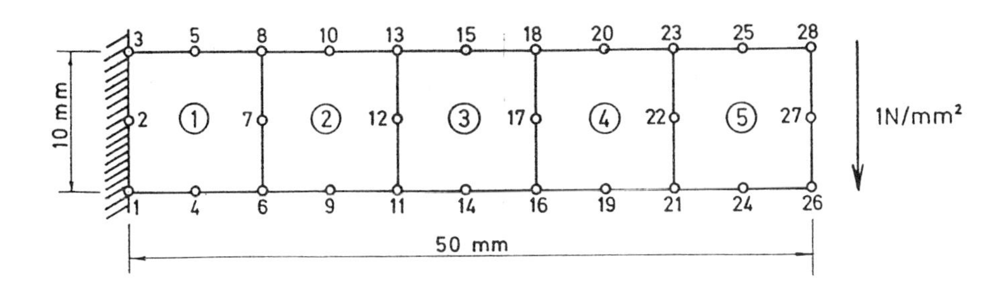

Fig. 8.16 A cantilever beam is divided into five Q8 elements

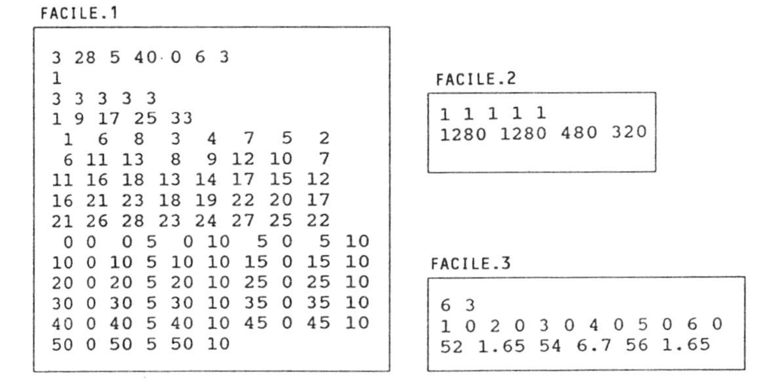

FACILE.1

```
3 28 5 40 0 6 3
1
3 3 3 3 3
1 9 17 25 33
 1  6  8  3  4  7  5  2
 6 11 13  8  9 12 10  7
11 16 18 13 14 17 15 12
16 21 23 18 19 22 20 17
21 26 28 23 24 27 25 22
 0 0  0 5  0 10  5 0  5 10
10 0 10 5 10 10 15 0 15 10
20 0 20 5 20 10 25 0 25 10
30 0 30 5 30 10 35 0 35 10
40 0 40 5 40 10 45 0 45 10
50 0 50 5 50 10
```

FACILE.2

```
1 1 1 1 1
1280 1280 480 320
```

FACILE.3

```
6 3
1 0 2 0 3 0 4 0 5 0 6 0
52 1.65 54 6.7 56 1.65
```

Finite element solution of vertical displacement at the tip, $\delta_{FEM} = 4.2257$ mm

Example 4: The bending of a circular plate

The bending of a circular plate under a central point load can be considered as an axisymmetric analysis. A typical section taken from the centre of the plate to the circumference is divided into five Q8 elements as shown in Fig. 8.17. The radius and thickness of the circular plate are 50 mm and 2 mm respectively. The plate which is fixed along the circumference is acted upon at the centre by a concentrated force of 100 N. The material properties of the plate are $E = 1200$ N/mm^2 and $\nu = 0.25$.

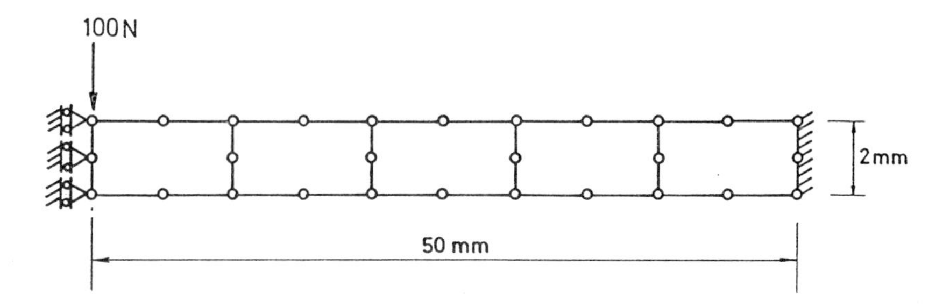

Fig. 8.17 Section of circular plate and finite element mesh for analysis

FACILE.1

```
4 28 5 40 0 9 1
1
3 3 3 3 3
1 9 17 25 33
 1  6  8  3  4  7  5  2
 6 11 13  8  9 12 10  7
11 16 18 13 14 17 15 12
16 21 23 18 19 22 20 17
21 26 28 23 24 27 25 22
0 0 0 1 0 2 5 0 5 2
10 0 10 1 10 2 15 0 15 2
20 0 20 1 20 2 25 0 25 2
30 0 30 1 30 2 35 0 35 2
40 0 40 1 40 2 45 0 45 2
50 0 50 1 50 2
```

FACILE.2

```
1 1 1 1 1
1440 1440 480 480 480
```

FACILE.3

```
9 1
1 0 3 0 5 0 51 0 52 0 53 0 54 0 55 0 56 0
6 100
```

The central deflection from Kirchhoff theory [22] is given by

$$\delta = \frac{Pa^2}{16\pi D}$$

where P = applied force, a = diameter, $D = \dfrac{Eh^3}{12(1-\nu^2)}$.

For the numerical constants used, $\delta = 5.8284$ mm.
The finite element solution, $\delta_{\mathrm{FEM}} = 5.8039$, differs by less than 1% from the thin plate solution.

Example 5: Temperature distribution on a circular disc

The temperature distribution on a circular disc is governed by the Laplace equation, for which an analytical solution exists. For a disc of unit radius and under the given boundary temperature distribution, the temperature at any point of the disc expressed in polar coordinates is given by

$$T = 5 + \frac{10}{\pi} \arctan \left(\frac{2r \sin \theta}{1 - r^2} \right)$$

In the finite element analysis, a mesh of five Q8 elements and three T6 elements is used as shown in Fig. 8.18.

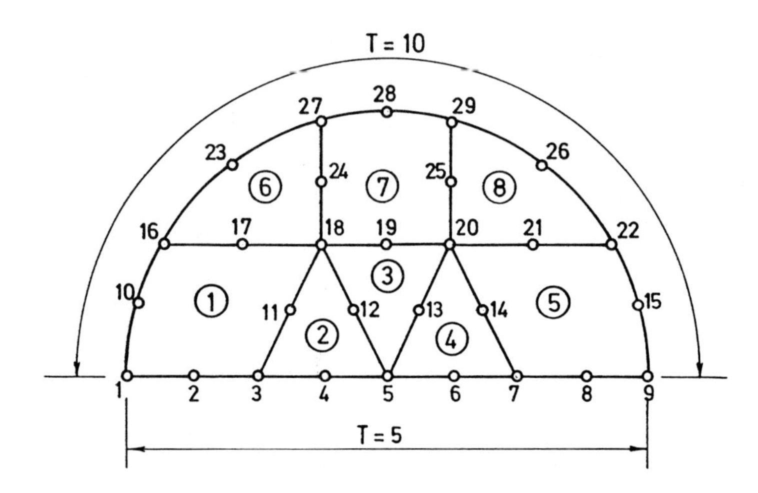

Fig. 8.18 Half of the circular disc divided into eight finite elements

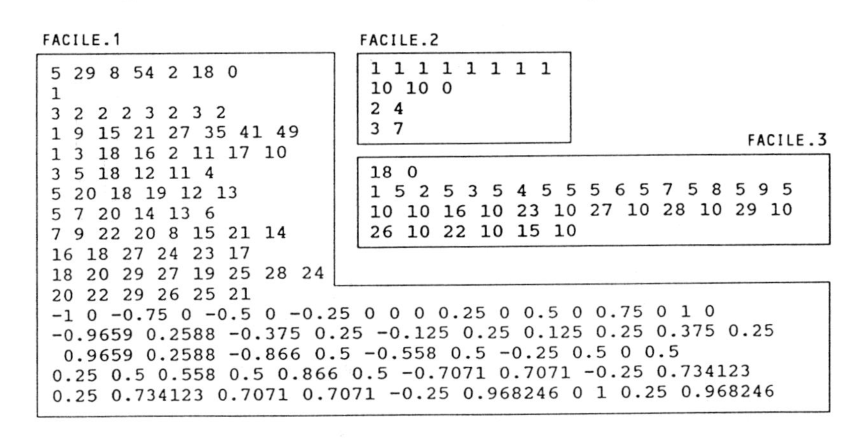

FACILE.1

```
5 29 8 54 2 18 0
1
3 2 2 2 3 2 3 2
1 9 15 21 27 35 41 49
1 3 18 16 2 11 17 10
3 5 18 12 11 4
5 20 18 19 12 13
5 7 20 14 13 6
7 9 22 20 8 15 21 14
16 18 27 24 23 17
18 20 29 27 19 25 28 24
20 22 29 26 25 21
-1 0 -0.75 0 -0.5 0 -0.25 0 0 0 0.25 0 0.5 0 0.75 0 1 0
-0.9659 0.2588 -0.375 0.25 -0.125 0.25 0.125 0.25 0.375 0.25
 0.9659 0.2588 -0.866 0.5 -0.558 0.5 -0.25 0.5 0 0.5
0.25 0.5 0.558 0.5 0.866 0.5 -0.7071 0.7071 -0.25 0.734123
0.25 0.734123 0.7071 0.7071 -0.25 0.968246 0 1 0.25 0.968246
```

FACILE.2

```
1 1 1 1 1 1 1 1
10 10 0
2 4
3 7
```

FACILE.3

```
18 0
1 5 2 5 3 5 4 5 5 5 6 5 7 5 8 5 9 5
10 10 16 10 23 10 27 10 28 10 29 10
26 10 22 10 15 10
```

Node	13	14	19	20	21	25
T_{FEM}	6.5479	6.5052	8.0259	8.2294	8.4263	9.2285
T_{EXACT}	6.5819	6.7837	7.9517	8.0829	8.6842	9.1563

Example 6: 3D analysis of a cantilever beam

The cantilever beam is considered as a 3D structure, and is modelled by two H20 hexahedron elements. The geometry of the beam, the boundary conditions and loading are shown in Fig. 8.19. The material constants used are $\lambda = 48\,000\,\text{kN/m}^2$

and $\mu = 48\,000\,\text{kN/m}^2$. A 14-point integration formula was used instead of the 8-point rule by default.

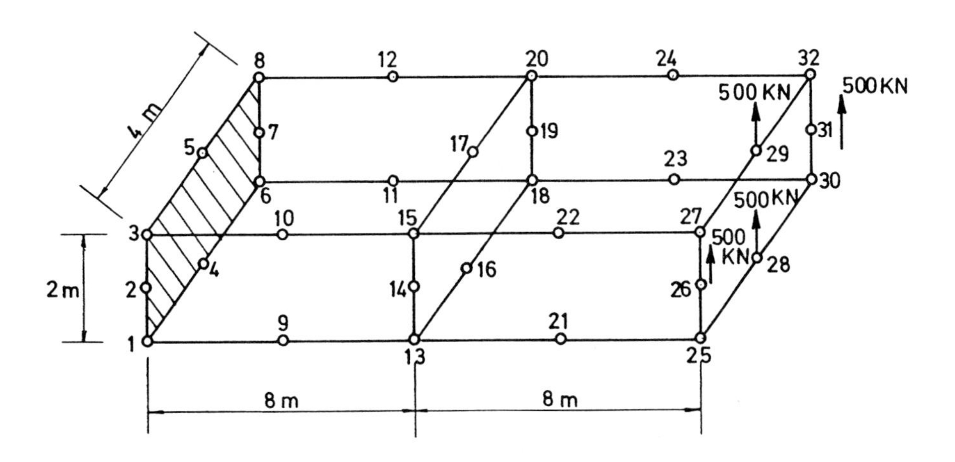

Fig. 8.19 3D analysis of a cantilever beam using two H20 elements

FACILE.1

```
6 32 2 40 1 60 30
1
7 7
1 21
1 13 18 6 3 15 20 8 9 16 11 4 2 14 19 7 10 17 12 5
13 25 30 18 15 27 32 20 21 28 23 16 14 26 31 19 22 29 24 17
0 0 0 0 0 1 0 0 2 0 2 0 0 2 2 0 4 0 0 4 1 0 4 2
4 0 0 4 0 2 4 4 0 4 4 2
8 0 0 8 0 1 8 0 2 8 2 0 8 2 2 8 4 0 8 4 1 8 4 2
12 0 0 12 0 2 12 4 0 12 4 2
16 0 0 16 0 1 16 0 2 16 2 0 16 2 2 16 4 0 16 4 1 16 4 2
```

FACILE.2

```
1 1
48000 48000
7 14
```

FACILE.3

```
24 4
1 0 4 0 7 0 10 0 13 0 16 0 19 0 22 0
11 0 14 0 2 0 5 0 8 0 17 0 20 0 23 0
6 0 21 0 3 0 12 0 18 0 9 0 15 0 24 0
78 500 84 500 87 500 93 500
```

Node	25	26	27	28
u	0.768 72	0	−0.768 72	0.766 98
v	0.003 99	0	−0.003 99	0
w	7.831 25	7.83167	7.831 25	7.829 17

Example 7: 3D heat conduction problem

This is a very simple example of the heat conduction problem, in which the two ends of a prismatic bar are maintained at different temperatures and no heat is allowed to be lost through the insulated lateral surfaces. The theoretical temperature distribution at any cross-section is uniform and is given by a linear interpolation defined by the temperatures at the ends. In the numerical analysis, the problem domain is divided into one H20 element and two P15 elements as shown in Fig. 8.20. The boundary conditions require that nodes 1 to 8 be given a temperature of 0°C and nodes 25 to 32 be given a temperature of 10°C. As expected, the finite element results coincide with the theoretical solution.

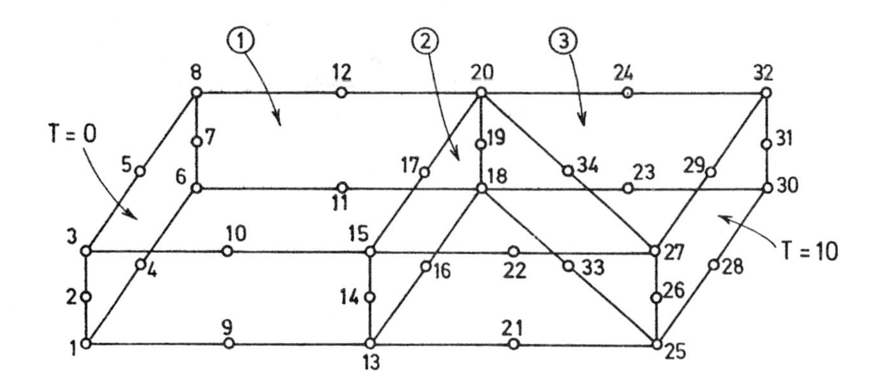

Fig. 8.20 The finite element mesh for the 3D heat conduction problem

FACILE.1

```
7 34 3 50 0 60 30
1
7 6 6
1 21 36
1 13 18 6 3 15 20 8 9 16 11 4 2 14 19 7 10 17 12 5
13 25 18 33 16 21 14 26 19 15 27 20 34 17 22
25 30 18 23 33 28 26 31 19 27 32 20 24 34 29
0 0 0 0 0 1 0 0 2 0 2 0 0 2 2 0 4 0 0 4 1 0 4 2
4 0 0 4 0 2 4 4 0 4 4 2
8 0 0 8 0 1 8 0 2 8 2 0 8 2 2 8 4 0 8 4 1 8 4 2
12 0 0 12 0 2 12 4 0 12 4 2
16 0 0 16 0 1 16 0 2 16 2 0 16 2 2 16 4 0 16 4 1 16 4 2
12 2 0 12 2 2
```

FACILE.2

```
1 1 1
10 10 10
```

FACILE.3

```
16 0
1 0 2 0 3 0 4 0 5 0 6 0 7 0 8 0
25 10 26 10 27 10 28 10 29 10 30 10 31 10 32 10
```

Example 8: The bending of a rectangular plate

A uniformly distributed load of intensity 5.76 kN/m^2 is applied on a 10 m $\times$ 10 m square plate encastré along its boundary as shown in Fig. 8.21. In the finite element analysis, one quarter of the plate is divided into four 9-node Lagrangian plate elements. The plate is assumed to be isotropic with material constants $E = 1.092 \times 10^9$ kN/m^2 and $\nu = 0.3$. The plate has a uniform thickness equal to 0.01 m.

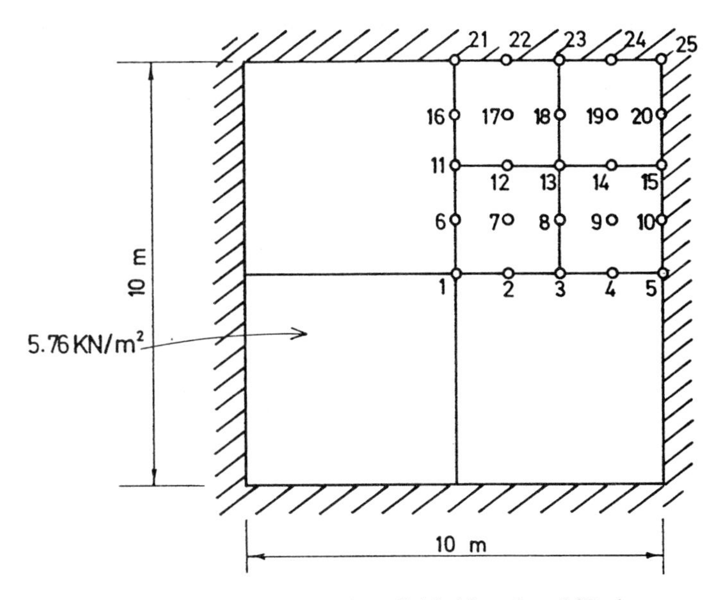

Fig. 8.21 One quarter of the plate divided into four L9P elements

FACILE.1

```
8 25 4 36 0 35 16
1
8 8 8 8
1 10 19 28
1 3 13 11 2 8 12 6 7
3 5 15 13 4 10 14 8 9
11 13 23 21 12 18 22 16 17
13 15 25 23 14 20 24 18 19
0 0    .01 1.25 0    .01 2.5 0    .01 3.75 0    .01 5 0    .01
0 1.25 .01 1.25 1.25 .01 2.5 1.25 .01 3.75 1.25 .01 5 1.25 .01
0 2.50 .01 1.25 2.50 .01 2.5 2.50 .01 3.75 2.50 .01 5 2.50 .01
0 3.75 .01 1.25 3.75 .01 2.5 3.75 .01 3.75 3.75 .01 5 3.75 .01
0 5.00 .01 1.25 5.00 .01 2.5 5.00 .01 3.75 5.00 .01 5 5.00 .01
```

FACILE.2

```
1 1 1 1
1.092E6 0.3
```

FACILE.3

```
35 16
13 0 14 0 15 0 30 0 29 0 28 0 45 0 44 0 43 0 60 0 59 0 58 0
61 0 62 0 63 0 64 0 65 0 66 0 67 0 68 0 69 0 70 0 71 0 72 0
73 0 74 0 75 0 2 0 3 0 6 0 9 0 12 0 17 0 32 0 47 0
1 1 4 4 7 2 10 4 16 4 19 16 22 8 25 16
31 2 34 8 37 4 40 8 46 4 49 16 52 8 55 16
```

The central deflection from the thin plate theory by Timoshenko [22] is

$$\delta = \frac{0.001\,26qa^4}{D} = 0.725\,76 \text{ m}$$

whereas from the finite element solution, $\delta_{\text{FEM}} = 0.735\,04\,\text{m}$.

Example 9: The pinching of a cylindrical shell

A cylindrical shell acted upon by a pair of equal and opposite point forces along a diameter of the middle plane is shown in Fig. 8.22. Both ends of the cylinder are covered with rigid diaphragms, which will allow displacement only in the axial direction of the cylinder. One-eighth of the cylinder is modelled by four 9-node Lagrangian shell elements. The material constants used are $E = 30\,000\,\text{N/mm}^2$ and $\nu = 0.3$.

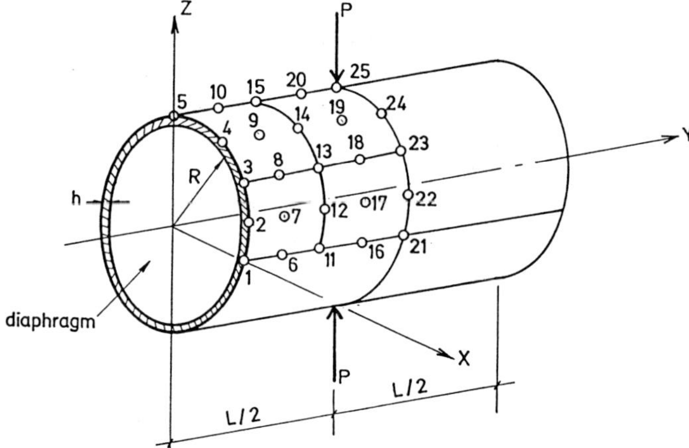

Fig. 8.22 The pinching of a cylindrical shell

FACILE.3

```
 51          1
  1 0    2 0    3 0    4 0    5 0
  6 0    7 0    8 0    9 0   10 0
 11 0   12 0   13 0   14 0   15 0
 16 0   17 0   18 0   19 0   20 0
 21 0   22 0   23 0   24 0   25 0
 28 0   29 0   46 0   49 0
107 0  110 0   53 0   54 0
 71 0   74 0  112 0  115 0
 78 0   79 0   96 0   99 0
117 0  120 0  102 0  103 0  104 0
105 0  121 0  122 0  124 0  125 0
123    2500.
```

FACILE.1

```
 9      25      4      36       0     51      1
 1
 9       9      9       9
 1      10     19      28
 1      11     13       3       6     12      8      2      7
 3      13     15       5       8     14     10      4      9
11      21     23      13      16     22     18     12     17
13      23     25      15      18     24     20     14     19
300.00     0.      .0      277.16     0.    114.81
212.13     0.    212.13    114.81     0.    277.16
  0.       0.    300.00    300.00   75.0      0.
277.16   75.0    114.81    212.13   75.0    212.13
114.81   75.0    277.16      0.     75.0    300.00
300.00  150.0      0.      277.16  150.0    114.81
212.13  150.0    212.13    114.81  150.0    277.16
  0.    150.0    300.00    300.00  225.0      0.
277.16  225.0    114.81    212.13  225.0    212.13
114.81  225.0    277.16      0.    225.0    300.00
300.00  300.0      0.      277.16  300.0    114.81
212.13  300.0    212.13    114.81  300.0    277.16
  0.    300.0    300.00
3.  3.  3.  3.  3.  3.  3.  3.  3.  3.  3.  3.  3.  3.  3.
3.  3.  3.  3.  3.  3.  3.  3.  3.  3.
```

FACILE.2

```
1 1 1 1
3.E6 0.3
```

For $R = 3000$ mm, $L = 6000$ mm, $h = 30$ mm, $P = 10\,000$ N, the analytical solution quoted in Reference 10 is $\delta = 1.8248$ mm.

From the finite element analysis using four L9S elements, $\delta_{FEM} = 1.3897$ mm.

Example 10: A cube under simple tensile force

This is a 3D elasticity problem, in which a uniform traction force of 600 N/cm^2 is applied on the top and bottom surfaces of a 10 cm $\times$ 10 cm $\times$ 10 cm cube. In the finite element analysis, the cube is divided into six 10-node tetrahedron T10 elements as shown in Fig. 8.23. The material is assumed to be isotropic, with $\lambda = 48\,000$ N/cm^2 and $\mu = 48\,000$ N/cm^2. In this simple tension problem, in spite of the coarse mesh used, the finite element analysis gives results identical to the theoretical solution.

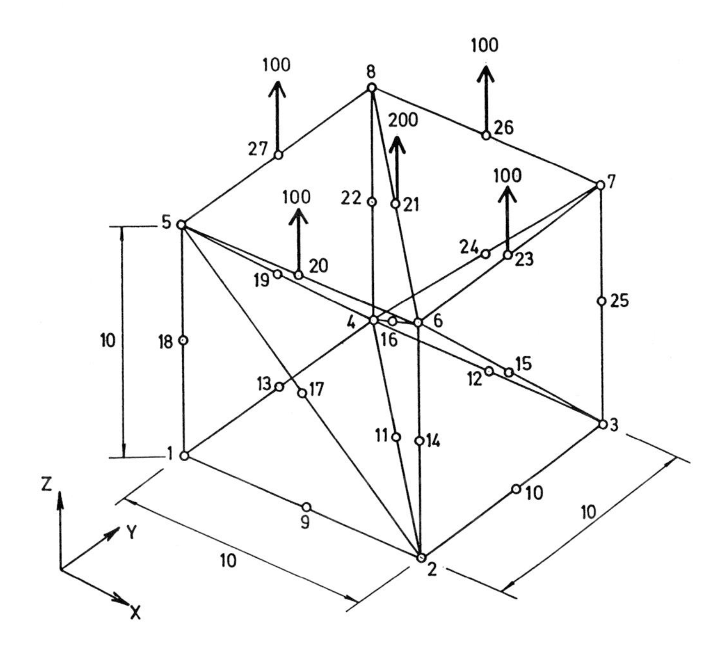

Fig. 8.23 A cube divided into six tetrahedral elements

FACILE.1

```
6 27 6 60 1 27 20
1
5 5 5 5 5 5
1 11 21 31 41 51
1 2 4 5 9 11 13 18 17 19
2 4 5 6 11 19 17 14 16 20
2 3 4 6 10 12 11 14 15 16
3 7 4 6 25 24 12 15 23 16
4 7 8 6 24 26 22 16 23 21
4 8 5 6 22 27 19 16 21 20
0 0 0 10 0 0 10 10 0 0 10 0
0 0 10 10 0 10 10 10 10 0 10 10
5 0 0 10 5 0 5 5 0 5 10 0
0 5 0 10 0 5 10 5 5 5 5 5
5 0 5 0 0 5 0 5 5 5 0 10
5 5 10 0 10 5 10 5 10 5 10 5
10 10 5 5 10 10 0 5 10
```

FACILE.2

```
1 1 1 1 1 1
48000 48000 48000
5 4
```

FACILE.3

```
15 5
3 0 27 0 6 0 39 0 33 0 30 0 12 0 36 0 9 0
25 0 34 0 29 0 38 0 31 0 32 0
60 1.E6 81 1.E6 63 2.E6 69 1.E6 78 1.E6
```

8.6.7 Program listing

We include here the complete Fortran listing of the program FACILE. The program is written in Fortran 77, and has been tested on an IBM-PC, an IBM RISC System 6000 machine and Digital VAX/VMS systems. FACILE consists of 26 routines, which are divided into groups and stored in six different files:

1. File **FACILE.FOR**
   Main control program.
2. File **EXECUT.FOR**
   Execution segment.
3. File **LIB1.FOR**
   Routines directly called by routine EXECUT.
4. File **LIB2.FOR**
   Routines of element interpolation functions.
5. File **LIB3.FOR**
   Routines for the evaluation of element stiffness matrices.
6. File **LIB4.FOR**
   Routines of special finite elements.

In the present version, the length LMAX of the global linear array V in the main program FACILE has been set equal to 80 000, which makes the size of the program a little bit less than 512 kilobytes. Although the exact number depends on the actual element connectivity of the structure and many other factors, with LMAX = 80 000, FACILE can handle approximately 1000 3D beam elements, 200 Q8 isoparametric elements, 50 L9P plate elements or 30 H20 hexahedron elements. When running under a virtual memory system or if more memory is available, the capacity of the program can be enhanced simply by changing LMAX for array V to a larger value.

Data input is taken from file FACILE.1, FACILE.2 and FACILE.3 in free format; and results are output to files FACILE.7 and FACILE.8. For users working on an IBM-PC or a compatible machine, the routines can be compiled and linked using the Microsoft Fortran Compiler and Linker. If the executable program file is given the name FACILE.EXE then the program is run by issuing the command "FACILE" at the DOS command level.

```
$LARGE
      PROGRAM FACILE
C
C     FINITE-ELEMENT ANALYSIS CODE IN LINEAR ELASTICITY (FACILE) by S. H. LO
C     Stored in file FACILE.FOR, last updated on 22 June 1991.
C
C     Spatial frame analysis, Plane stress/strain, Axisymmetric stress,
C     2D field problem, 3D elasticity, 3D field problem, plate bending,
C     Shell structures, etc.
C
      PARAMETER (LMAX=80000)
      DIMENSION MF(9),MM(9),MV(9),MIP(9),V(LMAX)
      CHARACTER*30 TOP(9)
      COMMON JP,NN,NE,NF,NEQ,LAW,NM,NC,N1,N2,N3,NVM
      DATA MV/2,6,8,6,10,15,20,9,9/
      DATA MM/0,4,4,5,3,2,3,2,2/
      DATA MIP/0,4,4,4,5,8,8,4,4/
      DATA MF/6,2,2,2,1,3,1,3,5/
      DATA TOP/'SPATIAL FRAME ANALYSIS','PLANE STRAIN ANALYSIS',
     -         'PLANE STRESS ANALYSIS','AXISYMMETRIC STRESS ANALYSIS',
     -'2D FIELD PROBLEM','3D ELASTICITY PROBLEM','3D FIELD PROBLEM',
     -'PLATE BENDING PROBLEM','SHELL STRUCTURE PROBLEM'/
      OPEN (1,FILE='FACILE.1')
      OPEN (2,FILE='FACILE.2')
      OPEN (3,FILE='FACILE.3')
      OPEN (7,FILE='FACILE.7')
      OPEN (8,FILE='FACILE.8')
C
C     ARRAYS :  MV(I) = NUMBER OF NODE IN ELEMENT TYPE I
C              MIP(I) = NUMBER OF INTEGRATION POINT FOR ELEMENT TYPE I
C              MF(JP) = NUMBER OF D.O.F. PER NODE FOR PROBLEM TYPE JP
C              MM(JP) = NUMBER OF MATERIAL PARAMETERS FOR PROBLEM TYPE JP
C             TOP(JP) = NAME OF PROBLEM TYPE JP
C
C     READ IN PROBLEM DEFINITION PARAMETERS JP,NN,NE,NM,NC,NBD,NPL,N1,N2,N3
C
      CALL ELAPTI (0,'CPU TIME COUNTER INITIALIZATION ')
      READ (1,*) JP,NN,NE,NM,NC,NBD,NPL
      WRITE (7,10) JP,TOP(JP)
   10 FORMAT (/' PROBLEM TYPE =',I2,10X,A30)
      WRITE (7,20) NN
   20 FORMAT (/' NUMBER OF NODES =',I5)
      WRITE (7,30) NE
   30 FORMAT (/' NUMBER OF ELEMENTS =',I5)
      WRITE (7,40) NM
   40 FORMAT (/' LENGTH REQUIRED FOR ARRAY ME =',I5)
      WRITE (7,130) NC
  130 FORMAT (/' NO. OF EL. TYPES REQUIRING A CHANGE IN',
     -         ' THE DEFAULT NO. OF INT. POINTS =',I3)
      WRITE (7,110) NBD
  110 FORMAT (/' MAXIMUM NUMBER OF SPECIFIED DISPLACEMENT B.C. =',I5)
      WRITE (7,120) NPL
  120 FORMAT (/' MAXIMUM NUMBER OF NON-ZERO FORCES IN A SINGLE LOADING',
     -         ' CASE =',I5)
      IF (JP.EQ.1) THEN
      READ (1,*) N1,N2,N3
      NVM=3*N1+2*N2+6*N3
      WRITE (7,140) N1,N2,N3
  140 FORMAT (/' NUMBER OF SECOND BASE VECTORS, N1 =',I5
     -        /' NUMBER OF MATERIAL TYPES, N2 =',I5
     -        /' NUMBER OF SECTION PROPERTIES, N3 =',I5)
      ELSE
      READ (1,*) N1
      N2=MM(JP)
      NVM=N1*N2
      WRITE (7,150) N1,N2
  150 FORMAT (/' NUMBER OF MATERIAL TYPES, N1 =',I5
     -        /' NUMBER OF PARAMETERS PER MAT. TYPE, N2 =',I5)
      ENDIF
C
C     CALCULATION OF PROBLEM SIZE
C
```

```
      NF=MF(JP)
      NEQ=NN*NF
      LMP=1+NM
      LMQ=LMP+NEQ+1
      LMC=LMQ+NN
      LMT=LMC+NE
      LMEP=LMT+NE
      LMBD=LMEP+NE
      LMPL=LMBD+NBD
      LX=LMPL+NPL
      LY=LX+NN*2
      LZ=LY+NN*2
      LTH=LZ+NN*2
      LD=LTH+NN*2
      LVBD=LD+NEQ*2
      LVPL=LVBD+NBD*2
      LVM=LVPL+NPL*2
      LGF=LVM+NVM*2
      LU=LGF+NEQ*2
      LA=LU+NEQ*2
      LAW=(LMAX-LA+1)/2
      WRITE (7,50) LMAX
   50 FORMAT (/' MAXIMUM TOTAL ARRAY DIMENSION =',I8)
      WRITE (7,60) NF
   60 FORMAT (/' NUMBER OF D.O.F. PER NODE =',I2)
      WRITE (7,70) NEQ
   70 FORMAT (/' TOTAL NUMBER OF EQUATIONS =',I6)
      WRITE (7,80) LA
   80 FORMAT (/' ARRAY SIZE REQUIRED TO HOLD INPUT DATA =',I7)
      WRITE (7,90) LAW
   90 FORMAT (/' SPACE REMAINED FOR MATRIX DECOMPOSITION =',I8)
C
      CALL EXECUT (V,V(LMP),V(LMQ),V(LMC),V(LMT),V(LMEP),
     -             V(LMBD),V(LMPL),V(LX),V(LY),V(LZ),V(LTH),V(LD),
     -             V(LVBD),V(LVPL),V(LVM),V(LGF),V(LU),V(LA),MV,MIP)
      STOP
      END
```

```
$LARGE
      SUBROUTINE EXECUT (ME,MP,MQ,MC,MT,MEP,MBD,MPL,X,Y,Z,
     -                   TH,D,VBD,VM,VPL,GF,U,A,MV,MIP)
C
C        FACILE - EXECUTION SEGMENT updated by S. H. LO on 22 June 1991.
C
      IMPLICIT DOUBLE PRECISION (A-H,O-Z)
      DIMENSION ME(*),MP(*),MQ(*),MC(*),MT(*),MEP(*),VM(*),
     -          MBD(*),MPL(*),X(*),Y(*),Z(*),TH(*),D(*),VBD(*),
     -          VPL(*),GF(*),U(*),A(*),MV(*),MIP(*),EK(60,60)
      COMMON JP,NN,NE,NF,NEQ,LAW,NM,NC,N1,N2,N3,NVM
      COMMON /BK1/C1(5)
      COMMON /BK2/C2(6)
      DATA BIG/1.D20/
C
C     READ IN MESH DATA X(*),Y(*),Z(*),MT(*),MEP(*),ME(*) FROM FILE FACILE.1
C     READ IN ELEMENT MATERIAL PROPERTIES FROM FILE FACILE.2
C
      READ (1,*) (MT(I),I=1,NE)
      READ (1,*) (MEP(I),I=1,NE)
      READ (1,*) (ME(I),I=1,NM)
      IF (JP.GE.2.AND.JP.LE.5) THEN
      READ (1,*) (X(I),Y(I),I=1,NN)
      ELSE
      READ (1,*) (X(I),Y(I),Z(I),I=1,NN)
      ENDIF
      IF (JP.EQ.9) THEN
      READ (1,*) (TH(I),I=1,NN)
      CALL NORMAL (NE,MEP,ME,MT,MV,X,Y,Z,GF,U)
      ENDIF
      READ (2,*) (MC(I),I=1,NE)
      READ (2,*) (VM(I),I=1,NVM)
C
C     CHANGE THE NUMBER OF INTEGRATION POINTS FOR CERTAIN ELEMENT TYPES
C     IT = TYPE OF ELEMENT, NP = NUMBER OF INTEGRATION POINTS
C
      DO 55 I=1,NC
      READ (2,*) IT,NP
   55 MIP(IT)=NP
      CALL ELAPTI (1,'INPUT MESH AND MATERIAL DATA      ')
C
C     CALCULATE ELEMENT STIFFNESS MATRICES AND ASSEMBLE SYSTEM VECTORS A & D
C
      CALL POINTN (NF,NE,MT,MEP,ME,MV,NN,MP,MQ)
      NA=MP(NEQ)
      WRITE (7,10) NA
   10 FORMAT (/' PROFILE OF THE GLOBAL STIFFNESS MATRIX =',I8)
      IF (NA.GT.LAW) STOP '*** NA > LAW ***'
      LT=0
      DO 66 I=1,NE
      IT=MT(I)
      IC=MC(I)
      IP=MEP(I)
      NV=MV(IT)
      NP=MIP(IT)
      NEF=NV*NF
      IF (JP.EQ.1) THEN
      K1=IC/100
      K2=(IC-100*K1)/10
      K3=MOD(IC,10)
      I1=3*K1-3
      I2=3*N1+2*K2-2
      I3=3*N1+2*N2+6*K3-6
      C1(1)=VM(I1+1)
      C1(2)=VM(I1+2)
      C1(3)=VM(I1+3)
      C1(4)=VM(I2+1)
      C1(5)=VM(I2+2)
      DO 77 J=1,6
   77 C2(J)=VM(I3+J)
      ELSE
      K=N2*(IC-1)
```

```
      DO 44 J=1,N2
   44 C1(J)=VM(K+J)
      ENDIF
      CALL ESTIMA (JP,LT,IT,IP,ME,NV,NP,X,Y,Z,TH,NEF,EK,GF,U)
      CALL ASSEMB (IP,ME,NV,NF,NEF,EK,MP,A,D)
   66 LT=IT
      CALL ELAPTI (2,'FORMING GLOBAL STIFFNESS MATRIX  ')
C
C     IMPOSING DISPLACEMENT BOUNDARY CONDITIONS STORED IN FILE FACILE.3
C     DECOMPOSITION OF GLOBAL STIFFNESS MATRIX A & D
C
      READ (3,*) NBD,NPL
      READ (3,*) (MBD(I),VBD(I),I=1,NBD)
      DO 33 I=1,NEQ
   33 U(I)=0.D0
      DO 11 I=1,NBD
      D(MBD(I))=BIG
   11 U(MBD(I))=BIG*VBD(I)
      CALL REDUC1 (NEQ,MP,D,A)
      CALL ELAPTI (3,'DECOMPOSE GLOBAL STIFFNESS MATRIX')
C
C     READ IN LOADING CONDITION AND CONSTRUCT RIGHT-HAND VECTOR U
C     CALCULATE AND OUTPUT GLOBAL DISPLACEMENT VECTOR U TO FILE FACILE.8
C
      READ (3,*) (MPL(I),VPL(I),I=1,NPL)
      DO 22 I=1,NPL
   22 U(MPL(I))=U(MPL(I))+VPL(I)
      CALL SOLVE1 (NEQ,MP,D,A,U)
      IF (JP.EQ.9) THEN
      WRITE (8,30) (U(I),I=1,NEQ)
   30 FORMAT (5F15.5)
      ELSE
      WRITE (8,20) (U(I),I=1,NEQ)
   20 FORMAT (6F13.5)
      ENDIF
      CALL ELAPTI (4,'SOLVING FOR DISPLACEMENT VECTOR  ')
      RETURN
      END
```

```
$LARGE
C         File LIB1.FOR, last updated on 11 July 1991
C
      SUBROUTINE ELAPTI (K,CH)
      IMPLICIT DOUBLE PRECISION (A-H,O-Z)
      CHARACTER*33 CH
      IH1=IH
      IM1=IM
      IS1=IS
      I101=I100
      CALL GETTIM (IH,IM,IS,I100)
      IF (K.EQ.0) RETURN
      TIME=3600*(IH-IH1)+60*(IM-IM1)+(IS-IS1)+0.01*(I100-I101)
      WRITE (*,10) K,CH,TIME
      WRITE (7,10) K,CH,TIME
   10 FORMAT (/' STAGE :',I2,5X,A33,5X,'CPU TIME =',F8.2)
      RETURN
      END

      SUBROUTINE REDUC1 (NEQ,MP,D,A)
      IMPLICIT DOUBLE PRECISION (A-H,O-Z)
      DIMENSION MP(*),D(*),A(*)
C
C     FUNCTION : CROUT DECOMPOSITION A = LDU
C
C     INPUT :    NEQ = NUMBER OF EQUATIONS
C              MP(I) = NUMBER OF OFF-DIAGONAL ELEMENTS UP TO COLUMN I
C               D(I) = ARRAY OF DIAGONAL ELEMENTS
C               A(I) = ARRAY OF OFF-DIAGONAL ELEMENTS
C
C     OUTPUT : D(I), A(I) reduced.
C
      DO 11 K=2,NEQ
      K1=K-1
      LK=MP(K)-K1
      KH=MP(K1)-LK+1
      S=D(K)
C
      DO 22 J=KH+1,K1
      J1=J-1
      LJ=MP(J)-J1
      JH=MAX(MP(J1)-LJ+1,KH)
      T=A(LK+J)
      DO 33 M=JH,J1
   33 T=T-A(LJ+M)*A(LK+M)
   22 A(LK+J)=T
C
      DO 44 J=KH,K1
      L=LK+J
      T=A(L)
      A(L)=T/D(J)
   44 S=S-T*A(L)
   11 D(K)=S
      RETURN
      END

      SUBROUTINE SOLVE1 (NEQ,MP,D,A,B)
      IMPLICIT DOUBLE PRECISION (A-H,O-Z)
      DIMENSION MP(*),D(*),A(*),B(*)
C
C     FUNCTION : SOLVE FOR X, LDUx = b      where U = transpose of L
C
C     1. FORWARD SUBSTITUTION : Lz = b
C
      DO 11 J=2,NEQ
      J1=J-1
      LJ=MP(J)-J1
      JH=MP(J1)-LJ+1
      T=B(J)
      DO 22 M=JH,J1
   22 T=T-A(LJ+M)*B(M)
   11 B(J)=T
```

```
C
C     2. DIVIDING BY DIAGONAL ELEMENTS : Dy = z
C
      DO 33 K=1,NEQ
   33 B(K)=B(K)/D(K)
C
C     3. BACKWARD SUBSTITUTION : Ux = y
C
      DO 44 K=NEQ,2,-1
      K1=K-1
      LK=MP(K)-K1
      KH=MP(K1)-LK+1
      T=B(K)
      DO 44 J=KH,K1
   44 B(J)=B(J)-T*A(LK+J)
      RETURN
      END

      SUBROUTINE POINTN (NF,NE,MT,MEP,ME,MV,NN,MP,MQ)
      DIMENSION MT(*),MEP(*),ME(*),MV(*),MP(*),MQ(*)
C
C     FUNCTION : TO CALCULATE POINTER VECTOR MP(NN*NF) FROM MESH INFORMATION
C                STORED IN ARRAY ME.
C
C     INPUT : NF = NUMBER OF DEGREES OF FREEDOM PER NODE
C             NE = NUMBER OF ELEMENTS INVOLVED
C          ME(*) = ARRAY CONTAINING THE NODE NUMBERS OF THE ELEMENTS
C          MT(I) = TYPE OF ELEMENT I
C         MEP(I) = START POSITION OF ELEMENT I ON VECTOR ME
C          MV(J) = NUMBER OF NODES IN ELEMENT TYPE J
C             NN = NUMBER OF NODES IN THE SYSTEM
C
C    OUTPUT : MP(J) = NUMBER OF NON-ZERO OFF-DIAGONAL ELEMENTS UP TO COLUMN J
C
C     WORKING ARRAY : MQ(NN)
C
      DO 11 I=1,NN
   11 MQ(I)=0

C     For each element, calculate the max. difference in node number, MQ(*)

      DO 22 I=1,NE
      IT=MT(I)
      NV=MV(IT)
      I1=MEP(I)
      I2=I1+NV-1
      DO 22 J=I1,I2-1
      NJ=ME(J)
      DO 22 K=J+1,I2
      NK=ME(K)
      IF (NK.GT.NJ) THEN
      MQ(NK)=MAX(MQ(NK),NK-NJ)
      ELSE
      MQ(NJ)=MAX(MQ(NJ),NJ-NK)
      END IF
   22 CONTINUE

C     Based on vector MQ(*), construct vector MP(*)

      K=1
      MP(1)=0
      DO 44 J=2,NF
      K=K+1
   44 MP(K)=MP(K-1)+J-1
      DO 33 I=2,NN
      L=NF*MQ(I)-1
      DO 33 J=1,NF
      K=K+1
   33 MP(K)=MP(K-1)+L+J
      RETURN
      END
```

```
      SUBROUTINE ESTIMA (JP,LT,IT,IP,ME,NV,NP,X,Y,Z,TH,NEF,EK,P,Q)
      IMPLICIT DOUBLE PRECISION (A-H,O-Z)
      DIMENSION ME(*),X(*),Y(*),Z(*),TH(*),P(*),Q(*),EK(NEF,*),
     -          GG(54),DG1(54),DG2(54),DGX(36),DGY(36),DT4(4),
     -          DHX(81),DHY(81),DH3(280),DET(18),DHG(3,280)
      COMMON /BK9/WT(14),HH(280),DH1(280),DH2(280)
C
C     FUNCTION : CALCULATE ELEMENT STIFFNESS MATRIX FOR VARIOUS PROBLEMS
C                AND ELEMENT TYPES
C
C     INPUT :        LT = LAST ELEMENT TYPE
C                    IC = MATERIAL CHARACTERISTIC GROUP NUMBER
C                    IP = ELEMENT NODE NUMBER START POSITION ON ARRAY ME
C                    NV = NUMBER OF NODES IN AN ELEMENT
C                    NP = NUMBER OF INTEGRATION POINTS
C                   NEF = TOTAL NUMBER OF D.O.F. OF THE ELEMENT
C                 ME(*) = ARRAY CONTAINING THE NODE NUMBERS OF THE ELEMENTS
C       X(*),Y(*),Z(*) = X,Y,Z NODAL COORDINATES
C                 TH(K) = THICKNESS OF PLATE OR SHELL AT NODE K
C
C    OUTPUT : EK(NEF,NEF) = ELEMENT STIFFNESS MATRIX
C
C     PROBLEM TYPE IDENTIFIER, JP        ELEMENT TYPE IDENTIFIER, IT
C     1. Spatial Frame                   1. Spatial beam element
C     2. Plane Strain                    2. Isoparametric 6-node triangle
C     3. Plane Stress                    3. Isoparametric 8-node quadrilateral
C     4. Axisymmetric Analysis           4. 6-node infinite quadrilateral
C     5. 2D Field Problem                5. Isoparametric 10-node tetrahedron
C     6. 3D Elasticity                   6. Isoparametric 15-node pentahedron
C     7. 3D Field Problem                7. Isoparametric 20-node hexahedron
C     8. Plate Bending Problem           8. 9-node Lagrangian plate element
C     9. Shell Structure                 9. 9-node Lagrangian shell element
C
C     CALCULATE THE VALUES AND THE DERIVATIVES OF INTERPOLATION FUNCTIONS
C     AT INTEGRATION POINTS FOR DIFFERENT TYPES OF ELEMENTS.
C
      IF (IT.NE.LT) THEN
      IF (IT.EQ.2) CALL DHT6 (NP,WT,HH,DH1,DH2)
      IF (IT.EQ.3) CALL DHQ8 (NP,WT,HH,DH1,DH2)
      IF (IT.EQ.4) CALL DHI6 (NP,WT,HH,DH1,DH2,GG,DG1,DG2)
      IF (IT.EQ.5) CALL DHT10 (NP,WT,HH,DH1,DH2,DH3)
      IF (IT.EQ.6) CALL DHP15 (NP,WT,HH,DH1,DH2,DH3)
      IF (IT.EQ.7) CALL DHH20 (NP,WT,HH,DH1,DH2,DH3)
      IF (IT.EQ.8) THEN
      CALL DHL9  (4,WT,GG,DG1,DG2)
      CALL DHL9  (9,WT,HH,DH1,DH2)
      ENDIF
      IF (IT.EQ.9) CALL DHL9 (NP,WT,HH,DH1,DH2)
      ENDIF
C
C     EVALUATION OF ELEMENT STIFFNESS MATRIX ACCORDING TO PROBLEM TYPE
C
      GOTO (1,2,2,2,2,3,3,4,5),JP
C
C     SPATIAL FRAME ANALYSIS
C
    1 J1=ME(IP)
      J2=ME(IP+1)
      CALL EKBEAM (X(J1),Y(J1),Z(J1),X(J2),Y(J2),Z(J2),EK)
      RETURN
C
C     TWO-DIMENSIONAL PROBLEMS
C
    2 IF (IT.EQ.4) THEN
      CALL JAC2D (IP,ME,NV,X,Y,NP,DG1,DG2,DH1,DH2,DHX,DHY,DET)
      IF (JP.EQ.4) CALL EKAXIS (NV,NP,GG,DHX,DHY,DET,WT,NEF,EK,IP,ME,X)
      ELSE
      CALL JAC2D (IP,ME,NV,X,Y,NP,DH1,DH2,DH1,DH2,DHX,DHY,DET)
      IF (JP.EQ.4) CALL EKAXIS (NV,NP,HH,DHX,DHY,DET,WT,NEF,EK,IP,ME,X)
      ENDIF
      IF (JP.LE.3) CALL EK2D   (NV,NP,DHX,DHY,DET,WT,NEF,EK)
      IF (JP.EQ.5) CALL EK2DF  (NV,NP,HH,DHX,DHY,DET,WT,EK)
```

```
      RETURN
C
C     THREE-DIMENSIONAL PROBLEMS
C
    3 CALL JAC3D (IP,ME,NV,X,Y,Z,NP,DH1,DH2,DH3,DH1,DH2,DH3,DHG,DET)
      IF (JP.EQ.6) CALL EK3D  (NV,NP,DHG,DET,WT,NEF,EK)
      IF (JP.EQ.7) CALL EK3DF (NV,NP,DHG,DET,WT,EK)
      RETURN
C
C     PLATE ELEMENT
C
    4 CALL JAC2D (IP,ME,NV,X,Y,4,DG1,DG2,DG1,DG2,DGX,DGY,DT4)
      CALL JAC2D (IP,ME,NV,X,Y,9,DH1,DH2,DH1,DH2,DHX,DHY,DET)
      CALL EKL9P (IP,ME,HH,DHX,DHY,DET,WT,GG,DGX,DGY,DT4,Z,EK)
      RETURN
C
C     SHELL ELEMENT
C
    5 CALL EKL9S (IP,ME,NV,X,Y,Z,TH,NP,HH,DH1,DH2,WT,DET,P,Q,EK)
      RETURN
      END


      SUBROUTINE ASSEMB (IP,ME,NV,NF,NEF,EK,MP,A,D)
      IMPLICIT DOUBLE PRECISION (A-H,O-Z)
      DIMENSION ME(*),EK(NEF,*),MP(*),A(*),D(*)
C
C     FUNCTION : Assemble element stiffness matrix EK(NEF,NEF) to global
C                stiffness vectors A and D.
C
C     INPUT :       IP = Element node number start position on array ME
C                ME(*) = Array of element node numbers
C                   NV = Number of nodes in the element
C                   NF = Number of d.o.f. per node
C                  NEF = Total number of d.o.f. within the element
C              EK(*,*) = Element stiffness matrix
C                MP(*) = Pointer array for global stiffness vector A
C
C     OUTPUT : A(*) = Array of off-diagonal elements
C              D(*) = Array of diagonal elements
C
      KK=IP-1
      DO 11 J1=1,NV
      J2=ME(KK+J1)
      DO 22 I1=1,NV
      I2=ME(KK+I1)
      IF (I2.GT.J2) GOTO 22
      IF (I2.EQ.J2) THEN

C     Assemble diagonal block coefficients

      DO 55 J=1,NF
      L1=NF*(J1-1)+J
      L2=NF*(J2-1)+J
      DO 33 I=1,J-1
      K1=NF*(I1-1)+I
      L=MP(L2)+I-J+1
   33 A(L)=A(L)+EK(K1,L1)
   55 D(L2)=D(L2)+EK(L1,L1)
      ELSE

C     Assemble off-diagonal block coefficients

      DO 44 J=1,NF
      L1=NF*(J1-1)+J
      L2=NF*(J2-1)+J
      DO 44 I=1,NF
      K1=NF*(I1-1)+I
      K2=NF*(I2-1)+I
      L=MP(L2)+k2-L2+1
   44 A(L)=A(L)+EK(K1,L1)
      ENDIF
```

```
   22 CONTINUE
   11 CONTINUE
      RETURN
      END

      SUBROUTINE NORMAL (NE,MEP,ME,MT,MV,X,Y,Z,P,Q)
      IMPLICIT DOUBLE PRECISION (A-H,O-Z)
      DIMENSION ME(*),MEP(*),MT(*),MV(*),X(*),Y(*),Z(*),P(*),Q(*)
      COMMON /BK9/WT(14),HH(280),DH1(280),DH2(280)
C
C     FUNCTION : Calculate the base vectors at each node of a Lagrangian
C                element. (Stored temporarily in arrays GF and U)
C
      NP=0
      CALL DHL9 (NP,WT,HH,DH1,DH2)

C     Loop over elements 1 to NE

      DO 11 I=1,NE
      IT=MT(I)
      NV=MV(IT)
      IP=MEP(I)-1
      L=0

C     Loop over element nodes 1 to NV

      DO 22 J=1,NV
      K=ME(IP+J)
      KK=5*K-5
      U1=0.D0
      U2=0.D0
      U3=0.D0
      V1=0.D0
      V2=0.D0
      V3=0.D0

C     Calculate the derivatives of (x,y,z) w.r.t. element coordinates

      DO 33 JJ=1,NV
      II=ME(IP+JJ)
      L=L+1
      U1=U1+DH1(L)*X(II)
      U2=U2+DH1(L)*Y(II)
      U3=U3+DH1(L)*Z(II)
      V1=V1+DH2(L)*X(II)
      V2=V2+DH2(L)*Y(II)
   33 V3=V3+DH2(L)*Z(II)

C     Forming orthonormal basis (u,v,w)

      U=DSQRT(U1*U1+U2*U2+U3*U3)
      U1=U1/U
      U2=U2/U
      U3=U3/U
      W1=U2*V3-U3*V2
      W2=U3*V1-U1*V3
      W3=U1*V2-U2*V1
      W=DSQRT(W1*W1+W2*W2+W3*W3)
      W1=W1/W
      W2=W2/W
      W3=W3/W
      W=W/DSQRT(V1*V1+V2*V2+V3*V3)
      IF (W.LT.0.5) WRITE (7,20) I,J,K,W
   20 FORMAT (/' ELEMENT',I4,'   VERTICE',I2,'   NODE',I4,'   W =',F9.5)
      IF (Q(KK+5).GT.0.0001) THEN
      DD=(Q(KK+2)-W1)**2+(Q(KK+3)-W2)**2+(Q(KK+4)-W3)**2
      IF (DD.GT.0.01) THEN
      WRITE (7,10) K,DD
   10 FORMAT (/' INCONSISTENCY OF NORMAL AT NODE',I4,5X,'DD =',G12.4)
      WRITE (7,30) Q(KK+2),Q(KK+3),Q(KK+4),W1,W2,W3
   30 FORMAT (/' N1 =',3F9.4,10X,'N2 =',3F9.4)
      ENDIF
```

```
      ENDIF

C     Store the (u,v,w) vectors temporarily in arrays P(*) and Q(*)

      IF (W.GT.Q(KK+5)) THEN
      P(KK+1)=U1
      P(KK+2)=U2
      P(KK+3)=U3
      P(KK+4)=W2*U3-W3*U2
      P(KK+5)=W3*U1-W1*U3
      Q(KK+1)=W1*U2-W2*U1
      Q(KK+2)=W1
      Q(KK+3)=W2
      Q(KK+4)=W3
      Q(KK+5)=W
      ENDIF
   22 CONTINUE
   11 CONTINUE
      RETURN
      END
```

```
$LARGE
C
C     File LIB2.FOR, last updated on 11 July 1991.
C
      SUBROUTINE DHT6 (NP,WT,HH,DH1,DH2)
      IMPLICIT DOUBLE PRECISION (A-H,O-Z)
      DIMENSION U(7),V(7),HH(*),WT(*),DH1(*),DH2(*)
C
C     FUNCTION  : To calculate the values of the interpolation functions
C                 and their derivatives at integration points for 6-node
C                 isoparametric triangular element T6.
C
C     INPUT : NP = NUMBER OF INTEGRATION POINTS (3,4,7)
C
C     OUTPUT : WT(I)  = Weight at integration point I
C        HH(6*I-6+J)  = Value of interpolation function for node J at
C                       integration point I.
C      DH1(*),DH2(*)  = Derivatives of interpolation functions w.r.t.
C                       element natural coordinates
C
      IF (NP.EQ.3) THEN
      T=1.D0/6.D0
      WT(1)=T
      WT(2)=T
      WT(3)=T
      U(1)=T
      V(1)=T
      U(2)=T*4.D0
      V(2)=T
      U(3)=T
      V(3)=T*4.D0
      END IF
C
      IF (NP.EQ.4) THEN
      WT(1)=-27.D0/96.D0
      WT(2)=25.D0/96.D0
      WT(3)=25.D0/96.D0
      WT(4)=25.D0/96.D0
      U(1)=1.D0/3.D0
      V(1)=1.D0/3.D0
      U(2)=.2D0
      V(2)=.2D0
      U(3)=.6D0
      V(3)=.2D0
      U(4)=.2D0
      V(4)=.6D0
      END IF
C
      IF (NP.EQ.7) THEN
      WT(1)=.1125D0
      C=(1.55D2+DSQRT(1.5D1))/2.4D3
      WT(2)=C
      WT(3)=C
      WT(4)=C
      D=3.1D1/2.4D2-C
      WT(5)=D
      WT(6)=D
      WT(7)=D
      U(1)=1.D0/3.D0
      V(1)=1.D0/3.D0
      A=(6.D0+DSQRT(1.5D1))/2.1D1
      U(2)=A
      V(2)=A
      U(3)=1.D0-2.D0*A
      V(3)=A
      U(4)=A
      V(4)=1.D0-2.D0*A
      B=4.D0/7.D0-A
      U(5)=B
      V(5)=B
      U(6)=1.D0-2.D0*B
      V(6)=B
```

```
      U(7)=B
      V(7)=1.D0-2.D0*B
      END IF
C
      J=0
      DO 11 I=1,NP
      R=U(I)
      S=V(I)
      T=1.D0-R-S
      HH(J+1)=R*(2.D0*R-1.D0)
      HH(J+2)=S*(2.D0*S-1.D0)
      HH(J+3)=T*(2.D0*T-1.D0)
      HH(J+4)=4.D0*S*T
      HH(J+5)=4.D0*T*R
      HH(J+6)=4.D0*R*S
      DH1(J+1)=4.D0*R-1.D0
      DH1(J+2)=0.D0
      DH1(J+3)=1.D0-4.D0*T
      DH1(J+4)=-4.D0*S
      DH1(J+5)=4.D0*(T-R)
      DH1(J+6)=4.D0*S
      DH2(J+1)=0.D0
      DH2(J+2)=4.D0*S-1.D0
      DH2(J+3)=1.D0-4.D0*T
      DH2(J+4)=4.D0*(T-S)
      DH2(J+5)=-4.D0*R
      DH2(J+6)=4.D0*R
   11 J=J+6
      RETURN
      END

      SUBROUTINE DHQ8 (NP,WT,HH,DH1,DH2)
      IMPLICIT DOUBLE PRECISION (A-H,O-Z)
      DIMENSION U(9),V(9),HH(*),WT(*),DH1(*),DH2(*)
      DATA ZERO,ONE,TWO,FOUR/0.D0,1.D0,2.D0,4.D0/
C
C     FUNCTION : Calculate the interpolation functions and their derivatives
C                at integration points for an isoparametric 8-node
C                quadrilateral element.
C
C     INPUT : NP = Number of integration points (4,7,9)
C
C     OUTPUT : WT(I)  = Weight at integration point I
C              HH(*)  = Values of interpolation functions at integration points
C      DH1(*),DH2(*)  = Derivatives of interpolation functions w.r.t. element
C                       coordinates at integration points.
C
      IF (NP.EQ.4) THEN
      WT(1)=ONE
      WT(2)=ONE
      WT(3)=ONE
      WT(4)=ONE
      T=ONE/DSQRT(3.D0)
      U(1)=T
      V(1)=T
      U(2)=-T
      V(2)=T
      U(3)=-T
      V(3)=-T
      U(4)=T
      V(4)=-T
      END IF
C
      IF (NP.EQ.7) THEN
      WT(1)=8.D0/7.D0
      WT(2)=20.D0/63.D0
      WT(3)=20.D0/63.D0
      WT(4)=20.D0/36.D0
      WT(5)=20.D0/36.D0
      WT(6)=20.D0/36.D0
      WT(7)=20.D0/36.D0
      U(1)=ZERO
```

```
      V(1)=ZERO
      U(2)=ZERO
      V(2)= DSQRT(14.D0/15.D0)
      U(3)=ZERO
      V(3)=-V(2)
      S=DSQRT(.6D0)
      T=1.D0/DSQRT(3.D0)
      U(4)=S
      V(4)=T
      U(5)=S
      V(5)=-T
      U(6)=-S
      V(6)=-T
      U(7)=-S
      V(7)=T
      END IF
C
      IF (NP.EQ.9) THEN
      WT(1)=25.D0/81.D0
      WT(2)=40.D0/81.D0
      WT(3)=25.D0/81.D0
      WT(4)=40.D0/81.D0
      WT(5)=64.D0/81.D0
      WT(6)=40.D0/81.D0
      WT(7)=25.D0/81.D0
      WT(8)=40.D0/81.D0
      WT(9)=25.D0/81.D0
      T=DSQRT(.6D0)
      DO 22 I=1,3
      DO 22 J=1,3
      U(3*I-3+J)=(J-2)*T
   22 V(3*I-3+J)=(I-2)*T
      END IF
C
      J=0
      DO 11 I=1,NP
      R=U(I)
      S=V(I)
      RM=ONE-R
      RP=ONE+R
      SM=ONE-S
      SP=ONE+S
      HH(J+1)=-RM*SM*(RP+S)/FOUR
      HH(J+2)=-RP*SM*(RM+S)/FOUR
      HH(J+3)=-RP*SP*(RM-S)/FOUR
      HH(J+4)=-RM*SP*(RP-S)/FOUR
      HH(J+5)=RP*RM*SM/TWO
      HH(J+6)=RP*SM*SP/TWO
      HH(J+7)=RP*RM*SP/TWO
      HH(J+8)=RM*SM*SP/TWO
      DH1(J+1)=SM*(R+R+S)/FOUR
      DH1(J+2)=SM*(R+R-S)/FOUR
      DH1(J+3)=SP*(R+R+S)/FOUR
      DH1(J+4)=SP*(R+R-S)/FOUR
      DH1(J+5)=-R*SM
      DH1(J+6)= SM*SP/TWO
      DH1(J+7)=-R*SP
      DH1(J+8)=-SM*SP/TWO
      DH2(J+1)=RM*(S+S+R)/FOUR
      DH2(J+2)=RP*(S+S-R)/FOUR
      DH2(J+3)=RP*(S+S+R)/FOUR
      DH2(J+4)=RM*(S+S-R)/FOUR
      DH2(J+5)=-RM*RP/TWO
      DH2(J+6)=-RP*S
      DH2(J+7)= RM*RP/TWO
      DH2(J+8)=-RM*S
   11 J=J+8
      RETURN
      END
C
      SUBROUTINE DHL9 (NP,WT,HH,DH1,DH2)
      IMPLICIT DOUBLE PRECISION (A-H,O-Z)
```

```
      DIMENSION U(9),V(9),HH(*),WT(*),DH1(*),DH2(*)
C
C     FUNCTION : Calculate the interpolation functions and their derivatives
C                at integration points for a 9-node Lagrangian biquadratic
C                element.
C
C     INPUT : NP = Number of integration points (4,7,9)
C             If NP=0, calculate HH(*),DH1(*),DH2(*) at nodal points.
C
C     OUTPUT : WT(I) = Weight at integration point I
C              HH(*) = Values of interpolation functions at integration points
C      DH1(*),DH2(*) = Derivatives of interpolation functions w.r.t. element
C                      coordinates at integration points.
C
      IF (NP.EQ.4) THEN
      WT(1)=1.D0
      WT(2)=1.D0
      WT(3)=1.D0
      WT(4)=1.D0
      T=1.D0/DSQRT(3.D0)
      U(1)=T
      V(1)=T
      U(2)=-T
      V(2)=T
      U(3)=-T
      V(3)=-T
      U(4)=T
      V(4)=-T
      END IF
C
      IF (NP.EQ.7) THEN
      WT(1)=8.D0/7.D0
      WT(2)=20.D0/63.D0
      WT(3)=20.D0/63.D0
      WT(4)=20.D0/36.D0
      WT(5)=20.D0/36.D0
      WT(6)=20.D0/36.D0
      WT(7)=20.D0/36.D0
      U(1)=0.D0
      V(1)=0.D0
      U(2)=0.D0
      V(2)= DSQRT(14.D0/15.D0)
      U(3)=0.D0
      V(3)=-V(2)
      S=DSQRT(.6D0)
      T=1.D0/DSQRT(3.D0)
      U(4)=S
      V(4)=T
      U(5)=S
      V(5)=-T
      U(6)=-S
      V(6)=-T
      U(7)=-S
      V(7)=T
      END IF
C
      IF (NP.EQ.9) THEN
      WT(1)=25.D0/81.D0
      WT(2)=40.D0/81.D0
      WT(3)=25.D0/81.D0
      WT(4)=40.D0/81.D0
      WT(5)=64.D0/81.D0
      WT(6)=40.D0/81.D0
      WT(7)=25.D0/81.D0
      WT(8)=40.D0/81.D0
      WT(9)=25.D0/81.D0
      T=DSQRT(.6D0)
      DO 22 I=1,3
      DO 22 J=1,3
      U(3*I-3+J)=(J-2)*T
   22 V(3*I-3+J)=(I-2)*T
      END IF
```

```
C
      IF (NP.EQ.0) THEN
      U(1)=-1.D0
      V(1)=-1.D0
      U(2)= 1.D0
      V(2)=-1.D0
      U(3)= 1.D0
      V(3)= 1.D0
      U(4)=-1.D0
      V(4)= 1.D0
      U(5)= 0.D0
      V(5)=-1.D0
      U(6)= 1.D0
      V(6)= 0.D0
      U(7)= 0.D0
      V(7)= 1.D0
      U(8)=-1.D0
      V(8)= 0.D0
      U(9)= 0.D0
      V(9)= 0.D0
      NP=9
      END IF
C
      J=0
      DO 11 I=1,NP
      R=U(I)
      S=V(I)
      F1=0.5D0*R*(R-1.D0)
      F2=1.D0-R*R
      F3=0.5D0*R*(R+1.D0)
      DF1=R-0.5D0
      DF2=-2.D0*R
      DF3=R+0.5D0
      G1=0.5D0*S*(S-1.D0)
      G2=1.D0-S*S
      G3=0.5D0*S*(S+1.D0)
      DG1=S-0.5D0
      DG2=-2.D0*S
      DG3=S+0.5D0
      HH(J+1)=F1*G1
      HH(J+2)=F3*G1
      HH(J+3)=F3*G3
      HH(J+4)=F1*G3
      HH(J+5)=F2*G1
      HH(J+6)=F3*G2
      HH(J+7)=F2*G3
      HH(J+8)=F1*G2
      HH(J+9)=F2*G2
      DH1(J+1)=DF1*G1
      DH1(J+2)=DF3*G1
      DH1(J+3)=DF3*G3
      DH1(J+4)=DF1*G3
      DH1(J+5)=DF2*G1
      DH1(J+6)=DF3*G2
      DH1(J+7)=DF2*G3
      DH1(J+8)=DF1*G2
      DH1(J+9)=DF2*G2
      DH2(J+1)=F1*DG1
      DH2(J+2)=F3*DG1
      DH2(J+3)=F3*DG3
      DH2(J+4)=F1*DG3
      DH2(J+5)=F2*DG1
      DH2(J+6)=F3*DG2
      DH2(J+7)=F2*DG3
      DH2(J+8)=F1*DG2
      DH2(J+9)=F2*DG2
   11 J=J+9
      RETURN
      END

      SUBROUTINE DHI6 (NP,WT,HH,DH1,DH2,GG,DG1,DG2)
      IMPLICIT DOUBLE PRECISION (A-H,O-Z)
```

```
      DIMENSION U(9),V(9),HH(*),WT(*),DH1(*),DH2(*),GG(*),DG1(*),DG2(*)
      DATA ZERO,ONE,TWO,FOUR/0.D0,1.D0,2.D0,4.D0/
C
C     FUNCTION : To calculate at Gaussian points of infinite element I6
C                Shape Function GG and its derivatives DG1, DG2, and
C                Interpolation Function HH and its derivatives DH1, DH2.
C
C     INPUT : NP = Number of integration points (4,7,9)
C
C     OUTPUT : WT(*) = Weights at integration points
C              HH(*),DH1(*),DH2(*),GG(*),DG1(*),DG2(*)
C
      IF (NP.EQ.4) THEN
      WT(1)-ONE
      WT(2)=ONE
      WT(3)=ONE
      WT(4)=ONE
      T=ONE/DSQRT(3.D0)
      U(1)= T
      V(1)= T
      U(2)=-T
      V(2)= T
      U(3)=-T
      V(3)=-T
      U(4)= T
      V(4)=-T
      END IF
C
      IF (NP.EQ.7) THEN
      WT(1)=8.D0/7.D0
      WT(2)=1.D2/1.68D2
      WT(3)=1.D2/1.68D2
      WT(4)=2.D1/4.8D1
      WT(5)=2.D1/4.8D1
      WT(6)=2.D1/4.8D1
      WT(7)=2.D1/4.8D1
      R=DSQRT(7.D0/1.5D1)
      S=DSQRT((7.D0+DSQRT(2.4D1))/1.5D1)
      T=DSQRT((7.D0-DSQRT(2.4D1))/1.5D1)
      U(1)=ZERO
      V(1)=ZERO
      U(2)=-R
      V(2)=-R
      U(3)= R
      V(3)= R
      U(4)= S
      V(4)=-T
      U(5)=-S
      V(5)= T
      U(6)= T
      V(6)=-S
      U(7)=-T
      V(7)= S
      END IF
C
      IF (NP.EQ.9) THEN
      WT(1)=25.D0/81.D0
      WT(2)=40.D0/81.D0
      WT(3)=25.D0/81.D0
      WT(4)=40.D0/81.D0
      WT(5)=64.D0/81.D0
      WT(6)=40.D0/81.D0
      WT(7)=25.D0/81.D0
      WT(8)=40.D0/81.D0
      WT(9)=25.D0/81.D0
      T=DSQRT(.6D0)
      DO 22 I=1,3
      DO 22 J=1,3
      U(3*I-3+J)=(J-2)*T
   22 V(3*I-3+J)=(I-2)*T
      END IF
C
```

```
      J=0
      DO 11 I=1,NP
      R=U(I)
      S=V(I)
      RM=ONE-R
      RP=ONE+R
      SM=ONE-S
      SP=ONE+S
      HH(J+1)= R*RM*S*SM/FOUR
      HH(J+2)=-R*RM*SM*SP/TWO
      HH(J+3)=-R*RM*S*SP/FOUR
      HH(J+4)=-RM*RP*S*SM/TWO
      HH(J+5)= RM*RP*SM*SP
      HH(J+6)= RM*RP*S*SP/TWO
      DH1(J+1)=S*SM*(RM-R)/FOUR
      DH1(J+2)=SM*SP*(R-RM)/TWO
      DH1(J+3)=S*SP*(R-RM)/FOUR
      DH1(J+4)=R*S*SM
      DH1(J+5)=-TWO*R*SM*SP
      DH1(J+6)=-R*S*SP
      DH2(J+1)=R*RM*(SM-S)/FOUR
      DH2(J+2)=R*RM*S
      DH2(J+3)=-R*RM*(SP+S)/FOUR
      DH2(J+4)=-RM*RP*(SM-S)/TWO
      DH2(J+5)=-TWO*RM*RP*S
      DH2(J+6)=RM*RP*(SP+S)/TWO
      GG(J+1)=R*S*SM/RM
      GG(J+2)=-TWO*R*SM*SP/RM
      GG(J+3)=-R*S*SP/RM
      GG(J+4)=-.5D0*RP*S*SM/RM
      GG(J+5)=RP*SM*SP/RM
      GG(J+6)=.5D0*RP*S*SP/RM
      R2=RM*RM
      DG1(J+1)=S*SM/R2
      DG1(J+2)=-TWO*SM*SP/R2
      DG1(J+3)=-S*SP/R2
      DG1(J+4)=-S*SM/R2
      DG1(J+5)=TWO*SM*SP/R2
      DG1(J+6)=S*SP/R2
      DG2(J+1)=R*(SM-S)/RM
      DG2(J+2)=FOUR*R*S/RM
      DG2(J+3)=-R*(SP+S)/RM
      DG2(J+4)=-.5D0*RP*(SM-S)/RM
      DG2(J+5)=-TWO*RP*S/RM
      DG2(J+6)=.5D0*RP*(SP+S)/RM
   11 J=J+6
      RETURN
      END

      SUBROUTINE DHT10 (NP,WT,HH,DH1,DH2,DH3)
      IMPLICIT DOUBLE PRECISION (A-H,O-Z)
      DIMENSION U(5),V(5),W(5),HH(*),WT(*),DH1(*),DH2(*),DH3(*)
C
C     FUNCTION : Calculate the interpolation functions and their
C                derivatives at integration points for an
C                isoparametric 10-node tetrahedron element.
C
C     INPUT : NP = Number of integration points (4,5)
C
C     OUTPUT :          WT(I)  = Weight at integration point I
C                       HH(*)  = Values of interpolation functions
C                                at integration points
C     DH1(*),DH2(*),DH3(*)     = Derivatives of interpolation
C                                functions at integration points
C
      IF (NP.EQ.4) THEN
      A=(5.D0-DSQRT(5.0D0))/2.D1
      B=(5.D0+DSQRT(4.5D1))/2.D1
      WEIGHT=1.D0/2.4D1
      U(1)=A
      V(1)=A
      W(1)=A
```

```
      WT(1)=WEIGHT
      U(2)=A
      V(2)=A
      W(2)=B
      WT(2)=WEIGHT
      U(3)=A
      V(3)=B
      W(3)=A
      WT(3)=WEIGHT
      U(4)=B
      V(4)=A
      W(4)=A
      WT(4)=WEIGHT
      END IF
C
      IF (NP.EQ.5) THEN
      A=0.25D0
      B=1.D0/6.D0
      C=0.5D0
      WEIGHT=3.D0/4.D1
      U(1)=A
      V(1)=A
      W(1)=A
      WT(1)=-2.D0/1.5D1
      U(2)=B
      V(2)=B
      W(2)=B
      WT(2)=WEIGHT
      U(3)=B
      V(3)=B
      W(3)=C
      WT(3)=WEIGHT
      U(4)=B
      V(4)=C
      W(4)=B
      WT(4)=WEIGHT
      U(5)=C
      V(5)=B
      W(5)=B
      WT(5)=WEIGHT
      ENDIF
C
      J=0
      DO 11 I=1,NP
      R=U(I)
      S=V(I)
      T=W(I)
      Q=1.D0-R-S-T
      R4=R*4.D0
      S4=S*4.D0
      T4=T*4.D0
      Q4=Q*4.D0
      HH(J+1)=R*(2.D0*R-1.D0)
      HH(J+2)=S*(2.D0*S-1.D0)
      HH(J+3)=T*(2.D0*T-1.D0)
      HH(J+4)=Q*(2.D0*Q-1.D0)
      HH(J+5)=R4*S
      HH(J+6)=S4*Q
      HH(J+7)=R4*Q
      HH(J+8)=R4*T
      HH(J+9)=T4*S
      HH(J+10)=Q4*T
      DH1(J+1)=R4-1.D0
      DH1(J+2)=0.D0
      DH1(J+3)=1.D0-Q4
      DH1(J+4)=0.D0
      DH1(J+5)=S4
      DH1(J+6)=-S4
      DH1(J+7)=Q4-R4
      DH1(J+8)=T4
      DH1(J+9)=0.D0
      DH1(J+10)=-T4
```

```
      DH2(J+1)=0.D0
      DH2(J+2)=S4-1.D0
      DH2(J+3)=1.D0-Q4
      DH2(J+4)=0.D0
      DH2(J+5)=R4
      DH2(J+6)=Q4-S4
      DH2(J+7)=-R4
      DH2(J+8)=0.D0
      DH2(J+9)=T4
      DH2(J+10)=-T4
      DH3(J+1)=0.D0
      DH3(J+2)=0.D0
      DH3(J+3)=1.D0-Q4
      DH3(J+4)=T4-1.D0
      DH3(J+5)=0.D0
      DH3(J+6)=-S4
      DH3(J+7)=-R4
      DH3(J+8)=R4
      DH3(J+9)=S4
      DH3(J+10)=Q4-T4
   11 J=J+10
      RETURN
      END

      SUBROUTINE DHP15 (NP,WT,HH,DH1,DH2,DH3)
      IMPLICIT DOUBLE PRECISION (A-H,O-Z)
      DIMENSION U(8),V(8),W(8),HH(*),WT(*),DH1(*),DH2(*),DH3(*)
C
C     FUNCTION : Calculate the interpolation functions and
C                their derivatives at integration points for
C                an isoparametric 15-node pentahedron element.
C
C     INPUT : NP = Number of integration points (6,8)
C
C     OUTPUT :          WT(I)  = Weight at integration point I
C                       HH(*)  = Values of interpolation functions
C                                at integration points
C     DH1(*),DH2(*),DH3(*)     = Derivatives of interpolation
C                                functions at integration points
C
      Z=1.D0/DSQRT(3.D0)
      IF (NP.EQ.6) THEN
      U(1)=.5D0
      V(1)=.5D0
      W(1)=-Z
      WT(1)=1.D0/6.D0
      U(2)=0.D0
      V(2)=.5D0
      W(2)=-Z
      WT(2)=1.D0/6.D0
      U(3)=.5D0
      V(3)=0.D0
      W(3)=-Z
      WT(3)=1.D0/6.D0
      U(4)=.5D0
      V(4)=.5D0
      W(4)=Z
      WT(4)=1.D0/6.D0
      U(5)=0.D0
      V(5)=.5D0
      W(5)=Z
      WT(5)=1.D0/6.D0
      U(6)=.5D0
      V(6)=0.D0
      W(6)=Z
      WT(6)=1.D0/6.D0
      END IF
C
      IF (NP.EQ.8) THEN
      U(1)=1.D0/3.D0
      V(1)=1.D0/3.D0
      W(1)=-Z
```

```
      U(2)=.2D0
      V(2)=.2D0
      W(2)=-Z
      U(3)=.6D0
      V(3)=.2D0
      W(3)=-Z
      U(4)=.2D0
      V(4)=.6D0
      W(4)=-Z
      U(5)=1.D0/3.D0
      V(5)=1.D0/3.D0
      W(5)=Z
      U(6)=.2D0
      V(6)=.2D0
      W(6)=Z
      U(7)=.6D0
      V(7)=.2D0
      W(7)=Z
      U(8)=.2D0
      V(8)=.6D0
      W(8)=Z
      WT(1)=-27.D0/96.D0
      WT(2)= 25.D0/96.D0
      WT(3)= 25.D0/96.D0
      WT(4)= 25.D0/96.D0
      WT(5)=-27.D0/96.D0
      WT(6)= 25.D0/96.D0
      WT(7)= 25.D0/96.D0
      WT(8)= 25.D0/96.D0
      END IF
C
      J=0
      DO 11 I=1,NP
      R=U(I)
      S=V(I)
      T=W(I)
      Q=1.D0-R-S
      TM=1.D0-T
      TP=1.D0+T
      H=0.5D0
      HTP=H*T+1.D0
      HTM=H*T-1.D0
      R2=R*2.D0
      S2=S*2.D0
      Q2=Q*2.D0
      HH(J+ 1)=R*TM*(R-HTP)
      HH(J+ 2)=S*TM*(S-HTP)
      HH(J+ 3)=Q*TM*(Q-HTP)
      HH(J+ 4)=S2*Q*TM
      HH(J+ 5)=Q2*R*TM
      HH(J+ 6)=R2*S*TM
      HH(J+ 7)=R*TM*TP
      HH(J+ 8)=S*TM*TP
      HH(J+ 9)=Q*TM*TP
      HH(J+10)=R*TP*(R+HTM)
      HH(J+11)=S*TP*(S+HTM)
      HH(J+12)=Q*TP*(Q+HTM)
      HH(J+13)=S2*Q*TP
      HH(J+14)=Q2*R*TP
      HH(J+15)=R2*S*TP
      DH1(J+ 1)=TM*(R2-HTP)
      DH1(J+ 2)=0.D0
      DH1(J+ 3)=TM*(HTP-Q2)
      DH1(J+ 4)=-S2*TM
      DH1(J+ 5)=TM*(Q2-R2)
      DH1(J+ 6)=S2*TM
      DH1(J+ 7)=TM*TP
      DH1(J+ 8)=0.D0
      DH1(J+ 9)=-TM*TP
      DH1(J+10)=TP*(R2+HTM)
      DH1(J+11)=0.D0
      DH1(J+12)=-TP*(HTM+Q2)
```

```
      DH1(J+13)=-S2*TP
      DH1(J+14)=TP*(Q2-R2)
      DH1(J+15)=S2*TP
      DH2(J+ 1)=0.D0
      DH2(J+ 2)=TM*(S2-HTP)
      DH2(J+ 3)=TM*(HTP-Q2)
      DH2(J+ 4)=TM*(Q2-S2)
      DH2(J+ 5)=-R2*TM
      DH2(J+ 6)=R2*TM
      DH2(J+ 7)=0.D0
      DH2(J+ 8)=TM*TP
      DH2(J+ 9)=-TM*TP
      DH2(J+10)=0.D0
      DH2(J+11)=TP*(S2+HTM)
      DH2(J+12)=-TP*(HTM+Q2)
      DH2(J+13)=TP*(Q2-S2)
      DH2(J+14)=-R2*TP
      DH2(J+15)=R2*TP
      DH3(J+ 1)=R*(T-R+H)
      DH3(J+ 2)=S*(T-S+H)
      DH3(J+ 3)=Q*(T-Q+H)
      DH3(J+ 4)=-S2*Q
      DH3(J+ 5)=-Q2*R
      DH3(J+ 6)=-R2*S
      DH3(J+ 7)=-R2*T
      DH3(J+ 8)=-S2*T
      DH3(J+ 9)=-Q2*T
      DH3(J+10)=R*(T+R-H)
      DH3(J+11)=S*(T+S-H)
      DH3(J+12)=Q*(T+Q-H)
      DH3(J+13)=S2*Q
      DH3(J+14)=Q2*R
      DH3(J+15)=R2*S
   11 J=J+15
      RETURN
      END

      SUBROUTINE DHH20 (NP,WT,HH,DH1,DH2,DH3)
      IMPLICIT DOUBLE PRECISION (A-H,O-Z)
      DIMENSION U(14),V(14),W(14),HH(*),WT(*),DH1(*),DH2(*),DH3(*)
      DATA ZERO,ONE,TWO,THREE,FOUR,EIGHT/0.D0,1.D0,2.D0,3.D0,4.D0,8.D0/
C
C     FUNCTION : Calculate the interpolation functions and
C                their derivatives at integration points for
C                an isoparametric 20-node hexahedron element.
C
C     INPUT : NP = Number of integration points (8,14)
C
C     OUTPUT :        WT(I)  = Weight at integration point I
C                     HH(*)  = Values of interpolation functions
C                              at integration points
C     DH1(*),DH2(*),DH3(*)  = Derivatives of interpolation
C                              functions at integration points
C
      IF (NP.EQ.8) THEN
      B=DSQRT(ONE/THREE)
      U(1)= B
      V(1)= B
      W(1)= B
      WT(1)=ONE
      U(2)= B
      V(2)= B
      W(2)=-B
      WT(2)=ONE
      U(3)= B
      V(3)=-B
      W(3)= B
      WT(3)=ONE
      U(4)= B
      V(4)=-B
      W(4)=-B
      WT(4)=ONE
```

```
      U(5)=-B
      V(5)= B
      W(5)= B
      WT(5)=ONE
      U(6)=-B
      V(6)= B
      W(6)=-B
      WT(6)=ONE
      U(7)=-B
      V(7)=-B
      W(7)= B
      WT(7)=ONE
      U(8)=-B
      V(8)--B
      W(8)=-B
      WT(8)=ONE
      ENDIF
C
      IF (NP.EQ.14) THEN
      A=DSQRT(19.D0/30.D0)
      B=DSQRT(19.D0/33.D0)
      W1=3.20D2/3.61D2
      W2=1.21D2/3.61D2
      U(1)= A
      V(1)=ZERO
      W(1)=ZERO
      WT(1)=W1
      U(2)=-A
      V(2)=ZERO
      W(2)=ZERO
      WT(2)=W1
      U(3)=ZERO
      V(3)= A
      W(3)=ZERO
      WT(3)=W1
      U(4)=ZERO
      V(4)=-A
      W(4)=ZERO
      WT(4)=W1
      U(5)=ZERO
      V(5)=ZERO
      W(5)= A
      WT(5)=W1
      U(6)=ZERO
      V(6)=ZERO
      W(6)=-A
      WT(6)=W1
      U(7)= B
      V(7)= B
      W(7)= B
      WT(7)=W2
      U(8)= B
      V(8)= B
      W(8)=-B
      WT(8)=W2
      U(9)= B
      V(9)=-B
      W(9)= B
      WT(9)=W2
      U(10)= B
      V(10)=-B
      W(10)=-B
      WT(10)=W2
      U(11)=-B
      V(11)= B
      W(11)= B
      WT(11)=W2
      U(12)=-B
      V(12)= B
      W(12)=-B
      WT(12)=W2
      U(13)=-B
```

```
      V(13)=-B
      W(13)= B
      WT(13)=W2
      U(14)=-B
      V(14)=-B
      W(14)=-B
      WT(14)=W2
      ENDIF
C
      J=0
      DO 11 I=1,NP
      R=U(I)
      S=V(I)
      T=W(I)
      RM=ONE-R
      RP=ONE+R
      SM=ONE-S
      SP=ONE+S
      TM=ONE-T
      TP=ONE+T
      R4=RM*RP/FOUR
      S4=SM*SP/FOUR
      T4=TM*TP/FOUR
      F1=(-R-S-T-TWO)/EIGHT
      F2=( R-S-T-TWO)/EIGHT
      F3=( R+S-T-TWO)/EIGHT
      F4=(-R+S-T-TWO)/EIGHT
      F5=(-R-S+T-TWO)/EIGHT
      F6=( R-S+T-TWO)/EIGHT
      F7=( R+S+T-TWO)/EIGHT
      F8=(-R+S+T-TWO)/EIGHT
      HH(J+ 1)=RM*SM*TM*F1
      HH(J+ 2)=RP*SM*TM*F2
      HH(J+ 3)=RP*SP*TM*F3
      HH(J+ 4)=RM*SP*TM*F4
      HH(J+ 5)=RM*SM*TP*F5
      HH(J+ 6)=RP*SM*TP*F6
      HH(J+ 7)=RP*SP*TP*F7
      HH(J+ 8)=RM*SP*TP*F8
      HH(J+ 9)=SM*TM*R4
      HH(J+10)=TM*RP*S4
      HH(J+11)=SP*TM*R4
      HH(J+12)=TM*RM*S4
      HH(J+13)=RM*SM*T4
      HH(J+14)=RP*SM*T4
      HH(J+15)=RP*SP*T4
      HH(J+16)=RM*SP*T4
      HH(J+17)=SM*TP*R4
      HH(J+18)=TP*RP*S4
      HH(J+19)=SP*TP*R4
      HH(J+20)=TP*RM*S4
      DH1(J+ 1)=-SM*TM*(F1+RM/EIGHT)
      DH1(J+ 2)= SM*TM*(F2+RP/EIGHT)
      DH1(J+ 3)= SP*TM*(F3+RP/EIGHT)
      DH1(J+ 4)=-SP*TM*(F4+RM/EIGHT)
      DH1(J+ 5)=-SM*TP*(F5+RM/EIGHT)
      DH1(J+ 6)= SM*TP*(F6+RP/EIGHT)
      DH1(J+ 7)= SP*TP*(F7+RP/EIGHT)
      DH1(J+ 8)=-SP*TP*(F8+RM/EIGHT)
      DH1(J+ 9)=-R*SM*TM/TWO
      DH1(J+10)= TM*S4
      DH1(J+11)=-R*SP*TM/TWO
      DH1(J+12)=-TM*S4
      DH1(J+13)=-SM*T4
      DH1(J+14)= SM*T4
      DH1(J+15)= SP*T4
      DH1(J+16)=-SP*T4
      DH1(J+17)=-R*SM*TP/TWO
      DH1(J+18)= TP*S4
      DH1(J+19)=-R*SP*TP/TWO
      DH1(J+20)=-TP*S4
      DH2(J+ 1)=-TM*RM*(F1+SM/EIGHT)
```

```
      DH2(J+ 2)=-TM*RP*(F2+SM/EIGHT)
      DH2(J+ 3)= TM*RP*(F3+SP/EIGHT)
      DH2(J+ 4)= TM*RM*(F4+SP/EIGHT)
      DH2(J+ 5)=-TP*RM*(F5+SM/EIGHT)
      DH2(J+ 6)=-TP*RP*(F6+SM/EIGHT)
      DH2(J+ 7)= TP*RP*(F7+SP/EIGHT)
      DH2(J+ 8)= TP*RM*(F8+SP/EIGHT)
      DH2(J+ 9)=-TM*R4
      DH2(J+10)=-S*TM*RP/TWO
      DH2(J+11)= TM*R4
      DH2(J+12)=-S*TM*RM/TWO
      DH2(J+13)=-RM*T4
      DH2(J+14)=-RP*T4
      DH2(J+15)= RP*T4
      DH2(J+16)= RM*T4
      DH2(J+17)=-TP*R4
      DH2(J+18)=-S*TP*RP/TWO
      DH2(J+19)= TP*R4
      DH2(J+20)=-S*TP*RM/TWO
      DH3(J+ 1)=-RM*SM*(F1+TM/EIGHT)
      DH3(J+ 2)=-RP*SM*(F2+TM/EIGHT)
      DH3(J+ 3)=-RP*SP*(F3+TM/EIGHT)
      DH3(J+ 4)=-RM*SP*(F4+TM/EIGHT)
      DH3(J+ 5)= RM*SM*(F5+TP/EIGHT)
      DH3(J+ 6)= RP*SM*(F6+TP/EIGHT)
      DH3(J+ 7)= RP*SP*(F7+TP/EIGHT)
      DH3(J+ 8)= RM*SP*(F8+TP/EIGHT)
      DH3(J+ 9)=-SM*R4
      DH3(J+10)=-RP*S4
      DH3(J+11)=-SP*R4
      DH3(J+12)=-RM*S4
      DH3(J+13)=-T*RM*SM/TWO
      DH3(J+14)=-T*RP*SM/TWO
      DH3(J+15)=-T*RP*SP/TWO
      DH3(J+16)=-T*RM*SP/TWO
      DH3(J+17)= SM*R4
      DH3(J+18)= RP*S4
      DH3(J+19)= SP*R4
      DH3(J+20)= RM*S4
   11 J=J+20
      RETURN
      END
```

```
$LARGE
C
C     File LIB3.FOR, last updated on 11 July 1991.
C
      SUBROUTINE JAC2D (IP,ME,NV,X,Y,NP,DG1,DG2,DH1,DH2,DHX,DHY,DET)
      IMPLICIT DOUBLE PRECISION (A-H,O-Z)
      DIMENSION DH1(*),DH2(*),X(*),Y(*),DHX(*),DHY(*),DET(*),ME(*),
     -          DG1(*),DG2(*),XX(9),YY(9)
C
C     FUNCTION : CALCULATE DHX,DHY,DET OF A 2D ELEMENT
C
C     INPUT : IP = ELEMENT NODE NUMBER START POSITION ON ARRAY ME
C             NV = NUMBER OF NODES IN THE ELEMENT
C          ME(*) = ARRAY OF ELEMENT NODE NUMBERS
C      X(*),Y(*) = NODAL COORDINATES
C             NP = NUMBER OF INTEGRATION POINTS
C  DG1(*),DG2(*) = DERIVATIVES OF SHAPE FUNCTIONS w.r.t. ELEMENT COORDINATES
C  DH1(*),DH2(*) = DERIVATIVES OF INTERPOLATION FUNCTIONS w.r.t. ELEMENT COOR.
C
C     OUTPUT :          DET(I) = DETERMINANT OF JACOBIAN AT INTEGRATION POINT I
C                DHX(*),DHX(*) = DERIVATIVES OF INTERPOLATION
C                                FUNCTIONS w.r.t. GLOBAL COORDINATES
C
      K=IP-1
      DO 44 I=1,NV
      XX(I)=X(ME(K+I))
   44 YY(I)=Y(ME(K+I))
      J=0
      DO 11 K=1,NP
      D11=0.D0
      D12=0.D0
      D21=0.D0
      D22=0.D0
      DO 22 I=1,NV
      L=J+I
      D11=D11+DG1(L)*XX(I)
      D12=D12+DG1(L)*YY(I)
      D21=D21+DG2(L)*XX(I)
      D22=D22+DG2(L)*YY(I)
   22 CONTINUE
      DET(K)=D11*D22-D12*D21
      D11=D11/DET(K)
      D12=D12/DET(K)
      D21=D21/DET(K)
      D22=D22/DET(K)
      DO 33 I=1,NV
      L=J+I
      DHX(L)=D22*DH1(L)-D12*DH2(L)
      DHY(L)=D11*DH2(L)-D21*DH1(L)
   33 CONTINUE
   11 J=J+NV
      RETURN
      END

      SUBROUTINE EK2D (NV,NP,DHX,DHY,DET,WT,NEF,EK)
      IMPLICIT DOUBLE PRECISION (A-H,O-Z)
      DIMENSION DHX(*),DHY(*),DET(*),WT(*),EK(NEF,*)
      COMMON /BK1/D1,D2,D3,D4,D5
C
C     FUNCTION : Evaluation of 2D elasticity element stiffness matrix by
C                numerical integration.
C
C     INPUT : NP = Number of integration points
C             NV = Number of nodes in the element
C            NEF = Total number of d.o.f. within the element
C         DET(I) = Determinant of Jacobian at integration point I
C          WT(I) = Weights at integration point I
C  DHX(*),DHY(*) = Derivatives of interpolation functions w.r.t. global coor.
C
C     OUTPUT : EK(NEF,NEF) = Element stiffness matrix
C
      DO 22 I=1,NV
```

```
      II=I+I
      DO 22 J=I,NV
      JJ=J+J
      S11=0.D0
      S12=0.D0
      S21=0.D0
      S22=0.D0
      DO 33 K=1,NP
      C=DET(K)*WT(K)
      L=NV*(K-1)
      FI=DHX(L+I)
      FJ=DHX(L+J)
      GI=DHY(L+I)
      GJ=DHY(L+J)
      S11=S11+C*FI*FJ
      S12=S12+C*FI*GJ
      S21=S21+C*GI*FJ
      S22=S22+C*GI*GJ
   33 CONTINUE
      EK(II-1,JJ-1)=D1*S11+D3*S22
      EK(II-1,JJ  )=D4*S12+D3*S21
      EK(II  ,JJ-1)=D3*S12+D4*S21
      EK(II  ,JJ  )=D3*S11+D2*S22
   22 CONTINUE
C
      DO 11 J=1,NEF-1
      DO 11 I=J+1,NEF
   11 EK(I,J)=EK(J,I)
      RETURN
      END

      SUBROUTINE EKAXIS (NV,NP,HH,DHX,DHY,DET,WT,NEF,EK,IP,ME,X)
      IMPLICIT DOUBLE PRECISION (A-H,O-Z)
      DIMENSION DHX(*),DHY(*),DET(*),WT(*),EK(NEF,*),HH(*),X(*),ME(*),
     -          R(9)
      COMMON /BK1/D1,D2,D3,D4,D5
C
C     FUNCTION : Calculate element stiffness matrix for axisymmetric analysis
C
C     INPUT : NV = Number of nodes in the element
C             NP = Number of integration points
C            NEF = Number of d.o.f. within the element
C         DET(I) = Determinant of Jacobian at integration point I
C          WT(I) = Weights at integration point I
C  DHX(*),DHY(*) = Derivatives of interpolation functions w.r.t. global coord.
C
C     OUTPUT : EK(NEF,NEF) = Element stiffness matrix

C     At each integration point, calculate the distance R from the axis

      J=IP-1
      DO 44 K=1,NP
      T=0.D0
      L=(K-1)*NV
      DO 55 I=1,NV
   55 T=T+HH(L+I)*X(ME(J+I))
   44 R(K)=T
C
      DO 22 I=1,NV
      II=I+I
      DO 22 J=I,NV
      JJ=J+J
      S11=0.D0
      S12=0.D0
      S13=0.D0
      S21=0.D0
      S22=0.D0
      S23=0.D0
      S31=0.D0
      S32=0.D0
      S33=0.D0
```

```
C     Sum over the integration points

      DO 33 K=1,NP
      C=DET(K)*WT(K)*R(K)*6.28318530718
      I1=(K-1)*NV+I
      J1=(K-1)*NV+J
      FI=DHX(I1)
      FJ=DHX(J1)
      GI=DHY(I1)
      GJ=DHY(J1)
      HI=HH(I1)/R(K)
      HJ=HH(J1)/R(K)
      S11=S11+C*FI*FJ
      S12=S12+C*FI*GJ
      S13=S13+C*FI*HJ
      S21=S21+C*GI*FJ
      S22=S22+C*GI*GJ
      S23=S23+C*GI*HJ
      S31=S31+C*HI*FJ
      S32=S32+C*HI*GJ
      S33=S33+C*HI*HJ
   33 CONTINUE
      EK(II-1,JJ-1)=D2*S11+D3*S22+D5*S13+D5*S31+D2*S33
      EK(II-1,JJ  )=D4*S12+D3*S21+D4*S32
      EK(II  ,JJ-1)=D3*S12+D4*S21+D4*S23
      EK(II  ,JJ  )=D3*S11+D1*S22
   22 CONTINUE
C
      DO 11 J=1,NEF-1
      DO 11 I=J+1,NEF
   11 EK(I,J)=EK(J,I)
      RETURN
      END

      SUBROUTINE EK2DF (NV,NP,HH,DHX,DHY,DET,WT,EK)
      IMPLICIT DOUBLE PRECISION (A-H,O-Z)
      DIMENSION HH(*),DHX(*),DHY(*),DET(*),WT(*),EK(NV,*)
      COMMON /BK1/D1,D2,D3,D4,D5
C
C     FUNCTION : Evaluation of 2D field problem element stiffness matrix
C                (without edge convection terms) by numerical integration.
C
C     INPUT  : NP = Number of integration points
C              NV = Number of nodes in the element
C          DET(I) = Determinant of Jacobian at integration point I
C           WT(I) = Weights at integration point I
C           HH(*) = Values of interpolation functions at integration points
C  DHX(*),DHY(*) = Derivatives of interpolation functions w.r.t. global coor.
C
C     OUTPUT : EK(NV,NV)  = Element stiffness matrix
C
      DO 22 I=1,NV
      DO 22 J=I,NV
      S1=0.D0
      S2=0.D0
      SS=0.D0
      DO 33 K=1,NP
      C=DET(K)*WT(K)
      L=NV*(K-1)
      S1=S1+C*DHX(L+I)*DHX(L+J)
      S2=S2+C*DHY(L+I)*DHY(L+J)
   33 SS=SS+C*HH(L+I)*HH(L+J)
   22 EK(I,J)=D1*S1+D2*S2+D3*SS
C
      DO 11 J=1,NV-1
      DO 11 I=J+1,NV
   11 EK(I,J)=EK(J,I)
      RETURN
      END

      SUBROUTINE JAC3D (IP,ME,NV,X,Y,Z,NP,DG1,DG2,DG3,DH1,DH2,DH3,
     -                  DHG,DET)
```

```
      IMPLICIT DOUBLE PRECISION (A-H,O-Z)
      DIMENSION DH1(*),DH2(*),DH3(*),X(*),Y(*),Z(*),DHG(3,*),DET(*),
     -          DG1(*),DG2(*),DG3(*),XX(20),YY(20),ZZ(20),ME(*)
C
C   FUNCTION : Calculate the derivatives of interpolation functions and
C              determinants of Jacobian at integration points of a 3D element
C
C   INPUT :           IP = Element node number start position on array ME
C                  ME(*) = Array of element node numbers
C                     NV = Number of nodes in the element
C                     NP = Number of integration points
C      X(*),Y(*),Z(*) = Nodal coordinates
C DG1(*),DG2(*),DG3(*) = Derivatives of shape functions w.r.t. element coor.
C DH1(*),DH2(*),DH3(*) = Derivatives of interpol. func. w.r.t. element coor.
C
C   OUTPUT :  DHG(3,*) = Derivatives of interpolation functions w.r.t. global
C                        coordinates.
C               DET(*) = Determinants of Jacobian at integration points
C
      K=IP-1
      DO 44 I=1,NV
      J=ME(K+I)
      XX(I)=X(J)
      YY(I)=Y(J)
   44 ZZ(I)=Z(J)
C
      J=0
      DO 11 K=1,NP
      D11=0.D0
      D12=0.D0
      D13=0.D0
      D21=0.D0
      D22=0.D0
      D23=0.D0
      D31=0.D0
      D32=0.D0
      D33=0.D0
      DO 22 I=1,NV
      L=J+I
      D11=D11+DG1(L)*XX(I)
      D12=D12+DG1(L)*YY(I)
      D13=D13+DG1(L)*ZZ(I)
      D21=D21+DG2(L)*XX(I)
      D22=D22+DG2(L)*YY(I)
      D23=D23+DG2(L)*ZZ(I)
      D31=D31+DG3(L)*XX(I)
      D32=D32+DG3(L)*YY(I)
      D33=D33+DG3(L)*ZZ(I)
   22 CONTINUE
      DT=D11*(D22*D33-D23*D32)+D12*(D23*D31-D21*D33)
     +  +D13*(D21*D32-D22*D31)
      C11=(D22*D33-D32*D23)/DT
      C12=(D32*D13-D12*D33)/DT
      C13=(D12*D23-D22*D13)/DT
      C21=(D23*D31-D33*D21)/DT
      C22=(D33*D11-D13*D31)/DT
      C23=(D13*D21-D23*D11)/DT
      C31=(D21*D32-D31*D22)/DT
      C32=(D31*D12-D11*D32)/DT
      C33=(D11*D22-D21*D12)/DT
      DET(K)=DT
      DO 33 I=1,NV
      L=J+I
      DHG(1,L)=C11*DH1(L)+C12*DH2(L)+C13*DH3(L)
      DHG(2,L)=C21*DH1(L)+C22*DH2(L)+C23*DH3(L)
      DHG(3,L)=C31*DH1(L)+C32*DH2(L)+C33*DH3(L)
   33 CONTINUE
   11 J=J+NV
      RETURN
      END

      SUBROUTINE EK3D (NV,NP,DHG,DET,WT,NEF,EK)
```

```
      IMPLICIT DOUBLE PRECISION (A-H,O-Z)
      DIMENSION DHG(3,*),DET(*),WT(*),EK(NEF,*),S(3,3),IDM(3,3)
      COMMON /BK1/D1,D2,D3,D4,D5
      DATA IDM/1,0,0,0,1,0,0,0,1/
C
C     FUNCTION : Evaluation of 3D elasticity element stiffness matrix by
C                numerical integration.
C
C     INPUT : NP = Number of integration points
C             NV = Number of nodes in the element
C            NEF = Total number of d.o.f. within the element
C         DET(I) = Determinant of Jacobian at integration point I
C          WT(I) = Weights at integration point I
C       DHG(3,*) = Derivatives of interpolation functions w.r.t. global coor.
C
C     OUTPUT : EK(NEF,NEF) = Element stiffness matrix
C
      DO 22 KA=1,NV
      K1=(KA-1)*3
      DO 22 KB=KA,NV
      K2=(KB-1)*3
      DO 44 I=1,3
      DO 44 J=1,3
   44 S(I,J)=0.D0
      DO 33 K=1,NP
      C=DET(K)*WT(K)
      LA=(K-1)*NV+KA
      LB=(K-1)*NV+KB
      DO 55 I=1,3
      DO 55 J=1,3
   55 S(I,J)=S(I,J)+C*DHG(I,LA)*DHG(J,LB)
   33 CONTINUE
      T=D2*(S(1,1)+S(2,2)+S(3,3))
      DO 66 I=1,3
      DO 66 J=1,3
   66 EK(K1+I,K2+J)=D1*S(I,J)+D2*S(J,I)+T*IDM(I,J)
   22 CONTINUE
C
      DO 11 J=1,NEF-1
      DO 11 I=J+1,NEF
   11 EK(I,J)=EK(J,I)
      RETURN
      END

      SUBROUTINE EK3DF (NV,NP,DHG,DET,WT,EK)
      IMPLICIT DOUBLE PRECISION (A-H,O-Z)
      DIMENSION DHG(3,*),DET(*),WT(*),EK(NV,*)
      COMMON /BK1/D1,D2,D3,D4,D5
C
C     FUNCTION : Evaluation of 3D field problem element stiffness matrix
C                (without surface convection terms) by numerical integration.
C
C     INPUT : NP = Number of integration points
C             NV = Number of nodes in the element
C         DET(I) = Determinant of Jacobian at integration point I
C          WT(I) = Weights at integration point I
C       DHG(3,*) = Derivatives of interpolation functions w.r.t. global coor.
C
C     OUTPUT : EK(NV,NV) = Element stiffness matrix
C
      DO 22 I=1,NV
      DO 22 J=I,NV
      S1=0.D0
      S2=0.D0
      S3=0.D0
      DO 33 K=1,NP
      C=DET(K)*WT(K)
      L=NV*(K-1)
      S1=S1+C*DHG(1,L+I)*DHG(1,L+J)
      S2=S2+C*DHG(2,L+I)*DHG(2,L+J)
   33 S3=S3+C*DHG(3,L+I)*DHG(3,L+J)
   22 EK(I,J)=D1*S1+D2*S2+D3*S3
```

```
C
      DO 11 J=1,NV-1
      DO 11 I=J+1,NV
   11 EK(I,J)=EK(J,I)
      RETURN
      END
```

```
$LARGE
C
C     File LIB4.FOR, last updated on 11 July 1991.
C
      SUBROUTINE EKBEAM (X1,Y1,Z1,X2,Y2,Z2,EK)
      IMPLICIT DOUBLE PRECISION (A-H,O-Z)
      REAL*8 L,I2,I3,J
      DIMENSION EK(12,*)
      COMMON /BK1/A2,B2,C2,E,G
      COMMON /BK2/A,I2,I3,J,F2,F3

C     Evaluation of the element stiffness matrix (12x12) of a spatial beam
C
C     (X1,Y1,Z1) = Spatial coordinates of node 1
C     (X2,Y2,Z2) = Spatial coordinates of node 2
C     (A1,B1,C1) = Principal axis (unit vector) of element local basis
C     (A2,B2,C2) = Second axis (unit vector) of element local basis
C              A = Cross-sectional area
C             I2 = Second moment of area about local axis 2
C             I3 = Second moment of area about local axis 3
C              J = Equivalent polar moment of area
C             F2 = Shear factor along local axis 2
C             F3 = Shear factor along local axis 3
C
C     Calculate element local orthonormal basis

      L=DSQRT((X2-X1)**2+(Y2-Y1)**2+(Z2-Z1)**2)
      A1=(X2-X1)/L
      B1=(Y2-Y1)/L
      C1=(Z2-Z1)/L
      IF (A2*A2+B2*B2+C2*C2.LT.0.01) THEN
      IF (A1*A1+B1*B1.GT.0.000001) THEN
      T=DSQRT(A1*A1+B1*B1)
      A3= B1/T
      B3=-A1/T
      C3=0.D0
      ELSE
      T=DSQRT(A1*A1+C1*C1)
      A3=-C1/T
      B3=0.D0
      C3= A1/T
      ENDIF
      A2=B3*C1-B1*C3
      B2=C3*A1-C1*A3
      C2=A3*B1-A1*B3
      ELSE
      A3=B1*C2-B2*C1
      B3=C1*A2-C2*A1
      C3=A1*B2-A2*B1
      ENDIF

C     Calculate element stiffness coefficients

      T2=E*I3/(L+L*F2)
      T3=E*I2/(L+L*F3)
      S1=E*A/L
      S2=12.D0*T2/L**2
      S3=12.D0*T3/L**2
      EK(1,1)=S1*A1*A1+S2*A2*A2+S3*A3*A3
      EK(2,1)=S1*A1*B1+S2*A2*B2+S3*A3*B3
      EK(3,1)=S1*A1*C1+S2*A2*C2+S3*A3*C3
      EK(2,2)=S1*B1*B1+S2*B2*B2+S3*B3*B3
      EK(3,2)=S1*B1*C1+S2*B2*C2+S3*B3*C3
      EK(3,3)=S1*C1*C1+S2*C2*C2+S3*C3*C3
      EK(7,7)=EK(1,1)
      EK(8,7)=EK(2,1)
      EK(9,7)=EK(3,1)
      EK(8,8)=EK(2,2)
      EK(9,8)=EK(3,2)
      EK(9,9)=EK(3,3)
      EK(7,1)=-EK(1,1)
      EK(8,1)=-EK(2,1)
```

```
      EK(9,1)=-EK(3,1)
      EK(8,2)=-EK(2,2)
      EK(9,2)=-EK(3,2)
      EK(9,3)=-EK(3,3)
      EK(7,2)=EK(8,1)
      EK(7,3)=EK(9,1)
      EK(8,3)=EK(9,2)
C
      S1=G*J/L
      S2=(4.D0+F3)*T3
      S3=(4.D0+F2)*T2
      EK(4,4)=S1*A1*A1+S2*A2*A2+S3*A3*A3
      EK(5,4)=S1*A1*B1+S2*A2*B2+S3*A3*B3
      EK(6,4)=S1*A1*C1+S2*A2*C2+S3*A3*C3
      EK(5,5)=S1*B1*B1+S2*B2*B2+S3*B3*B3
      EK(6,5)=S1*B1*C1+S2*B2*C2+S3*B3*C3
      EK(6,6)=S1*C1*C1+S2*C2*C2+S3*C3*C3
      EK(10,10)=EK(4,4)
      EK(11,10)=EK(5,4)
      EK(12,10)=EK(6,4)
      EK(11,11)=EK(5,5)
      EK(12,11)=EK(6,5)
      EK(12,12)=EK(6,6)
      S1=-G*J/L
      S2=(2.D0-F3)*T3
      S3=(2.D0-F2)*T2
      EK(10,4)=S1*A1*A1+S2*A2*A2+S3*A3*A3
      EK(11,4)=S1*A1*B1+S2*A2*B2+S3*A3*B3
      EK(12,4)=S1*A1*C1+S2*A2*C2+S3*A3*C3
      EK(11,5)=S1*B1*B1+S2*B2*B2+S3*B3*B3
      EK(12,5)=S1*B1*C1+S2*B2*C2+S3*B3*C3
      EK(12,6)=S1*C1*C1+S2*C2*C2+S3*C3*C3
      EK(10,5)=EK(11,4)
      EK(10,6)=EK(12,4)
      EK(11,6)=EK(12,5)
C
      S1= 6.D0*T2/L
      S2=-6.D0*T3/L
      EK(4,1)=S1*A2*A3+S2*A3*A2
      EK(5,1)=S1*A2*B3+S2*A3*B2
      EK(6,1)=S1*A2*C3+S2*A3*C2
      EK(4,2)=S1*B2*A3+S2*B3*A2
      EK(5,2)=S1*B2*B3+S2*B3*B2
      EK(6,2)=S1*B2*C3+S2*B3*C2
      EK(4,3)=S1*C2*A3+S2*C3*A2
      EK(5,3)=S1*C2*B3+S2*C3*B2
      EK(6,3)=S1*C2*C3+S2*C3*C2
      EK(10,1)=EK(4,1)
      EK(11,1)=EK(5,1)
      EK(12,1)=EK(6,1)
      EK(10,2)=EK(4,2)
      EK(11,2)=EK(5,2)
      EK(12,2)=EK(6,2)
      EK(10,3)=EK(4,3)
      EK(11,3)=EK(5,3)
      EK(12,3)=EK(6,3)
      EK(10,7)=-EK(4,1)
      EK(11,7)=-EK(5,1)
      EK(12,7)=-EK(6,1)
      EK(10,8)=-EK(4,2)
      EK(11,8)=-EK(5,2)
      EK(12,8)=-EK(6,2)
      EK(10,9)=-EK(4,3)
      EK(11,9)=-EK(5,3)
      EK(12,9)=-EK(6,3)
      S1= 6.D0*T3/L
      S2=-6.D0*T2/L
      EK(7,4)=S1*A2*A3+S2*A3*A2
      EK(8,4)=S1*A2*B3+S2*A3*B2
      EK(9,4)=S1*A2*C3+S2*A3*C2
      EK(7,5)=S1*B2*A3+S2*B3*A2
      EK(8,5)=S1*B2*B3+S2*B3*B2
```

```
      EK(9,5)=S1*B2*C3+S2*B3*C2
      EK(7,6)=S1*C2*A3+S2*C3*A2
      EK(8,6)=S1*C2*B3+S2*C3*B2
      EK(9,6)=S1*C2*C3+S2*C3*C2
      DO 11 I=1,11
      DO 11 K=I+1,12
   11 EK(I,K)=EK(K,I)
      RETURN
      END

      SUBROUTINE EKL9P (IP,ME,HH,DHX,DHY,DET,WT,GG,DGX,DGY,DT4,TH,EK)
      IMPLICIT DOUBLE PRECISION (A-H,O-Z)
      DIMENSION ME(*),HH(*),DHX(*),DHY(*),DET(*),WT(*),GG(*),DGX(*),
     -          DGY(*),DT4(*),TH(*),EK(27,*),H4(4),H9(9)
      COMMON /BK1/D1,D2,D3,D4,D5
C
C     FUNCTION : Evaluation of element stiffness matrix of a 9-node Lagrangian
C                biquadratic plate bending element (isotropic case)
C
C     INPUT  : HH(*),DHX(*),DHY(*), = Values and derivatives of interpolation
C              DET(*),WT(*)           functions, determinant of Jacobian,
C                                     and weight at 3x3 Gaussian points.
C              GG(*),DGX(*),DGY(*), = Values and derivatives of interpolation
C              DT4(*)                 functions, and determinant of Jacobian
C                                     at 2x2 Gaussian points.
C              TH(J)  = Thickness of plate at node J
C
C     OUTPUT : EK(27,27)  = Element stiffness matrix

C     Calculate the plate thickness H at integration points

      DO 111 K=1,4
      H=0.D0
      L=9*(K-1)
      DO 222 I=1,9
  222 H=H+GG(L+I)*TH(ME(IP-1+I))
  111 H4(K)=H
      DO 333 K=1,9
      H=0.D0
      L=9*(K-1)
      DO 444 I=1,9
  444 H=H+HH(L+I)*TH(ME(IP-1+I))
  333 H9(K)=H
C
      T1=D1/(2.4D0+2.4D0*D2)
      T2=D1/(12.D0-12.D0*D2*D2)
      DO 22 KA=1,9
      K1=(KA-1)*3
      DO 22 KB=KA,9
      K2=(KB-1)*3
      S11=0.D0
      S12=0.D0
      S13=0.D0
      S21=0.D0
      S22=0.D0
      S23=0.D0
      S31=0.D0
      S32=0.D0

C     Summing over integration points for shear stiffness

      DO 33 K=1,4
      C=DT4(K)*T1*H4(K)
      LA=(K-1)*9+KA
      LB=(K-1)*9+KB
      S11=S11+C*DGX(LA)*DGX(LB)+C*DGY(LA)*DGY(LB)
      S12=S12-C*DGX(LA)*GG(LB)
      S13=S13-C*DGY(LA)*GG(LB)
      S21=S21-C*DGX(LB)*GG(LA)
      S31=S31-C*DGY(LB)*GG(LA)
   33 S22=S22+C*GG(LA)*GG(LB)
      S33=S22
```

```
C      Summing over integration points for bending stiffness

       DO 55 K=1,9
       C1=WT(K)*DET(K)*T2*H9(K)**3
       C2=D2*C1
       C3=(0.5D0-0.5D0*D2)*C1
       LA=(K-1)*9+KA
       LB=(K-1)*9+KB
       S22=S22+C1*DHX(LA)*DHX(LB)+C3*DHY(LA)*DHY(LB)
       S23=S23+C2*DHX(LA)*DHY(LB)+C3*DHY(LA)*DHX(LB)
       S32=S32+C2*DHY(LA)*DHX(LB)+C3*DHX(LA)*DHY(LB)
    55 S33=S33+C1*DHY(LA)*DHY(LB)+C3*DHX(LA)*DHX(LB)
       EK(K1+1,K2+1)=S11
       EK(K1+1,K2+2)=S12
       EK(K1+1,K2+3)=S13
       EK(K1+2,K2+1)=S21
       EK(K1+2,K2+2)=S22
       EK(K1+2,K2+3)=S23
       EK(K1+3,K2+1)=S31
       EK(K1+3,K2+2)=S32
       EK(K1+3,K2+3)=S33
    22 CONTINUE
       DO 11 J=1,26
       DO 11 I=J+1,27
    11 EK(I,J)=EK(J,I)
       RETURN
       END

       SUBROUTINE EKL9S (IP,ME,NV,X,Y,Z,TH,NP,HH,DH1,DH2,WT,DET,P,Q,EK)
       IMPLICIT DOUBLE PRECISION (A-H,O-Z)
       DIMENSION ME(*),X(*),Y(*),Z(*),TH(*),HH(*),DH1(*),DH2(*),DET(*),
      -          P(*),Q(*),WT(*),EK(45,*), XX(9),YY(9),ZZ(9),TT(9),
      -          U1(9),U2(9),U3(9),V1(9),V2(9),V3(9),W1(9),W2(9),W3(9),
      -DHX(162),DHY(162),DHZ(162),DZHX(162),DZHY(162),DZHZ(162),R11(18),
      -R12(18),R13(18),R21(18),R22(18),R23(18),R31(18),R32(18),R33(18)
       COMMON /BK1/D1,D2,D3,D4,D5
C
C      FUNCTION : Evaluation of element stiffness matrix of a 9-node Lagrangian
C                 biquadratic shell element (isotropic case)
C
C     U1(*),U2(*),... W3(*) = Base vectors at element nodes (Fibre)
C      DHX(*),DHY(*),DHZ(*) = Derivatives of interpolation functions w.r.t.
C                             global coordinates
C   DZHX(*),DZHY(*),DZHZ(*) = Derivatives of (zeta * interpolation functions)
C                             w.r.t global coordinates
C R11(*),R12(*),... R33(*) = Base vectors at integration points (Laminar)
C
       K=IP-1
       DO 44 I=1,NV

C      At each element node I, retrieve the coordinates XX(I),YY(I),ZZ(I),
C      thickness TT(I) and local basis vectors (u,v,w)

       J=ME(K+I)
       XX(I)=X(J)
       YY(I)=Y(J)
       ZZ(I)=Z(J)
       TT(I)=TH(J)
       JJ=5*J-5
       U1(I)=P(JJ+1)
       U2(I)=P(JJ+2)
       U3(I)=P(JJ+3)
       V1(I)=P(JJ+4)
       V2(I)=P(JJ+5)
       V3(I)=Q(JJ+1)
       W1(I)=Q(JJ+2)
       W2(I)=Q(JJ+3)
    44 W3(I)=Q(JJ+4)
C
       NP2=NP*2
       KK=0
```

```
      DO 33 M=1,2
      ZETA=(2*M-3)/DSQRT(3.D0)
      J=0

C     Loop over integration points

      DO 11 K=1,NP
      KK=KK+1
      C11=0.D0
      C12=0.D0
      C13=0.D0
      C21=0.D0
      C22=0.D0
      C23=0.D0
      C31=0.D0
      C32=0.D0
      C33=0.D0

C     Calculate the derivatives of (x,y,z) at integration points w.r.t.
C     element coordinates xi,eta,zeta

      DO 22 I=1,NV
      L=J+I
      T=TT(I)/2.D0
      ZT=ZETA*T
      C11=C11+DH1(L)*(XX(I)+ZT*W1(I))
      C12=C12+DH1(L)*(YY(I)+ZT*W2(I))
      C13=C13+DH1(L)*(ZZ(I)+ZT*W3(I))
      C21=C21+DH2(L)*(XX(I)+ZT*W1(I))
      C22=C22+DH2(L)*(YY(I)+ZT*W2(I))
      C23=C23+DH2(L)*(ZZ(I)+ZT*W3(I))
      C31=C31+HH(L)*T*W1(I)
      C32=C32+HH(L)*T*W2(I)
      C33=C33+HH(L)*T*W3(I)
   22 CONTINUE

C     Construction of orthonormal basis at integration points

      C1=C12*C23-C22*C13
      C2=C13*C21-C23*C11
      C3=C11*C22-C21*C12
      C=DSQRT(C1*C1+C2*C2+C3*C3)
      C1=C1/C
      C2=C2/C
      C3=C3/C
      C=DSQRT(C11*C11+C12*C12+C13*C13)
      A1=C11/C
      A2=C12/C
      A3=C13/C
      B1=C2*A3-A2*C3
      B2=C3*A1-A3*C1
      B3=C1*A2-A1*C2
      R11(KK)=A1
      R12(KK)=A2
      R13(KK)=A3
      R21(KK)=B1
      R22(KK)=B2
      R23(KK)=B3
      R31(KK)=C1
      R32(KK)=C2
      R33(KK)=C3

C     Compute the inverse of the Jacobian matrix

      D11=A1*C11+A2*C12+A3*C13
      D12=B1*C11+B2*C12+B3*C13
      D13=C1*C11+C2*C12+C3*C13
      D21=A1*C21+A2*C22+A3*C23
      D22=B1*C21+B2*C22+B3*C23
      D23=C1*C21+C2*C22+C3*C23
      D31=A1*C31+A2*C32+A3*C33
      D32=B1*C31+B2*C32+B3*C33
```

```
      D33=C1*C31+C2*C32+C3*C33
      DT=D11*(D22*D33-D23*D32)+D12*(D23*D31-D21*D33)
     +   +D13*(D21*D32-D22*D31)
      C11=(D22*D33-D32*D23)/DT
      C12=(D32*D13-D12*D33)/DT
      C13=(D12*D23-D22*D13)/DT
      C21=(D23*D31-D33*D21)/DT
      C22=(D33*D11-D13*D31)/DT
      C23=(D13*D21-D23*D11)/DT
      C31=(D21*D32-D31*D22)/DT
      C32=(D31*D12-D11*D32)/DT
      C33=(D11*D22-D21*D12)/DT
      DET(KK)=DT*WT(K)

C     Calculate the derivatives of the interpolation functions w.r.t. x,y,z.

      DO 55 I=1,NV
      L=J+I
      LL=NV*(KK-1)+I
      DHX(LL)=C11*DH1(L)+C12*DH2(L)
      DHY(LL)=C21*DH1(L)+C22*DH2(L)
      DHZ(LL)=C31*DH1(L)+C32*DH2(L)
      DZHX(LL)=ZETA*DHX(L)+C13*HH(L)
      DZHY(LL)=ZETA*DHY(L)+C23*HH(L)
      DZHZ(LL)=ZETA*DHZ(L)+C33*HH(L)
   55 CONTINUE
   11 J=J+NV
   33 CONTINUE

C     Evaluate the element stiffness coefficients

      F1=D1/(1.D0-D2*D2)
      F3=(1.D0-D2)/2.D0
      F4=5.D0*F3/6.D0
      DO 222 KA=1,NV
      K1=(KA-1)*5
      T=TT(KA)/2.D0
      V1A=-T*V1(KA)
      V2A=-T*V2(KA)
      V3A=-T*V3(KA)
      U1A= T*U1(KA)
      U2A= T*U2(KA)
      U3A= T*U3(KA)
      DO 222 KB=KA,NV
      K2=(KB-1)*5
      T=TT(KB)/2.D0
      V1B=-T*V1(KB)
      V2B=-T*V2(KB)
      V3B=-T*V3(KB)
      U1B= T*U1(KB)
      U2B= T*U2(KB)
      U3B= T*U3(KB)
      S11=0.D0
      S12=0.D0
      S13=0.D0
      S14=0.D0
      S15=0.D0
      S21=0.D0
      S22=0.D0
      S23=0.D0
      S24=0.D0
      S25=0.D0
      S31=0.D0
      S32=0.D0
      S33=0.D0
      S34=0.D0
      S35=0.D0
      S41=0.D0
      S42=0.D0
      S43=0.D0
      S44=0.D0
      S45=0.D0
```

```
      S51=0.D0
      S52=0.D0
      S53=0.D0
      S54=0.D0
      S55=0.D0
C     Loop over integration points
      DO 333 K=1,NP2
      C=DET(K)*F1
      A1=R11(K)
      A2=R12(K)
      A3=R13(K)
      B1=R21(K)
      B2=R22(K)
      B3=R23(K)
      C1=R31(K)
      C2=R32(K)
      C3=R33(K)
      L=NV*(K-1)+KA
      DX=DHX(L)
      DY=DHY(L)
      DZ=DHZ(L)
C     Compute coefficients B11A-B53A and C11A-C52A
      B11A=A1*DX
      B21A=B1*DY
      B31A=A1*DY+B1*DX
      B41A=B1*DZ+C1*DY
      B51A=C1*DX+A1*DZ
      B12A=A2*DX
      B22A=B2*DY
      B32A=A2*DY+B2*DX
      B42A=B2*DZ+C2*DY
      B52A=C2*DX+A2*DZ
      B13A=A3*DX
      B23A=B3*DY
      B33A=A3*DY+B3*DX
      B43A=B3*DZ+C3*DY
      B53A=C3*DX+A3*DZ
      W11=A1*V1A+A2*V2A+A3*V3A
      W21=B1*V1A+B2*V2A+B3*V3A
      W31=C1*V1A+C2*V2A+C3*V3A
      W12=A1*U1A+A2*U2A+A3*U3A
      W22=B1*U1A+B2*U2A+B3*U3A
      W32=C1*U1A+C2*U2A+C3*U3A
      C11A=W11*DZHX(L)
      C21A=W21*DZHY(L)
      C31A=W11*DZHY(L)+W21*DZHX(L)
      C41A=W21*DZHZ(L)+W31*DZHY(L)
      C51A=W31*DZHX(L)+W11*DZHZ(L)
      C12A=W12*DZHX(L)
      C22A=W22*DZHY(L)
      C32A=W12*DZHY(L)+W22*DZHX(L)
      C42A=W22*DZHZ(L)+W32*DZHY(L)
      C52A=W32*DZHX(L)+W12*DZHZ(L)
      L=NV*(K-1)+KB
      DX=C*DHX(L)
      DY=C*DHY(L)
      DZ=C*DHZ(L)
C     Compute coefficients B11B-B53B and C11B-C52B
      B11B=A1*DX+D2*B1*DY
      B21B=D2*A1*DX+B1*DY
      B31B=F3*(A1*DY+B1*DX)
      B41B=F4*(B1*DZ+C1*DY)
      B51B=F4*(C1*DX+A1*DZ)
      B12B=A2*DX+D2*B2*DY
      B22B=D2*A2*DX+B2*DY
      B32B=F3*(A2*DY+B2*DX)
```

```
      B42B=F4*(B2*DZ+C2*DY)
      B52B=F4*(C2*DX+A2*DZ)
      B13B=A3*DX+D2*B3*DY
      B23B=D2*A3*DX+B3*DY
      B33B=F3*(A3*DY+B3*DX)
      B43B=F4*(B3*DZ+C3*DY)
      B53B=F4*(C3*DX+A3*DZ)
      W11=A1*V1B+A2*V2B+A3*V3B
      W21=B1*V1B+B2*V2B+B3*V3B
      W31=C1*V1B+C2*V2B+C3*V3B
      W12=A1*U1B+A2*U2B+A3*U3B
      W22=B1*U1B+B2*U2B+B3*U3B
      W32=C1*U1B+C2*U2B+C3*U3B
      DX=C*DZHX(L)
      DY=C*DZHY(L)
      DZ=C*DZHZ(L)
      C11B=W11*DX+D2*W21*DY
      C21B=D2*W11*DX+W21*DY
      C31B=F3*(W11*DY+W21*DX)
      C41B=F4*(W21*DZ+W31*DY)
      C51B=F4*(W31*DX+W11*DZ)
      C12B=W12*DX+D2*W22*DY
      C22B=D2*W12*DX+W22*DY
      C32B=F3*(W12*DY+W22*DX)
      C42B=F4*(W22*DZ+W32*DY)
      C52B=F4*(W32*DX+W12*DZ)

C     Multiply and sum up the contribution from each integration point

      S11=S11+B11A*B11B+B21A*B21B+B31A*B31B+B41A*B41B+B51A*B51B
      S21=S21+B12A*B11B+B22A*B21B+B32A*B31B+B42A*B41B+B52A*B51B
      S31=S31+B13A*B11B+B23A*B21B+B33A*B31B+B43A*B41B+B53A*B51B
      S41=S41+C11A*B11B+C21A*B21B+C31A*B31B+C41A*B41B+C51A*B51B
      S51=S51+C12A*B11B+C22A*B21B+C32A*B31B+C42A*B41B+C52A*B51B
      S12=S12+B11A*B12B+B21A*B22B+B31A*B32B+B41A*B42B+B51A*B52B
      S22=S22+B12A*B12B+B22A*B22B+B32A*B32B+B42A*B42B+B52A*B52B
      S32=S32+B13A*B12B+B23A*B22B+B33A*B32B+B43A*B42B+B53A*B52B
      S42=S42+C11A*B12B+C21A*B22B+C31A*B32B+C41A*B42B+C51A*B52B
      S52=S52+C12A*B12B+C22A*B22B+C32A*B32B+C42A*B42B+C52A*B52B
      S13=S13+B11A*B13B+B21A*B23B+B31A*B33B+B41A*B43B+B51A*B53B
      S23=S23+B12A*B13B+B22A*B23B+B32A*B33B+B42A*B43B+B52A*B53B
      S33=S33+B13A*B13B+B23A*B23B+B33A*B33B+B43A*B43B+B53A*B53B
      S43=S43+C11A*B13B+C21A*B23B+C31A*B33B+C41A*B43B+C51A*B53B
      S53=S53+C12A*B13B+C22A*B23B+C32A*B33B+C42A*B43B+C52A*B53B
      S14=S14+B11A*C11B+B21A*C21B+B31A*C31B+B41A*C41B+B51A*C51B
      S24=S24+B12A*C11B+B22A*C21B+B32A*C31B+B42A*C41B+B52A*C51B
      S34=S34+B13A*C11B+B23A*C21B+B33A*C31B+B43A*C41B+B53A*C51B
      S44=S44+C11A*C11B+C21A*C21B+C31A*C31B+C41A*C41B+C51A*C51B
      S54=S54+C12A*C11B+C22A*C21B+C32A*C31B+C42A*C41B+C52A*C51B
      S15=S15+B11A*C12B+B21A*C22B+B31A*C32B+B41A*C42B+B51A*C52B
      S25=S25+B12A*C12B+B22A*C22B+B32A*C32B+B42A*C42B+B52A*C52B
      S35=S35+B13A*C12B+B23A*C22B+B33A*C32B+B43A*C42B+B53A*C52B
      S45=S45+C11A*C12B+C21A*C22B+C31A*C32B+C41A*C42B+C51A*C52B
      S55=S55+C12A*C12B+C22A*C22B+C32A*C32B+C42A*C42B+C52A*C52B
  333 CONTINUE

C     Assign the computed value to the element stiffness matrix

      EK(K1+1,K2+1)=S11
      EK(K1+2,K2+1)=S21
      EK(K1+3,K2+1)=S31
      EK(K1+4,K2+1)=S41
      EK(K1+5,K2+1)=S51
      EK(K1+1,K2+2)=S12
      EK(K1+2,K2+2)=S22
      EK(K1+3,K2+2)=S32
      EK(K1+4,K2+2)=S42
      EK(K1+5,K2+2)=S52
      EK(K1+1,K2+3)=S13
      EK(K1+2,K2+3)=S23
      EK(K1+3,K2+3)=S33
      EK(K1+4,K2+3)=S43
```

```
      EK(K1+5,K2+3)=S53
      EK(K1+1,K2+4)=S14
      EK(K1+2,K2+4)=S24
      EK(K1+3,K2+4)=S34
      EK(K1+4,K2+4)=S44
      EK(K1+5,K2+4)=S54
      EK(K1+1,K2+5)=S15
      EK(K1+2,K2+5)=S25
      EK(K1+3,K2+5)=S35
      EK(K1+4,K2+5)=S45
      EK(K1+5,K2+5)=S55
  222 CONTINUE
      DO 111 J=1,44
      DO 111 I=J+1,45
  111 EK(I,J)=EK(J,I)
      RETURN
      END
```

8.7 References to Chapter 8

1. S. H. Lo (1985) A new mesh generation scheme for arbitrary planar domains, *Int. J. Num. Meth. Engng*, **21**, 1403–26.
2. I. S. Duff, J. K. Reid & J. A. Scott (1989) The use of profile reduction algorithms with a frontal code, *Int. J. Num. Meth. Engng*, **28**, 2555–68.
3. N. E. Gibbs, W. G. Poole Jr & P. K. Stochmeyer (1976) An algorithm for reducing the bandwidth and profile of a sparse matrix, *Siam J. Num. Anal.*, **13**, No. 2, 236–50.
4. S. W. Sloan (1986) An algorithm for profile and wavefront reduction of sparse matrices, *Int. J. Num. Meth. Engng*, **23**, 239–51.
5. G. C. Everstine & E. H. Cuthill (1983) The optimal ordering of tree networks, Technical Note, *Computers & Structures*, **17**, No. 4, 621–2.
6. S. W. Sloan (1989) A FORTRAN program for profile and wavefront reduction, *Int. J. Num. Meth. Engng*, **28**, 2651–79.
7. R. L. Taylor, E. L. Wilson & S. J. Sackett (1981) Direct solution of equations by frontal and variable band, active column method, in *Nonlinear Finite Element Analysis in Structural Mechanics*, ed. by K. J. Bathe & E. Stein, Springer, New York.
8. K. J. Bathe (1982) *Finite Element Procedures in Engineering Analysis*, Prentice-Hall, Englewood Cliffs, New Jersey, USA.
9. T. J. Chung (1988) *Continuum Mechanics*, Prentice-Hall International Editions.
10. E. N. Dvorkin & K. J. Bathe (1984) A continuum mechanics based four-node shell element for general nonlinear analysis, *Engng Comp.*, **7**, 77–88.
11. O. C. Zienkiewicz & R. L. Taylor (1989) The Finite Element Method, 4th edn, Vol. 1 – *Basic Formulation and Linear Problems*, McGraw-Hill International Editions.
12. R. D. Cook (1981) *Concepts and Applications of Finite Element Analysis*, John Wiley & Sons.
13. T. J. R. Hughes (1987) The Finite Element Method in *Linear Static and Dynamic Finite Element Analysis*, Prentice-Hall International Editions.
14. B. Irons & S. Ahmad (1984) *Technique of Finite Elements*, Ellis Horwood Series in Engineering Science.
15. I. M. Smith & D. V. Griffiths (1982) *Programming the Finite Element Method*, 2nd edn, John Wiley & Sons.
16. R. H. Gallahger (1975) *Finite Element Analysis – Fundamentals*, Prentice-Hall, Englewood Cliffs, New Jersey, USA.
17. E. Hinton & D. R. J. Owen (1980) *Finite Element Programming*, Academic Press.
18. J. S. Przemieniecki (1968) *Theory of Matrix Structural Analysis*, McGraw-Hill.
19. G. Dhatt & G. Touzot (1984) *Une Présentation de la méthode des éléments finis*, S. A. Maloine (ed.), 27 rue de l'Ecole de Médecine, 75006 Paris.
20. *Micro Computers in Engineering Applications* (1987) ed. by B. A. Schrefler & R. W. Lewis, John Wiley & Sons.
21. H. C. Martin & G. F. Carey (1973) *Introduction to Finite Element Analysis*, Tata McGraw-Hill Publishing.
22. S. P. Timoshenko & S. Woinowsky-Krieger (1959) *Theory of Plates and Shells*, McGraw-Hill Kogakusha.
23. P. G. Ciarlet (1982) *Introduction à l'analyse numérique matricielle et à l'optimisation*, Masson, Paris.

Index